Materials modification by electronic excitation

Electronic excitation is a means to change materials properties. This book analyses the important features of the changes induced by electronic excitation, identifies what is critical, and provides a basis from which materials modification can be developed successfully. Electronic excitation by lasers or electron beams can change the properties of materials. In the last few years, there has been a mix of basic science, of new laser and electron beam tools, and of new needs from microelectronics, photonics and nanotechnology. This book extends and synthesises the science, addressing ideas like energy localisation and charge localisation, with detailed comparisons of experiment and theory. It also identifies the ways this understanding links to technological needs, like selective removal of material, controlled changes, altering the balance between process steps, and possibilities of quantum control. This book will be of particular interest to research workers in physics, chemistry, electronic engineering, and materials science.

Noriaki Itoh has been Emeritus Professor of Physics at Nagoya University since 1995. Until recently he was Professor of Physics at the Faculty of Information Science, Osaka Institute of Technology. He spent much of his career at Osaka and Nagoya Universities. After completing his PhD, he was Research Associate, Lecturer and Associate Professor of Nuclear Engineering at the Faculty of Engineering, Nagoya University. At Nagoya University, he became in turn Associate Professor and Professor of Nuclear Engineering at the Faculty of Engineering, Professor of Physics in the Faculty of Science and finally Emeritus Professor. He served as department heads occasionally. He has been Emeritus Professor of Nagoya University since 1995. He was a Council Member of the Japanese Society of Applied Physics. He was Editor of *Radiation Effects and Defects in Solids*, and a Member of the Advisory Boards of *Journal of Physics Condensed Matter*, *Nuclear Instruments and Methods B* and *Review of Solid State Science*. His main interests in materials science and in experimental physics are in the ways in which materials are modified by irradiation with photons, electrons and ions. His publications include books on radiation effects on solids as well as about 300 papers. As an experimental physicist he has been stimulated by theorists whose main interest is modelling of atomic processes.

Marshall Stoneham is Massey Professor of Physics and Director of the Centre for Materials Research at University College London. In 1989, he was elected a Fellow of the Royal Society. He is a Fellow of the Institute of Materials, of the Institute of Physics, and of the American Physical Society. He spent much of his career at Harwell, becoming in turn Leader of the Solid State and Quantum Physics Group, Head of Materials Physics and Metallurgy Division, Director of Research for AEA Industrial Technology, and finally AEA Technology's Chief Scientist. He is a Director of Oxford Authentication Ltd and, as Vice President of the Institute of Physics, chairs the Board of Directors of Institute of Physics Publishing. He is a Senior Fellow of Corning. His wide interests in materials science and in basic and applied physics include the ways in which materials modelling can be used to advantage. His publications include books on defects in solids and on the reliability of non-destructive inspection, as well as nearly 400 papers. As a theorist, he particularly enjoys the stimulus of working alongside good experimenters.

Preface

Electronic excitation is a means to change materials properties. It offers both novel types of change and better-controlled changes. The new developments in laser technologies, in electron beam methods, and also in the use of particle beams have led to a wide range of new opportunites, both scientific and technological.

The basic ideas are not new. Photography is over 150 years old. Photochromics have been on sale for decades. Photolithography, electron lithography, and ablation are standard tools. Photosynthesis is nearly as old as life, and incorporates some of the ideas we shall describe. What has changed in the last few years has been the gradual growth in insight in areas which are not usually connected, and a vast increase in the technology available to manipulate materials by excitation in a controlled way. Our aim is to analyse the important features of the changes induced by electronic excitation, to identify what is critical, and to provide a basis from which materials modification can be developed successfully.

We shall emphasise the mechanisms underlying materials modifications in inorganic non-metals following electronic excitation. Metals and organics are mentioned mainly when they show interesting parallels with the inorganic insulators which we emphasise. It would take at least another book to cover the photochemistry of organic and biological systems. We shall not (and could not) cover every aspect of electronic excited states, which would need to include vast areas of photonics. We shall not include every type of excited state; purely thermal effects, for instance, are covered in many texts. Radiation damage associated with atomic collisions will normally be ignored, except when it sheds light on materials modification processes associated with electronic excitation. For example, radiation damage and ion-beam studies show that outcomes can be good or bad. Semiconductor implantation is a route to high-quality products which would be hard to realise in any other way; long-term reactor irradiation may embrittle structural steels, and countermeasures need understanding.

The aims of materials modification can be categorised. The main themes are these: *removal of surface material,* whether to deposit it elsewhere (e.g. sputter coating) or for analysis (e.g. secondary ion mass spectrometry, SIMS); *modification of surface layers,* as in the creation of diamond-like carbon films by using low-energy nitrogen ions to remove hydrogen from surface hydrocarbons; *modification of the bulk of a material,* as in neutron transmutation doping of silicon; *altering the rate of a process,* as in the effects of low-energy electrons on silicon oxidation; *switching from one process to another,* by changing branching ratios between radiative and non-radiative channels; *damage production,* whether desirable, as for isolation of regions of a semiconductor device, or undesirable, as in embrittlement which degrades mechanical properties.

For these aims to be realised, certain basic solid-state processes are needed. Again, they can be categorised. The first is *energy localisation*, whether ionic or electronic. If atoms are to be displaced, some minimum energy is needed, and this energy must be associated with specific local atomic motions. Electronic excited states often have features qualitatively different from the ground state, notably their equilibrium geometry and their degeneracy or near-degeneracy. Their excess energy can be dissipated by radiative or non-radiative transitions. A second process is *charge localisation.* This is less necessary for materials modification, but is one simple means of guiding energy localisation. A third process is *energy transfer.* The displacements do not need to occur at the sites originally excited, but can occur at more distant sites. A common example, if not yet fully understood, is photosynthesis. Fourthly, there is *energy storage.* Energy sinks can delay damage processes, and sometimes change their character. Finally, there are the effects of *charge transfer* and space charge. Charge build-up can be important for indirect reasons, e.g. because a macroscopic field can influence subsequent damage events.

Excitation can take many forms. It may excite primarily nuclear motion, as by an ion beam, or it can involve selective excitation of the electrons, as by electromagnetic radiation. The excitation can be concentrated in regions of high energy density, or spread over larger regions; it can be sudden, or drawn out in time. The transfer of energy between electronic and ionic subsystems has many subtle features, and these will be a major theme of this book. Such transfer between subsystems is most apparent in non-metals, where there are spectroscopic means to probe the electronic system, but there are also parallel effects in metals. There are thus certain common challenges. First, we shall often need to discuss highly non-equilibrium systems. It is not sufficient to understand electronic ground-state, adiabatic energy surfaces. The second challenge concerns timescales, which range from femtoseconds to tens of years. The third challenge concerns length scales, and the link between microscopic (atomistic) and mesoscopic (microstructural) scales. There is useful progress, with an important and continuing collaboration between theory and experiment. It is this collaboration which we wish to stress. Experiment continues to throw challenges to theory, and to need from theory both a conceptual framework and a validated quantitative understanding. We hope to demonstrate progress in meeting the many challenges of materials modification by electronic excitation: the removal of material, the modification of bulk or surface material, the altering of rates of processes, the changing of branching ratios between alternative processes, and the control of damage.

Acknowledgments

We are grateful to our colleagues, collaborators and friends for discussions, comments and insights over many years. Mr Charles Barnard prepared the figures, and we are greatly indebted to him. We are grateful to Maureen Storey, whose excellent copy editing was very valuable. Even more, we thank our wives, Ayako and Doreen, for their support and encouragement during the writing of this book.

Noriaki Itoh
Osaka, Japan
Marshall Stoneham
Dorchester-on-Thames, UK

Chapter One

Concepts: Excitation, polarons and electronic structure

1.1 Basic ideas about the localisation of charge and energy

Our theme is excited state phenomena in materials, and the way that excitation can affect the nature and performance of materials. The emphasis on excited states means that we are primarily concerned with non-metals, although certain of the interesting features do not depend on the existence of a bandgap. It is natural that excited states evolve with time as equilibrium is reestablished. Our concern includes the ways in which the direction of this evolution can be controlled, for example, so as to produce chosen defect populations. How this happens, and the degree of control which is possible, depends on how the excitation energy is provided as well as on the intrinsic features of the excited states.

The components of materials modification by excitation can be classified in a few categories, although the mechanisms may be very varied. First, there are the transport processes: *energy transport* and *charge transport.* These are the processes by which excitation reaches specific regions of a solid. Control of which zones are affected and which are unaffected is important in any proposed application. Secondly, there are the localisation processes: *energy localisation* and *charge localisation.* Energy localisation can often be described as the trapping or self-trapping of an exciton. In some cases, it will be preceded by charge localisation, as in the trapping or self-trapping of a hole followed by capture of an electron. Normally, we shall be referring to localisation on a particular ion or small group of ions; occasionally, we shall mean localisation in a confined region, perhaps within one or two atomic layers of a surface. Localisation can begin on the mesoscopic scale, such as on a large molecule, like a polymer, or a particular grain in a ceramic, before localisation on the atomic scale. Thirdly, there is some process involving ionic displacement which results from the energy localisation. There are usually several alternative channels for dissipating the energy without ion displacements, so the efficiency of this displacement process is a matter of keen interest in applications. In the most useful processes, the available energy is directed into chosen regions of space and chosen channels of behaviour. The photographic process is a good example, where very few photons can generate an image which is sharp and accurate. But, before any of these processes, we must look at the ways by which the initial excitation can occur, and especially the intensity, timescales and region sizes of excitation.

If localisation provides a means for such a change to happen, the resulting processes are central to any systematic use of excitation. These processes involve a wide range of timescales and length scales. In certain cases, the nuclear motion, not only the electronic system, exhibits quantum behaviour. It can be useful to classify the processes (Stoneham 1990) according to whether they occur near thermal equilibrium or not, and also as to whether they are adiabatic or not. Thus diffusion processes often occur near thermal equilibrium, and involve motion on an adiabatic energy surface. The response to a sudden perturbation, perhaps following an electronic or nuclear (Mössbauer) transition is likewise adiabatic, but far from equilibrium. Many non-radiative transitions are ones between two adiabatic energy surfaces, and hence are explicitly non-adiabatic.

The point of materials modification is the control and improvement of materials properties. The properties to be enhanced can be remarkably varied. They might be mechanical properties, whether strength or tribological. Improvements may be in electrical properties, such as in conductivity, resistivity, or control of breakdown. It may be advantageous to alter optical properties, including colour changes or refractive index changes. There may also be compatibility issues which make changes desirable, and these could be environmental (as in electrochemical corrosion) and possibly even biological (e.g. for *in situ* sensors).

We shall be concentrating primarily on electronic excitation by photons and by electrons and ions, although we shall not ignore other forms of radiation. We shall find that the outcome of excitation can be essentially the same for quite a variety of electronic excitation routes. Yet even a single type of excitation can be exploited to have different consequences. It is possible to control the photon energy, the photon flux, the area over which the energy is applied, the degree of coherence, of polarisation, or of divergence, and so on. The sources of light may vary from synchrotron radiation to the near-field microscope, from a laser to a lamp. The sources of electrons might involve an electron microscope, charge injection from an electrode, a particle accelerator, a plasma, or a scanning tunnelling microscope. These all affect the excited states created and the way these develop.

Simply because we are concerned with excited states, we have a range of phenomena which we must understand. First, there will be relaxation processes, since the equilibrium geometry for the excited state will not normally be the same as that for the initial (ground) state. The relaxation energy associated with this motion is often not available for defect production. Secondly, there may be transfer of energy between states. These may be other electronic states, or vibrationally-distinct states; there may be excitation of plasmons, or magnons, or other electrons, or the emission of light. This potential variety of these excited states demands a descriptive framework which is suitably general.

The band descriptions of solids, and the descriptions in terms of the component atoms or ions, provide complementary ways of regarding the electronic structure. Our emphasis is on excited states and on the phenomena of energy and charge localisation, which lie outside the range of some of the most popular approaches. We shall find that the real-space description is usually the most helpful, especially for those excited states which show localisation by self-trapping. Some situations, especially highly non-equilibrium transient ones, are hard to model or describe adequately using any simple picture. The crystalline or amorphous nature of a solid is far less troublesome. Many features, like the existence of a bandgap, can occur in both crystalline and amorphous systems; indeed, the most transparent solids available are probably silica-based glass fibres, which are amorphous and obviously have a wide bandgap. Needless to say, there are many ways in which amorphous solids do behave differently, notably in the fact that all sites are not identical. For atomic vibrations, there are similar issues. For any harmonic system (whether crystalline or not) one can define normal modes, and describe their excitation in terms of phonon occupation numbers. We must also remember that solid-state phenomena can involve several different length scales. The structure at the mesoscopic scale (grain structure, surface roughness, inclusions of other phases, screening of charge by charged defects, . . .) can all be important. This presents further problems of characterisation, especially in applications.

1.1.1 The polaron concept

It is obvious that localised charge or energy could result from some local structural difference in a region of a solid, perhaps an impurity, a defect, a surface, a fluctuation in topology. Yet the most important mechanism for the processes of charge and energy localisation needs no such inhomogeneity, apart perhaps from the thermal motion of the ions in a crystal. *Self-trapping* is sufficient. This term carries strong associations with the *polaron* concept (a carrier plus its associated polarisation and host lattice deformation field). These ideas will be discussed alongside experiment later, but it is helpful to recall how they began.

In 1932 Landau observed that there were two quite distinct situations for an electron coupled to an otherwise perfect (but deformable) lattice. In one case, the electron moved 'freely' with an effective mass somewhat modified by coupling to lattice polarisation. In the other case, the electron would be localised, *'self-trapped'* in a region strongly distorted by the electron itself through the electron–lattice coupling. Kanzig in 1955 was the first to see such localised carriers unambiguously (V_k centres, self-trapped holes). Exciton immobilisation (self-trapping) at low temperatures was appreciated by Frenkel in 1926; Kabler & Patterson (1967) and Kink *et al.* (1967) observed self-trapped excitons in KI in 1967. Landau also remarked that the electron might have to surmount a barrier before being self-trapped. This *barrier to self-trapping* was verified only much later for excitons in alkali-metal iodides (Kuusmann *et al.* 1976; see Rashba 1982 for a review), although the idea was developed in 1957 by Rashba (1957), who stressed the important role of short-range interactions.

The strong electron–phonon coupling assumed by Landau implied very short carrier mean free paths. J Yamashita & Kurosawa (1958) and Holstein (1959) showed that the carriers would move by an incoherent *'hopping'* transport mechanism in most cases of interest. Instead of the traditional behaviour of semiconductors like silicon, or like metals, where the phonon scattering becomes stronger as vibrational amplitudes increase, so the carrier mobility decreases when the solid grows hotter, the small polaron mobilities *increase* with temperature as incoherent hopping from one site to another becomes easier. This gives an important experimental clue in identifying small-polaron systems, although further tests are needed (Chapter 4). However, thermally-activated behaviour can have several different origins, including impurity trapping. Further, if self-trapping is only just stable, thermal excitation to the large-polaron form may dominate over the hopping of small polarons in the mobility observed.

Landau's ideas have been generalised in several ways. One route is phenomenological (Section 3.3); other studies use new, powerful, methods (Chapter 3). The main interests lie in three areas, all of which will be important in this book. First, there are developments in the *criteria for self-trapping*. Such approaches will be a guide as to whether charge localisation is stable. Secondly, there are attempts to simulate the *process of self-trapping* from delocalised initial states. Here, one issue is the timescale for localisation, and how it compares with competing processes. Thirdly, the studies of the *microscopic structure and properties* of localised polarons, including activation energies for motion, are a major test of the accuracy and capability of theories. Reviews of polarons include those by Stoneham (1989), Gerlach & Lowen (1991), Fisher *et al.* (1989), Shluger & Stoneham (1993), and Stoneham & Smith (1990).

These ideas have become increasingly important, especially with the recent development of femtosecond optical spectroscopic techniques (Keimer 1998).

These experiments began a new class of studies of the mechanisms of dynamical processes in condensed phases after electronic excitation. Examples include clarification of the relaxation dynamics of the triplet self-trapped exciton in NaCl (Iwai *et al.* 1991), of (hole) polaron self-trapping in KBr (Tanimura *et al.* 1993), KI and RbI (Iwai *et al.* 1994), and of the relaxation of the solvated electron in water (Rentzepis *et al.* 1973; Migus *et al.* 1987) and in alcohol systems (Kanney-Wallace 1981). Such studies of the dynamics of early stages of self-trapping of polarons and excitons in real systems, rather than model systems, have created a new class of very fascinating problems.

We remark how widespread the application of the small-polaron concept is. A large range of materials shows features characteristic of small polarons. The polaron ideas have many parallels with conventional defect issues, where lattice deformation and polarisation is a crucial component of defect processes and spectroscopy. The polarisation and lattice relaxation which drives self-trapping is similar to the lattice relaxation which stabilises impurites in a range of charge states; similar relaxations determine defect optical spectra, and contribute to characteristic activation energies. The similarity of polaron and defect characteristics is at its clearest for cases like the self-trapped hole bound to an Na ion in CsI, which may be regarded as a small polaron trapped by an impurity (Ong *et al.* 1979). These interactions of small polarons with defects, or with other small polarons, are especially important in polaron dynamics at finite concentrations, as in non-stoichiometric solids. But, just as the small-polaron ideas give a new way of looking at certain defect states, so do defect ideas point to small-polaron ideas which were not evident in Landau's original paper. The electronic excited states of self-trapped excitons are very varied: the electron or hole (or both) can be excited, leading to a rich area of spectroscopy. Trapped small polarons (like the V^- centre in MgO, a hole is localised on an oxygen next to a cation vacancy) show two classes of excited states: charge transfer (the hole moving to other oxygens) and electronic excitation of the oxygen ion itself.

A further point is more general: why do small polarons matter? Are they a mere scientific curiosity? We believe that there will be a growing exploitation of electronic excitation as a tool in materials modification. At the time of writing, applications are varied but isolated, the result of independent opportunities. In the halides, the technological impact is present but not dominant: there are links to the understanding of the photographic process for silver halides; there are polaron aspects to the operation of X-ray phosphors like CsI:Na; the sputtering and selective desorption processes have associations with small-polaron processes. The self-trapping of excitons is seen in a wide variety of insulating solids with strong exciton–lattice coupling, especially in halides and oxides (N Itoh 1982; Williams & Song 1990; Lushchik & Lushchik 1989). This energy localisation is responsible for various photochemical and radiation-induced processes, including defect formation and optically-induced desorption (Lushchik & Lushchik 1989; N Itoh & Tanimura 1990). The localisation of charge or energy can stimulate reactions which are harmful, as well as beneficial. For example, we believe that one of the key steps in electrical breakdown involves a localisation stage.

1.1.2 Excitation of metals and insulators: What is special about insulators?

Radiation damage, even collisional damage, shows systematic differences between insulators, semiconductors, and metals. Why is this? A first reason is that most of the insulators which we shall discuss involve more than one element: NaCl contains Na and Cl, whereas Cu contains only Cu. Clearly, there are exceptions, like diamond and like ordered alloys. There will be effects simply because of the different masses of the species. Moreover, there is a strong rule which asserts that defects in ionic crystals tend to keep each species on its own sublattice. Thus Na tends to remain on the Na sublattice, Cl on the Cl sublattice. Exceptions are rare but significant. In the radiation damage of alkali halides, perfect dislocation loops form by a process in which neutral halogen molecules occupy anion–cation divacancies. This gives a defect which is almost impossible to observe, but is essentially an antisite defect. In III–V semiconductors, antisite defects form with relatively low energies. This affects dislocation climb strongly, since simple defect reactions can transfer disorder from one sublattice to the other (so a Ga interstitial can displace As to form an As interstitial and a Ga antisite) without the need for the complicated mechanisms once assumed. Another example concerns iodine, a key fission product of UO_2. In addition to the obvious sites, such as I^- at the oxygen site, in certain situations (Grimes *et al.* 1989) I^+ is stable at the cation site.

A second difference between metals and non-metals is that defects in insulators or semiconductors can have a range of possible charge states. For oxides like MgO and $SrTiO_3$ transition-metal dopants can occur in three or four different states, and equilibrium over these states can take days to establish. The variety of charge states has several effects: on defect reactions, on carrier transport, and, through the image interaction, on the energetics of interfaces and radiation-enhanced adhesion.

A third difference between metals and non-metals is that excited states last longer, and are far more important, in non-metals. This leads to a number of major themes of our book. Thus, when the bandgap is large enough, it may be possible to localise energy sufficient to create defects. Electronic excitation can also affect defect reactions, after the creation of the primary defects; this can be by directly affecting the jump probability through the provision of extra vibrational excitation (which we shall call *local heating*) or through a change in electronic state (possibly, but not necessarily, a change of charge state), which we shall call *local excitation*. The relatively long life of excited states means also that the spectroscopy of defects is possible, and this gives a wealth of information and understanding about the transient processes of materials modification. Optical measurements are, in any case, a valued source of information on defect populations emerging under excitation. Sapphire, for instance, proves to be an effective radiation dosimeter up to relatively high temperatures; quartz is the basis of the thermoluminescent authentication of art and archaeological ceramics.

1.2 Methods of excitation

For many purposes, the key factors which determine what happens during excitation are the energy spectrum of the excitation, the total energy input, the energy density, the energy flux, and whether or not there is charge injection as well as energy injection. We now describe some aspects of the main classes of excitation, drawing attention to features which may be important for materials modification.

1.2.1 Excitation by electromagnetic radiation

Radiofrequencies (energies of order 1 cm^{-1} or less, as used in spin resonance)

In the commonest form, the radiofrequency excitation induces transitions among the Zeeman levels of ions in an applied magnetic field. The energy is often transferred to electron spins, which may redistribute energy among the various magnetic ions (spin–spin relaxation) or relax to equilibrium by excitation of lattice vibrations (spin–lattice relaxation). The Zeeman energy splittings can have sharply-defined values, so that the vibrations excited may be modes in a narrow frequency range for which the density of states is small. The release of energy to these low-energy phonon modes can lead to a highly non-equilibrium distribution in the phonon energy spectrum, which may reduce the rate at which the energy is being transferred; this situation is known as a *phonon bottleneck*.

Microwave and infrared frequencies (energies of order 1 cm^{-1} to 1 eV; domestic microwaves operate at 2.45 GHz, or about 0.08 cm^{-1})

Microwaves readily excite the motions of ions or of molecules with dipole moments (like water). Conduction electrons, whether in a metal or excited to the conduction band of an insulator, absorb electromagnetic radiation for energies below the plasmon energy. The skin depth is a measure of the penetration of the radiation and, for a free-electron metal of conductivity σ, is $(2\epsilon_0 c^2/c\omega)^{1/2}$ (Feynman 1963, §II.32.11). For radiation of 10^6 MHz, the skin depth is 670 nm, so that the energy is deposited close to the surface.

Bel'kov *et al.* (1999) have reported that circular spots of enhanced free-electron density are formed in thin n-GaAs subjected to a uniform microwave field. The spots, which can be of order 1 mm across, have sharp boundaries, and form spontaneously when the microwave power is above a threshold. Dopant densities have to be above a threshold, typically $10^{15}/\text{cm}^3$. No current passes.

Multiphoton excitation and dissociation of molecules by infrared excitation is a well-known feature of non-linear optics. Infrared excitation can be sufficiently intense for the electric field to enhance ionisation processes. The generation of visible light by infrared excitation is also possible, e.g. by resonant absorption of CO_2 light by gas molecules (Isenor & Richardson 1971). In such molecular cases, the relaxation processes and energy transfer between the vibrationally-excited states are critical. The characteristic times are microseconds in certain cases. For molecules in or on solids, relaxation processes tend to be much faster. For molecules on metal surfaces, the molecular dipolar vibrations couple strongly to the electron gas, and relaxation is fast. Even for the Si–H stretching mode on Si(111), the coupling to surface phonons gives relaxation in nanoseconds. However, for porous silicon, excitation of the Si–O stretching vibrations by 1076 cm^{-1} infrared pulses leads to emission at the characteristic 1.8 eV energy of porous silicon luminescence. In this and other special cases, intense low-energy excitation can provide much larger energies for subsequent processes (e.g. Diener *et al.* 1995). These cumulative processes are not always understood.

Rapid imaging of long wavelength electromagnetic radiation is difficult. However, the waveform $E(t)$ of terahertz infrared pulses has been measured by encoding it onto a stretched chirped optical pulse (Jiang & Zhang 1998), a method which may prove useful.

Optical frequencies (energies from the infrared to the ultraviolet, typically 1–5 eV)

At these energies, electrons in semiconductors and insulators can be excited. For photon energies less than the bandgap, unless there are strongly-absorbing defects (colour centres), the absorption is likely to be weak, and due to impurities or lattice vibrations. Defect spectroscopy is possible, sometimes with very high resolution, and this is a major tool in the experiments we shall describe. As the photon energy is increased, there is often a clear exciton absorption just below the bandgap, and this may merge with band-to-band absorption. Just above the bandgap, the absorption rises rapidly with energy. If the gap is indirect, there may be a further threshold corresponding to a higher-energy direct gap. The bandgaps depend on temperature, and this can be significant for small gaps; the gap for Si is about 1.2 eV at ambient temperatures, falling to about 0.6 eV at the melting point. The optical absorption cross-section σ is typically 10^{-17} cm^2 for photon energies above the bandgap. The optical absorption coefficient $\mu = \sigma N$, where N is the number of atoms per cubic centimetre, is about 10^{+5} cm^{-1}, so that photons are absorbed within 100 nm of the surface.

Since the bandgap is a measure of the energy available in a number of solid-state processes then, at least in principle, this can have a significant effect on behaviour. At energies just above the bandgap, electrons and holes are created, and these may react to form an exciton. Electrons and holes may recombine, either with the hole being filled by the electron excited (geminate recombination; see Section 2.6) or by processes involving other traps. For higher photon energies, there are various channels for energy loss, and typically an energy of about three bandgaps is needed to generate a measured electron–hole pair. The 'missing' energy is lost primarily through the generation of secondary carriers with energies too small to cause further electron–hole pair excitation. This is discussed further in Chapter 2.

Light sources

Conventional light sources, which were used widely before lasers became available, include the tungsten lamp for the visible range and the mercury lamp, the Xe lamp, and the deuterium lamp for the ultraviolet range. These lamps, whose input power is typically in the range of 100–1000 W, emit of the order of 10^{20} photons per second. After passing through a monochromator, typically 10^{15} photons cm^{-2} s^{-1} reach a specimen. Xe flash lamps are available, and emit 1 μs pulses.

Lasers, pulsed or continuous wave (CW)

Progress in producing short pulses has been dramatic, starting from special techniques like mode-locking, pulse-compression, and pulse-shaping. Pulses as short as 10 fs (less than 10 cycles of visible light) have been achieved for several laser systems. The production of very short light pulses is only one component of what is needed. Ultrafast clocking is also needed, so that the consequences of an ultrashort pulse can be monitored. Some methods use a second, delayed, pulse; others use intensity autocorrelation (the 'correlation clock'), streak camera methods, Kerr upconversion gates, and other methods.

Lasers currently available provide intense light sources for photons ranging from the infrared to ultraviolet. Excimer lasers are particularly useful as ultraviolet light sources. Solid-state lasers, like the YAG laser, emit infrared light and, by using the second and third harmonics, can be used for the visible and ultraviolet ranges as well. Tunable lasers are particularly useful for investigating the excitation spectra of photo-induced processes. Dye lasers, Ti sapphire lasers, and optical parametric systems can cover the spectrum continuously from the infrared to the visible. The quality and effectiveness of a laser for materials modification will involve a number of factors. The *photon energy*, *maximum peak power*, and *beam profile* will matter in almost all applications. In many cases, the *pulse duration*, *repetition rate,* and *scan rate* over a chosen area will all be crucial. In more special circumstances, the *pointing stability* (avoidance of unwanted beam defelections) will be essential, especially if the process involves interference at some distant point. In the more advanced applications, *spatial coherence* and *temporal coherence* are critical. For most current optical fibre applications, lateral spatial coherence of about 1 mm is needed, and a temporal coherence length of 10 mm or more will be required.

Laser sources can be either continuous (CW) or pulsed. A few laser systems can be used for both CW and pulsed lasers, but only pulsed beams can be emitted from those laser systems in which the lower lasing level has a lifetime longer than the upper level. Techniques such as Q-switching, changing the Q value of the resonator, or mode-locking, resonating pulses lying within a frequency interval, are employed to generate pulses of nanosecond and picosecond ranges. Femtosecond laser pulses can be generated by a pulse compression technique. The characteristics of lasers commonly used for materials modification are listed in Table 1.1.

The primary interaction of laser beams in non-metals above bandgap energies is the excitation of valence electrons. A typical cross-section per ion at the surface, or for the bulk of a direct gap material, is of order 0.1 $Å^2$. This initial excitation is followed by several secondary effects, because of the high intensity (Allmen 1987). The secondary effects include self-focusing, free-carrier heating, and thermal ionisation, especially for semiconductors. Self-focusing occurs if the real part of the dielectric constant varies as a function of beam intensity. Since laser beams are more intense near the beam axis, there is a focusing somewhat similar to a lens placed in the beam path. The generation of free carriers in non-metals alters the optical absorption coefficient significantly. Free carriers gain energy by absorption of photons. The high-energy electrons cause impact ionisation, and increase the number of free electrons (e.g. Epifanov *et al.* 1976) at a rate linear in the instantaneous amplitude of the electric field. For pulsed lasers, these secondary effects are more important at a given fluence as the pulse width is reduced. Because of these processes,

Table 1.1. *Characteristics of lasers for materials modification.* Values are representative only.

Laser type	Pulse or CW?	Wavelength (nm)	Typical maximum output	Remarks
Gas lasers				
excimer ArF	pulse	193	1 J/pulse	
excimer KrF	pulse	249	1 J/pulse	
excimer XeCl	pulse	308	1 J/pulse	
Ar^+	CW	488, 514	10 W	
CO_2	CW	10600	10 kW	
Solid state				
ruby	pulse	694.3	1 J/pulse	
Nd YAG	pulse	1064	0.5 J/pulse	
Tunable lasers				
Ti: saphire	CW	660–1200	1 kW	
	pulse		1 J/pulse	
colour centre	CW	990–2500	0.1 W	using several colour centres in several halides
dye lasers	CW	420–800	1 W	using several dyes
	pulse	320–970	0.1 J/pulse	output depends on wavelength
parametric	pulse	220–2700	0.1 J/pulse	output depends on wavelength

heating is inevitable when pulsed laser beams are incident on solids. A ruby laser pulse of 1 J/cm^2 has 3.5×10^{18} photons/cm^2; if photons are absorbed in 10^{-4} cm, the energy absorbed per atom is approximately 3.5 eV, about 10 times larger than the latent heat. Clearly, thermal conduction tends to reduce the temperature rise. The kinetic equations have been solved for a variety of systems, and indicate melt thresholds for many materials in the range of 0.01–1 J/cm^2, decreasing as the pulse width decrease.

Photons below bandgap energies are absorbed by multiphoton band-to-band transitions or by transitions where there is lattice disorder. Multiphoton absorption, the simultaneous absorption of n photons, requires that those n photons are incident upon its cross-sectional area within a time interval δt of order $2\pi/\omega$. The probability for one photon is proportional to the photon flux $I/\hbar\omega$, implying a rate of multiphoton excitation given by $\sigma^{(n)}(I/\hbar\omega)^n$. For each extra photon involved, the cross-section is smaller by a factor typically in the range 10^{-33}–10^{-17} cm^2/s; this factor can be larger by a factor of perhaps 10^{+4}–10^{+8} if there is a resonant effect (Haglund & Itoh 1994). Two-photon transitions occur between states of the same parity. Multiphoton excitation of vibrational motion underlies one route to isotope separation. When one of the intermediate states of the multiphoton transition is close to a resonant state, the multiphoton cross-section is enhanced. Photoexcitation at defect sites involves single-photon transitions, or possibly multiphoton transitions needing a smaller number of photons than the multiphoton band-to-band transitions. One-photon transitions have cross-sections per defect which are typically in the range 10^{-4}–1 Å^2. The excitation of defects can be as significant as band-to-band transition, even though the number of atoms directly involved in the transitions is less. Photoexcitation at the surface of insulators often alters stoichiometry and consequently enhances the optical absorption coefficient. Irradiation with laser beams on surfaces causes laser ablation, which causes the loss of large numbers of atoms. Pre-existing defects often influence laser breakdown and the laser ablation of wide-bandgap materials (Dickinson *et al.* 1996; Ribeiro *et al.* 1997a,b, 1998a,b). The emitted atoms are ionised by electrons emitted from the surface, and form a plasma; the plasma formed in front of the surface is known as the plume. When a plume forms, it can absorb or reflect light, and so alter energy absorption by the target. The evolution of the plume has been analysed using analytical models (e.g. Phipps *et al.* 1988) and computer simulations (e.g. Vertes *et al.* 1988; Dreyfus 1993). The models assume that vaporisation at the surface follows laser irradiation immediately and that there is local thermodynamic equilibrium. We discuss the solid-state processes of ablation, in Chapter 11.

Scanning near-field optical microscopy (SNOM)

The light source is an optical fibre terminating in an aperture with dimensions of the order of the wavelength of light, or less. By appropriate design, the regions of high intensity can be localised to dimensions determined more by the aperture than by the wavelength. Another approach uses the near field around the tungsten tip of an illuminated scanning probe microscope, where the electromagnetic effect of the sample and probe tip intensify the field by a large factor, perhaps as much as 10^6. This method enables high-precision ablation and micromachining. Related approaches can be used to deposit material on the substrate, or to intensify a laser for microlithography.

Confocal and other special optical systems

These allow attention to be confined to a localised region, perhaps as small as 50–100 nm across, with three-dimensional sectioning and possible femtosecond excitation (Gruber *et al.* 1997; Hell *et al.* 1996). Individual molecules can be observed. There are several approaches (Orrit & Bernard 1990; Betzig &

Chichester 1993; Nie *et al.* 1994; see also Brown *et al.* 1997, especially their Fig. 2). These include: (a) a monomode non-pulled optical fibre, with light collection by a parabolic mirror; (b) a confocal setup, with a microscope objective both bringing in and collecting light; and (c) near-field excitation via a pulled optical fibre, with fluorescence collected (at the far side) by a microscope objective.

Higher photon energies (energies from perhaps 5 eV to 200 eV)

For the lower-energy part of this range, non-linear processes can be used to double laser photon energies. This is a powerful technique, since it retains many of the advantages of laser systems. Developments using a hollow, gas-filled waveguide instead of a non-linear crystal (Rundquist *et al.* 1998) have extended the range of energies to the soft X-ray range (40–70 eV).

Synchrotron radiation provides an intense, highly coherent, directed source. The electronic excitations are of several types. There are one-electron excitations from valence or core levels to conduction bands. There are also collective excitations (plasmons), essentially modulations in the charge density. The cross-section for excitation by photons decreases rapidly as the excitation energy falls. The excitation of core electrons dominates when the photon energy exceeds the core excitation energy, producing an energetic electron and a core hole. Usually Auger processes follow. These yield Auger electrons, which cause further ionisation, creating electron–hole pairs, which deposit energy near the site of the initial core hole. An example of the spatial distribution of energy deposition of Auger electrons generated from a core hole in alkali halides is given by Elango *et al.* (1983).

X-rays and gamma rays (energies from 100 eV upwards)

Conventional X-ray sources provide a spectrum with sharp features, and with intensities usually far lower than a synchrotron source. The X-rays are incoherent and unpolarised. However, standard sources are favoured over synchrotrons for many applications, partly because they are portable and can be cheaper, but generally because they are more readily available.

At higher energies, the main effects of electromagnetic radiation interacting with a solid can be divided into: (a) the *photoelectric effect* at lower energies (up to 0.1–1 MeV, the energy of the divide increasing with atomic number), (b) the *Compton effect*, and (c) *pair production*, at the highest energies (a few MeV or above, taking over at higher energies for lower atomic numbers).

The photoelectric effect produces an electron with energy $h\nu - E_c$ and a core hole, where $h\nu$ is the photon energy and E_c is the core electron energy. The cross-section of the photoelectric effect decreases rapidly as the photon energy increases beyond E_c, so the cross-section has a sharp maximum at the core electron binding energies. The cross-section of the photoelectric effect for each shell increases roughly as Z^4, where Z is the atomic number. Core holes produced by X-rays and γ-rays are filled by emitting photons (fluorescence X-rays) or Auger electrons. The probability of emitting fluorescence X-rays increases with increasing the core energy. Auger electrons, as well as photoelectrons, cause further excitation of core and valence electrons. The Compton effect produces energetic scattered electrons, which cause further excitation of core electrons. Thus the processes that follow the absorption of X-ray photons are similar to those induced under electron irradiation: the number of electron–hole pairs produced is approximately equal to the energy absorbed divided by 2–3 times the energy per electron–hole pair (see Section 1.5.2). The inhomogeneity in the distribution of energy deposited for synchrotron radiation becomes less significant as the photon energy increases.

γ sources

Gamma sources are often associated with the nuclear industry, where they are used to understand and develop the technology of that industry. When spent fuel is being reprocessed, this exposes equipment and glass viewing windows to relatively intense gamma sources. Spent fuel can itself be a useful source of gammas, for example during storage (as in a spent fuel pond) awaiting some further processing or treatment. Other, weaker, sources (like ^{60}Co) are often used to mimic the effects of gammas in a controlled way, so that the behaviour can be understood. One special gamma source comes from the Mössbauer effect. Here, a nucleus emits a γ and the recoil is taken up by the lattice as a whole, with the result that the γ has a very sharp energy. The recoil following γ emission by a nucleus in a solid can cause local damage.

Dosimetry is important for hard X-rays and γ-rays. The conventional unit of absorbed dose for those radiations which mainly cause electronic excitation, such as γ-rays and electron beams, is the gray (Gy, the SI unit, 1 joule per kilogram). The unit used previously, the rad (100 erg per gram) is 10^{-2} Gy. The SI unit of activity is the becquerel (Bq), corresponding to a radionuclide decaying at an average rate of one spontaneous nuclear transition per second. Thus $1\ \text{Bq} = 1\ \text{s}^{-1}$. The former unit, the curie (Ci), is equal to 3.7×10^{10} Bq. The curie was originally chosen to approximate the activity of 1 g of ^{226}Ra.

A γ-ray source emitting photons of energy E [J] of Q [Bq] emits Q photons or an energy of EQ [J/s] per second. The energy flux at a distance r is $EQ/4\pi r^2$ [J/m^{2}s]. If the absorption coefficient for the γ-rays of the absorbing materials is μ [m^{-1}] and its density is ρ [kg/m^3], the absorbed dose is $\mu EQ/4\pi\rho r^2$. Thus, for 10^4 [Ci] ^{60}Co, the absorbed dose of a specimen with absorption coefficient

24 cm^{-1} and density 4×10^{-3} kg/m^3, placed at a distance of 1 m, is 1×10^3 Gy or 3×10^{13} [eV/(cm^3s)]. The absorbed dose is significantly modified if the specimen is bulky because of scattered γ-rays (Goldstein 1959).

As with X-rays, the gammas are incoherent, and the spectrum can have sharp features. Unlike X-rays, gamma sources cannot be switched off. This is inconvenient both scientifically and because of the associated legislative controls. In many applications, gamma radiation is accompanied by neutrons. The two types of radiation are not simply additive in their effects, partly because the neutrons contribute nuclear displacement damage and the gammas contribute electron redistribution. For example, neutron displacement damage might be stabilised by a change in charge state of a radiation product through the capture of a carrier generated by the gammas. Studies of these mixed effects include the darkening of optic fibres, in relation to communications in a nuclear radiation environment.

1.2.2 *Excitation by electrons*

The electronic excitation of solids by particle beams can take several forms, depending on whether single-particle or collective excitation dominates. One important mechanism of energy loss is plasmon excitation, through the collective excitation of valence electrons. Plasmon excitation (see Section 1.5.3) is most commonly identified under electron beam excitation, but is identifiable in quite different circumstances. For example, the specular reflection of 0.5 MeV protons by the (100) surface of KCl excites surface plasmons. In this case, there is almost complete conversion of excited surface plasmons to electron–hole pairs (Kimura *et al.* 1998). The important parameters for electron excitation of plasmons are the plasmon energy $\langle E \rangle$, the centroid of the energy loss function $\mathrm{Im}(\epsilon^{-1})$, and the inelastic mean free path, λ. The escape depths for secondary electrons are especially important. The slowing down of fast electrons with kinetic energy K is determined by $\mathrm{d}K/\mathrm{d}s = -\langle E \rangle/\lambda$, with s the total path length. Typically, $\langle E \rangle$ is a few tens of an electron volt. For K above about 50–100 eV, λ is given by empirical expressions such as this (Szajman *et al.* 1981):

$$\lambda\ [\mathrm{nm}] = 0.18 \langle E \rangle K^{3/4} / (\hbar \omega_{\mathrm{p}})^2;$$

typical values are a few nanometres. If the electron energy is lost only in discrete amounts of order $\langle E \rangle$, the total path length s will be much longer than λ. The inelastic mean free path for energies K less than the bandgap will be determined by interactions with lattice vibrations. This is discussed in Section 1.5.2, where we also discuss the energy needed to create an electron–hole pair.

Excitation of a sample will create charge within it, whether by the deposition

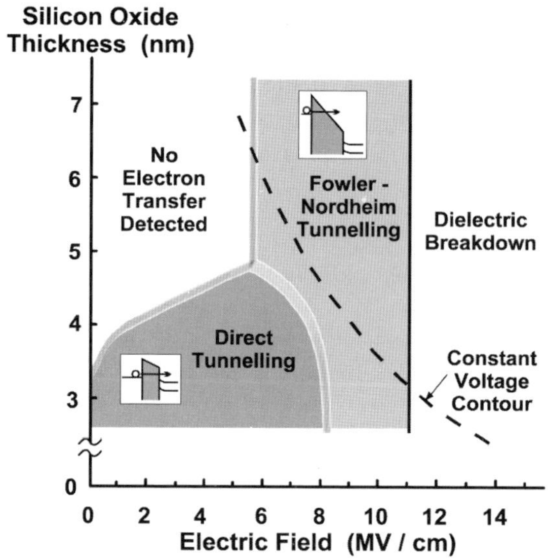

Fig. 1.1. Regimes of behaviour for thin oxide on silicon (extended from Hori 1997). The inserts show the distinction between Fowler–Nordheim and direct tunnelling. The constant voltage contour is shown because defects cannot be created unless there is a sufficient voltage; breakdown fields are only appropriate for thicker oxides.

of charged particles or by causing the emission of secondary electrons. Cazaux (1999) discussed a double-layer model to describe the secondary yield from electron-irradiated insulators; he also remarked on the role of dislocations. The effects of excitation will be complicated by the electric fields from this charge and from the redistribution of this charge during the recovery to the initial charge state.

Injection from an electrode

Injection can be a relatively gentle process, for energies from 0.1–1 eV are typical. It takes place by tunnelling, whether direct or Fowler–Nordheim (Fig. 1.1) below the dielectric breakdown threshold. The distribution of electrons in space (and hence the injection intensity) can be very inhomogeneous, especially since there will usually be asperities both at the typical grain size (a few microns) and on smaller scales. When electrical breakdown occurs, it can also be highly localised. This localisation can involve several stages. In a polymer, there may be some low-density region which leads to charge localisation, followed by energy localisation involving only a few atoms. In MgO, the process appears to involve dislocation climb, whereas in SiO_2 defect generation by an excitonic process may be involved.

The several key processes are clearly discussed in Mott & Gurney (1948, Chapter V sections 6–9). Among the issues for metal/non-metal contacts are these. First, are there shallow traps on the non-metal side? If there are no shallow traps (Mott & Gurney's insulator case) the total number of electrons in the insulator is thermally activated, and the potential is characterised by a scale length which falls as temperature rises. Space charge is important when the field is small. Injection into a film of thickness d provides a current $j = AV^2/d^3$ (where A is $\mu(9/32\pi)$, with μ the mobility). When there are shallow traps (Mott & Gurney's semiconductor case), the characteristic length rises with temperature. Secondly, under an applied field, do electrons tunnel through the potential barrier between metal and insulator, or are they thermally activated over the barrier? Thirdly, to what extent does the image interaction (the effect of the polarisation of the metal by a carrier in the non-metal) affect the behaviour?

The ideas given in Mott & Gurney need supplementing in several respects. First, Bardeen recognised that there could be interface states, and that these could trap charge which has profound consequences. Secondly, as we shall see in later chapters, carriers can be self-trapped, and the relationship between the injection process (which may be to a non-self-trapped state) and the self-trapping step will often be complicated (see, e.g., Ramos *et al.* 1994).

Injection from a scanning tunnelling microscope (STM) tip

The electron source is highly localised, so there will be an intense local beam. In special cases, this can be highly directional. The excitation can be confined to a region a few nanometres across. Energies are typically 1–5 eV; currents are small, although the current density can be large.

Electron microscope

The energy, energy density, and the current can be varied within quite large ranges. For transmission electron microscopy, the standard operating voltage is as high as 200 keV (Flower 1993), with 300–400 keV for medium-energy microscopes, and over 500 keV for high-voltage operation. Surface-sensitive electron diffraction, on the other hand, exploits electrons with energies of less than 200 eV (0.2 keV). As with the STM, the source of electrons can be used to image the specimen, and hence to monitor the results of the excitation. Dose rates can be very high. This has advantages, especially in rapid assessment of large doses. It can also be a problem, for instance by making the observation of radiation effects in organic materials tricky, since damage occurs as observation continues.

Electron accelerators

The electron accelerators used for materials modification work in an energy range of 1–10 MeV. At higher energies, the penetration depth increases beyond 1 cm and the energy deposition rate is reduced. Electrostatic accelerators are used routinely for modifying materials such as organics, and can generate electron beams below 2 MeV and 1 mA. There are the classical methods to generate the high voltage, such as van de Graaff and Cockcroft–Walton accelerators, as well as the more recent resonant transformer types. Linear accelerators (linacs) use radiofrequency electric fields, synchronised with the passage of the electron beam, and can generate much higher energies. Pulsed electron beams are often used for time-resolved measurements in materials modification. The application of a pulsed electric field to a cold needle cathode generates an electron pulse of a few hundreds of amperes in nanoseconds. Shorter electron pulses can be generated by linear accelerators using pulse compression techniques. One successful application of electron linear accelerators is the controlled coloration of semiprecious gemstones, like topaz.

Electrons with energy in excess of 0.6 MeV lose kinetic energy in a solid by electronic excitation within a range R, which is given approximately by a universal equation (Birkhoff 1958):

$$R = 0.526\,(E\,[\mathrm{MeV}])^{-0.094}\ [\mathrm{g/cm^2}].$$

The energy coming from the incident electrons is absorbed relatively uniformly within the range. Thus, the energy input per unit volume to a specimen irradiated with an electron beam of V [V] and current I [A/cm^2] for time t is roughly VI/R.

Electrons emitted from solids

In Auger spectroscopy, or in X-ray photoelectron spectroscopies, there is interest in the energy spectrum of the electrons emitted following excitation of the surface at a well-defined energy. The use of these spectra for chemical analysis is clear. However, there is another reason for us to be interested in the spectra. These emitted electrons can originate at different depths, and they will lose energy on their way to the surface. Only with special techniques is there good reason to believe that the outermost surface is being sampled. An example is metastable ion emission spectroscopy (MIES), in which an excited He atom is incident and an Auger process leads to emission of an electron from a surface atom. In other approaches, electron energy loss needs attention. The energy loss processes are essentially the same as those which occur when excited electrons move within a solid, redistributing charge and energy, and these processes are a component of materials modification. Ultraviolet photoelectron spectroscopy (UPS) can be analysed to reveal surface band structures.

When the primary spectrum of electron excitation is narrow, the spectrum of energies of emitted electrons can be divided into three sectors. There is a component with the primary energy, from those electrons emitted from very close to the surface, and for which there is very little energy loss or momentum transfer. A second, near-peak, component is dominated by the electrons from shallow layers which may have lost energy, but which have not been deflected greatly. The third, off-peak, component consists of those electrons which have lost some energy and which have also been deflected significantly. The key parameters are three characteristic lengths (Tougaard & Sigmund 1982). First, there is the inelastic mean free path λ, which is often estimated using a dielectric theory of the electron gas (Lindhard 1954) and which has a characteristic minimum at about 50 eV (Seah & Dench 1979). Secondly, there is a transport mean free path λ_1 defined in terms of the inverse differential mean free path for scattering, which is expressed in terms of a cross-section σ and a number N of scatterers per unit volume:

$$1/\lambda_1 = N \int d\sigma(\theta)\,[1-\cos\theta],$$

where θ is the scattering angle. This transport mean free path is similar to the inelastic mean free path at low energies, but becomes much larger at higher energies, above the minimum in $\lambda(E)$. Thirdly, there is the escape depth $\Lambda = (\lambda\lambda_1/3)^{1/2}$, and it is this which should be compared with observed escape depths; Λ also has a minimum at about 50 eV.

There have been systematic studies of secondary electron emission. Some, like Grais & Bastawros (1982) used the analysis of Dionne (1973), which assumes electron emission follows a two-step process of production and emission. The equation expresses the cross-section in the form [(escape probability)/(bandgap + electron affinity)]$\cdot f(\alpha_s d, \alpha_p d)$, with d the penetration depth, α_s the absorption coefficient of the secondary electrons, and α_p the penetration depth of the primary electrons. Empirically, for incident electrons of a few keV, the secondary yield from insulators is an order of magnitude larger than that for metals. Values for alkali halides are typically 10–20 (Grais & Bastawros 1982); for MgO the value is 22 (Whether 1964) while for diamond it is as high as 84 (Shih *et al.* 1997).

Photoemission, as described, can be regarded as the analogue of Einstein's photoelectric effect, generalised to allow for energy losses within the solid. But there is a quite different form of emission, closer to thermionic emission. This happens when there are many electrons excited, and when they interact sufficiently to establish a temperature (the timescales for this and competing processes are discussed in Chapter 2). The electrons then evaporate from the solid.

1.2.3 Other forms of particle excitation

Ion-beam excitation

Our main emphasis is on the consequences of the controlled electronic excitation of a material. Radiation damage by particle beams can be a more violent process. Its initial effect will often be nuclear displacement, with electronic excitation a secondary consequence. Even when nuclear motions are the key events, there is complex and interesting physics, since the processes occur far from equilibrium, and often involve several length scales and several timescales. These phenomena become more important at high excitation levels, as in the situations discussed in Chapter 11.

It would be wrong to assume that the only role of electrons arises because they give rise to an interatomic potential or to the screening of charges, as in conventional molecular dynamics or even self-consistent molecular dynamics. Neither approach gives a complete picture of what is happening. In any damage events, there will certainly be excited electrons as well as fast-moving nuclei. The transient (highly non-equilibrium) behaviour will indeed affect the way that radiation modifies the host matcrials. The importance of electronic excited states depends on their lifetimes. As we note elsewhere, excited states can persist long enough to be important even in metals.

If descriptions in terms of adiabatic nuclear motion are incomplete, it is true nevertheless that some broad general ideas can be obtained from very simple descriptions of collision dynamics. These concern the total amount of energy transferred in typical collisions. Clearly, a projectile which loses its energy in many small energy transfers will cause damage which is very different from one which transfers large parts of its energy in single events (see Fig. 1.2). Such qualitative differences in the damage produced by electrons, by neutrons, and by heavy ions, relate to several separate factors.

The first factor arises because the scatter cross-section can take significantly different forms. If it is dominated by long-range Coulomb interactions (Rutherford scattering, especially important for protons or alpha particles), the total cross-section will be large, but with a large fraction of the collisions having small energy transfer. For neutrons and slow-moving heavy ions, the scatter is closer to that of hard spheres. A second factor concerns the relative masses of the projectile and target atom. The energy transfer is largest for equal projectile and target masses, and small when the mass ratio is very large or very small (Table 1.2). This has important consequences. In a plasma, the electrons and ions may have different temperatures, and the equipartition of energy between ions and electrons may be slow.

A third factor concerns the effects of secondaries. If a large energy is given to a target nucleus, that target nucleus also generates damage. This can be seen from the collision dynamics of classical hard-sphere atoms, ignoring chemical effects. For instance, if a 1 MeV neutron is incident on an atom similar in mass to Cu (for simplicity, we consider an element instead of a compound), the average energy transfer per primary event is 30 keV. A 30 keV Cu atom has a

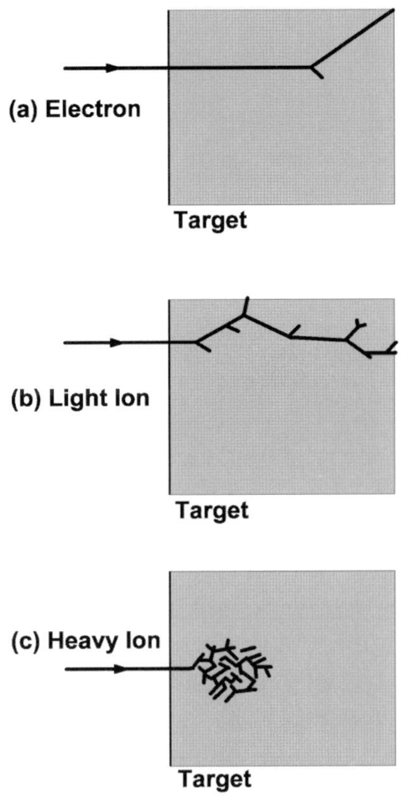

Fig. 1.2. Contrasted damage distributions arising from irradiation with: (a) electrons, (b) light ions, and (c) heavy ions.

range of about 100 nm, and will give a compact, highly damaged region containing a few hundred vacancies and interstitials. Likewise, fission fragments (typically heavy ions with atomic weights in the range 80–160, and with far higher energies of 100 MeV) leave damage tracks, typically 10 microns long.

When we come to interpret the actual tracks, we can no longer use just classical collision theory. Electronic processes do matter, and Cu would be an unhelpful analogy. The standard models of track formation are of two sorts (see Chapter 11). One model argues that there is electrical breakdown, driven by the electric field associated with electrons scattered from the centre of the track. A second model argues that sufficient energy is given to nuclei that there is melting, leaving an amorphous or damaged region. We shall argue later that electronic excitation, followed by energy localisation through self-trapping, can be the main cause of track formation. Clearly, in two of these models, the ideas go beyond simple adiabatic dynamics.

Table 1.2. *Energy transfer in a binary collision: Energy transferred in a binary collision between a 1 MeV projectile (electron, neutron, proton, Cu atom) and a target atom (H or Cu). In these binary collisions, the particles behave like hard spheres; we have chosen an element (Cu), rather than a compound, for simplicity.*

	H atom target	Cu atom Target
1 MeV electron	0.043% max, i.e. 430 eV	0.007%, max i.e. 70 eV
1 MeV neutron	100% max	6.1% max, i.e. 61 keV (30 keV average, hard sphere limit)
1 MeV proton	100% max	6.1% max, i.e. 61 keV (230 eV average, Rutherford scatter)
1 MeV Cu atom	6.1% max, i.e. 6.1 keV	100% max

Plasma treatments and gas-phase excitation

It is a standard technique in microelectronic processing to use an Ar or SF_6 plasma to remove oxide from silicon (e.g. Y H Lee 1992). An ideal plasma would remove the oxide very precisely, and would ignore the silicon. What are the mechanisms, and what can be done to control what happens? Control of the plasma involves such parameters as the chemical composition of the plasma, the applied bias voltage, and the plasma potential or the plasma density (which are interdependent). We shall not consider processes in the plasma phase directly, but only the energy input through a plasma source.

Clearly, several processes are involved, and these may be physical or chemical. For instance, an SF_6 plasma acting on Si improves the minority carrier lifetime at first, possibly through passivation of the surface dangling bonds by fluorine. Further exposure removes material in a way consistent with ionic radiation damage, with a standard value for the displacement energy. An Ar plasma creates a rough surface as well as removing material. On the oxide, the physical sputtering mechanism seems to apply only at higher biases; at lower energies, fluorine-containing plasmas appear to show chemical sputtering, which allows etching at ion energies which cause far less damage to the underlying silicon.

These two mechanisms – chemical and physical sputtering – are only two of a number of processes. The plasma has electrons which could cause damage, and also produces photons. These can cause electronic excitation and, possibly, lead to materials modifications by mechanisms we discuss in Chapters 9 and 11. Charging effects are of especial concern, since these can degrade silicon oxide gates during any plasma step after the oxide has been formed. Charging seems to involve several distinct components, partly from plasma non-uniformity, and partly from a current imbalance associated with the fact that the ion bombardment is more directional than the electron component.

There is a further plasma-based approach, which (in essence) exploits the plasma as a source of extreme ultraviolet light. A 1996 Sandia Laboratories patent described an approach in which a laser is focused on a beam of xenon gas clusters expanding at supersonic speeds; the plasma so created is the photon source. The source has demonstrated proof of principle in the generation of lines and spaces as fine as 0.15 μm for microelectronic devices.

Photo-induced chemical vapour deposition uses lasers or even synchrotron radiation to control chemical vapour deposition. The excitation is not strictly a plasma, but the excited species in the gas phase have some analogies. As an example, Matsui *et al.* (1992) discussed growth of silicon dioxide from O_2 and Si_2H_6 on Si or MgF_2 at rates of up to 13 nm/min. They reported Si–H and Si–OH species in the films.

1.2.4 Other forms of excitation

The most important further forms of excitation involve heat or chemical energy. We note, but shall not discuss, two of the special techniques involving lattice vibration. The first is *ultrasonics*, where there is the potential for special effects from high intensities, coherence, or directional effects. At still higher intensities are *shock waves*. Some of the consequences of strong local excitation may be the *generation* of shock waves. Nanosecond shocks can be achieved by using high-energy, short laser pulses to excite a dye embedded in a binder coated onto the sample. Pulsed shock waves can deliver transient temperatures of several hundred degrees and peak pressures of tens of thousands of atmospheres.

Chemical reactions

Chemical reactions between atomic or molecular species occur in the solid state as well as on surfaces or in free space. There are the obvious processes of oxidation and reduction, and these can be exploited in potential electrochemical energy sources. In the heterogeneous catalysis of exothermic reactions, significant energies can be released at a small particle during operation, and the particle temperature can be raised to close to its melting temperature. In self-propogating synthesis, a solid-state reaction can occur throughout a bulk system. This process lies outside our scope, except that it might be nucleated by electron excitation.

Thermal excitation

Heat is, of course, completely standard as a means to control processes, and we shall not regard standard thermal processing as one of our excitation mechanisms for materials modification. Nevertheless, there are some aspects which we shall need to mention. First, lasers allow processes which are far from thermal equilibrium, such as in laser cutting (Prokhorov *et al.* 1990). Prokhorov *et al.* also discussed laser-induced oxidation. Secondly, we note that, in systems with very narrow bandgaps, electronic excitation by thermal means is possible. Even with wide-gap materials, heat can change the charge states of shallow impurities or have other effects which enhance diffusion (see Chapter 7). Thirdly, we note that we shall often find situations in which the temperature changes rapidly. We may wish to compare a thermally-activated process, like diffusion, with some other phenomenon. If so, it is convenient to define a simple guide as to how much diffusion will occur. One such guide is the idea of a cooling time τ, such that the number of jumps made during cooling is the same as the number which would have been made at the initial temperature T_i in time τ. This can be estimated assuming linear cooling at rate $\theta = -\mathrm{d}T/\mathrm{d}t$. If $\gamma = (1 - \theta t/T_i)$ is essentially constant, then τ is of order $\gamma k T_i^2 / \theta E_m$ where E_m is the activation energy for diffusion. As an example, this type of expression is helpful in deciding when non-equilibrium segregation occurs (T M Williams *et al.* 1976).

In special systems, electronic excitation may lead to a subsystem being heated to high effective temperatures. Some examples are these:

The phonon bottleneck The distribution over spin states of an electron spin in a magnetic field can be controlled by an electromagnetic field. If the fraction of spins in excited states is raised, then it is often appropriate to describe the situation as one in which there is a spin temperature higher than that of the host crystal (which constitutes a thermal bath). The spin interacts with this host lattice, with the absorption or emission of phonons whose energy equals the Zeeman splitting (this is the so-called direct process of spin–lattice relaxation). Since these Zeeman energies are very low, typically 0.3 cm^{-1}, the phonon density of states is very low. Those modes which have the Zeeman energy can have far higher phonon occupancies than expected for thermal equilibrium. Indeed, the effective temperature for these phonons can become sufficiently close to the spin temperature that the energy transfer between spins and phonons is slowed down. This is the phonon bottleneck, a consequence of a non-equilibrium phonon distribution (van Vleck 1941; Scott & Jeffries 1962; Stoneham 1965, 1975).

Excitation of clusters In catalysis, small particles often play a key role. Because they are small this offers a large surface area for a relatively small amount of a possibly high-cost material. If there is an exothermic reaction on the surface (releasing say 5 eV) and the cluster is small (say 500 atoms), then the energy per atom could be as much as 0.01 eV, corresponding to a temperature of about 110 K. We shall not normally be concerned with such cases, but we should be aware of the increased importance of excitation in small clusters. An analogous example is the thermal isomerisation of isolated CsCl, CsBr, CsI clusters [$(CsX)_{27}$ or $(CsX)_{27}Cs$] (Fatemi *et al.* 1996).

Rapid thermal processing (RTP), which competes with some excitation methods. In a conventional furnace, temperature ramp rates are a few tens of degrees per minute. Fast-ramp furnaces have rates approaching 100 degrees per minute. RTP exploits tungsten halogen lamps, with a typical ramp-up of perhaps 25 °C/s and a cool down of say 50 °C/s, but may ramp up as fast as 50–200 degrees per *second.* Most of the processes operate between 400 °C and 900 °C; low temperatures are those below 800 °C, and medium temperatures from 800 to 950 °C. Short timescales would be less than 20 minutes; medium timescales would be 20 minutes or somewhat longer.

The excitation is used solely to heat, rather than to selectively excite specific species. The main aim is annealing, and hence to control diffusion processes or solid-state reactions in microelectronic processing operations. About 70% of production systems operate in the low-temperature regime, especially for silicide or nitride formation. However, the range of processes is wide: pre-cleaning and de-oxidising silicon wafers, oxidation or nitridation and, in combination with chemical vapour deposition, the adjustment of oxide dielectric thickness. The faster temperature rise permits chosen thermal diffusion processes, without so large an effect on other processes (we shall discuss thermal diffusion further in Section 1.5). A shorter time at a high temperature leads to a lower thermal budget.

1.3 Structure at the atomic scale

1.3.1 Structural issues: Where do crystalline and amorphous materials differ?

Some features of amorphous solids (or nanocrystalline solids) are significantly different from those for crystalline solids. We now consider some examples.

First, amorphous solids do not have dislocations in the usual sense, which has consequences for deformation and fracture (plasticity may need rebonding if there are directed bonds; in diamond-like carbons, a-C:H, for instance, it is likely that hydrogen motion is an important ingredient of plastic deformation). In nanocrystals, there is likely to be only a single dislocation, if any. Such a dislocation may account for the dependence of diffraction spot widths on momentum transfer known as paracrystallinity. Similarly, there is no natural analogue of grain boundaries in amorphous materials (and certainly not a clear distinction between tilt and twist boundaries) although, of course, there are interfaces (surfaces, amorphous/crystalline boundaries, boundaries between different compositions or topologies) and there can be medium-range order.

Secondly, there are no special close-packed directions. Special radiation damage phenomena, like focused collision sequences (important in separating vacancies from interstitials and hence in limiting recombination) and channelling are eliminated. The lack of long-range order means that diffraction experiments give somewhat different information for crystalline and non-crystalline materials.

Thirdly, all sites are not equivalent in amorphous materials, whereas the average positions of atoms in crystals do have translational symmetry. In an amorphous solid, the variations in energy from site to site have a correlation length which is often small, perhaps a few interatomic spacings. For rare-earth ions in an oxide glass, for instance, the precise energy for one of the sharp optical transitions will vary essentially randomly from ion to ion whenever the correlation lengths for the glass structure are smaller than the typical spacings between rare-earth ions. This has consequences for spectral energy transfer. In a crystalline solid, there will also be variations in energy for sharp spectral lines. These, however, will be caused by random strains or random electric fields of defects. In such cases, the correlation length will be the typical spacing between the defects causing the energy shifts. This can be quite long, typically a micron or so for dislocations. There will therefore be significant correlations in energy between the sharp lines of one ion and of its neighbours. In amorphous solids, like glasses, such long-range strain fields (not due to dislocations, of course, but perhaps due to inclusions) tend to be eliminated by topological changes in the structure.

Many models concentrate exclusively on short-range geometry or topology. Some of the standard models are shown in Fig. 1.3. However, medium-range order could be important. Medium-range order can take various forms. In crystalline solids, grains (or even smaller units) can be ordered to some extent. In amorphous solids, there are competing mechanisms which will be one source of medium-range ordering. Thus coordination constraints (Fig. 1.3(b), e.g. Si preferences for four-fold coordination, O preferences for two-fold) favour a continuous random network, whereas close packing of oxygens (Fig. 1.3(a)) with irregular interstitial atoms favours another arrangement.

When sites are inequivalent, there will be special sites which may make damage easier to initiate, or may make charge localisation easier. Thus electronic holes can localise at special sites in vitreous silica, but do not seem to localise in quartz. The special site may have a favourable electrostatic potential, or a favourable relaxation energy. The inequivalence from site to site means that the weighting used in estimating any 'average' quantities must be done especially carefully. In nanocrystalline materials, electric fields may be large on a microscopic scale, even when there is no macroscopic field. Among the 'special' sites must be mentioned the so-called tunnelling-level systems (two-

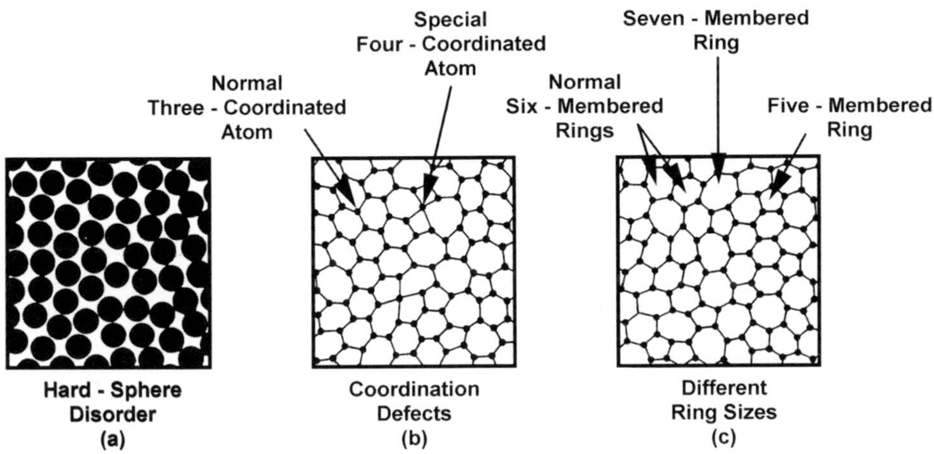

Fig. 1.3. Some idealised models of amorphous solids. (a) Hard-sphere disorder involves only short-range repulsion between the atoms. (b) Coordination disorder shows cases where there is a widespread pattern of neighbours (here three-fold), but with exceptions (e.g. four-fold coordination). (c) Topological disorder shows through altered numbers of atoms in the rings (five-fold and seven-fold, as well as six-fold rings).

level systems). The vibrational tunnelling shows up through characteristic soft modes, which alter the thermal conductivity and specific heat at low temperatures. Such features are seen as well in disordered crystalline solids, like doped fluorite systems (Kazanskii 1985). These sites have not been identified unambiguously, although there are several suggestions, mainly involving some cooperative motion of several atoms. What is surprising is that these 'soft' modes have not been associated systematically with sites at which defects are readily formed or where carriers easily trapped.

When there are small variations from one site to another, this can lead to trapping, since there will be individual sites at which carriers can be bound. For significantly higher disorder, a new phenomenon can occur. Anderson localisation (Anderson 1958) describes a situation in a disordered material when the disorder is sufficient to remove all delocalised states. This remarkable phenomenon should not be devalued by confusing it with trapping, when the disorder merely causes some of the states to become localised, and where delocalised states continue to exist.

1.3.2 The varied forms of 'amorphous'

Amorphisation is one of the possible results of excitation. However, operationally, it is usually enough that a system shows diffraction rings, rather than the spots expected of crystalline systems. However, diffraction rings can result from systems which have structures which differ widely (e.g. Stoneham 1987a). In particular, the scale of the amorphous features can vary considerably.

At one extreme, there are *polygrannular* systems. These, which include many oxides, consist of small crystalline fragments, randomly oriented; they can be thought of as a powder which has been fully consolidated. The opposite extreme might include continuous random networks. Some of the amorphous silicas and related glasses are close to this idealised structure. For the continuous random networks, there is a fixed topology, with clear short-range order: in an amorphous silica, each Si will have four O neighbours, each O will have two Si neighbours, and the Si–O distances will be fairly well defined. There will be limited medium range order, perhaps associated with the packing of the oxygens, and there will be no very discernible order at ranges beyond a few nanometres.

Another type of disorder is simply alloy disorder, as in mixed alkali halides like (Na, K)(Cl, Br). There is certainly an underlying crystal structure, but a random component is superimposed. There is still an underlying crystal structure when one of the species is molecular, and can reorient, like O_2^-, or like NO_3^-, or in a highly non-stoichiometric system, like $Fe_{1-x}O$. These cases are not normally described as amorphous, yet they may have similar behaviours in some experiments.

One must realise that, even for a single well-defined chemical composition, there need not be a single amorphous structure. In the case of Si, for instance, the liquid phase (and presumably the same system after infinitely rapid quenching) is metallic, whereas the lower-density forms of a-Si, like that formed by irradiation of c-Si, are semiconductors. There is plenty of evidence that a-SiO_2 can have a spectrum of structures, each with a well-defined density and enthalpy (Chapter 8). Different forms can be created reproducibly by choosing different formation routes: radiation-induced amorphisation, rapid quenching, or creation under external constraints. When Si is oxidised, for example, the constraints at the Si/SiO_2 interface influence the local oxide structure. Moreover, subsequent thermal or stress treatments appear to alter the topology of the Si–O network.

1.3.3 Mesostructure

Characteristic lengths

Materials are specified not only by their composition and crystal structure, but also by the microstructure which affects many properties. There are other characterisitic lengths which influence the performance of a material and its evolution under electronic excitation.

Grain sizes, and the dimensions of other well-defined domains like distinct phases, provide one obvious scale. Typical dislocation spacings provide another significant length scale. The roles of the boundaries and dislocations are varied. They can provide fast ion transport paths. They may act as sources or sinks of charges or of ions. Any sources or sinks of charge, whether dislocation or interface, will generate another characteristic length, namely a screening length. Typical space-charge layers extend for about 10–100 nm from a surface. With charge densities of the order of 1 electron for every 10 000 surface atoms, this corresponds to a difference in potential of perhaps 0.2 eV and a corresponding electric field of 10^4–10^5 V/cm. One can be more precise by combining the Poisson equation with a Boltzmann factor (see, e.g., Hayes & Stoneham 1985 p. 360) to obtain a characteristic length:

$$L = \{8\pi Z^2 e^2 c_0(T)/\epsilon kT\}^{1/2},$$

in which charges Ze are present in concentration c_0 in the absence of the boundary. If the dielectric constant $\epsilon = 10$, $T = 300$ K and c_0 is $10^{16}/\text{cm}^3$, then L is about 2.5 nm.

Interface roughness provides another scale. There are at least two characteristic lengths. One is the roughness amplitude. For surfaces prepared for scientific experiments, the root mean square deviation from an average plane may be of atomic dimensions. For surfaces prepared in technologies other than microelectronics, substantially larger root mean square amplitudes are possible. A second characteristic length is the correlation length across the interface. This gives the typical dimensions of protrusions or islands, and typical interstep distances. The two characteristic lengths together determine the electric field concentration. A final class of scale lengths of significance includes the mean free paths for electrons, holes, or excitons. It is these scale lengths which determine how far energy or charge is redistributed following excitation.

Evolutionary changes can happen without excitation: Ostwald ripening of colloids or voids, sintering, grain growth. We shall need to be able to characterise the effects of excitation on these processes, for instance when we contrast dislocation climb in the presence or absence of electron–hole pairs (Section 7.6.2). Some of the approaches to evolving microstructure are described in Section 2.6.

Length scales and time scales in radiation damage

In radiation damage, the key lengths will depend on the type of radiation and on the timescale. Electron damage produces isolated defects spread over fairly large volumes. As the incident particles increase in mass, localised cascades become more important. As noted, the range and the dimension of the energy deposition are related in standard radiation damage models (Sigmund 1972b). Sometimes there are special factors, such as rare precursor species whose spacing defines another further length scale. For instance, the rare Si–Si species is believed to be crucial in writing gratings in silica optical fibres.

On a microscopic scale, the key issues are ones like whether an interstitial recombines with the vacancy it has left (geminate recombination) or instead finds some other defect or trap. Clearly, there is a parallel for excited electrons and their recombination or trapping (Section 2.6). However, damage can generate quite new structures on a mesoscopic scale. One familiar example is seen in the cone structures which build up when there is sputtering by ion beams at non-normal incidence. The void lattice, comprising void arrays with long-range order, is another instance.

One important phenomenon, often with consequences on a mesoscopic scale, arises when electrons are excited and move away from some zone, leaving a net positive charge. The zone can be the surface, when there is exoelectron emission; this can be extremely important in the microscopy of insulators, and also in laser ablation. Electron redistribution can also be from the surface into the bulk when Compton scatter of incident gammas occurs. The positive zone can be a linear region, perhaps round a fission track, where the electric fields can lead to permanent damage. In materials with very low electrical conductivity, substantial energies may be stored as electrostatic energy, and can be released by mechanical impact. An example is a Ce-doped glass used for windows exposed to radiation. In such glasses, since Ce can occur in 3+ or 4+ charge states, electronic conduction is all but eliminated by trapping, and the slower ionic conduction of interstitial alkali ions gradually eliminates the net charge.

The positively-charged region decays most rapidly in metals, although one must be cautious in saying how fast, since this depends on the electron mean free path as well as on the plasma frequency (Bochove & Walkup 1990). For Cu and Ni, the timescale is 0.2 fs for the period of the plasma oscillations, with 0.6 fs (Ni) or 1.5 fs (Cu) as the electron collision time in the liquid metal. The lattice vibrational periods are typically 70 fs (0.07 ps). The characteristic times of transfer of energy between nuclei and electrons were given as 0.1 ps (Ni) and 4.3 ps (Cu) by Finnis *et al.* (1991), or as 10 ps from the analysis of Stoneham (1990) based on the general theory of plasmas due to Landau. These are similar to the 1–10 ps predicted by Sigmund (1974) for energy transfer at the end of a collision cascade; in his model, electrons are not explicit. Laser measurements for the recovery of hot electrons in Cu show a value of 1.8 ps (Golov *et al.* 1995). The energy loss time in metals following ultrashort optical pulses is proportional to $(\tau_{ee}\tau_{eph}^{2})^{1/3}$ (Gusev & Wright 1998), which involves both electron–electron and electron–phonon scattering times.

1.4 Basic issues of electronic structure

In this section we draw together some of the aspects of electronic structure which we shall need in later parts of the book. The discussion is thus selective, not comprehensive. However, we shall avoid restrictive assumptions as far as possible.

Our discussion is specifically not confined to macroscopic crystalline solids and traditional band structure methods. Defect production and amorphisation are, after all, a common consequence of excitation. At a time when there is much activity involving nanocrystalline materials and nanostructures, such as thin films and quantum dots, we cannot ignore the structure of solids in smaller units or on finer scales. Since we shall be discussing both crystalline and amorphous materials, we shall begin by identifying some general features common to all structures.

1.4.1 Band structures: General features for crystalline and amorphous solids

Bandgaps

The existence of bands and bandgaps is not related to crystallinity. The fact that one can see through glass and water should be reason enough to believe bandgaps do not need periodic structures. Certainly, most materials will become metallic when sufficiently compressed, but this requires conditions which, for many materials, are far from normal. Figure 1.4 shows some basic features of band structures. In particular, it identifies two key energies. The first is the *bandgap* δE. It is the gap which is a measure of how much energy is available from electron–hole recombination. The second key energy is the *bandwidth*. The bandwidth is a measure of the kinetic energy cost in localising a carrier on a single site. In non-crystalline systems, the band edge will be hard to define uniquely, and the lowest-energy excitations may vary substantially from site to site. Figure 1.5 shows the systematic trends in bandgap for many of the non-metals we shall consider.

At surfaces, there are several distinct effects which must be included. The surface breaks the three-dimensional translational symmetry of most band pictures. It is possible, but not always helpful, to think in terms of a band structure. The surface-related effects include the following. First, there will always be some surface relaxation or rumpling, causing deviations from the ideal

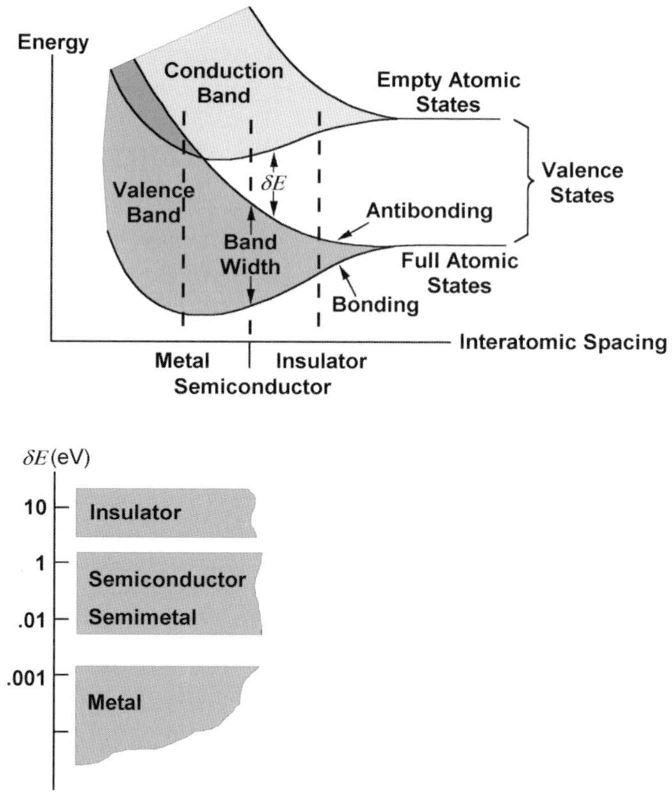

Fig. 1.4. Electronic energies as a function of interatomic spacing. Also shown is the energy δE needed to raise a solid to an electronically excited state, with typical values for metals, semimetals, semiconductors, and insulators.

termination of the basic crystal structure. Relaxation and rumpling will normally lead to a dipole layer at the surface, which will affect the energy needed to bring a charge into the crystal or to remove charge from the solid. There may be reconstruction as well, in which periodic surface structures are found. These changes will modify the electrostatic potentials and fields on surface and near-surface ions from the values expected for ideal terminations (which are, of course, different from those in the bulk). Spectra will be changed as well. Secondly, there may be charges trapped at interface states. Such charges are often represented by band bending: negative charges trapped bend the band upwards, meaning that more energy is needed to bring an electron in the conduction band to the surface. Another important term is the image potential, associated with the polarisation of the solid by charged species: the vacuum is unpolarisable, so that charged species will have lower energy away from a free surface. Conversely, metals have essentially infinite polarisability, so that

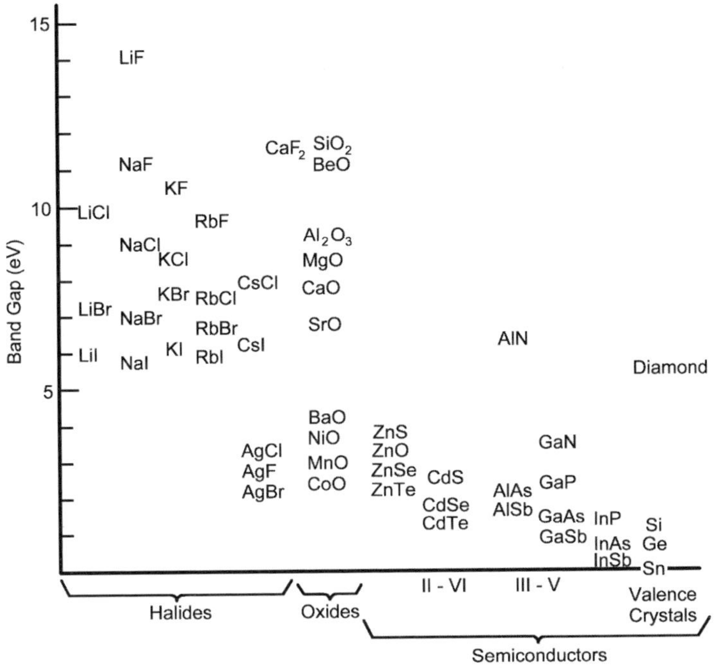

Fig. 1.5. Band gaps δE for some of the systems discussed in this book. Note the systematic trends with column of the periodic table, and with anion.

charged species in a non-metal gain energy from proximity to an interface with a metal. Thirdly, electrons and holes can be confined to regions close to surfaces or interfaces, either because there is some extra stabilisation, or through the combination of a band edge discontinuity and an applied electric field.

Bandwidths

When energy is localised, or charge is localised, there is a competition between two energy terms. One is the energy gain on localisation, for example from polarisation or distortion of the lattice, or from the formation of chemical bonds. This we shall discuss further in Chapter 3. The other energy term favours delocalisation. For a free electron, this is the kinetic energy. In the simplest tight-binding monatomic lattices (for which the band is symmetric about its centroid), it is easy to show that the energy cost in localising a charge from a band edge delocalised state to a single Wannier function is half the bandwidth. Thus, for the self-trapping of holes, the valence bandwidth is a critical quantity. The precise quantity needs to be defined with care when there are several atoms per unit cell, since the localisation may be onto only one of these atoms, and localisation to the Wannier function (which relates to the cell) is not quite right. Further, the bands may have some substructure, and be reasonably described as overlapping bands. These are some of the several reasons why comparisons of localised and delocalised states of carriers cause problems.

Another challenging issue concerns transition energies between bound and ionised states. For shallow levels, especially in those non-polar semiconductors where effective mass theory works readily, there are no problems unless there are high impurity concentrations. For shallow levels in polar hosts, the complications come for the extent to which lattice distortion or polarisation affects these energies. The situation is complex and can be tricky, but the issues are understood (see, e.g., Stoneham (1975), Chapter 8). For deep levels, accurate (rather than approximate) estimates are very hard. The problems include especially the electron correlation terms and the electron–lattice coupling (polaron) corrections for the two (localised and delocalised) states. In the case of those non-stoichiometric oxides for which the carriers are self-trapped, the relevant states are both localised, and their energy difference is more straightforward to estimate.

Absolute energies: Electron affinities and work functions

In some situations, it is necessary to know electron energies in one material relative to those in vacuum or in another material. There are several potential complications, often because there will usually be a dipole layer associated with the surface or interface. This means that there can be a dependence on interface (which crystallographic planes) even for the same materials. Experimental measurements of band alignment in Group IV or III–V semiconductors are in a relatively satisfactory state. Estimates of electron affinities (the energy released on moving an electron from the vacuum to the bottom of the conduction band) for polar solids like MgO range widely. In such cases, it is important to be sure precisely which experiment has been done, and whether it relates closely to the electronic processes being considered.

1.4.2 Approaches to electronic structure

Many descriptions of the electronic structure of a solid concentrate on two components: the charge density in the ground state, and the electronic band structure. It is interesting that the development of density functional theory, a powerful method especially suited to ground-state behaviour, has led to the interests of theorists moving away from spectroscopy (which involves excited states), despite the continued – or even growing – needs of experimenters. Understanding the ground-state behaviour is often sufficient to predict cohesive energies (and hence energies of alternative structures), elastic constants,

Compton profiles, and so on. Yet, for our purposes, approaches emphasising the ground state provide a very incomplete description. In our discussions here, we shall emphasise the simpler ideas, since we shall be discussing very complex behaviour. However, we shall certainly take advantage of the best available techniques when appropriate.

1.4.3 Special cases

Empirical and parameter-free methods

The ideas which underlie standard descriptions are often more helpful than the specific realisations when we come to analyse the behaviour of excited states. For example, a variety of simpler theories all recognise three distinct factors in understanding chemical behaviour. We can see this from theories like those of Miedema: the semiempirical chemistry approaches (e.g. complete neglect of differential overlap, CNDO) and the tight-binding methods (e.g. Horsfield *et al.* 1996). All these simpler theories invoke at least three common ideas. These are: (a) an orbital radius for the outermost electrons actually involved in chemical behaviour, (b) an electronegativity, which determines the natural direction for charge transfers between unlike atoms, and (c) a more elusive parameter which decides whether charge builds up to form bonds or is transferred, as in an ionic system.

These ideas are framed here in a way convenient for developing empirical parameters. The distinction between such *empirical* approaches and so-called *a priori* or *'first principles'* methods is not as great as the names imply (not least because most '*a priori*' methods are often a long way from the first principles often claimed). The choice of gradient corrections (or choosing to omit them), or of which pseudopotential should be used, or the size and shape of the cell for periodic boundary conditions, all can introduce uncertainties which are not always avoidable. Substantial errors can arise because of inadequate handling of ionic polarisation in small supercell models, making it especially hard to compare defect energies for states of different charge. Experienced workers can often avoid the difficulties. Those who regard the codes as a 'black box' may not be aware of their opportunities to make gross errors.

We shall not be making a general value judgement between the empirical and the parameter-free (often wrongly called *a priori*) methods. They are both important, and the best choice for a particular problem depends on that problem, on the system, and on the questions being asked. Some of the questions we shall need to ask are still too difficult for either approach at present. It is simply not true that self-consistent molecular dynamics means that electronic structure problems have been solved apart from details. The challenges are especially problematic for excited states, or when correlation is crucial and where the simpler models, like the Hubbard model, are insufficient.

The usual way to develop an empirical method is to start from a rather general approach, and make systematic approximations to identify certain key quantities; these key quantities may use some empirical information in their evaluation. The empirical method can, in fact, be regarded as a way to extend more fundamental methods to larger and more complex systems. We note two examples of simple methods which have given much insight: semiempirical calculation of the adiabatic potential energy surfaces and molecular dynamics. Issues concerning these simple methods follow.

Interatomic potentials for shell model calculations

When interatomic potentials are used, the electrons are implicit in the calculation, built into certain simple models. Simplicity does not imply inaccuracy, for some of the most accurate defect calculations have been achieved by these methods. The shell model provides a rather general framework at the dipole level (i.e. where monopoles and dipoles, but not higher multipoles, are associated with each ion). In this approach, the shell is virtually massless, and corresponds (not too closely) to the valence electrons responsible for short-range repulsive forces; the shell is coupled harmonically to a massive core which includes the nucleus and relatively inert inner electrons. The motion of the shell relative to the core gives the dipole moment, but – of course – the polarisability depends on the short-range interactions with neighbouring ions. In many cases (notably the halides and the closed-shell oxides like MgO), the potentials show both high accuracy and successsful transferability. The accuracy comes in part from the fact that Coulomb and polarisation energies are often the largest energies in some observable characteristic, and the experimental static dielectric constant is often an input to the potentials. The transferability is surprisingly wide. Potentials fitted only to perfect crystal bulk properties often give good predictions for surfaces, for behaviour under shock wave compression, for highly-distorted geometries like diffusion saddle points, and even for certain electronic (optical) charge transfer transition energies (see, e.g., Stoneham *et al.* 1981, Stoneham 1987c).

Zero differential overlap methods (ZDO; CNDO and more sophisticated approaches)

Formally, these can be derived by systematic simplification of either Hartree–Fock theory or of density-functional theory (Lindholm & Lundqvist 1985). The electronegativity and orbital radius parameters are fairly tightly constrained by known atomic and ionic information. The third ('bonding') parameter can be fitted in many ways. In our experience, a good strategy is to regard the ZDO methods as a means of extrapolation or interpolation. Thus, if one is interested in predicting vibrational properties, it can be useful to obtain bonding parameters from systematic fits to diatomic molecular vibrational data. It goes without saying that the parameters should be fixed at the start of a series of calculations, and not be changed during the calculations: failure to agree with experiment gives information, about either the model or one's interpretation of what is happening.

Embedding versus periodic boundary conditions

Even the massive calculations which handle 10^8 atoms explicitly are a long way from a full treatment of a solid of 10^{23} atoms. There are two broad ways in which significant results can be obtained from a more manageable number, perhaps 100 atoms. The first method is *embedding*, which includes cluster methods as a special case. The second approach uses *periodic boundary conditions.* Both approaches have virtues, and both do have problems if used in a naive way.

In a cluster approach, a modest number of atoms or ions is taken. The cluster may have special atoms or pseudo-atoms at the surface to terminate dangling bonds. For an ionic crystal, there may be extra point ions outside it, so as to ensure the correct Madelung potential, or the right electric field. The embedding approaches go further in the way that the central group of ions is handled. There will normally be a way to make sure that the region far from the central cluster (which is often defective) correctly reproduces the bulk properties of the solid. Thus the band structure will be correct. It is usually more important to ensure that the ionic and electronic polarisation is correct. For charged defects, errors in the polarisation energy can be very large (5–10 eV in some cases known to us). The way in which the outer region and the central cluster are matched is a problem, with many attempts at solution. In most cases, useful solutions are possible when there is understanding of the physics of the specific system.

In a periodic boundary condition approach, there is no external boundary, simply a series of large unit cells, each of which may contain a defect. One problem concerns the intercell defect–defect interactions. These are especially a problem when the defect is charged, both because of the polarisation problem and since it may not be sufficient merely to subtract off monopole terms using the common trick a uniform charge backgound. Indeed, many workers seem wary of studying charged defects within a periodic boundary conditions framework, and some published results contain large errors (as much as 5 eV) through inadequate attention to large energies.

It is usually a matter of personal choice as to which approach one uses. In practice, neither approach is easy to apply to shallow defects with a wavefunction which spreads over many atoms. Excited states are a problem, but marginally easier in cluster methods; unpaired spins cause difficulties. Careful calculations can achieve good results by either route.

Molecular dynamics

Molecular dynamics, in various forms, is an extremely important tool for simulating atomic motion after excitation: it solves the classical Newton equations numerically for the motion of ions. The differences from one form to another are primarily in the way that the forces on the ions are calculated at each timestep. Most work uses interatomic potentials, usually in a rigid ion model. Other approaches solve the electronic structure problem in some way at each timestep, and use this as the basis for an estimate of the forces. This is done at several levels: from tight-binding calculations; from semiempirical quantum chemistry; by use of the Car–Parrinello approach, in which the description of the electron density is included in new dynamical variables.

The strength of molecular dynamics is that, often, very few assumptions need be put in as to which approaches are important. It is often not necessary to guess a saddle point, or to decide which defect configuration is the most stable. The limitations are primarily those of system size and of the length of real time modelled. Some calculations have successfully modelled up to 10^8 atoms for useful periods. However, certain problems remain. First, rare events need special treatment. If a process has a high activation energy, then some trick must be chosen to ensure that the event occurs within a molecular dynamics run. Secondly, each molecular dynamic run starts from a particular set of momenta and positions. Some processes will be sensitive to the precise conditions, so there may be a need to make runs for many such realisations of a system, i.e. for many different starting coordinates and momenta. Thirdly, and especially important for us, it is very hard to handle excited states in self-consistent molecular dynamics. The underlying reason is that self-consistent methods seek a minimum energy state. For the completely general symmetries met in molecular dynamics, it is hard to prevent the system slipping from the desired excited-state energy surface onto the ground-state energy surface. One standard way is to force the excited state to have a different spin from the ground state, but this is not always an appropriate solution.

1.4.4 Localising charge

Kanzig's dramatic spin resonance observation of the self-trapped hole (V_k centre) in KCl (Kanzig 1955) caused theorists special concern, especially those whose background was band theory. Bloch had proved that, for a perfectly periodic solid, the eigenstates must also have periodicity; surely one should not find apparently localised states? The problem is more apparent than real. One must consider the whole system (electrons and phonons) and recognise that the small polaron might be found on any translationally-equivalent site with equal probability. However, there are important points to note. First, the point symmetries of the free carrier and of the localised polaron states (say, the one-centre hole and the V_k centre) are different. Secondly, for the localised state, the transition matrix elements of the Hamiltonian between different sites determine the polaron transport properties, and these will not be the same for the self-trapped polarons (for which there will be a lattice overlap component) and the free polaron. Thirdly, special attention should be paid to precisely which experiment is done, and to the characteristic timescale relative to the timescale of polaron relaxation from a free state and the time between transitions from site to site.

The electronic polaron

Electrons respond more rapidly than ions to the creation of a carrier, and the electronic polarisation of the host lattice ions affects the way the localisation process evolves. This suggests a description as an electronic polaron, a quasiparticle of an electron or hole together with its associated electronic polarisation (Toyozawa 1954). This description has been used often in interpreting the interband optical properies of insulators, photoelectron emission spectra, X-ray photoemission spectra (XPS), X-ray absorption spectra (XAS), electron spectroscopy for chemical analysis (ESCA), and other experiments (see Siegbahn *et al.* (1967) and references therein). The same ideas are also relevant to electrochemical processes involving thermalisation and solvation of electrons.

The concept leads to two basic questions. First, what is the wavefunction of the electronic polaron? This determines matrix elements of the electronic excitations in spectroscopy, and the localisation rate in the polaron studies. Secondly, what is the energy of the electronic polaron with respect to the ground state of the system before excitation? This determines the ionisation and excitation spectra, and also the relaxation energy into a final polaron state.

For many systems, the dynamical effects of electronic relaxation and correlation can have a large effect on the extent of localisation and the energy of the quasiparticle. This is so for polar solids and liquids, and for the hydrated electron (del Buono *et al.* 1992). In some other systems, such as electrons in liquid helium, the situation is simpler. In these simpler systems, one can neglect the effects of electronic relaxation of the atoms induced by the polaron, and the polaron localisation can be obtained using a one-electron model and classical polarisation of the medium (Barnett *et al.* 1988). The need for the more sophisticated approach can be seen for core-level spectroscopies where, in the final state of X-ray photoemission or absorption spectra, a core hole is left behind, and this couples with valence electrons. There are similar problems for excitations from different states in the valence band of crystals with mixed bands, such as AgCl. In systems with partly-filled f- or d-shells, the coupling between the core hole and the d- or f- electrons is strong enough to give characteristic spectral splitting in XPS and XAS (see Kotani & Ogasawara (1992) for a review).

Electronic relaxation following excitation

The energy of electronic relaxation following localised excitation or the ionisation of core states can be larger than those interactions which determine the valence bandwidth. These many-electron effects are often treated using quantum-chemical methods. Some form of embedded cluster model is also common; this can give reasonable ionisation and excitation energies, but there are some important technical issues.

Many of the many-electron calculations use a one-electron (but self-consistent) approximation, such as Hartree–Fock theory. With such methods, the eigenfunction of the Fock operator is not necessarily an eigenfunction of the point group operator representing the total group of the system. Bagus & Schaefer (1972) pointed out that symmetry-unrestricted or 'broken symmetry' Hartree–Fock solutions gave a better first-order description in such cases (see also Fukutome (1983) and Davidson & Borden (1972) and references therein). A simple model analysis (Broer & Nieuwpoort 1981; Sawatzky & Lenselink 1980) indicates the conditions favouring broken symmetry solutions for hole states. The criterion depends on the relative magnitudes of the orbital energy splitting (a measure of delocalising interactions) and the relaxation energy associated with the electron ionisation. If the local relaxation accompanying ionisation is larger than the orbital energy splitting, the lowest energy has a wavefunction corresponding to the lower symmetry. The hole will be able to localise on one of the geometrically equivalent atoms. Since all atoms remain at their sites for the electronic polaron, both point and translational symmetry should be preserved in its exact wavefunction, at least for the time-averaged atomic positions. This can be achieved by using a more rigorous treatment. Thermal vibrations mean that, on timescales short compared with vibrational periods (typically 10^{-13} s), individual atomic sites have lower symmetries. The variations from one site to another can initiate self-trapping.

Why does a broken symmetry Hartree–Fock wavefunction (with the hole localised at a single site in a small cluster) have a lower energy? This is partly due to an increase of the electronic relaxation energy (the relaxation energy decreases as the charge is spread out), and partly due to electronic polarisation of surrounding atoms (see, for example, Snyder (1971) and p. 35 of Hayes & Stoneham (1985)).

However, the correlation between electrons promotes the delocalisation of the hole and restoration of the symmetry of the wavefunction, and this correlation is only partly taken into account in the broken-symmetry Hartree–Fock state. The correlation can be treated by projecting proper symmetry states out of broken-symmetry wavefunctions and by using a configuration interaction technique. The wavefunction of the system can be constructed as a linear combination of wavefunctions corresponding to the hole localised onto different equivalent sites. The elements of the Hamiltonian between these states are many-electron transfer integrals. If the off-diagonal matrix elements are large, the total energy will be reduced significantly, and the electron density may be quite different from that of the broken-symmetry solution. The broken-symmetry solution gives quite a good description of photoionisation and optical absorption spectra, to a few tenths of an electron volt (Broer & Nieuwpoort 1988). Even this accuracy is not enough for the barriers for self-trapping of polarons and excitons, and a fuller account of electron correlation is needed.

The electronic polarisation of the lattice accompanying the electronic polaron has little effect on transport properties. The slowly-moving quasiparticle can be considered as a 'free' electron. In simple cases, such as alkali halides, the electronic polaron state for the valence hole may be treated sensibly as localised on one of the p-orbitals of anion. The dispersion of the free hole band is then the same as that for the valence band. As Fowler (1966) has shown, in these crystals, the classical Mott–Littleton method is acceptable for calculating electronic polarisation effects.

Calculation of the localisation energy

Localising an electron on a single site clearly causes an increase in kinetic energy relative to a delocalised state. If the localised state is the Wannier function corresponding to a single band for which the energy is $E(\mathbf{k})$ for wavevector $\mathbf{k}$, then the expectation value of the Hamiltonian is $\langle E(\mathbf{k})\rangle$, the mean energy of the states of that band. For a single electron in an otherwise empty band, the lowest energy delocalised state is that at the bottom of the band. For a single hole in an otherwise full band, the lowest energy state is for a hole in a state at the top of the band (the band of free holes corresponds to the valence band inverted, so the bottom of the hole band corresponds to the top of the valence band (see Stoneham (1975), Section 2.2.5). For a symmetrical band,

$\langle E(\mathbf{k})\rangle$ is in the middle of a band of width $2B$, then the energy cost E_{loc} in localising the carrier is simply B, or half the bandwidth. This result is especially helpful in the simpler cases, such as the s-like anion core states in oxides.

There are several reasons why this result cannot be better than an order of magnitude estimate in many cases. Any errors in E_{loc} can be important, since the stability of the small polaron can involve a delicate balance of energies. The first problem is that the usual Wannier function corresponds to a unit cell, not a single atom or ion, whereas we are often concerned with localisation at a single site. A second problem is that the band may have quite a complex structure. In many crystals, the valence band has p-like or d-like orbital character. For these cases, using the simple approach substantially overestimates the localisation energy (Kantorovich *et al.* 1993; Shluger *et al.* 1992a,b). The situation is even more complicated when the valence band is a mixture of the states derived from two or more sublattices, as for AgCl, Al_2O_3, SiO_2 and transition metal oxides. For the mixed band case, a method based on a model Hamiltonian technique and the tight-binding approximation has been proposed (Heifets & Shluger 1992), which takes account of the hole polaron states and of orbitals of surrounding ions.

Most approaches assume that the matrix elements of the tight-binding Hamiltonian may be obtained from matrix elements from a band structure calculation for the perfect crystal. This link is possible if, at least, two conditions are fulfilled. First, the atomic orbitals of anions and cations may be taken as Wannier functions, which corresponds to a strong tight-binding limit (cf. Stoneham 1975, Section 2.2.3). Secondly, when we exploit an intermediate localised reference state, Φ_i, in our later cluster calculations of self-trapping, the wavefunction of this reference state is *physically equivalent* to the Wannier function localised at a certain perfect lattice site. These assumptions may be too simple. In many interesting cases, the electronic relaxation for the electronic polaron can differ from that for the valence band electrons, a difference only partly taken into account in the methods of Heifets & Shluger (1992), Kantorovich *et al.* (1993), Shluger *et al.* (1992a,b). The localisation energy calculated using the frozen wavefunctions of the perfect crystal may be inaccurate, and a more flexible variational approach is desirable to obtain accurate localisation and relaxation energies.

An alternative approach uses cluster calculations. The wavefunction of the intermediate polaronic state in the perfect lattice, which should correspond to the wavefunction of the localised polaron state, is often a broken-symmetry solution of the Hartree–Fock equations. Broer & Nieuwpoort (1988) suggested how the Hamiltonian matrix elements between broken-symmetry wavefunctions might be calculated with holes localised on different equivalent sites.

Their variational method is useful in investigations of 'electronic polaron' bands (Broer & Nieuwpoort 1988; Janssen & Nieuwpoort 1988; Broer *et al.* 1991), and thus for evaluation of the polaron localisation energy. The important advantage of their approach is that the matrix elements needed to evaluate both the localisation and the relaxation energies of polarons can be calculated using the same quantum-chemical technique and sets of parameters. For electron (as opposed to hole) polarons, there is a special problem because calculations of conduction bands and of the crystal electron affinity are very unreliable.

Atomistic modelling of polarons

For some parts of the calculation of small polaron stability and dynamics, it is possible to avoid explicit consideration of electrons altogether. The electrons are implicit, since they determine the key interatomic potentials. In this way, several workers have used the major methods of atomistic modelling (combining electronic structure theories and shell-model-based approaches) to predict what might be expected in real solids. This approach is valuable because it can be applied systematically to complicated crystals, and its accuracy can be surprisingly high.

Mott's early work was summarised in Mott & Gurney (1948), and led the way to the systematic use of Mott–Littleton methods (Mott & Littleton (1938); see also the articles in Catlow & Stoneham (1989)) in serious quantitative studies of polarons. The approaches of Mott & Gurney, Gilbert (1966), Toyozawa (1961), Tolpygo (1957, 1961) and Fowler (1968) have been developed by many other workers, although challenges remain (see Shluger & Stoneham (1993), Stoneham (1989a)). The uses of atomistic computer modelling fall into three main classes. First are the mechanisms and the dynamics of polaron self-trapping. Secondly, there are static properties, such as the relative energies of large and small polarons, the optical transitions expected, the positions of nearby ions, and vibrational entropies. Thirdly, there is small polaron hopping and diffusion. Analytic theories provide a basis for some parts of these computational studies.

Atomistic studies of polarons involve certain technical difficulties, as can be shown from examples. First, localised carriers mean ions in non-standard charge states. Some, like the 3+ states of Co or Ni, or the 1− state of O, are seen in nature, and so interatomic potentials and shell-model parameters can be validated. In other cases, like the −1/2 state of Cl (as in the V_k centre, essentially an X_2^- ion) known only in molecular ions, it is not easy to validate estimates of their polarisabilities and short-range interactions with other ions are problematic. Such ions have to be treated using quantum techniques. Secondly, we shall combine (and sometimes compare) energies from two quite distinct calculations: those from band structure for a rigid perfect lattice, and lattice distortion and polarisation from cluster calculations. Great caution is needed to avoid double-counting of energies. There are also problems in determining the correct bandwidth of the polaron band and the appropriate electron affinity for the crystal. Thirdly, some key energy differences are very small, only 0.1 eV or so, and useful accuracy is easier to claim than to demonstrate.

Phenomenological approaches

The phenomenological pictures (Ueta *et al.* 1986) focus on how the stability of different types of polarons depends on the strength of the electron (exciton)–lattice coupling, on the dimension of the lattice, and other parameters. Gerlach & Lowen (1991) reviewed much of this work. Emin & Holstein (1976) used a scaling approach based on an adiabatic approximation and a continuum model for the lattice. Depending on the nature of the electron–phonon coupling (e.g. deformation potential versus Frohlich coupling), they find several different categories of behaviour (see Section 3.3). Some systems might have large polarons only, others small polarons only. In other systems, polarons could be of either type, with a barrier between the configurations. For a three-dimensional continuum, a purely short-range electron–continuum interaction always yields two distinct cases: a small-polaron state and non-polaronic free state. The addition of a long-range component to the electron–continuum interaction can convert a non-polaronic state into a finite-radius polaron state. Both Emin & Holstein (1976) and Toyozawa & Sumi (1974) identified a dependence on the dimensionality of the lattice. In one dimension, only a single stable state is predicted, a localised polaron. This contrasts with the three-dimensional case, where two minima of the adiabatic potential, corresponding to delocalised and strongly localised situations, are always predicted.

Toyozawa and his coworkers looked at the role of electron correlation too, in an '*STU*' model (*S*, a measure of electron–phonon coupling; *T*, a measure of the bandwidth of a bare electron in the absence of such coupling; *U*, the Hubbard *U*, measures the on-site electron–electron interaction and so includes electron correlation). The theory and results of extensive experimental studies reveal a much richer picture of possible polaron and exciton states, which we summarize in Table 1.3. The terms 'large polaron' and 'small polaron' are convenient but, as Ueta *et al.* (1986) have noted, should be used with care. The 'small' polaron is not just the strong coupling limit with electron–polarisation (optical phonon) interaction. Short-ranged interactions with acoustic phonons can trigger a self-trapped state which can coexist with a large radius or delo-

Table 1.3. *Types of polaron. This table (based on Shluger & Stoneham (1993)) gives examples of some of the polaron systems observed experimentally.*

Large polaron	e in alkali halides; e, h in II–V and Group IV semiconductors.
Small polaron	(i) One species only (e small or h small, not both) on one ion: h in AgCl, solvated e in water and alcoholic systems. (ii) Both electrons and holes small; centred on one ion only: e, h on cation for certain transition metal oxides. (iii) One species on two sites: h in halides (V_k centre). (iv) One species on many sites: 'polaron' in trans-polyacetylene. (v) Exciton: excitons in alkali halides, SiO_2.
Trapped polaron	(i) Trapped onto impurity site: transition metal ion impurity. (ii) Trapped onto impurity/defect sublattice: V_k centre next to Br in KCl. (iii) Trapped onto another sublattice: V_k next to Na in KCl; h near Li in MgO. (iv) Both large and small polarons trapped by the same defect: (a) bistability: CdF_2:In; (b) CsI:Na; (e large, h small).
Bipolaron	(i) Electrochromics WO_3 etc. (ii) Oxide superconductors.

calised state (Rashba 1957; Toyozawa 1961). It is the short-range interaction which makes the self-trapping a type of critical phenomenon, giving rise to an abrupt change of a parameter like the effective mass for a small change in the electron–phonon coupling constant. Optical phonons do have a big effect on the nature of the polaron state, but interaction with them alone leads only to a smooth change of the effective mass. Experimentally, it is very hard to alter the coupling constant, so it is difficult to validate those phenomenological theories which emphasise behaviour as a function of coupling constant. Their value lies mainly in providing a framework for the atomistic theories.

1.5 Excitation and excited states

Both crystalline and amorphous materials have a similar variety of excited states. These excitations include: the correlated motion of electrons and holes, as in excitons and aggregates of excitons; collective excitations, like plasmons; energy localisation, as for bound (localised) excitations; coupled carrier and lattice deformation, as for polarons; charge or energy localisation from self-trapping, and so on. The similarity of crystalline and non-crystalline systems does not mean that crystallinity is irrelevant. It means more that the meaning of selection rules which are based on symmetry is secondary and, in particular, that a 'direct' or 'indirect' bandgap description can become unhelpful.

1.5.1 Optical excitation

In a perfect crystalline solid, the lowest-energy electronic excitations create excitons (Table 1.4) below the band-to-band transitions which generate electron–hole pairs. In absorption, the exciton lies below the band edge by an energy corresponding to the Coulombic binding of the electron and hole; in emission, there are further factors, such as the consequences of lattice deformation and possible self-trapping.

In some semiconductors, especially those with small bandgaps and large effective-mass Bohr radii, molecular excitons and other combinations of electrons and holes are formed. These include plasmas or electron–hole drops (see, e.g., Hayes & Stoneham (1985) Section 1.3.3 and Fig. 1.19 for an outline). The phase diagram of the exciton gas has regions corresponding to an exciton gas, to an exciton gas with droplets of liquid, and to an electron–hole liquid at higher densities. High exciton densities correspond to a highly-excited state, yet there is very little energy localisation. These excited states usually have no effect on materials modification, except in those cases where a change in charge state can enhance a diffusion rate.

Optical transitions have selection rules. The most significant for the perfect periodic solid corresponds to conservation of quasimomentum, and results from translational symmetry. There are different asymptotic behaviours for the tails of systems with direct bandgaps (conduction band minimum and valence band maximum at the same point of the Brillouin zone) and with indirect band structures (see Knox (1963)). In this section, we consider those primary processes which do not involve the lattice relaxations of self-trapping and consequent defect production.

The rate of single-photon band-to-band processes can be described as a cross-section σ, which is typically 10^{-17} cm^2 for direct bandgaps, and smaller for indirect gaps. Similar cross-sections hold for many defects and impurities, but energy absorption peaks at different energies (for a fuller discussion, see Stoneham (1975)). The number of excitations per second per site in a solid is then:

$$N = \sigma I / h\nu = \sigma F$$

where $h\nu$ is the photon energy, I the *irradiance*, usually quoted in W/cm^2, and F the photon *flux density*, usually in units cm^{-2} s^{-1}. Typical optical attenuations are of order 10^5/cm for light with photon energy greater than the bandgap in a

Table 1.4. *A glossary of the main species of excitons.*

Exciton	*In essence, an exciton is an electron and a hole moving with a correlated motion as an electron–hole pair. The word also is used with various more specific meanings.*
Wannier exciton	Electron and hole both move in extended orbits. Energy levels related to hydrogen atom levels by scaling using effective masses and dielectric constant. Occurs in covalent solids such as silicon.
Frenkel exciton	Electron and hole both move in compact orbits, usually essentially localised on adjacent ions. Seen in ionic solids, such as KCl, in absorption.
Self-trapped exciton	One or both carriers localised by the lattice distortion they cause. Observed in ionic solids, such as KCl, in emission. Self-trapped excitons can take several forms, e.g. one-centre (hole on one halogen ion) or two-centre (hole forming a halogen molecular ion). A perturbed self-trapped exciton includes at least one impurity ion next to the halogen ion(s) involved.
Bound exciton	A useful idea for cases in which a defect merely prevents translational motion of an exciton and does not otherwise cause significant perturbation. The description has often been misused in semiconductor systems
Core exciton	Lowest-energy electronic excitation from a core state, leaving an unoccupied core orbital (e.g. the 1s level of a heavy atom) and an electron in the conduction band whose motion is correlated with that of the core hole.
Excitonic molecule	A complex involving two holes and two electrons.
Multiple bound excitons	Complex of many holes and a similar number of electrons, apparently localised near impurities. Some controversy exists, but up to six pairs of localised carriers have been suggested.
Exciton gas	High concentration of electrons and holes in which each electron remains strongly associated with one of the holes (an insulating phase).
Electron–hole drops	High concentration of electrons and holes in which the motions are plasma-like (a metallic phase), not strongly correlated as in excitons and included here only for comparison.

Table 1.5. *Energies (in units of the bandgap) needed to create an e–h pair.*

Classes of system	System	*W* value
Halides	NaI	2.7
	CsI	2.5
	CeF_3	3.8
Oxides	$Bi_4Ge_3O_{12}$	3.3
	Y_2O_3	4.6
	YVO_4	7.5
Sulphides/oxysulphides	ZnS	2.9
	La_2O_2S	4.0
Phosphides and related	$LuPO_4$	6.1
	CeP_5O_{14}	7.8
Semiconductors	c-Si	3.2
	a-Si	3.5
	Ge	4
	GaAs	2.7

direct gap material; values for indirect gap materials are somewhat less. Haglund & Itoh (1994) note there are typically 100 band-to-band excitations per site in a nanosecond laser pulse of 1 J/cm^2. The energy absorbed is ultimately coverted into heat. If absorption is primarily at the surface, or at an internal interface like a grain boundary, the heating will also be local. This is important in laser ablation (Chapter 11).

1.5.2 Excitation by ionising radiation

The number of electron–hole pairs produced in non-metals is approximately proportional to the energy deposited by irradiation as electronic excitation:

Number of electron–hole pairs = (local energy deposited in electronic collisions)/(*W* value).

The *W* value is typically 2–3 times the bandgap energy, irrespective of the incident energy or type of incident particle (Table 1.5). The *W* values are used in health physics (ICRU 1979), where they are one aspect of *degradation spectra*. *W* values have been calculated for many solids and evaluated experimentally. The reason that *W* is greater than unity is partly that there are energy losses from the generation of optical phonons by the primary fast carriers, and partly that any fast carriers will cause ionisations, ultimately yielding carriers with too little energy to create new ionisations. Plasmons may be another mode of energy loss, but plasmons can also excite electrons and holes. Robbins (1980) gave a discussion of the proportions of energy going to phonons and to ionisation, and also analysed the energy lost through the creation of slow carriers. Dubeau *et al.* (1996 and references therein) evaluated the energy loss of secondary electrons against impact ionisation and the energy possessed by electrons and holes which have energies less than the bandgap. Alternatively, a statistical method (the so-called crazy carpenter method) can be used (van Roosbroek 1965), in which the key step is to divide the input energy into small components, each lying between the bandgap energy and twice the bandgap energy. Blasse (1993) remarked that the Shockley–Robbins description is used in the West, but Eastern European scientists use a different description, based on fast electrons losing energy mainly to plasmons; this also gives W values of about 3.

1.5.3 Excitation at higher energies

Excitation with higher energy photons

As the photon energy increases above the bandgap energy, the kinetic energy imparted to the electron increases and the cross-section for excitation falls. When the photon energy exceeds the core excitation energy, the cross-section for core-electron excitation becomes bigger than that for valence-electron excitation, the largest cross-section being for excitation of those core electrons whose binding energy is closest to the incoming photon energy. The kinetic energy imparted to the electrons gives rise to impact ionisation and to heat (lattice vibrations) via the electron–lattice interaction. The core holes become filled by electrons from higher levels, transferring the holes to upper core levels, mostly by the Auger process.

Electrons excited during the Auger process lose their kinetic energy in the same way as the photoelectrons. They cause further ionisation, and create electron-hole pairs. Electrons in the 50–1000 eV energy range have relatively small mean free paths. It follows that energy possessed by low-energy electrons is deposited near the site where the core hole was produced. For example, Kikas & Elango (1985) made Monte Carlo calculations of the electron–hole pairs produced by Auger electrons with several energies in NaCl. They showed that the electron–hole pairs were distributed within a few nanometres of the core hole. Thus, the ionisation of core electrons leads to dense electron–hole pairs as well as multiple holes. The number of electron–hole pairs in Si produced per photon changes by less than 1% as the photon energy crosses the core excitation energy, probably because of a change in the distribution of the Auger-electron energies (Sholtz *et al.* 1996). The creation of a core hole is accompanied by the creation of n holes in the valence band and $(E_K - nE_V)/W$ electron–hole pairs, where E_K is the core energy and E_V the valence band energy with respect to vacuum.

If we include the electron–hole pairs produced both by the Auger processes and by the photoelectrons from the initial photoexcitation, the total number of electron–hole pairs per photon is approximately the photon energy absorbed divided by the *W* value, the average energy needed to generate an electron–hole pair under uniform excitation (Section 1.5.2). For irradiation with photons of energy below 10 keV, the *W* value can be affected by the non-uniform distribution of electron–hole pairs generated by core excitation. A change in *W* value is seen when the photon energy crosses a core excitation energy, but the deviation is less than 3% (Scholz *et al.* 1996). We will see later that the yield of some radiation products is altered when the photon energy crosses a core excitation energy.

Electronic correlation is especially important in the very narrow core-hole bands and, for this reason, core holes are relatively immobile, and can be localised on a single site more readily (see Chapter 3). In special cases, the migration of core holes can be observed. The 4p hole on Rb^+ in (Rb/Cs)F can be monitored as it moves and is trapped as a 5p hole on Cs^+ (M Itoh *et al.* 1992). The diffusion constant is estimated to be 1.5×10^{-5} cm^2/s.

Holes in the core band nearest to the valence band can be also filled by an Auger process if the bandgap energy E_G is smaller than the energy difference between the core and the top of the valence band. This condition is satisfied in most insulators, except for BaF_2, CsF, CsCl, CsBr, and RbF. In these materials, electron transitions from the valence band to the core band are radiative, emitting valence-to-core-band luminescence. This luminescence is not quenched by phonons, and can be observed even above room temperature. The transition is usually allowed, with a lifetime of order 1 ns (see, e.g., Kubota *et al.* (1986) and references therein). These properties make the luminescence suitable for scintillators.

Plasmon excitation

The collective excitations of an electronic system are seen experimentally, for instance, in the energy loss of an electron beam. The simplest understanding of behaviour is obtained by considering the recovery of a classical uniform free-electron gas following a displacement of charge. The use of Poisson's equation leads quickly to a solution in which there are excitations whose energy is simply related to the electron density. The quantum generalisation is given less often (see, e.g., Haken (1976), Section 27). We shall be concerned with two main types of plasmon. First, there are the plasma oscillations of the valence electrons. It is these which give the plasmon energies measured in electron microscopy and which are quoted as materials parameters. Secondly, there are the plasmons associated with excited electrons, and perhaps with an electron–hole plasma. Such plasmons have been measured in special cases, such as confined excited states of multilayer semiconductor systems (see, e.g., Ryan (1998)).

The effectiveness of the simple approaches is deceptive for two reasons. First, it is usually very tricky to distinguish experimentally between one-electron and plasmon excitations for any complex system. Secondly, calculations based on a single tight-binding band do not yield the correct plasmon behaviour for subtle reasons. There is no essential direct problem with the limited basis set; indeed, the Haken (1976) analysis carries through without problems even for delta function charges. However, a one-band tight-binding model has an implicit limit on how fast charge can redistribute, and it is this implicit limit which causes problems in estimating plasmon behaviour.

The plasma frequency, as the natural frequency of response of charge density fluctuations, is one natural timescale for the recovery of an electronic system after excitation. Another important timescale for free carriers is the time associated with electron–electron collisions. The decay of an electrical perturbation involves three stages (Bochove & Walkup 1990; Ohanian 1983): charge density relaxation; expulsion of electromagnetic fields to the exterior and of currents to the surface; and damping by ohmic and radiative damping if there are surface currents.

In an electron–hole plasma in a non-metal, five main types of process occur. First, screening will evolve as electrons begin to screen hole–hole interactions (and vice versa). There is a screening length which varies as $n^{-1/6}$ in simple cases: the Poisson equation leads to a length proportional to the density of states. For a free-electron gas, the density of states is proportional to the reciprocal of the square root of the Fermi energy, or to the reciprocal of the Fermi wavevector. Since the number of electrons per unit volume is proportional to cube of Fermi wavevector, the $n^{-1/6}$ form results. We remark that the cutoff in Fermi energy implies Friedel oscillations. Secondly, the kinetic energy of free electrons and holes will be converted into heat, primarily through coupling to longitudinal optic phonons in polar crystals (since these are associated with an electric field which couples to the charges). Thirdly, there will be Auger processes, as electron–hole recombination occurs, with energy and momentum conserved by interactions with other carriers. Fourthly, some of the electrons and holes will combine to form excitons, and this will affect the subsequent dynamics. Finally, there are localisation processes. Some of the excitons and possibly some of the other carriers will self-trap; some of the carriers and possibly some of the excitons will be trapped by defects or impurities.

Femtosecond laser pulses with energy similar to the surface plasmon resonance of Ag particles in glass have permanent effects on the observed colour and dichroism (Kaempfe *et al.* 1999). This materials modification appears to result from changes in the Ag particle shape.

1.5.4 Excitation at higher intensities

When the intensity of illumination is high, two-photon (or many-photon) absorption can become appreciable, in addition to the standard one-photon excitation. The selection rules are different when several photons are involved, and there is a stronger dependence on intensity. Thus Haglund & Itoh (1994) remarked that a nanosecond pulse with a fluence of 1 J/cm^2 gives only of order 10^{-7}–10^{-10} excitations per site, compared with 100 for one-photon excitation. Of course, values can be higher for resonant transitions.

Some of the situations which concern us involve very large, localised, energy inputs, as when there is a collision cascade. There is then a mix of timescales and of length scales, and there is excitation of the nuclear system as well as of the electronic system. The transfer of energy between electronic and nuclear systems is slow, because of the difference in masses; it is complicated still further by the Coulomb interactions. Landau's analysis (Landau 1936) is still helpful. In Chapter 11, we shall discuss collision cascades, fission tracks, laser annealing, laser ablation, and laser breakdown. Under laser excitation, the optical absorption of the free electrons of an electron–hole plasma can be considerable. This may lead to heating of the host lattice which, in turn, can generate mechanical stresses (for NaCl and SiO_2, see N Itoh & Haglund (1994) p. 17) or cause phase changes.

1.6 Excitation of defects and recovery after excitation

We now identify some of the processes which occur following excitation, and group them in ways which identify common aspects.

Energetic free carriers

These may be carriers injected from an electrode or other electron source, created by impurity ionisation, or generated by excitation to energies in excess of the bandgap. For the highest energies, atomic displacements are possible simply by knock-on collisions, but the large disparity in mass between electrons and nuclei makes this relatively inefficient. Energy can be lost by light emission, notably by bremsstrahlung, although this is not relevant in most situations which we shall discuss. Important energy loss mechanisms come from the electron–electron interaction and the electron–phonon interaction. The electron–electron interaction can simply redistribute electron kinetic energy; it may also generate collective plasmon excitations.

The electron–phonon interaction is extremely important (see Chapter 3). It is this which limits the mobility of carriers at higher temperatures in many systems. As the temperature rises, the scatter increases (as do the vibrational amplitudes of the lattice ions) and the carrier mobility decreases. The several types of interaction are usually classed as Frohlich coupling (to the electric fields of longitudinal optic phonons), deformation potential coupling, and piezoelectric coupling (for which the screening of the electric field by other carriers must be treated seriously). The usual descriptions are helpful but simplistic; normally, the analysis of carriers interacting with defects is handled more fully. In many systems which we shall discuss, Frohlich coupling is the important component. Since the longitudinal optic phonons are relatively dispersionless (so they have an energy almost independent of wavevector), energy loss by free carriers is often in units of this phonon energy, i.e. typically 0.1 eV quanta.

In the limit of strong coupling, self-trapping can occur (see Chapters 3, 4). The carrier is effectively immobilised, and moves by an incoherent hopping motion. The mobility rises with temperature (contrary to the case for other free carriers) with an activation energy associated with the deformation energy of self-trapping.

Excited carriers associated with defects: the isolated defect

It is common to treat three related situations as distinct. The first is *carrier capture* by a defect. This process may proceed in many steps (an initial capture to a shallow state followed by a subsequent cascade through many excited states) or directly to a deep level. The second situation is *recombination,* including the recapture of an electron which has been ionised from a defect. The third situation, and the one with which we shall be most concerned, is *recovery after excitation.* Here a defect might be excited optically, and recover to its ground state via radiative (light-emitting) or non-radiative (heat generating) channels. We shall discuss later what determines the branching ratio between these alternatives. The recovery process can be complex. The main reasons are associated with the fact that the equilibrium sites for the neighbouring ions are different in the ground and excited states: the electron–lattice coupling depends on the state. When the electron switches state, the forces on the nearby ions change (they do so rapidly compared with typical vibration periods). However, the ions will start to vibrate about their new equilibria only after a time of the order of the vibrational period. How effectively the lattice can take up energy as a function of frequency when these forces change is described by the *lineshape function,* which determines both the optical lineshape (as the name implies) and the non-radiative transition behaviour.

It is useful to illustrate what happens for a single oscillator of force constant K and frequency ω, for which the coordinate is $x(t)$ and the velocity $\mathrm{d}x/\mathrm{d}t = v(t)$. Suppose there is an electronic change of state at time $t=0$ which leads to an extra constant force F applied to the oscillator. In the static limit, the force would cause a displacement $Q=F/K$, and there would be a relaxation energy $R=F^2/2K$. In the dynamic case, the oscillator adjusts its motion to vibrate about Q, rather than about $x=0$. The time dependence of the displacement $y(t)$ from the new mean position is given by:

$$y(t)=x(t)-Q=[v(0)/\omega]\sin \omega t+[x(0)-Q]\cos \omega t$$

and the time for the first crossing of the new mean position ($y=0$) is

$$t_1=[1/\omega]\tan^{-1}\{[Q-x(0)]/[v(0)/\omega]\}.$$

There is always a time less than π/ω for which there is such a crossing, although the motion can be in either direction. Clearly, the characteristic time for responding to a constant force (rather than a large impulse, as in a collision) is determined by the oscillator period, and not strongly by F, which relates to electron–lattice coupling in the new state. The other factor of interest is the energy transferred. This proves to be $-Fx(0)$ (the sign being chosen so that $Q=+F/K$). We can look at the ensemble average of this energy for a (classical) thermal distribution of oscillators. This shows that the mean square energy transfer is $2RkT$, in agreement with the relevant limit of Huang–Rhys theory (Section 3.2.4 here, or Chapter 10 of Stoneham (1975)). The fact that the energy transfer depends directly on $x(0)$ shows that *the phase of the oscillation at the time of the transition is important.* If we are to achieve detailed control of the states of some system during excitation, we must do so in a way which takes account of this phase. This is already recognised in other examples of quantum control, as in the laser control of how readily atoms absorb or emit light (Meshulach & Silberberg 1998; Bucksbaum 1998).

We may distinguish two basic types of behaviour at this stage. One concerns the establishment of vibrational equilibrium in the excited state. The transitions are primarily the transfer of vibrational energy between degrees of freedom, including the relaxation processes as ions shift their mean positions. Such transitions we describe as *cooling transitions,* for their main effect is to reduce the energy in some reaction coordinate associated with the defect. The second class of behaviour leads to transitions to a second electronic state, often the ground state. These can be optical transitions, giving luminescence; if the luminescence occurs before the cooling process is complete, it is known as hot luminescence. They may also be non-radiative transitions, where the energy is transferred to other degrees of freedom (phonons, magnons, electronic kinetic energy through Auger processes, and so on).

Metastability

Many excited states have rather short lifetimes, e.g. a few nanoseconds for optically-allowed transitions. Yet some systems can survive for very long periods in states which are not the lowest-energy ones. Diamond is metastable relative to graphite, for instance, yet it lasts a long time. Stored charge can last days in many good insulators.

Hierarchical processes

It is natural that there will be processes with different timescales. Some will be fast, with a femtosecond timescale, especially those involving closely-coupled electrons. Others will be far slower, perhaps involving diffusion processes. When the processes on these different timescales are coupled, the observed time dependence can be complex or even misleading.

A first example concerns a steel component in a nuclear reactor. The collision cascades are complex processes, where the important steps take only a few femtoseconds (10^{-15} s, say), and most significant steps are over in a few picoseconds. The consequences of these cascades, and the slower diffusion processes which follow, affect the ductility of the steel. This ductility can decide the safety case for a 30-year-old reactor to run for another 10 years, i.e. events on the billion second timescale (10^{+9} s) which dramatically affect reactor economics (Stoneham *et al.* 1998a). This is an example of hierarchical processes. A second example is the photographic process (Chapter 5), where the latent image is created, and then a development stage is needed. A third example involves a crystalline oxide which amorphises under irradiation. This shows hierarchical behaviour, for the radiation response is affected strongly by whether the sample is in the amorphous or crystalline state. In crystals, channelling can be possible, and there may be focused collision sequences which separate interstitials from vacancies. In amorphous solids, the topology might be changed, e.g. by modifying the ring structures in silicas; the density may change as well. A more complex case is the evolution of a glass containing radioactive waste, where there will be transmutation as well as excitation-induced mixing. In organic systems, cross-linking or scission will lead to an evolving series of molecules, and often to hydrogen loss or oxidation. Their response will change as the bonding patterns are modified (Milinchuk & Tupikov 1989). A final example of a hierarchical situation concerns the effects of a laser on the roughness of a surface. Whether the effects are of smoothing or of roughening, the energy deposition will be affected.

Chapter Two

Energy deposition and redistribution in solids

Energetic particles in solids lose their energy partly to atoms and ions, by elastic collisions, and partly to electronic excitations. Most of the present chapter describes the way in which energetic particle beams generate electronic excitation. We shall begin by considering the interaction of charged particles with solids. We include the effects of X-ray, γ-ray and neutron irradiation, since these lead to secondary electrons or ions. In discussing the basics of particle beam interactions with solids, our emphasis is on identifying situations where there is reason to believe that electronic excitation is an essential feature. We shall say relatively little about those situations which are described by elastic collisions, and for which the electrons merely provide a potential to define the interactions between atoms. Thus we shall not discuss certain areas of major industrial significance, such as the doping of semiconductors by ion implantation, for which electronic excitation is a secondary feature. However, we should be aware of the energies and doses which are used in these applications, since they will influence the extent to which excitation is an issue. In ion implantation, the critical parameters are chosen to produce a specific concentration profile for a specific need. For example, doping polysilicon uses high doses (typically 10^{16} ions/cm^2) whereas the adjustment of threshold voltages uses low doses (perhaps 10^{12} ions/cm^2), even though both use similar energies, perhaps 50 keV. The production of buried layers of B or P might need ion energies of a few 1000 keV; other applications are moving to lower energies of 1 keV or less. Using such low energies emphasises that almost a continuum of ion energies can be selected for applications. These range from deposition, as in molecular beam epitaxy (MBE) and chemical vapour deposition (CVD) through to implantation methods.

The primary reasons for using particle beam methods are varied. The incident particles may be to change composition, perhaps by implantation or deposition. They might remove material by sputtering, whether to deposit on a substrate, or for analysis, as in secondary ion mass spectrometry (SIMS). The beam may be part of a preparation process for further processing, as in certain forms of lithography, or for imaging, as in electron microscopy. It can cause mixing of the different species, especially across an interface. Sometimes particle bombardment is simply an undesirable source of damage or degradation. Thus oxygen ions with about 5 eV energy are a source of damage in the space environment. Underlying all of these phenomena, however, are some general features of collision science and its dependence on projectile energy and mass, and on specific features of the target. These features provide a guide as to the energy input, how the energy is divided between electronic and nuclear excitation, and the timescales and region sizes affected. We shall need to distinguish between primary and secondary phenomena. For example, the main initial effect of the irradiation of polymers may be the removal of H atoms, a phenomenon which is not central to our theme. The secondary effect, in which the polymer responds by electronic cross-linking processes, is potentially important to us, to the extent that electronic excitation has a role.

Even though classical collisions are not our main concern, there is one basic result from collision theory which is important for our purposes (see Section 1.2.3). This result gives the maximum energy transferred in a classical collision between a projectile of mass M and energy E and a stationary target atom of mass M_T:

$$E_{max} = E\,[4M\,M_T/(M + M_T)^2].$$

Collisions between equal masses are the most effective in transferring energy. The first consequence is that electron beams ($M = m_e$) are much better at transferring energy to elctrons than to atoms, and will often cause electronic excitation and its consequences. The second consequence is that, for light incident particles like electrons, only those with very high energies will be able to displace atoms ($M = m_e << M_T$) in classical, collisions. These atoms displaced will have such a small recoil that they are unlikely to produce secondary displacements. One should expect to find well-separated point defects. Thirdly, energy transfer in collisions between particles with very different masses will be very ineffective. Systems of electrons and nuclei will not equilibrate very fast. In a plasma, for example, there may be a well-defined electron temperature and a different, equally well-defined, nuclear temperature.

2.1 Interactions of charged particles with solids

Charged particles include electrons, protons, and other ions, both as primary particles from an incident beam and as secondary particles. Electrons are generated as secondary particles in several ways; as photoelectrons, Compton electrons, and Auger electrons, and their energies can range from a few electron volts to even megaelectron volts. Atoms ejected from target atom sites can also have a wide range of energies, and can cause further interaction with solids. They behave in the same way as ions, since atoms traversing a solid are ionised, having higher charge states for higher energies.

Energy thresholds for atomic displacement

We now turn to the idea of a displacement energy. For a specific ion to be displaced by momentum transfer in a collision, there must be a critical momentum transferred to that ion in a collision. Electron beams are a common means of materials modification by electronic excitation. Sometimes electrons are chosen for defect generation by elastic encounters, since they produce uniformly distributed defects in solids. If atomic displacements are wanted, the energy of electrons should be higher than about 100 keV, depending on the target, since the energy transferred must exceed the displacement energy.

For displacement following electronic excitation, rather than through collisional momentum transfer, there should be an electronic force large enough to drive the displacement. That electronic force may be generated after other intermediate processes. For example, the excitation of at least two plasmons (requiring about 40 eV) is needed to fragment C_{60} (Cheng *et al.* 1996; Hunsche *et al.* 1996); in that case, there are distinct channels of ionisation, of loss of C_2 fragments, and of multiple fragmentation.

The momentum transfer or force of minimum magnitude is defined by both a magnitude and a direction, for there will be easy and difficult directions in which the ion can be moved. The ionic displacement needs localised energy greater than that for defect generation. Further, the displacement should take a form which does not make immediate recombination easy. There are thus several thresholds to be achieved if there is to be damage. The variation with direction of the momentum transfer could be a factor of 2 or 3. The displacement energy averaged over the direction of the momentum transfer angle is usually used for analysis.

As remarked, the displacement energy is determined by monitoring Frenkel pairs or other simple defects in the bulk. Values for four-fold coordinated compounds are listed in Table 2.1. The corresponding displacement energy is lower on surfaces, where the atoms can be emitted (Yamamura & Bohdanski 1985): an analysis of experimental data of the sputtering yield of monoatomic metals reveals that the threshold energy of all metals is approximately five times the vaporisation energy, 10–20 eV, while the displacement energy in the bulk is 25–40 eV. Exit-surface sputtering of thin films by electron irradiation indicates

Table 2.1. *Values of displacement energies (eV) for four-fold coordinated compounds. Where experiment does not give a definitive answer, a range is shown. When there are two host species, the value for the cation is shown first (so 9/9.4 for GaAs means 9 for Ga and 9.4 for As).*

II–VI compounds	III–V compounds	Group IV elements
ZnO 30–60/60–120	GaAs 9/9.4	C 25 graphite
ZnS 7–9/15–20		C 35–80 diamond
ZnSe 7–10/6–8	InP 6.7/8.7	Si 13
CdS 2–7/8–25	InAs 6.7/8.7	Ge 13–16
CdSe 6–8/5–8	InSb 5.7/6.6	Sn 12
CdTe 5.6–9/5–8		

that the threshold energy for the sputtering of Au to the forward direction is only 5 eV (Cherns 1979), while defect formation energy in the bulk is 35 eV (Bauer & Sosin 1964). A molecular dynamics simulation for crystallisation at crystalline/amorphous interfaces indicates that, when 15–20 eV recoils are initiated at the amorphous side of the interface, the crystal regrows by solid-phase epitaxy (Caturla & Diaz de la Rubia 1995), even though the displacement threshold energy in the bulk has the larger value of 23 eV using the same molecular dynamics code (Stock *et al.* 1995).

A parameter like the displacement energy is good phenomenology, but awkward to make precise. For how long must the ion remain displaced before recombination? How can one define a 'displacement' for an amorphous solid, where there are no defined sites or reference structure for comparison and, at best, there is a defined topology? Knowing the topology is not enough. Changes from a 'before the event' structure to an 'after the event' structure can be followed in calculations, but are not normally identifiable in experiment.

Many models of collision sequences exploit the idea of a displacement energy, and use some classical mechanics approach to estimate the minimum kinetic energy to be given to an atom to cause a displacement such that the atom and the vacancy do not instantly recombine. The number of atoms displaced in a collision cascade is often written in a form slightly generalised from that given originally by Kinchin & Pease (1955):

Number of displacements in a cascade $=(\eta/2)$ (local energy deposited in nuclear collisions)/(displacement energy).

Here η is an efficiency factor, typically 0.8, which allows for some displaced atoms to recombine immediately with vacancies. The expression suggests that the density of displacements is directly proportional to the energy deposited in nuclear collisions, at least in cases when the threshold is not a major factor.

Stopping power and its dependence on projectile velocity and mass

The excitation of a bound electron by charged particles can be analysed in various ways. For fast particles, when particle velocity V is higher than that of the orbital electrons, the standard approach uses the Born approximation. In classical terms, the Born approximation is equivalent to ejection of an orbital electron by Rutherford collisions with the incident particles. The cross-section is proportional to the square of the atomic number of the incident particle and inversely proportional to the binding energy E_B of the electron excited. In the Born approximation, the cross-section increases with increasing projectile energy E when the energy exceeds the binding energy ($E>E_B$) and starts to decrease after a maximum which occurs nearly at 3 times the binding energy. The *stopping power* or linear energy transfer dE/dx for a monatomic target is given by:

$$dE/dx = [(4\pi Z_1{}^2 e^4)/(MV^2)]NZ_2 \ln(2MV^2/I),$$

where Z_1 and Z_2 are the atomic numbers of the projectile ions and target ions, respectively. I is the logarithmic average of the binding energies E_B of orbital electrons.

The Born approximation fails at energies lower than an energy E which is slightly less than the maximum in the energy dependence of the stopping power. In this low-energy range, another method is used in which the electrons are regarded as a viscous medium, which removes and redistributes kinetic energy.

The electronic system will modify the damage, as can be seen from a simple model (Stoneham 1990) or from molecular dynamics (Jakas & Harrison 1984, 1985). Suppose an ion of mass M enters a solid with velocity V, and so with energy $E=\frac{1}{2}MV^2$. The rate of energy loss is AV, linear in particle velocity for simple mechanisms. If the energy is shared over N ions, then the energy loss to electrons increases. For equal sharing, each ion has energy E/N, and hence velocity $v=V/N^{\frac{1}{2}}$; each ion has energy loss Av with A a constant independent of velocity, so the total rate of energy loss to the electrons is now $NAv=AVN^{\frac{1}{2}}$. Clearly, the electronic viscosity will affect the subsequent damage. This approach, with related effects, has been incorporated in molecular dynamics (Finnis *et al.* 1990)

Using classical electrodynamics, the 'viscous' energy loss associated with the electrons can be related to the frequency-dependent dielectric constant $\epsilon(\mathbf{k}\cdot\mathbf{v})$,

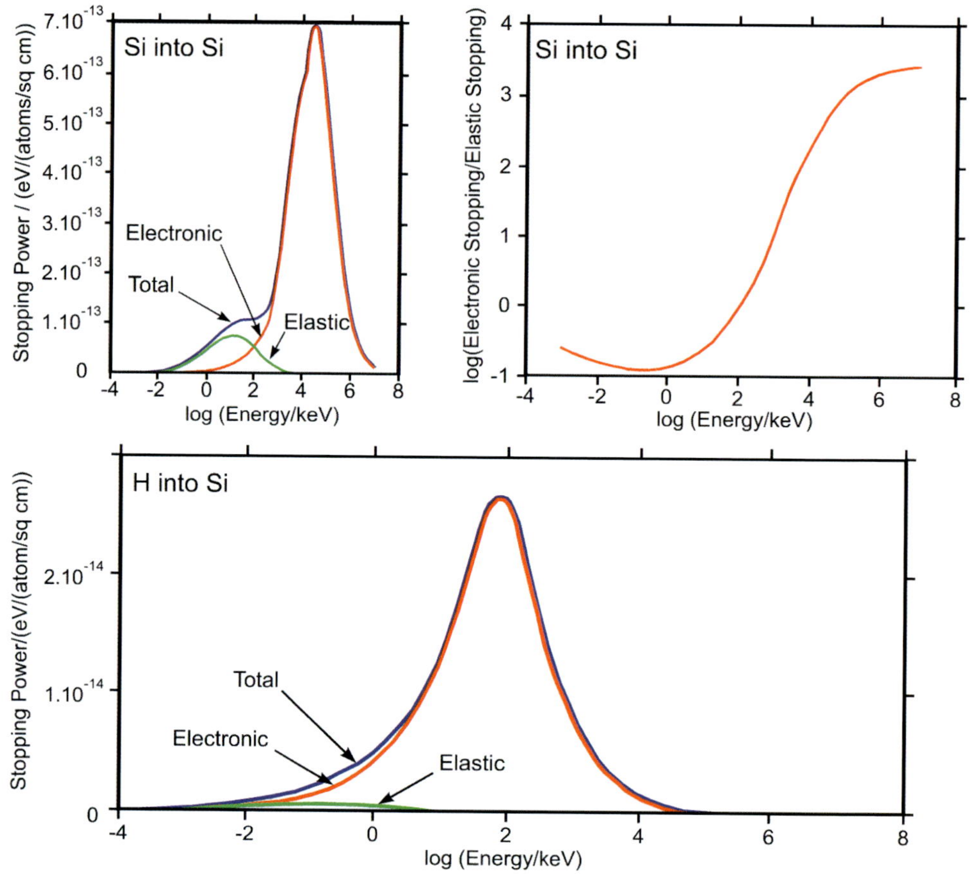

Fig. 2.1. Stopping powers of Si for incident Si atoms and protons from the TRIM code (for which we thank Dr A H Harker). The projectile energies are in keV, and the stopping powers in eV/(atom/cm^2); to convert to dE/dx, one should multiply by the number of Si atoms per unit volume, 4.94×10^{22} per cm^3; the maximum value for Si into Si is thus about 3 keV/nm.

where **k** is the wavenumber and **v** is the velocity. We shall define $\epsilon^{-1} = \eta = \eta' + i\eta''$. The Poisson equation for a charge Ze moving with velocity **v** gives a potential whose Fourier components are:

$$\phi_k = 4\pi Z / k^2 \epsilon(\mathbf{k} \cdot \mathbf{v}).$$

Associated with this potential is the retarding electric field:

$$\mathbf{E}_\mathrm{k} = -\mathrm{i}\, \mathbf{k}\, \phi_k$$

and hence a retarding force $\mathbf{F} = Z\mathbf{E}$, which is given by:

$$\mathbf{F} = -[4\pi e^2/2\pi^2] \int \mathrm{d}^3\mathbf{k}\, \mathbf{k} / [k^2 \epsilon(k, \omega)].$$

The differential cross-section for a particle losing energy $\hbar\omega$ and momentum $\hbar k$ is given in the dielectric approach by:

$$\sigma(v, \omega, k) = (2/\pi)(Z^2/v^2k)\ \mathrm{Im}\,(-1/\epsilon(k, \omega)).$$

Several model expressions exist for the dielectric function. Ashley (1991) extended optical data $\epsilon(0, \omega)$ to finite k in a way which ensured that the f-sum rule was obeyed for each k. Penn (1987) gave an alternative route to a response function from optical data. Tung *et al.* (1979) adopt a somewhat different model, exploiting ideas of Lindhard *et al.* (1963a,b) and of Rezvani & Friauf (1993). Bartholomé *et al.* (1997, 1998) have made predictions of proton-induced electron emission from alumina using these methods. In this process, it is necessary to calculate both the energy loss by the proton and the mean free path of the electrons scattered. It is reassuring that all three methods give largely similar results.

The simplest form of the frequency-dependent dielectric function is that for an electron gas, which provides a useful reference case:

$$\epsilon(\omega) = 1 - [4\pi Ne^2/m\omega^2]$$

at large frequencies ω, then we can define an average frequency ω_{av} by:

$$\ln(\omega_{av}) = [m/2\pi^2Ne^2] \int d\omega\ \omega\,|\eta''(\omega)|\ \ln(\omega).$$

More general models of the dielectric function can be used instead. These are often expressed in terms of the plasma frequency ω_p and bandgap G, for which limiting cases of more general expressions are (Barerra & Duke 1976):

$$\omega_p^2 = N\,e^2/m\epsilon_0,$$

$$\epsilon(\omega) = 1 + \omega_p^2/(G^2 - \omega^2).$$

The stopping power takes the form:

$$F(q_0) = (4\pi Ne^2/mv^2)\ \ln(q_0v/\omega_{av}).$$

Other generalisations which bring in the band structure do so through a dielectric function $\epsilon(\omega)$, and lead to an electron energy loss proportional to $-\mathrm{Im}\,[1/\epsilon(\omega)]$. This can be compared with the energy absorbed from photons, which is proportional to $\mathrm{Im}\,[\epsilon(\omega)]$. The difference between the electron and the photon cases is mainly because the electric displacement **D**, not the field **E**, enters into the Poisson equation (Landau & Lifshitz 1984; Kittel 1996). For a uniform electron gas, the plasma frequency (and hence the plasmon energy) is a function of the electron density.

The plasma oscillations are a collective excitation of the electron gas and can be an effective energy sink. Some of the effects of plasmons can be subtle, e.g., it is possible that a plasmon process, through its effect on energy spectra in a silicon detector, was responsible for the apparent detection of massive neutrinos (Choy & Stoneham 1993).

Balance of energy between electrons and nuclei

What are the relative proportions of energy losses to electronic excitation and to atomic motion? At low energies, where the Born approximation fails, the balance between electrons and nuclei can be evaluated from the ratio of the stopping powers for electronic excitation and for elastic encounters with nuclei.

At high energies, a simplistic argument is helpful. We regard the transfer of energy from a fast particle to a nucleus of charge Z and its Z asociated electrons as resulting from a force F per electron charge, acting for a period t, equal to a characteristic distance (like a screening length) divided by the projectile velocity. We assume that the incident particle is moving sufficiently fast that its own velocity changes by a negligible amount. Such a force F acting on a stationary particle of mass M and charge Qe gives it a kinetic energy $Q^2(Ft)^2/2M$. In the model, t will be inversely proportional to the projectile velocity, and the energy transfer will be inversely proportional to the projectile energy in the non-relativistic case. There are Z electrons per nucleus, so the ratio *(energy transferred to the Z electrons)/(energy transferred to the nucleus)* $=(A/Z)(m_p/m_e)$, where $A=M/m_p$ is the atomic weight. Clearly, since the proton mass m_p is much larger than the electron mass m_e, the energy is primarily given to the electrons. This model implies, therefore, that high-energy charged particles lose energy primarily by electronic excitation.

More detailed analyses suggest that the energy loss (dE/dx, where x is the distance along the particle trajectory) can be split into two components. The electronic loss term is roughly linear in velocity, as if the moving ion is in a viscous medium (see below). The nuclear loss has a peak, and falls off at large energies. For a silicon target, the nuclear loss (decreasing with E) and the electronic loss (increasing with E) are equal at an energy which depends on the particle mass: 10 keV for B, 135 keV for P, 33 keV for As, and above 1000 keV for Sb. Typical examples of the elastic and electronic stopping powers of Si for several particles are shown in Fig. 2.2.

Since the electronic loss is the dominant term at high energies, a fast particle entering a solid will lose energy electronically at first, with nuclear loss taking over when some slowing has occurred. The particle will change direction of motion frequently. This means that the actual length of the trajectory (total track length) is greater than the projected range. It means also that there will be a statistical distribution of ranges and of lateral deviations for particles entering a solid with a given momentum vector.

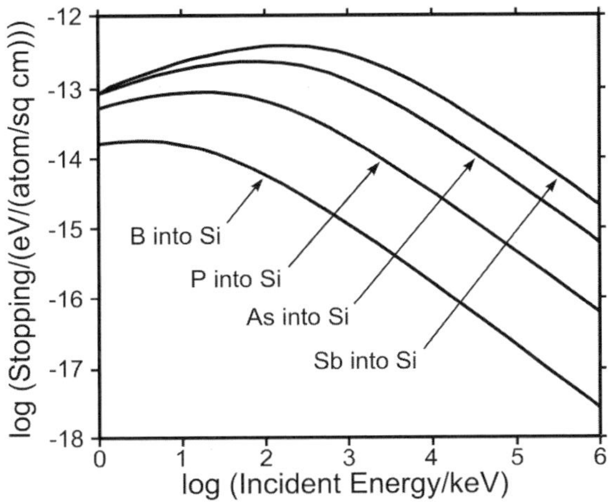

Fig. 2.2. The elastic stopping powers for B, P, As, and Sb ions in Si, calculated using the TRIM code (for which we thank Dr A H Harker). The incident energy is in keV, and the stopping power is in units eV/(atom/cm^2); to convert to dE/dx, one should multiply by the number of Si atoms per unit volume, 4.94×10^{22} per cm^3. The maximum value for Sb into Si is thus about 5 keV/nm.

Excitation of valence and core electrons

In the Born approximation, the cross-section for electronic excitation is inversely proportional to the binding energy of the electron to be excited. This implies that a core electron will be excited with much lower probability than a valence electron. Further, when an electron fills the core hole, the energy released is given to Auger electrons or X-ray photons. For relatively shallow core holes, the probability of the Auger process is significantly larger than that for photon emission. Since a core hole leads to a series of Auger transitions, many low-energy electrons are emitted, and these dissipate energy nearby. In discussing the causes of materials modifications, we need to distinguish between those changes induced by a core hole and those induced by changes in the chemically-important valence electrons (see Section 5.2.2).

Timescales

There are implicit timescales in the evolution of materials modification. Once energy has been transferred to the target, we can identify a number of typical timescales for subsequent events. First, there are *fast processes,* shown in Table 2.2. Bonds can be broken in times of order of femtoseconds, since we can relate an energy uncertainty of, say, 0.5 eV to 3 fs. On the femtosecond timescale, atoms are frozen in the positions to which thermal vibration has carried them. In semiconductors, the timescale for changes due to electron–phonon coupling is at least of the order of the vibrational period (typically 0.1–1 ps; see Chapter 1), and cooling of carriers to the lattice temperature can take much longer

Table 2.2 *Fast processes under irradiation. These results are mainly for metals, since comparable data are lacking for insulators. The characteristic times of Finnis and of Landau relate to the transfer of energy from nuclei to electrons. The characteristic time of Sigmund relates to energy transfer at the end of a collision cascade; electrons are not explicit.*

Faster than ps	
0.2 fs	Typical plasma oscillation period
0.6 fs (Ni) 1.5 fs (Cu)	Electron collision time for liquid metal
70 fs (0.07 ps)	Typical lattice vibration period
100 fs–10 ps	Self-trapping (for zero barrier)
100 fs (Ni)	Characteristic time (Finnis *et al.* 1990)
100 fs–1 ps	Loss of memory for excited electron momenta
Times of a few ps	
1–10 ps	Fast non-radiative processes
1–10 ps	Fast photochemical processes
1.8 ps (Cu)	Laser recovery data (Govolov *et al.* 1996)
1–10 ps	Characteristic time (Sigmund 1974)
4.3 ps (Cu)	Characteristic time (Finnis *et al.* 1990)
10 ps	Characteristic time (Landau 1936)

in certain circumstances. The establishment of an electron temperature will normally take a time determined either by the plasmon frequency or by the electron–electron collision time (Bochove & Walkup 1990). For quantum wells (where the subband behaviour introduces further complexities) Ryan (1998) cite measurements of the loss of polarisation correlations in GaAs which suggest thermalisation in about 200 fs for carriers of density $2\times10^{16}/\mathrm{cm}^3$ and 30 fs at a density of order $5\times10^{17}/\mathrm{cm}^3$. There is a plasmon frequency corresponding to 1 ps at about 10^{16} carriers/cm^3.

Another group of times relates to somewhat slower processes: allowed optical transitions (1–1000 ns, typically 10 ns), typical spin-forbidden optical transitions (1 ms), and diffusion-controlled processes (1 s to geological times). It is important to recognise that some processes on the femtosecond (10^{-15} s) timescale can influence what happens on the engineering timescales (30 years, or 10^9 s) and even geological timescales (10^{17} s since the natural fission reactors started in the Gabon).

Sputtering

Sputtering induced by elastic encounters is measured by a coefficient S, which is inversely proportional to a surface binding energy, and directly proportional to the nuclear stopping cross-section at the incident energy and to a function of the ratio of projectile and target masses (Sigmund 1969). The surface binding energy is not always well defined, but is often approximated by the cohesive energy multiplied by a geometric factor to correct for the fewer neighbours for surface atoms. Fuller investigation shows that the sputtering threshold energy is smaller than the bulk displacement energy (Yamamura & Bohdanski 1985), as discussed in Chapter 10. There is clearly scope for better models

Mixing

At internal interfaces between different phases, intermixing can result from bombardment. There are several different mechanisms. Recoil mixing results from momentum transfer which drives a particle across the interface. The thermodynamic properties of the interface are known to influence the parameters involved in mixing. For example, the amount of mixing at Cu/W interfaces is close to that expected from recoil mixing, whereas that at Cu/Au interfaces is an order of magnitude higher (Wang *et al.* 1982). There is a higher mixing rate for those interfaces which gain free energy by mixing. This is interpreted as thermodynamically-driven recoil mixing, and its rate increases with temperature. Diffusion can occur, perhaps enhanced by local increase in temperature or vacancy concentration, or influenced by radiation-induced segregation. A further discussion of this problem is given in Chapter 11.

2.2 Theory of the interaction of charged particles with solids

Many of the important features of radiation effects can be understood from more sophisticated, yet still simple models, which do not discuss electrons explicitly. One approach is to concentrate on classical collisions, with the electrons implicit in displacement probabilities, or alternatively in the interatomic potentials, ionic charges, and ionic radii.

Strategies for modelling radiation effects

There are several strategies, depending on the results required. There are various analytical models, often based on the *Boltzmann transport equation*, whose value lies in the way they give insight to trends. *Monte-Carlo* approaches can embody empirical information as well as more basic calculations in their (scientifically-based) assumptions about the probabilities of events, which are then built into a stochastic code. The TRIM code (Biersack & Haggmark 1980) describes various aspects of the interaction of ions with solids: the range, range distribution, the lateral spread of ion paths, the energy deposited to electrons and atoms, the number of defects produced, and so on. In *molecular dynamics*, which is based on calculated interatomic potentials, Newtonian mechanics is used to follow the motion of individual atoms.

Certain phenomena depend in detail on features of the lattice structure. These include *channelling* and *focused collision sequences.* Channelling involves the motion of a particle down one of the empty channels which occur in some structures. The collisions are ones in which only small deflections occur, guiding the particle down the channel. The channelling particle can move relatively long distances, but its passage may be interrupted by defects or obstructions. This means that channelling can be used both as a means of achieving deeper distributions in implantation, and also as a means to identify the site occupied by some defect which may affect several channels. In a focused collision sequence, one member of a row of atoms is struck so that it moves at a small angle θ_1 to the axis of the row. This atom, in turn, strikes another, which moves at a smaller angle $\theta_2 = \lambda\theta_1$ to the axis of the row. As more and more collisions occur, the momentum transfer becomes more and more axial, and the sequence of collisions is focused. For hard spheres of radius R separated by distance D, the necessary condition is $4R > D$. Focused collision sequences are especially important in photochemical damage, where the energy input is only just sufficient to create a vacancy–interstitial pair. In such cases (like the production of F–H pairs in alkali halides), rapid recombination is averted because there can be a focused collision sequence which takes the interstitial some distance from the vacancy.

Even the most fundamental of these approaches, molecular dynamics, has significant limitations. Two obvious limitations of molecular dynamics relate to computer power, namely the *timescale for runs* (the real time which is simulated), and *system size* (the region simulated). These limits are not trivial, and the simple evolution of computer hardware will not eliminate the problem. The solution will be some form of mesoscopic modelling. A third limitation concerns *potentials* (or, strictly, how one should best calculate total energies for chosen positions of nuclei). This area has developed dramatically. There are excellent potentials for many metals and insulators, and self-consistent molecular dynamics (whether based on density functional theory, as in the Car–Parrinello method, or in more traditional chemical approaches, as with the CHEMOS code (Wallace *et al.* 1991) extended so that it is practical for at least a few radiation damage studies (Stoneham 1997b)). However, these limitations are not the only ones. In much of this book, we shall be discussing highly inhomogeneous, excited systems which are far from equilibrium. In some cases, quantum effects cannot be ignored.

The energy given to electrons in solids alters their energy distribution. In metals, the energy imparted to conduction electrons will be dissipated rapidly among the conduction electrons, or may be transferred to the lattice via the electron–lattice interaction. These processes happen on the femtosecond to picosecond timescale, and the energy is deposited locally, not transferred over significant distances. The situation is completely different when there is a finite bandgap. For semiconductors and insulators, electrons and holes are excited only when energy larger than the bandgap is available. The electron–hole pairs so created are relatively long-lived. The electron and hole created move significant distances before recombination, so that the energy may be redistributed within the solid, e.g., providing energy for surface atoms to be desorbed. The importance of energy transport, probably involving electronic excitation, can be identified in cases such as fast ions incident on poly[phenyl sulphide], where there are deviations from the usual assumption that damage should be proportional to deposited energy density (Papaléo *et al.* 1996).

Lindhard theory

This formulation of radiation damage and of ion stopping in matter has been of major importance. Lindhard used as working assumptions a set of rules which are still central to this field (Lindhard *et al.* 1963a,b). Briefly stated, these assumptions assert:

(i) electrons produced in collision cascades do not produce recoil atoms with appreciable energies;
(ii) atomic binding energies can be ignored for heavy particle collisions at the energies for which electronic stopping has any appreciable significance;
(iii) energy transfers from nuclei to individual electrons are relatively small;
(iv) electronic and nuclear collision contributions to energy loss can be regarded as separate (i.e. they are additive).

Range and energy deposition in Lindhard theory

With these rules, and some simple but powerful approximations, Lindhard and others made predictions which have been at the core of radiation damage studies. Even with computer modelling methods available, the analytical descriptions provide an impressive framework for understanding. Many of the results exploit potentials proportional to (ion–target atom distance)$^{-m}$ characterised by a parameter m (typically between 1/3 and 1/2). A useful example gives the spatial distribution of deposited energy when the projectile range is $R(E)$ as a function of its initial energy E. The spread is characterised by a mean square length:

$$\langle x^2 \rangle - \langle x \rangle^2 = R^2[2m/(1+2m)]^2 / (1+4m).$$

so the spread is typically one quarter to one third of the range. Since $R(E)$ scales as E^{2m}, the energy density should scale as $E/(E^{2m})^3$. The energy per target atom (and hence the peak local temperature, as defined by the nuclear kinetic energies) will vary as E^{1-6m}. Approaches following Lindhard's ideas are also quantitative, so that one can compare the maximum target atom energies with those of other excitations, like plasmon energies and bandgaps, and with chemical bond energies.

Stopping power

In Lindhard theory, electronic stopping resembles stopping in a viscous medium, so the decelerating force is proportional to velocity:

$$(\mathrm{d}E/\mathrm{d}x)_{\mathrm{electronic}} = k_e E^{1/2},$$

where k_e has a weak dependence on the masses and charges of target and projectile ions. As noted above, electronic stopping becomes increasingly important at higher projectile energies. Nuclear stopping is normally written as the product of a target atom number density and a stopping power

$$(\mathrm{d}E/\mathrm{d}x)_{\mathrm{nuclear}} = N\, S(E).$$

It can be transformed into 'universal' forms by expressing variables in terms of reduced variables which involve ϵ, ρ, and a, defined by:

$$\epsilon = \{\text{Energy } E\}\ [\mathrm{M}_1/(M_1 + \mathrm{M}_2)]\ /\ (Z_1 Z_2 q^2/a),$$

$$\rho = \{\text{distance } x\}\ N\pi a^2 4M_1 M_2/(M_1 + M_2)^2,$$

$$a = \text{screening length} = 0.88\ a_0/(Z_1^{2/3} + Z_2^{2/3})^{1/2},$$

where a_0 is the Bohr radius. Accordingly, the stopping power $\mathrm{d}\epsilon/\mathrm{d}\rho$ takes the form of an integral over energy of a factor depending on the potential, divided by $1/\epsilon$. The nuclear component dominates at low energies, although this is not transparent from the form just given. The ratio of the electronic stopping power to the nuclear stopping power is important for much of this book. Examples of stopping power contributions as a function of energy are shown in Fig. 2.2; they have a characteristic limit as the energy increases beyond the energy for the maximum stopping power.

Stopping powers for ions in various materials have been tabulated by Ziegler *et al.* (1985). Their tables were obtained by fitting experimental data, using Lindhard theory for the low-velocity range and the Born approximation for the high-velocity range. The overall stopping power S is approximated by the sum:

$$1/S = 1/S_{\mathrm{LOW}} + 1/S_{\mathrm{HIGH}},$$

where S_{LOW} is the stopping power which follows Lindhard theory and S_{HIGH} is based on the relativistic Born approximation. The stopping power of Si for various incident ions (Fig. 2.1) was derived using this approach.

The stopping power depends on the projectile atomic number Z_1 and the target atomic number Z_2, being proportional to $Z_1Z_2/(Z_1^{2/3}+Z_2^{2/3})$ at low energies, and to $Z_1^2Z_2$ in the high-energy range. The deviation from these relations, sometimes called Z_1 or Z_2 oscillation, is significant again in the low-velocity range where the electron distribution around nuclei is critical.

Stopping powers for compounds or alloys can be estimated from Bragg's rule, namely that the stopping power S_C of a compound $C=A_nB_m$ is the concentration-weighted average of the stopping powers S_A and S_B of the component materials:

$$S_C = n\,S_A + m\,S_B.$$

Although this rule fits experimental results well in the high-velocity range, agreement is poor at low velocities, where the Z_1 or Z_2 oscillations are significant.

2.3 Issues: Beyond the standard models

Non-uniform energy deposition

The standard approaches give us information on the energy imparted in nuclear and electronic collisions, the numbers of displacements, and the number of electron pairs generated in bulk of a solid. Yet there are limitations in describing the non-uniformity of energy deposition and its time evolution. Moreover, although the concept of a displacement threshold is a useful measure for evaluating the number of defects produced, it is simplistic. Transient displacements and displacements near pre-existing defects are situations for which collisions transferring less energy than this displacement threshold may lead to lattice modification. Molecular dynamics, which has its own limits, can treat this extra complexity.

There are several related terms, used by different authors, all describing regions in which there is a large local energy deposition. The terms 'thermal spike', 'collision cascade', and 'hot spot' are sometimes given different meanings by different workers, so we shall often use alternative descriptions. In particular, 'thermal spike' suggests that some measure of thermal equilibrium has been achieved. In one class of energy deposition process, the energy is deposited in the electronic system (large electronic stopping); in the other, the energy is largely deposited through elastic collisions, with the electrons merely providing a potential. These two extremes are actually less different than first appears, at least when there is significant electron–phonon interaction, as there is for most of the systems which we consider in this book. Suppose, for example, that all the energy is given to electrons in some small region. There are then various consequences. Some of the electrons will move out of the region rapidly, leaving it with a net positive charge. There will be a highly non-equilibrium energy distribution for the electrons, with many excited from bonding orbitals, so that interatomic forces are altered. There may be strong forces towards self-trapping of carriers. These effects will cause rapid trasfer of energy from the electrons to the nuclei, so that the ion temperature will rise rapidly. Alternatively, suppose that the energy is given solely to the ions. There will be some transfer of energy to the electrons, as noted by Flynn & Averbach (1988), and also by Landau (1936), in his early discussion of plasmas. The electrons will be able to redistribute heat, and can also act as an energy reservoir, affecting the release of energy. Often we shall use descriptions like 'local energy deposition', plus an explanation, instead of the picturesque but less defined common terms.

Localisation of nuclear collisional energy

In primary collisions, clearly the energy is localised, for it is transferred to a specific atom. The energy deposited is redistributed to surrounding atoms, and may lead to a collision cascade. The standard theory describes mainly those atoms which gain an energy greater than the displacement energy. The atoms in the vicinity of the primary collision experience secondary collisions, from which they receive a variety of energies. If these atoms are in the bulk, most of them will not be displaced permanently from their sites. However, those atoms at interfaces or other special sites may move to some other configuration. This is one class of energy localisation. We shall see later (Chapter 9) that the threshold energy for sputtering from surfaces is much less than the bulk displacement energy. Another class of localisation of collisional energy occurs along the path of an ion. The secondary ions have low energies, and their range limits the track to a small radius. Experimentally, this shows in the fluence dependence of materials modification. So long as the overlap is negligible, the permanent modifications induced by ion beams are proportional to fluence, but the changes are less than this when the overlap between tracks is significant.

Localisation of electronic excitation energy

The localisation of individual electrons and holes occurs either by self-trapping or by trapping by defects, as discussed in Chapter 1. This type of localisation can be the cause of local lattice modification. Another type of localisation of electronic excitation energy stems from the non-uniform generation of electron–hole pairs. The generation and recombination of dense electron–hole pairs can enhance local lattice modification. We shall discuss such processes in the following chapters.

Dense electronic excitation can be induced in several ways. Low-energy secondary electrons have short ranges: a 100 eV electron has a range of 10 nm.

Even under radiation with X- or γ-rays, the secondary electrons produce dense electronic excitation, so there is quite a high probability of two neighbouring molecules being excited (Varley 1962). Perhaps the most significant dense electronic excitation due to low-energy electrons appears after Auger transitions to core holes. The internal energy stored by a core hole is released by emitting a series of Auger electrons, and these deposit energy close to the core hole. The effect of this dense excitation can be seen when the photon energy is scanned across a core ionisation edge (Kondo *et al.* 1993). This shows that the significant differences in the products result from the changes in the energy distribution of the secondary electrons, not from the presence of the core hole. Another example of the localisation of electronic excitation energy is observed along the tracks of mesoscopic scale (see Cahn & Lifshin (1993), p 468) created by energetic heavy ions. The secondary electrons produced by heavy ions can have only low energies because of the mass mismatch. Their stopping power is proportional to the square of the atomic number of incident ions. The density of electron–hole pairs along a track can be so high that every molecule is excited. The consequences of this type of excitation are treated in Chapter 11.

Defect formation from electronic excitation

In ionic crystals and amorphous materials, electron–hole pairs can create defects. This raises two important issues. First, there must be localisation of electronic excitation energy in excess of the formation energy for a defect pair, whether a vacancy and interstitial, or an electron centre and a hole centre. For the damage to persist, even more energy is needed to separate the vacancy and interstitial components; for example, in alkali halides, there can be a displacement sequence along a close-packed row which ensures that the vacancy and interstitial are too far apart to recombine. Secondly, the thermodynamic equilibrium formation energy is usually much smaller than the displacement energy. This is partly because the energy surfaces are altered by the electronic excitation, and partly because the timescales are different. Since the maximum energy reasonably available from an electron and hole will of the order of the bandgap, we expect defect formation by electronic excitation to occur only when, first, the bandgap exceeds the formation energy for a defect pair and, secondly, the displacement takes a form which makes rapid recombination difficult.

These ideas imply several qualitatively distinct types of threshold. A large bandgap (MgO, SiO_2, alkali halides) is necessary but not sufficient for defect creation. Small bandgaps (as for Si and many semiconductors) usually rule out defect formation; instead, certain defects may exist in metastable states. Early discussions of collision-induced radiation damage went further, and raised the question of whether the stopping power should fall to zero for projectiles with energy less than the bandgap. To our knowledge, there is no evidence that the stopping power, let alone the atomic displacements, show such a threshold. This may be because experiments always generate some carriers, and these carriers mask the direct effect of the gap.

The bandgap itself relates to the one-electron excitations of the solid. One should not ignore the role of the collective oscillations, the plasmons. These are significant in a few special cases. Thus photodesorption from alkali metal clusters is seen following plasmon excitation (Weidenauer *et al.* 1989; note also Choy & Stoneham 1993).

The defects introduced by elastic encounters can act as energy or charge localisation sites, and can enhance the defect yield by electronic excitation. An example of this interplay of electronic and elastic energy deposition is given by Macaulay-Newcomb & Thompson (1984). They showed by Rutherford backscattering experiments that SiO_2 crystals are damaged by simultaneous irradiation with heavy ions (50 keV Sb^+) and 1 MeV He ions, yielding five times as many defects as irradiation with the heavy ions only. The He ions by themselves produce no damage. Similarly, defects are generated in relatively few materials by electronic excitation alone. However, electronic excitation is effective far more widely once the defects have been created by some other means. In later chapters we shall note cases where electronic excitation plays a role, even though it is not the major cause of the modification.

2.4 Challenges: Non-equilibrium situations

Radiation damage leads to highly non-equilibrium situations, often with large amounts of energy associated with small regions of a solid. Kinetic energies of atoms and of electrons in such regions can be large, with individual particles having energies of several electron volts being common. These regions can have a net charge, since electrons will be scattered out of the central zone, and will return only after some delay. Standard statistical equilibrium expressions will not apply; in particular, there need not be a usefully-defined Fermi level in the non-equilibrium condition. It is to be expected that the condition will depend on just how excitation has occurred. Some of the common terms (Section 2.3), like 'thermal spike' or 'hot spot', suggest that some sort of temperature can be defined. There is no assurance that a single temperature can be identified. Even if the region of space is well defined, equilibration among particles of similar masses (like the electrons alone, or the nuclei alone) can happen much faster than the equalisation of any electron temperature with a corresponding lattice temperature. This identifies an important distinction between modes of excitation, namely whether the energy is given initially to the electrons or to the nuclei.

One possible mode of excitation could result from collisions of a projectile nucleus with target nuclei. In this case, energy will be predominantly nuclear kinetic energy. Some will be lost in defect creation; much will be redistributed over the vibrational degrees of freedom as thermalisation occurs. These parts can be predicted (in principle) by standard methods, including self-consistent molecular dynamics, since only adiabatic energy surfaces need be considered. A certain amount of energy will excite electrons (so nuclear energy becomes electronic energy), which can then act as an energy reservoir and as a means to transfer energy. There is evidence that these roles affect defect recovery in radiation damage. This component inevitably involves non-adiabatic transitions, and cannot be treated at present, even by methods such as self-consistent molecular dynamics (but see also Section 3.7.4).

The alternative situation provides energy primarily to the electrons. The electrons, holes, and any excitons (possibly after some recombination) will begin to self-trap, yielding charge localisation (such as self-trapped holes), energy localisation (self-trapped excitons), and probably damage, whether this occurs as part of the self-trapping process or afterwards. In these processes, as in photochemistry, electronic energy becomes nuclear energy of motion. A further question concerns the ionisation produced by particles. Do the electrons and holes generated affect the motion of defects, or influence the damage process? Certainly any charge redistribution generates electric fields, and these can have substantial effects (Chapter 11). However, the participation of free carriers in individual atomic motions or reactions is harder to identify. For example, studies of muon behaviour in liquid nitrogen under the influence of electric fields show whether or not the electrons generated by a muon interact with the muon itself (Storchak *et al.* 1997).

2.5 Thermal diffusion: Processes near thermal equilibrium

We shall discuss diffusion dynamics, i.e. the calculation of absolute diffusion rates, in Chapter 3. In this section we concentrate on the phenomenology of diffusion and on the ideas which are used in understanding the processes following radiation damage or other parallel situations. General discussions include Stoneham (1989c) and the books of Flynn (1972), of Allnatt & Lidiard (1993) and of Kotomin & Kuzovkov (1996).

Diffusion processes, sequences of incoherent atomic jumps, are described by Fick's laws. The usual form of Fick's first law relates a diffusion flux to a concentration gradient:

$$J = -D\, \partial c / \partial x$$

with D the diffusion constant, a measure of the absolute rate of diffusion. Fick's second law relates to the time dependence of a concentration:

$$\partial c/\partial t = \partial/\partial x\,(D\,\partial c/\partial x).$$

These two laws are usually combined with phenomenological descriptions for D, and with the Nernst–Einstein relations to link D to a mobility under an applied field.

2.5.1 The phenomenology of diffusion rates: The Arrhenius and Meyer–Neldel (compensation) expressions

The temperature dependence of the diffusion constant often (but not always) has a so-called Arrhennius form:

$$D = D_0 \exp(-E/kT).$$

This Arrhennius (thermal activation) expression for D is the first of two major empirical rules concerning D_0 and E. The second major rule is the so-called compensation rule, or Meyer–Neldel rule, which asserts that large values of D_0 are usually observed with large values of E:

$$\log D_0 = A + BE.$$

This rule applies when comparing a number of 'related' processes, e.g. a number of related systems, in each of which only a single mechanism dominates diffusion. There is a lack of precision in defining which processes are indeed related, and there are examples which show distinct limits (see Stoneham (1989c)). Thus, in a-Si:H deposited at different temperatures, one series of experiments (Irsigler *et al.* 1983) showed that the compensation rule applied for the dark conductivity in samples which had been fully annealed after illumination but, even in the same samples, showed no sign of compensation if illuminated before measurement.

There are several ways in which compensation can arise, and we give just one example. Thermal expansion will usually lead to reductions in energy barriers E. A linear dependence of E on temperature

$$E = E_0 - \alpha kT$$

simply modifies the prefactor:

$$D_0 \rightarrow D_0 \exp(\alpha)$$

without affecting the apparent Arrhenius energy. We can readily imagine that it is the larger energy barriers which are most reduced by thermal expansion.

If so, a large E will imply a large α, which will be the source of compensation. Detailed analysis shows this to hold for cation diffusion in some simple oxides (Harding *et al.* 1987). There is no way to separate $D_0\exp(\alpha)$ into its components; α is not measurable separately and, contrary to some assertions, the temperature $E_0/\alpha k$ has no physical significance.

The recognition that E might depend on temperature can be analysed more systematically by using thermodynamic relationships (Catlow *et al.* 1981; Harding 1985). Experimentally, the concern is usually with Gibbs free energies, g_p. Theoretically, it is usually easier to estimate constant volume (or, more strictly, constant lattice parameter) quantities, like the internal energy u_v. One of the standard approaches which we shall discuss, Vineyard theory, works with a constant volume ensemble. The key expressions for the enthalpy h_p and entropy s_p are these, if the thermal expansion is sensibly linear in the temperature range of interest:

$$h_p = u_v(0) + T^2\,(\mathrm{d}s_v/\mathrm{d}V)_T\,(\mathrm{d}V/\mathrm{d}T)_p;$$

$$s_p = s_v(0) - (\mathrm{d}V/\mathrm{d}T)_p\,[(\mathrm{d}u_v/\mathrm{d}V)_T - 2T\,(\mathrm{d}s_v/\mathrm{d}V)_T].$$

The fact that the leading term of h_p is $u_v(0)$ explains the relative success of those many calculations which merely estimate an activation energy at constant volume and zero temperature. For the vibrational entropy, however, the constant volume and constant pressure forms may even be opposite in sign. The correction from the volume dependence of u_v can be large, and is analogous to the α term above.

2.5.2 Special cases of diffusion

Applied fields and bias terms: The Nernst–Einstein relations

Fick's laws may need to be generalised substantially in practice, for example, if an electric field or temperature gradient contributes to the particle fluxes. The generalisation of the first law for particles of charge Z^*q is:

$$J = L\,[(-\nabla\zeta)_T - Q^*\,(\nabla T/T) + Z^*\,Eq],$$

with ζ the chemical potential; $-\nabla\zeta$ reduces to $kT\nabla c/c$ in simple cases, and the factor L becomes Dc/kT. Expressions for the Seebeck coefficient P follow at zero concentration gradient by choosing a field and temperature gradient which give zero current. The heat of transport Q^* is then given by PTZ^*q. Flynn (1972) and Allnatt & Lidiard (1993) discuss these issues in depth. The Seebeck coefficient is one diagnostic for small polaron behaviour (e.g., Mott & Davis 1979).

One important result relates ionic conductivity (diffusive motion under a potential gradient) to diffusion under a concentration gradient. The simplest forms of the Nernst–Einstein relations allow one to relate ionic conductivity σ, ionic mobility μ (ionic velocity per unit applied field) and diffusion constant D:

$$\mu = (q/kT)D,$$

$$\sigma = (Nq^2/kT)D$$

for N charges per unit volume, each of charge q, and electric field E. Clearly, these relationships mean that μ and D cannot both have a strictly Arrhenius temperature dependence, since there is a factor T to be incorporated. However, in many cases, an extra term, weakly dependent on temperature, can be subsumed into a modified activation energy.

Non-Arrhenius temperature dependences

Caution is needed in interpreting both the prefactor D_0 and the activation energy E. For example, E may include both a defect motion component (an energy associated with the energy surface for the ionic motion) and a defect formation process (whether from detrapping or from the thermal generation of defects). It may be hard to separate these terms unambiguously. In an amorphous or disordered system, there may be a variety of activation energies which combine in a complex way. Even in simpler systems, one must be careful about whether E is a thermodynamic internal energy (as in many theoretical papers) or an enthalpy (as usually measured). Quantitative modelling is now an important complement to experiment in understanding the contributing processes.

However, a more subtle problem is that the qualitative Arrhenius form may not hold. This is especially true for small activation energies, especially so when the experiments cover only a modest range of temperature. A power-law dependence of rate on temperature T^n will look like a process with activation energy nkT' from data near $T = T'$ (so T^4 at room temperature would give about 0.1 eV). A power-law dependence of the Arrhenius prefactor will introduce a similar energy into the apparent activation energy. These temperature dependences are typical of cases where a single phonon can induce a transition, or where two phonons combine in a Raman process. Such cases include quantum diffusion (Chapter 3) and multiple trapping. Other cases which show non-Arrhenius behaviour include variable range hopping (Mott & Davis 1979) in which, in disordered systems at low temperatures, carriers must tunnel to the nearest sites which also have the right energy; this leads to an $\exp(-A/T^{1/4})$ behaviour.

Behind this variety of dependences is the warning that transport of particles may not be the standard surmounting of a barrier. At the very least, the single particle on a unique energy surface can be misleading at times. In the quantum diffusion of light interstitial particles (H, muons μ), for instance, there is a light diffusing particle and sluggish, heavy, host lattice atoms. The light particle does not surmount a barrier created by atoms moving adiabatically. The light particle responds adiabatically to the moving atoms, and the jump corresponds to the *coincidence configuration*, when the particle would have the same energy on either the initial or final site, so tunnelling is unhampered. The same ideas apply to polaron motion (Chapter 3). Surprisingly, they apply with minor changes to the classical diffusion of light particles. The point is that the transfer of energy between particles of very different masses is inefficient (cf. Landau's discussions of electron–ion energy exchange in plasmas, cited above). Thus a light interstitial may find no barrier to moving from site A to a site B of lower energy, but will not be able to lose energy fast enough at site B to benefit from the lower energy; instead, it may return to interstice A. The idea of a coincidence configuration is again useful.

Characteristic energies

A natural assumption, which is often valid, is that most transport involves the movement of a single atom or ion. We note, however, that it is often the case that a dimer can move with a lower activation energy. In essence, when one component of the dimer is in a potential valley, the other may be near a peak in the energy; together, the dimer components average out the hills and valleys. Nor should we assume that large clusters of atoms must be immobile. On surfaces, in particular, large rapid motions are seen when there is an energy source, such as a chemical reaction which the particle catalyses. Geguzin & Krivoglaz (1973) give an extensive review of the motion of macroscopic inclusions.

2.6 Transport and capture processes

2.6.1 Geminate recombination

When an electron is excited, one critical question is whether it will recombine with the electronic hole left behind (geminate recombination), or whether it will escape, to recombine or be trapped elsewhere. The simple model of Smolouchowski provides a reference for discussion of more complicated systems (Freeman 1987). Experimentally, different behaviours are observed following the excitation of insulators by intense subpicosecond laser pulses (Petite *et al.* 1998). In SiO_2, geminate recombination of electron and hole occurs; in NaCl, recombination is primarily non-geminate. The difference appears to be primarily the large phonon energies in SiO_2, which mean that the carriers lose momentum faster and have less chance to escape. The creation process for primary radiation defects means that they (F and H centres, for instance) can be correlated in space. This correlation affects both the temperature dependence and impurity concentration dependence of defect accumulation (Kotomin & Kuzovkov 1992).

The idealised problem concerns diffusion of the electron from an initial site at distance r from the hole (or Coulomb trap). The equation describing the evolution of the probability $p(r)$ includes the standard diffusion process, with diffusion constant D, and a term from the Coulomb interaction, which we write as $v_0 = -C/r$:

$$\mathrm{d}p/\mathrm{d}t = D\,\nabla^2 p + D\,\nabla[(p/kT)\,\nabla v_0].$$

One length scale will be the so-called Onsager radius, that distance r_c at which the Coulomb energy is equal to kT:

$$r_c = C/kT,$$

and we may argue that the electron will escape with a probability given by

$$\phi_0 = \exp(-C/rkT) = \exp(-r_c/r).$$

When an electric field is applied as well, v_0 is altered to

$$v = v_0 + eEr\cos(\theta),$$

but the same equations apply with trivial generalisation. There is now a second characteristic length, determined by E, namely $b = 2kT/Ee$. The length b is often written as β^{-1}; the factor 2 is standard. Clearly, r_c/b $(=(Ee/2kT)(C/k\mathrm{T}) \propto E/T^2)$ is a dimensionless parameter which can be used to describe the change in escape probability in the presence of an electric field. When the field is weak, and the Coulombic binding is relatively strong, the solution yielded is:

$$\phi = \phi_0\,[1 + (r_c/b)] = \phi_0\,[1 + (eEC/2k^2T^2)].$$

Large fields suppress geminate recombination. It is less obvious why high temperature and large Coulomb attraction should seem to have the same effect and, of course, they do not, for their main effect is through ϕ_0.

The effect of an applied field on recombination luminescence has been analysed for F centres (Portnyagin *et al.* 1991) and excitons (Portnyagin *et al.* 1992) in Al_2O_3.

Donor–acceptor recombination

When a system of donors and acceptors is excited at low temperatures, there is often luminescence as equilibrium is reestablished. Tunnelling (i.e. the overlap of the electron and hole states) is the central process. The spectrum of the luminescence and the intensity both change with time. At first, it is the close pairs of donors and acceptors which recombine, and those at larger spacings become more important as recovery continues. The recombination energy depends on the separation. The intensity can be predicted, subject to some simple but reasonable assumptions (see, e.g., Stoneham (1975) Chapter 25, Dunstan (1984), Blumen & Klafter (1983)). These assumptions involve explicit expressions for the probability of recombination $W(r)$ for a spacing r, for the distribution of donors and acceptors in space, and for the recombination energy as a function of r. Thus $W(r)$ will fall exponentially with r if the electron–hole overlap is crucial. The distribution of donors and acceptors might be close to random. The r dependence of the recombination energy comes primarily from a Coulomb interaction between the charged donor and acceptor. Various generalisations are possible, but these simple ideas give a fruitful and respectably accurate description of behaviour.

2.6.2 Rate theory and defect aggregates

In materials modification, some of the most important effects can be the result of defect aggregation. Interstitial clusters can stabilise interstitials and vacancies against recombination in alkali halides at higher temperatures, for instance. Examples of aggregation processes through the accumulation of point defects include the formation of stable defect clusters, dislocation processes (the creation and growth of loops, or dislocation climb and multiplication), the growth of metal colloids through anion loss, the growth of voids, and the accumulation of charge at interfaces or other special sites. These defect aggregates have varied but important consequences. The dislocations may affect mechanical properties or even electrical breakdown; colloids will affect optical properties, such as reflectivity and energy absorption; the voids might be sinks for fission product gases. The major processes in defect aggregation include the motion of point defects and the motion of small clusters (Geguzin & Krivoglaz 1973). As these processes occur, the aggregates evolve, and we need a description of this class of evolution.

What are the options in analysing what is happening? It is most unlikely that we can use molecular dynamics for the whole process, for example. The timescales may be years, whereas picoseconds are long times for molecular dynamics. Even single diffusion jumps are sufficiently rare that ingenuity may be needed to see an event unless the initial state is specially chosen. The distances and length scales for aggregation are often microns, and a cubic micron may well contain 10^{12} atoms. So other methods are needed for understanding the evolution.

Clearly, the rate-determining point defect processes must be identified. In the simplest cases, this will be diffusion of one of the point defects, typically an interstitial or a vacancy. But in what charge state? And in which electronic state? There is evidence that anion vacancies in alkali halides have diffusion rates which depend on charge state (F centre or simple vacancy) and, for the F centre, are faster in the excited state (see Chapter 7). Is there recombination-enhanced diffusion, assisted by local energy release as an electron and hole recombine (or as an electron or hole is captured by a defect)? Such enhanced diffusion is also well known in semiconductors, and responsible for degradation of some light-emitting diodes in operation. In more complex cases, two or more defects may move together. Thus the motion of O in Si is enhanced by the involvement of H. On a surface, adatom pairs will often have lower activation energies than single ones: the dimer energy surface will be flatter, since one atom in a minimum will be compensated by the other being near a maximum. In non-equilibrium segregation, a vacancy–impurity pair moves faster than either a vacancy or an impurity alone so that, when the pair reaches a surface or grain boundary, the vacancy disappears, leaving an immobile impurity (T M Williams, *et al.* 1976).

The aggregates may nucleate spontaneously through the random coming together of a minimum immobile nucleus. In other cases, nucleation will occur at pre-existing features, like dislocations. The surface is often important, as are internal surfaces like grain boundaries. The nucleation of dislocation loops and of colloids or voids is especially important in radiation damage, for instance in understanding the evolution of oxide fuels in an operating reactor. We may wish to discuss colloids containing hundreds of atoms, so the simplest rate equations (master equations) will be impractical for all but the earliest stages. There is a need to understand the balance between nucleation and growth, and this has led to the use of *rate theory*, a variant of *effective medium theory.*

The basic ideas are these (Brailsford & Bullough 1972, 1973; Stoneham 1979a). We shall consider the evolution of different stationary defect aggregates (sinks, such as voids, loops, colloids) as they compete for different mobile defects (vacancies, interstitials, impurities). Rate theory goes beyond simple chemical kinetics by considering each sink in a self-consistently chosen average environment. The main emphasis is on competition for diffusing species, allowing the sinks and the average environment to evolve with time. The average environment therefore mimics the real, inhomogeneous environment as a competing sink for the mobile species.

The first stage is the calculation of a sink strength k_i^2 for species i; we may also write $k_i^2 = \rho_i \kappa_i^2$ to separate out the sink concentration ρ_i. There will be a total sink strength for a medium, which will be a sum k^2 of all the k_i^2, and we shall need to solve the diffusion equation for a single sink in a medium with this average sink density. The solution will yield a concentration $c(r, k^2)$ in the steady state, where r is the distance from the sink; the boundary condition at the sink is usually chosen to avoid the final jump being rate-determining. The second stage is to consider the medium with many sinks, but to subsitute another equation for the steady-state diffusion equation. In this, the possibility of reemission of mobile defects must be recognised, and an effective chemical potential at the sink surface is needed. In all, one has an expression involving (effective area) × (flux in diffusion model per unit area) of the form $D\rho_i\kappa_i^2(c_i - c_{iav})$. A third component is the rate of production of defects, and the possible annihilation of defects (vacancy–interstitial reconbination) and similar reactions independently of any sinks. With these components, the solution of the basic equations and conservation laws is then possible. The most important part of these rate equations is competition between alternative evolution routes. Will voids develop, or will there be dislocation loops? Will there be small defect clusters, or fewer large ones? One of the crucial factors proves to be bias: does a particular sink accept vacancies more readily than interstitials?

To give a specific example, we outline the discussion by Jain & Lidiard (1977) of the development of dislocation loops during the electron irradiation of an alkali halide. For a discussion of other aspects, including lattice discreteness and the consequences of further defect reactions, see Kuzovkov & Kotomin (1988) and Kotomin & Kuzovkov (1996). The main mobile defect species are neutral halogen vacancies (F centres) and interstitials (H centres). The F centres may be in their ground state, or in an excited state with a lower motion energy in their excited state. The rate equations which describe the evolution of the F and H centres are these:

$$dc_F/dt = K' - K_1 c_F - K_2 c_F c_H,$$

$$dc_H/dt = K - K_3 c_H - K_2 c_F c_H.$$

The right hand side of each equation consists of three terms: defect creation (K or K'), a loss term to aggregates or to dislocations (K_1, K_3), and a vacancy–interstitial annihilation term. K' differs from K in that there is a concentration c_F^e of F centres in equilibrium with the colloids. The loss terms can be written in terms of diffusion constants D_F, D_H, to colloid and dislocation concentrations (C_c and ρ_d; the dislocation density will include a term from H centre aggregation) and to quantities which characterise the aggregate strengths as sinks (r_c and z_F, z_H), where we recognise that bias may be important. The basic equations for the creation and loss terms are then:

$$K_1 = D_F\,(4\pi\, r_c\, C_c + z_F\, \rho_d),$$

$$K_3 = D_H\,(4\pi\, r_c\, C_c + z_H\, \rho_d),$$

$$K' = K + 4\pi\, r_c\, C_c D_F c_F{}^{e}.$$

If the H centres are mobile and effectively in a steady state then, near saturation, the prediction has a form very similar to that observed by Sonder & Templeton (1967):

$$c_F = c_F(\infty)\,[1 - \exp(-2K_1 t)]^{1/2}.$$

Under the conditions which are to be expected, Jain & Lidiard (1977) and Hughes & Lidiard (1989), the time constant for c_F is independent of K, the dose rate. The saturation value $c_F(\infty)$ is proportional to $K^{1/2}$ and inversely proportional to the square root of the F centre diffusion constant, so varying as $D_F{}^{-1/2}$(T). Colloid production depends on the relative bias for F centres and H centres $(z_F - z_H)/z_H$. There is a dose rate at which no colloids should be formed; in essence, $c_F(\infty)\,(z_F - z_H)/z_H$ must exceed the mole fraction of F centres in equilibrium with the colloids. The dose rate and time dependence for the colloids depend on the dislocation density, as might be expected.

The production of colour centres and colloids in alkali halides has certain parallels with the photographic process (Chapter 5). One important difference is that the photographic process seems to need sites either at a dislocation or at a grain surface. This affects the evolution of metal colloids profoundly. It should be no surprise that colloid evolution proceeds differently in different systems. The possible charge state of the moving species (F centres or anion vacancies, for instance) is one factor. Moreover, the excitation of the irradiation process modifies these diffusion rates (see Chapter 7). The extent of interaction between F and H centres to cause annihilation is a second factor. A third factor is just how the metal atoms are to come together to form the colloid. In the fluorite structure of CaF_2, for instance, the loss of fluorine atoms leaves a region of metal atoms with about the same density as the free metal.

Chapter Three

Electron–lattice coupling and its consequences

The linking of electronic and nuclear dynamics is of major importance in understanding the ways that electronic excitation can lead to materials modification. In particular, the exchange of energy between electronic excitation and lattice vibrational energy is crucial in the dynamics of excited states. This chapter summarises these unifying features, and provides a framework for fuller descriptions.

3.1 Basics of electron–lattice coupling

Some relatively simple ideas form the basis for understanding lattice vibrations. They can be generalised, so as to apply to transient, non-equilibrium systems and in other situations important to us.

Adiabatic approximations

The first basic assumption is used to separate the electronic and nuclear motions. The *Born–Oppenheimer approximation* asserts that the light-mass electrons respond so rapidly to the sluggish motion of the nuclei that one can assume the nuclei to be at rest when calculating the electronic wavefunctions. There is then an electronic term in the energy which determines the motion of the nuclei.

This approximation is excellent in many circumstances (for fuller references, see Stoneham (1975)). Some of its generalisations are extensions of the same idea, for instance when there is electronic degeneracy, and the Jahn–Teller effect is important. In other circumstances, as in non-radiative transitions, the breakdown of the adiabatic assumption is the driving force for the transitions. In other cases, the same ideas are exploited but in different ways. Thus, when there are light atoms (like hydrogen, or like muons) in a solid consisting of heavy atoms, then two distinct adiabatic approximations are needed: the first separates the electronic motion for that of all the nuclei; the second separates the light-nucleus motion from that of the heavier ones. The different dynamics of light and heavy particles also affects their non-equilibrium behaviour, since energy exchange is inefficient in collisions between particles of very different masses. This means, for example, that excited electrons may equilibrate amongst themselves to establish a temperature much faster than they can transfer energy to nuclear motion, which can be characterised by a different temperature (Chapter 2).

Harmonic approximation

Often, the nuclear motion consists of small vibrations about some well-defined equilibrium geometry (exceptions might include diffusion processes and certain tunnelling motions). This geometry does not need to be a crystal lattice, and clearly the equilibrium geometry will change when electrons change their states. If the displacements from equilibrium are written $\mathbf{x}_i$ for the *i*th atom, then it is usually sufficient to concentrate on the leading term in the potential energy:

$$(1/2)\,\mathbf{x}\cdot\mathbf{A}\cdot\mathbf{x} \equiv (1/2)\,\Sigma_{i,j}\,\mathbf{x}_i\cdot\mathbf{A}_{ij}\cdot\mathbf{x}_j.$$

The term linear in $\mathbf{x}$ vanishes because we have defined $\mathbf{x}$ relative to the equilibrium positions. $\mathbf{A}$ is a *force-constant matrix* describing the coupling between the many atoms of the solid. However, it would be much more convenient to represent the motion as a that of a set of *dynamically-independent oscillators*, or *normal modes.* This is always possible in the harmonic approximation, irrespective of whether the solid is crystalline or amorphous. What is needed is a linear transformation to these new modes $\mathbf{q} = \mathbf{a}\cdot\mathbf{x}$, with $\mathbf{a}$ the matrix describing the transformation. These new modes have their own vibrational frequencies, and are dynamically independent in the sense that there are not interactions between modes to this approximation: if one mode is excited, there is no means for it to transfer energy to the others, and it will remain excited indefinitely.

Dipole approximation

When a solid is ionic, or when an atom is deformed during vibrational displacement, there will normally be a long-range electric field because of induced multipole moments of the atom. It is rare for any term beyond the dipole term to be important, and the dipole approximation recognises this. The shell model is a general model to this approximation; the dipole moment has ionic and electronic parts, and with the polarisabilities influenced by the local environment of the ions.

Table 3.1. *Polaron coupling constants and effective Rydbergs R_0 is the Rydberg using the free electron mass, and screening by the static dielectric constant. $\hbar\omega_0$ is the longitudinal optic phonon energy. α is the Fröhlich coupling constant.*

Class of crystal	Actual crystal	α	$R_0/\hbar\omega_0$
Alkali halides	KCl	3.60	17.3
	RbCl	3.99	19.0
	KBr	3.15	19.1
	KI	2.15	27.7
Silver halides	AgCl	1.91	2.46
	AgBr	1.60	1.60
Oxides	SiO_2	1.9	4.3
II–VI semiconductors	CdS	0.65	0.74
	CdSe	0.39	0.49
III–V semiconductors	AlN	1.03	0.97
	AlSb	0.12	0.97
	GaN	1.1–1.4	0.33–0.37
	GaAs	0.031	0.22
	GaSb	0.021	0.08
	InP	0.08	0.13
	InAs	0.056	0.07
	InSb	0.015	0.025

Simple models for electron–lattice coupling

There are two broad approaches to estimating electron–lattice coupling. The first approach is direct, based on electronic structure calculations at some level. It is this approach which will usually be adopted in the present book. For most of our analysis, it is the terms linear in ionic displacements which are most significant, i.e. terms which can be written $-\mathbf{F}\cdot\mathbf{Q}$, where $\mathbf{Q}$ can be a specific ionic displacement, a combination of such displacements constituting a reaction coordinate (i.e. determined by some process), or a normal mode (and so determined by the energy terms quadratic in displacements). $\mathbf{F}$ is clearly a generalised force; when $\mathbf{F}$ is determined by a defect in some state, we shall refer to these forces as the *defect forces* associated with that defect in that state.

The second approach is to relate the coupling to macroscopic quantities. This approach includes *Fröhlich* coupling, based on the electric field of longitudinal optic phonons, *deformation potential* coupling, related to the shifts of band edges under uniform stress, and *piezoelectric* coupling. These couplings are characterised by *coupling constants,* such as the dimensionless Fröhlich polaron coupling constant, α (Table 3.1). Our emphasis in this book is on energy localisation and on charge localisation, so that macroscopically-defined quantities are of less value. Yet they do allow one to identify trends, especially from one crystal class to another.

3.2 The configuration coordinate diagram

For many reasons, not least clarity, it would be good to mimic by a single harmonic oscillator the lattice interacting with a defect system. In very special cases, of course, the defect may indeed only interact with a single local mode. Usually, this is not so, and the assumption of a single oscillator is just a helpful simple framework to represent complicated behaviour. Yet this *configuration coordinate model* is surprisingly effective, beyond this descriptive power. Its value comes partly from its simplicity: one can draw and interpret diagrams. In addition, there are some general and exact results for systems with just a single frequency. The one-mode (or even two-mode) assumptions fail when dephasing occurs, or when energy is redistributed over vibrational degrees of freedom in some effectively-irreversible way. In the diffusion of small polarons (Section 3.4.2), for instance, the incoherence of successive jumps relies on the fact that there is a negligible probability that the initial and final states will be strictly translationally equivalent, including those degrees of freedom associated with the lattice vibrations.

Any of a large number of atomic displacements might influence a defect in a solid. Often, a very specific combination of displacements is especially important, for example the radial displacement of the nearest neighbours to the defect, and this can be represented by a single coordinate Q. We emphasise that it is most unlikely that the coordinate Q will be one of the dynamically-independent normal modes $\mathbf{q}$. It is almost certain that normal modes of many different frequencies will be needed to make up Q. Moreover, the precise combination of modes making up Q will vary from defect to defect, from one electronic state to another of the same defect, and even from property to property for the same state (see Section 3.2.2).

Happily, this choice is not critical for some purposes. More important is the recognition that the reaction coordinate (configuration coordinate) comprises modes of many frequencies. This has implications for non-equilibrium dynamics, for an initially-coherent motion in Q will die away as the contributing normal modes get out of phase. However, for some equilibrium properties, the assumption of a single mode is effective. Suppose we know the mean square displacement $\langle x^2 \rangle$ or mean square momentum $\langle p^2 \rangle$ for some coordinate at temperature T. Then there are quite tight bounds on $\langle x^2 \rangle$ and $\langle p^2 \rangle$ for other

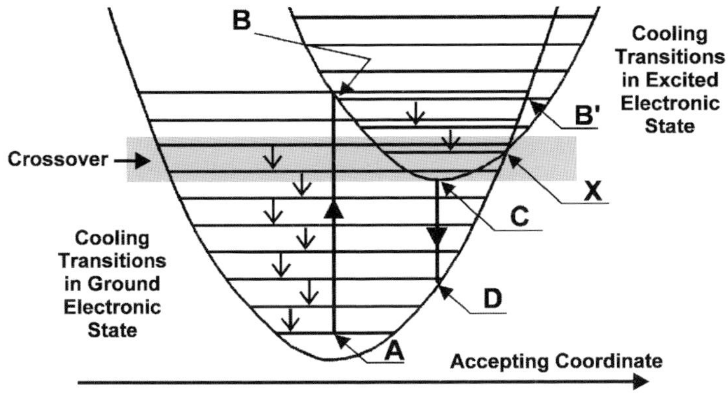

Fig. 3.1. Configuration coordinate diagram. The figure shows the main characteristic energies and transitions, both radiative and non-radiative. The key energies are the zero-phonon transition energy E_0 (the energy difference between C and A), the effective activation energy to the intersection of energy surfaces, E_A (the energy difference between X and C) and the relaxation energy E_R (the energy difference between B and C or between D and A).

temperatures, irrespective of the details of the vibrational spectrum. Quantities depending on averages like these can be modelled almost as well by a single frequency as by the full spectrum (see Stoneham (1975), Section 10.8.3).

The earliest descriptions of configuration coordinates applied to the situation where the breathing (radial) motion of the nearest neighbours was a reasonable choice. But, clearly, an asymmetric set of displacements will be appropriate for some systems. For the neutral vacancy in diamond, for example, the ^1E to $^1T_2$ transition couples primarily to a Jahn–Teller distortion of E symmetry.

3.2.1 The basic configuration coordinate model

The basic model is illustrated in Fig. 3.1. It shows harmonic energy surfaces for two electronic states. For each surface, the different levels of vibrational excitation are shown, with energies $(n+\frac{1}{2})\hbar\omega$ appropriate for a harmonic oscillator. In the ground state, the potential energy surface is quadratic in the displacement $(Q-Q_{g0})$ and proportional to a force constant K_g:

$$E_g = E_{g0} + \tfrac{1}{2} K_g (Q - Q_{g0})^2.$$

Similarly, the potential energy surface for the excited state is

$$E_e = E_{e0} + \tfrac{1}{2} K_e (Q - Q_{e0})^2 = (E_{e0} + \tfrac{1}{2} K_e Q_{e0}^2) - K_e Q_{e0} Q + \tfrac{1}{2} K_e Q^2.$$

Clearly, if a transition occurs at constant Q (a Franck–Condon transition) then a force of magnitude $K_e Q_{e0}$ is imposed by the change in electronic state. This force drives the system to oscillate about a new configuration centred on Q_{e0} rather than Q_{g0}. If the force constants K_g, K_e are equal, often a reasonable assumption, then the several energies can be obtained in a convenient form from the energies of states shown in Fig. 3.1: the zero-phonon line energy $E_0 = E_{e0} - E_{g0} = p\hbar\omega$ (E_C–E_A in the figure), the effective activation energy for deexcitation to the ground state from the relaxed excited state $E_A = (E_0 - E_R)^2/4\,E_R$ (which is E_X–E_C in the figure), and the relaxation energy, $E_R = S_0\,\hbar\omega$ (E_B–E_C or E_D–E_A in the figure).

The dimensionless constants S_0 (the Huang–Rhys factor) and p often appear in discussions of defect–lattice interactions. Our definitions contain some crucial assumptions: the system should be harmonic; if the coupling is strong, a single reaction coordinate should suffice somehow; and the frequencies of any modes involved (or for the reaction coordinate for strong coupling) should be the same in the ground and excited states. These assumptions can be avoided, within limits. Values for S_0 for Jahn–Teller systems will depend on which property (intensity of the zero-phonon line; Stokes shift; linewidth) is used to deduce them (see Stoneham (1975), Lannoo & Stoneham (1968)). There are expressions for different frequencies in ground and excited states, although these are rarely useful (see Stoneham (1981)). Analytical expressions exist for certain special anharmonic systems, such as those with Morse potentials, but the results are often too complicated to be really useful. Self-consistent molecular dynamics allows some treatment of anharmonic systems (e.g., Mainwood & Stoneham 1994). Yet the standard approaches concentrate on the use of the single-frequency theory of Huang & Rhys (1950) as an empirical framework.

3.2.2 Choices of configuration coordinate

Apart from rare special cases (e.g. interaction with one local mode only) there is always a compromise in choosing the 'best' combination of normal modes to represent Q. The aim is to include as much of the important electron–lattice coupling in this single coordinate. The choice depends on the problem of interest. Some options are these:

(a) Maximise the relaxation energy in the initial state relative to some selected 'perfect' state, such as the perfect crystal.
(b) Maximise the relaxation energy in the excited state from the equilibrium configuration for the ground state.
(c) Optimise the overlap of wavefunctions near the two minima, so as to predict zero-phonon line properties well.

(d) Optimise some other part of an energy surface, e.g. that of the excited state at energy corresponding to the Franck–Condon transition. This is, in fact, a useful choice for certain non-radiative transitions.
(e) Ensure the right rate or temperature dependence of a transition probability at one temperature extreme (a high temperature limit, or a low temperature limit).

All of these choices of Q will give slightly different frequencies or force constants. This is one of the basic limitations of the standard approaches, the other being the heavy algebra (and restrictive assumptions) needed for most calculations.

Just as there has been a move from simple approximations for the electron–lattice coupling to direct electronic structure calculations, so there is a move from the simpler uses of the configuration-coordinate diagram to more general approaches using molecular dynamics. However, neither move is completely straightforward. In many cases, the more advanced approach merely gives a *relaxation energy* or an *energy surface* which is then applied using formulae based on more simple pictures. This can cause problems. For instance, various expressions which are equivalent in the harmonic approximation can give wholly different answers (predictions of activation energy for small polarons are an example). Further, transition probabilities can cause problems: even if one has the atomic coordinates and electronic wavefunctions for each molecular dynamics step, one cannot easily calculate a non-adiabatic rate, since this usually needs nuclear wavefunctions. We shall discuss a way forward in Section 3.7.4.

3.2.3 Simple cases: The F centre

Most of the important features of the energy surfaces of localised states can be illustrated by a two-state system. To be specific, we can consider the F centre as a system with two electronic states, interacting with modes of a single frequency. The ground state is s-like (using an analogy with the H atom) and the excited state p-like, though we shall ignore the degeneracy for the present. In its simplest form, the energy surfaces are characterised by the phonon energy $\hbar\omega$, the Huang–Rhys factor S_0, and p, the energy separation of the two relaxed states in units of $\hbar\omega$. The energy surfaces correspond to the simple model described above. We can discuss the cycle of optical excitation and deexcitation.

In *optical absorption*, the main feature is likely to be the Franck–Condon transition (A to B in Fig. 3.1). In *optical emission*, the corresponding transition is C–D. If the electron–lattice coupling is weak (e.g. if there is very little change in charge distribution as a result of excitation), then there will be a sharp *zero-*

phonon line, corresponding to transition A–C. This energy will be similar to that for *thermal excitation*, which is dominated by the lowest-energy transitions. In any detailed quantitative discussions, however, one has to be careful as to which thermodynamic quantity is measured. Thus, most predicted configuration coordinate diagrams would give thermodynamic internal energies; most experiments provide a Gibbs free energy.

Cooling transitions (B–C)

The energy is lost from the motion in the reaction coordinate. If the configuration coordinate were a normal mode (dynamically independent) then there would be no cooling at all in the harmonic approximation. The loss of coherent motion in this configuration coordinate is an adiabatic process, and occurs by anharmonic processes and (often more important) harmonic processes because the coordinate is not an exact mode. We shall discuss the related cooling processes for free carriers in Section 3.5. In the optical case, the system may make radiative transitions (hot luminescence) or non-radiative transitions (non-adiabatic processes) as cooling occurs. Clearly, there is a question about what happens for states close to the intersection X of the two energy surfaces; we discuss this in Section 3.6.

The cooling transitions imply a finite speed of the lattice relaxation (see Section 3.5). It is conceivable that the rate-determining factor could be in the lattice vibrational system, rather than associated with electron–lattice coupling. The phonon bottleneck is well known in spin–lattice relaxation, for example (Van Vleck 1941; Stoneham 1975, Chapter 14). Wagner & Koengeter (1988), discussing the transition between free and localised states of an exciton, suggested there could be a 'phonon bottleneck' because the lattice modes coupled to the exciton cannot reestablish thermal equilibrium with the other modes sufficiently fast. The speed of the lattice relaxation plays a crucial role in all dynamic processes, including the polaron (exciton) localisation and transport (Lushchik & Lushchik 1989). It influences strongly the nature of the 'hot' luminescence of excitons in alkali halides which is emitted in the course of their self-trapping (Hizhnyakov *et al.* 1981). Two-colour femtosecond absorption methods show non-equilibrium hot-phonon effects associated with the photo-excited electron gas in GaAs (DelFatti *et al.* 1999).

Luminescence

In the absence of hot luminescence or non-radiative decay, cooling takes the system to the relaxed excited state. Luminescence may follow, again by a Franck–Condon transition (C–D). After this has occurred, there will be a further *cooling transition* (D–A), taking the system to its ground state.

3.2.4 Optical transitions

In free space, the optical absorption cross-section for a particular transition of a single atom can be written in terms of a cross-section σ_{ab}, where

$$\sigma_{ab} = (2\pi^2 e^2/mc)\, f_{ab}\, \delta(\omega - \omega_{ab}),$$

in which $\hbar\omega$ is the photon energy, $\hbar\omega_{ab}$ is the transition energy for the transition $a \rightarrow b$, f_{ab} is the oscillator strength

$$f_{ab} = (2m\, \omega_{ab}/3\hbar)\, M_{ab}^{\ 2},$$

and M is the matrix element of **r**, the position vector. The oscillator strength is especially useful, because it is constrained by sum rules (see Appendix I of Stoneham (1975)). In particular, the Thomas–Kuhn sum rule gives $1 = \Sigma_j f_{aj}$, which leads to constraints on the absolute rates of optical transitions. The absorption coefficient, which measures the attenuation of optical intensity with distance, is then $N\sigma$, so that the intensity falls off exponentially as $\exp(-N\sigma z)$. For a defect in a solid, these formulae are modified (Stoneham 1975, Chapter 10). There will be a multiplicative factor, from the effects of the refractive index on the velocity of light and on the photon density of states. Obviously, there will be detailed changes in the matrix elements and transition energy; there will also be an effective field correction, from the polarisation of the solid in which the defect is embedded. An especially important effect is the fact that energy can be transferred to or from lattice vibrations, which leads to an important change in the delta function, which ensures energy conservation.

The optical absorption spectrum, including zero-phonon lines, any phonon sidebands, and major features like the Franck–Condon transitions, is given by the *lineshape function* $G(\omega)$. If the energy-conserving factor in the optical absorption cross-section been a delta function in the imagined absence of electron–lattice coupling, then the lineshape function is what would replace it when the electron–lattice coupling was included. Clearly, the lineshape function recognises that the lattice vibrations, as well as the main electronic term, can take up energy during the excitation process. The excitation process provides a sudden perturbation to the lattice system, which can be written in the general form $(\mathbf{F}_x - \mathbf{F}_g)\cdot\mathbf{Q}$, depending on the change in the defect forces **F** on excitation from the ground state g to the excited state x. These forces act on any of the ions in the solid; in the shell model, it may be useful to distinguish between forces on cores and on shells. Just because this lineshape function describes how effectively the lattice can take up energy under a sudden perturbation, the function also appears in the rates of non-radiative transitions (Section 3.6).

Discussions of the lineshape function are given in several places (Stoneham 1975, Chapter 10; Hayes & Stoneham 1985, Section 4.2). Figure 4.3 of Hayes & Stoneham (1985) illustrates the predictions and compares them with observed spectra. We note two limiting cases for a single phonon frequency (in principle, any number of modes can be included by convolution, but this is rarely helpful). First, for very *weak coupling*:

$$G(\omega-\omega_{ab})=(1-S)\,\delta(\omega-\omega_{ab})+[(S+S_0)/2]\,\delta(\omega-\omega_{ab}-\omega_0)+ [(S-S_0)/2]\,\delta(\omega-\omega_{ab}+\omega_0)+\text{small terms},$$

where S is the Huang–Rhys factor, and S_0 is its value at zero temperature. The first term is the zero-phonon line; the second and third terms correspond, respectively, to the emission and absorption of a single phonon.

For strong coupling and low temperatures, a Poisson distribution is predicted:

$$G(\omega-\omega_{ab})=\Sigma_p[\exp(-S_0)\,S_0^{\,p}/p!]\,\delta(\omega-\omega_{ab}-\omega_0),$$

in which p is the net number of phonons emitted. At high temperatures (or, strictly, large mean values of p) this expression reduces to a Gaussian:

$$G(\omega-\omega_{ab})\propto\exp[-\alpha S_0\,(\omega-\omega_{ab}-S_0\omega_0)^2],$$

which is centred on the Franck–Condon transition energy.

Optical cross-sections can often be estimated quite well, partly because of sum rules (see Appendix 1 of Stoneham (1975)). Nevertheless, detailed estimates of absolute rates for optical transitions in solids are tricky because of the polarisability of the environment and hence the effective field corections E_{eff}/E which are needed (see Section 10.5 of Stoneham (1975)). Two useful relationships are these. The first relates the radiative lifetime τ in an excited state and the wavelength λ and oscillator strength of the transition:

$$f\tau=(c/2e^2)\,(1/2\pi^2)\,\lambda^2\,(nE_{\text{eff}}/E)^2=(4\times10^{-9}\,[\text{s}])(\lambda/500\,[\text{nm}])^2\,(nE_{\text{eff}}/E)^2,$$

with n the refractive index. The second expression is Smakula's formula. If N is the number of absorbing centres per cm^3 and $A=0.82\times10^{17}/\text{cm}^3$ is a constant, and α is a number depending on the actual lineshape (usually α lies between 1 and $\pi/2$) then

$$N\,(E_{\text{eff}}/E)^2\,f/A\,n\,\alpha=(\text{maximum absorption coefficient})(\text{full width at half maximum}).$$

This expression is often used to estimate defect concentrations.

The Principle of Detailed Balance and the Einstein relations do allow relationships to be made between optical absorption and emission rates. These are more complex in a solid, but can still be useful (Stoneham 1975, Chapter 10).

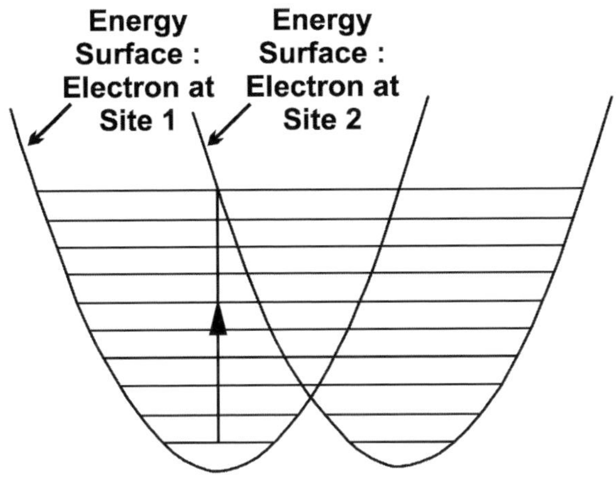

Fig. 3.2. Optical charge transfer transition between equivalent sites, as for the V^- centre in MgO. The diagram also emphasises that the fully-relaxed ground state would have the same energy as the initial state; in this case, there may be no luminescence.

3.2.5 Charge transfer transitions

Many of the transitions which we shall describe show obvious analogies with atomic excitations. However, when a transition involves charge transfer, the origin of the energy is significantly different. Consider, for example, the singly-charged cation vacancy in MgO. This can be regarded as a hole trapped by the vacancy charge onto the oxygens nearest to the vacancy. The hole is localised further onto a single one of these six oxygens (to give an O^- ion) in much the same way that self-trapping occurs in a perfect lattice. There are then two quite distinct types of transition, as shown in Fig. 3.2. The first is intra-atomic excitation of that O^- ion. The second involves transfer of the hole to one or more other oxygens, an interionic charge transfer process. In a rigid, unpolarisable lattice, this interionic charge transfer process would need no energy. In a polarisable, distortable lattice, there is a finite energy for a Franck–Condon transition, because the ionic polarisation is determined by the original ground-state charge density, and does not change in the transition. The transition energy, apart from a modest tunnelling term, is dominated by the polarisation energy, as can be seen from the configuration coordinate diagram. The dominance of the polarisation energy means that shell-model treatments can give excellent predictions, especially when they have been parameterised to ensure the correct static dielectric constant. Calculations which work with the electronic structure explicitly can have problems if they underestimate the ionic polarisation. The success of shell-model calculations has been demonstrated for MgO (Norgett *et al.* 1977, and for ZnSe (Harding & Stoneham 1982). The optical transition energies can be linked to stability energies through shell-model calculations, since a polarisation energy is involved once again. Charge transfer transitions, whether like this or strictly from one ion species A to another ion species B, are common for transition metal ions and gemstones. The low-lying excited states of V_k centres appear to be intramolecular, although certainly there will be some more extended excited states. For the analogous H centre system, where spin resonance shows some spin density on the next axial (110) halogen ions, theory and experiment (Shluger *et al.* 1995) identify charge transfer transitions from the inner two halogens to the outer two ions. These could, of course, be regarded as intramolecular within a four-halogen group.

3.3 Relaxation energies and defect stability

The deformation and polarisation of a host lattice by a defect or by localised charges can involve substantial energies. These energies are themselves important in determining which defect state is stable. When we are concerned with charge localisation, these energies are critical in answering questions like why Mn can be stable in MgO in three or even four different charge states, whether a particular defect can trap a carrier, whether self-trapping occurs, and even at which site an impurity will be found. The experimental data provide substantial corroborative evidence of predictions of relaxation energies.

One result which is widely applicable (but not universal) asserts that a charge will have lowest energy when localised on a single site. The argument is this (Hayes & Stoneham 1985, p. 35). Suppose a carrier is spread over N sites (for simplicity, we shall ignore interference effects, since detailed calculations lead to the same conclusion). In the simplest description, the defect forces associated with each site are F/N, since the Coulomb forces driving polarisation and distortion are proportional to the fraction of charge at each site. There are N contributions with their own relaxation energies $(F/N)^2/2K$, with K the relevant force constant, so that the total relaxation energy for the charge is $N(F/N)^2/2K$, which is proportional to $1/N$. For greatest stability, N is 1: localising a carrier stabilises it. As stated, the argument holds both for ionic and electronic polarisation. The argument applies to more complex situations (such as Jahn–Teller systems) as well, after some generalisation: concentrating the defect forces on a small number of atoms is usually favoured.

Even in continuum models, it is not sufficient to talk of 'electron–phonon coupling', as if all forms were equivalent. Scaling arguments (Emin & Holstein 1976) show that there are four classes of behaviour. These result from the different importances of specific energy terms as the scale L of the wavefunction is varied from some reference value L_0. The kinetic energy scales as $+L^{-2}$. Coulomb terms (including deformation-induced polarisation) scale as $-L^{-1}$.

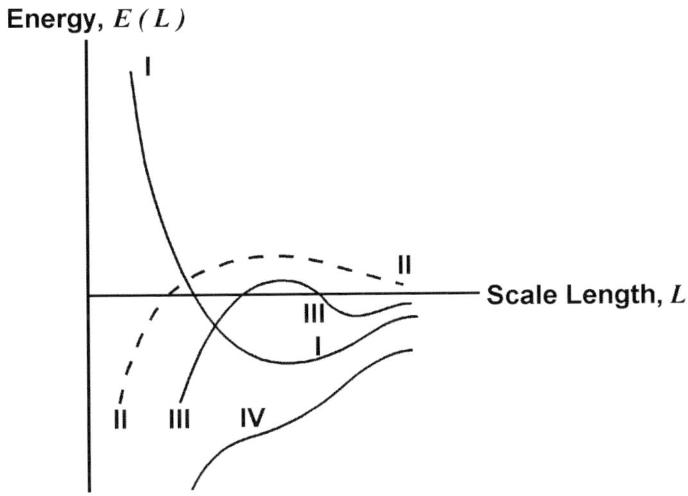

Fig. 3.3. Classes of behaviour for an electron coupled to a deformable polarisable lattice (following Emin & Holstein (1976)). The classes are discussed in the text. Note small polarons are stable in Classes II, III, IV. Coexistence of large and small forms is found in Class III. The dielectric polaron (Class I) does not show a clearly-defined switch in properties as the scale length L of the wavefunction is reduced.

Short-range terms, as from the deformation potential, scale as $-L^{-3}$. Finally, there are various cross-terms which scale as $-L^{-2}$. The four regimes which result are shown in Fig. 3.3. In Class I, there are only Coulomb and kinetic energy terms. Only a single minimum in energy is found as a function of L: there is only one class of polaron. For Class II, with only short-range and kinetic energy terms, the energy is lowest for L zero (which would be the small polaron, if atomic structure were included) or L infinite (the free carrier). When both short-range and Coulomb terms are present, with the kinetic energy, there is a small-polaron form, and a large-polaron form which is metastable (Class III) or unstable (Class IV). These continuum pictures need generalisation for real systems: there is an atomic structure, so L cannot fall below some limit; there is a dependence on the effective mass used to estimate the kinetic energy, and so on. Nevertheless, the Emin–Holstein analysis is illuminating.

3.3.1 Stability and instability

When describing stability, it is essential to ask 'stable against what?' We have already discusses the stability or instability of free carriers against self-trapping (Section 1.1). For the charge state stability of some impurity ion, there are two reactions which apply even for one isolated defect. The first is stability against loss of an electron to the conduction band $X^{N+} \to X^{(N+1)+} + e_c^-$; the second is stability against loss of a hole to the valence band $X^{N+} \to X^{(N-1)+} + h_v^+$ (or, equivalently, $e_v^- + X^{N+} \to X^{(N-1)+}$). In deciding stability in these cases, we should assume in energy calculations that the free carrier (e_c^-, e_v^-, h_v^+) can relax to its lowest energy state, which could be a small polaron. When there are at least two isolated defects, one needs to add a further condition, namely stability against charge disproportionation $2X^{N+} \to X^{(N+1)+} + X^{(N-1)+}$. When disproportionation does occur, it is often said to be a 'negative *U*' situation. Positively-charged vacancies V^+ in Si, for example, appear to disproportionate into V^{2+} and V^0 (Baraff *et al.* 1979). In monovalent ionic crystals, negative *U* behaviour is relatively rare (Stoneham & Sangster 1983).

These conditions refer to the specific processes by which charge states change. However, in real systems in thermal equilibrium, it is the thermodynamic chemical potential for electrons (i.e. the Fermi energy) which matters, rather than the band edge. If the Fermi energy can be controlled independently, then the stable charge state can often be manipulated. Thermodynamic equilibrium is often a valid working approximation for semiconductors. It is less useful for insulators, where the timescales for charge states to relax to equilibrium can be hours or days.

Our discussion has assumed that it is a carrier which is captured or emitted when a system is unstable. It is fundamental to materials modification by excitation that, at least for some systems, the instability which lowers the energy can involve a defect process in which an ion or atom moves to a new site. An example might be some system (perhaps a Cl^- ion on a KCl surface) for which the capture of one carrier (Cl^- becomes Cl^0) causes no instability, but the capture of a second carrier (Cl^0 becomes Cl^+ at a site which is stable for negative ions) brings about an instability.

We should add at this point that, normally, the energies should be thermodynamic free energies, since experiments are normally carried out at constant pressure and temperature. Theoretically, most calculations provide thermodynamic internal energies (Catlow *et al.* (1981); see also Hayes & Stoneham (1985), p. 110).

3.3.2 Examples of charge state stability

For simple impurities in oxides, like transition metal ions in MgO, the most important terms in the energies determining charge state stability are the following (Sangster & Stoneham 1980). First, there are terms which are essentially the ionisation potentials for the different charge states of the free ions. Secondly, there is a Madelung energy, of order $24(N-2)$ eV if the ion has charge $N+$ and we assume the Mg ion replaced has charge 2+. Thirdly, there is a term from polarisation of the lattice. For MgO, this energy is of order $6(N-2)^2$ eV. There are also some relatively small terms from covalency, size effects, Jahn–Teller distortions, crystal field corrections, and so on. Nevertheless, the three leading terms dominate, and are individually often larger than the bandgap. The existence of three or four stable charge states for 3d transition metal ions in MgO stems from the cancellation between the electrostatic terms and the free ion ionisation potentials which is important; charge transfer is largely negligible.

In III–V semiconductors, there are also three or four stable charge states for 3d ions. For the III–Vs, it is the charge transfer between the ions on different sublattices which is crucial. In several other systems, the change of charge state will be accompanied by some other change, possibly including ion migration. Photochromism and the photographic effect are examples. Another interesting example is that, in suitable circumstances, the fission product I in UO_2 is stable as a cation on the U site or as an anion on the O site (Grimes *et al.* 1989).

For certain amphoteric defects, elastic interactions can be important. Examples include CsI : Na and SiO_2 : Ge. For CsI : Na, an electron can be bound (giving Na^0, in effect) by the electronegativity difference; the hole (giving Na^{2+}, in effect) is bound primarily by the elastic interaction of the V_k centre with the small cation (Ong *et al.* 1979). For SiO_2 : Ge (Hagon *et al.* 1985), the electron is bound primarily by the extra polarisability of the Ge, whereas the hole is again bound primarily elastically (the free hole in this case is not self-trapped, but is very close to being so).

3.3.3 Stability of self-trapped polarons: Strategies

There are two types of test of our understanding of self-trapping. Here we assess whether or not we can predict correctly when self-trapping will occur, i.e. stability of the self-trapped state relative to a free state. Later, we ask whether the self-trapped state does indeed have the properties which we predict.

Whereas understanding the localisation process needs information about much of the energy surface, studies of the polaron stability are concerned with *extrema* of the energy surface and energy differences between such extrema. Usually, we shall have to consider both one-centre and two-centre polaron states. The phenomenological theory of small-radius polarons (see the opening paragraph of Section 3.3) suggests that one-centre polarons will be more stable than the two-centre state, and that the energy difference between these configurations is the activation barrier for the diffusion of the one-centre polaron (see I G Lang (1977)). However, in some crystals (alkali halides, alkaline-earth fluorides (Stoneham 1975), possibly alumina (Jacobs & Kotomin 1992; Jacobs *et al.* 1992) the two-centre hole polaron is stabilised by an anion–anion chemical bond. One has to go beyond phenomenology to quantum-mechanical calculations to predict the bonding and to make estimates of observables like magnetic resonance parameters for a quantitative comparison between theory and experiment.

The pioneering atomistic calculations of small-polaron versus large-polaron stability were those of Gilbert (see Fowler (1968)). He proposed a two-step procedure: first localise the polaron from a completely delocalized state onto one or two lattice sites; secondly, calculate the energy gain from lattice relaxation from this state to a final polaron state. The first calculations were primitive, but illuminating. They suggested the scheme taken up in many later approaches to polaron structure and stability. In particular, Gilbert showed why holes self-trap in alkali-metal halides but electrons do not. In essence, localisation of holes is favoured by relatively narrow valence bands and by chemical binding forces within the anion molecular ions; for electrons, there is no such bond, and the wide conduction bands mean a large energy cost for localisation.

Early analyses argued that polaron localisation caused ionic polarisation, which created a potential well (involving $\epsilon_{\rm ion}$, a dielectric constant defined by

$$1/\epsilon_{\rm ion} = 1/\epsilon_\infty - 1/\epsilon_0$$

in terms of high frequency and static dielectric constants). The potential well would be Coulombic, screened by the ionic polarisation outside a polaron radius $R_{\rm p}$ and roughly constant inside the well. Minimising the total energy, including kinetic energy, gives $R_{\rm p}$ in terms of $\epsilon_{\rm ion}$. For the small polaron, the effective radius becomes small, comparable with or less than interatomic spacings. There are variants on this approach. The scaling arguments of Emin & Holstein (1976) assert that, if we change the length scale of the (normalised) eigenfunction, the energy as a function of the scale factor should have a minimum at the scale corresponding to the actual eigenstate. Within the more sophisticated technique developed by Toyozawa and coworkers, both states (delocalised and localised polaron) can be treated using the same model Hamiltonian (Ueta *et al.* 1986; Toyozawa 1990).

None of these phenomenological methods allows us to take into account the microscopic structure of the polaron in the real solid. Fuller, atomistic approaches are needed. These might replace the kinetic energy by band parameters, and model the ionic polarisation using the shell model and the Mott–Littleton method. However, the different roles of electronic and ionic polarisations continue to be essential, and the simple dielectric picture is still of value.

3.3.4 Stability of small polarons: Static approaches

We now discuss generalisations and alternatives to Gilbert's strategy. A possible generalisation would be to assume the polaron is localised in a small region of, say, tens of ions or molecules, embedded in the rest of the medium. One could vary the parameters which describe the wavefunction, like its radius, whilst simultaneously minimising the total energy of the crystal with respect to the atomic coordinates. In this way, both short-range and long-range contributions to the total energy are taken into account. Such a static approach is particularly convenient for systems for which only one electron needs to be treated explicitly, and the others can be treated by their pseudopotentials. This has been used for defect states, like the excited states of the F centre (Stoneham & Bartram 1970), and for the simulation of the self-trapped excitons in alkali halides by Song & Leung (1987). An analogous approach has been used for the quantum molecular dynamic simulation of an excess electron in water by Rossky & Schnittke (1988) and Motakabbir & Rossky (1989) and in small clusters by Landman *et al.* (1985), Barnett *et al.* (1988) and Barnett *et al.* (1989). In the many-electron calculations, the localised and the delocalised state usually belong to different Hamiltonians. Sometimes one can obtain acceptable results by taking differences between energies calculated by the most appropriate method at each stage. The difficulties should not be underestimated.

As in almost all approaches, two energies are compared in these calculations: the kinetic energy cost E_{loc} from delocalising a carrier in an undistorted lattice, and the polarisation and distortion energy gain E_{rel} from having a localised carrier. The self-trapping energy is thus the difference:

$$E_{st} = E_{loc} - E_{rel}.$$

When one uses a standard stationary quantum-chemical technique, one must introduce an artificial intermediate state, F_i, as a common reference point for the calculation of E_{loc} and E_{rel}. Two steps are needed. First, one needs to construct an electronic state F_i in the perfect lattice, which is qualitatively the same as the localised configuration. Secondly, one assumes the magnitudes of the electronic polarisation produced by this state and by the completely delocalised polaron state are either the same or, more rarely, are related in some simple way. The validity of this second approximation was analysed by Fowler (1966). With these assumptions, the polaron localisation energy is given in terms of H_{ee}, the electronic Hamiltonian, by:

$$E_{loc} = \langle F_i | H_{ee} | F_i \rangle - \langle F_{del} | H_{ee} | F_{del} \rangle,$$

where F_{del} is the wavefunction of the delocalised state. The lattice relaxation energy is given in terms of the total Hamiltonian H by:

$$E_{rel} = \langle F_i | H | F_i \rangle - \langle F_{st} | H | F_{st} \rangle,$$

where F_{st} is the wavefunction of the self-trapped state. If the electronic polarisation of the crystal and the self-energy of the polaron in the delocalised and the intermediate states do indeed compensate each other, then the localisation energy is simply the change in the kinetic energy of the localising particle. The electronic polarisation of the perfect lattice in the intermediate state has to be taken into account in calculating the relaxation energy.

If E_{st} is positive, then the localised state is certainly unstable. The polaron by itself can exist only as the delocalised state. A localised state might form in the presence of a defect, or near an interface. If E_{st} is negative, the localised state should be stable. However, what is observed will also depend on kinetics: an exciton must self-trap before recombination for the localised state to be seen. The behaviour of the polaron also involves the barriers and the transfer matrix elements, $H_{LL'}$, for diffusion between different localised states.

3.3.5 Stability of small polarons: Microscopic calculation of the relaxation energy

Many model Hamiltonian methods used in studies of self-trapping assume that the harmonic approximation suffices, and that the energy of interaction between the particle and the medium is linear in coordinates representing the deformation of the medium from its equilibrium configuration (Rashba 1982; Ueta *et al.* 1986). These assumptions are dubious for strong lattice distortions. It is in calculating this lattice relaxation that the Mott–Littleton method and related techniques come into their own; indeed, this was recognised in early work (Gilbert 1966). Gilbert's calculations were the basis for further ones by Jette *et al.* (1969), who made extensive estimates for alkali-metal fluorides and chlorides. This work showed that self-trapped holes could be regarded rather well as halogen molecular ions in a polarisable, distortable matrix. The polarisable point-ion model used in this work has been superseded for most serious estimates by the shell model. Later work in the same spirit, but using the shell model, has been published for a wide range of systems: for the alkali-metal halides (including mixed alkali-metal halides) by Cade *et al.* (1984, 1986), for the Cs-structure alkali-metal halides by Monnier *et al.* (1977) and for the alkali-earth metal halides by Norgett & Stoneham (1973). The 'molecule in a crystal' model for the hole polarons in alkali-metal halides (V_k centre) has been examined by Stoneham (1975), by Tasker & Stoneham (1977) by Cade *et al.*

(1984), and by others. Here a diatomic molecular ion is studied quantum mechanically (Tasker & Stoneham 1977), and embedded into a classical ionic host. The strong covalent bonding within the halogen molecular ion X_2^- is included, but there are only ionic interactions with the rest of the crystal.

A more sophisticated approach combines a quantum-mechanical treatment of a small molecular cluster (larger than the two halogens) with Mott–Littleton calculations of the polarisation of the rest of the crystal (for reviews, see Catlow & Stoneham (1989), Shluger *et al.* (1993), Vail (1990), Kantorovich *et al.* (1991), Grimes *et al.* (1992)). As in conventional Mott–Littleton methods, the crystal is divided into two regions: an inner region (I) including a part treated by quantum methods, embedded in an outer region (II) which responds as a dielectric continuum (there can be an intermediate region in which there is lattice structure, but with displacements extrapolated from the continuum model). The region treated quantally is placed at the centre of region I.

This quantum-chemical approach has been used for studies of several alkali-metal halides (Shluger *et al.* 1993) and oxides (Vail *et al.* 1991), and provides satisfactory microscopic models for the self-trapped holes and excitons in these crystals. The quantum-chemical calculations of V_k centres in several alkali halides (Kantorovich *et al.* 1993; Shluger *et al.* 1992a; Shluger *et al.* 1991a,b) give atomic displacements similar to those obtained by Cade *et al.* (1984) using classical Mott–Littleton methods. Similar results were obtained with both techniques for the lattice distortion around the hole in MgO (Shluger *et al.* 1992b; Foot *et al.* 1988). However, the main advantage of the quantum-mechanical methods is that they provide a stronger connection between the localised and delocalised states.

The atomic structures of point defects in ionic crystal can be studied more or less routinely. Sadly, this is not the case for calculations of the relaxation energy. The biggest problem is the choice of the path for polaron localisation and, as a result, the choice of the intermediate state as the reference point for calculating of the relaxation energy (Kantorovich *et al.* 1993). A further point concerns accuracy, since the self-trapping energy rarely exceeds several tenths of an electron volt and is the difference of two large, similar values.

The situation is not simple even for the dielectric polaron, in which electrons or holes in a polar lattice interact mainly with long-range polarisation. The phenomenological theory indicates that the dielectric polaron will have a large radius, not a small radius. The problems can be illustrated for MgO, where there are extensive calculations of hole trapping. All these calculations of lattice relaxation predict the hole to be strongly localised on a single oxygen ion. The relaxation energy E_{rel} is the energy difference between the final, completely-relaxed state and an intermediate state for which there is a hole on one oxygen in an otherwise perfect lattice, but with electronic relaxation taken into account. Various methods give values of E_{rel} in the range 1.7–2.4 eV (Shluger *et al.* 1992b; Foot *et al.* 1988; Shluger *et al.* 1991c). This large energy is fully consistent with theoretical analysis of the observed charge-state stabilities of transition metal impurities in MgO. Estimates of the localisation energy E_{loc} are in the range 2.0–2.4 eV (Shluger *et al.* 1992b). The value of 2.4 eV is half the experimental valence band width (4.8 eV, Tjeng *et al.* 1990), and so gives an upper limit. These calculations use the same quantum methods to calculate both the relaxation and localisation energies, and suggest a self-trapping energy of about -0.4 eV. Simple treatments of small dielectric polarons indicate that the hopping energy for diffusion motion is less than this and probably less than half of the self-trapping energy, so the polaron should be very mobile. Indeed, despite strong localisation implicit in the cluster calculation, and despite the large relaxation energy, the small-polaron behaviour will differ little from that of the large polaron at any but the lowest temperatures. It appears that the transfer matrix elements $H_{LL'}$ are substantial, and the stationary localised state on one site is not a good approximation.

In the case of the 'molecular' polaron, self-trapping is caused by the short-range interaction. The situation is complicated by the possibility of coexistence of both types of polarons. The V_k centre is the classic example. In this case, it is not clear *a priori* if one should consider the localisation in the one-centre form first, and then calculate the additional energy gain due to the formation of the chemical bond between two anions, or whether two-centre localisation is to be preferred from the start (see Section 3.3.3).

Evidently, the relaxation energy only makes sense as a part of the self-trapping energy, since it depends on the intermediate state(s) chosen in its calculation. The relaxation energy cannot be measured directly, although it may be estimated using energy cycles, which combine available experimental and theoretical data (N Itoh *et al.* 1989a). Relaxation energies contribute to a wide range of observable quantities, including defect formation energies and optical charge transfer processes. The observed systematics of behaviour (e.g. which charge states of transition metals are stable in a given host) give strong support to the theoretical estimates of these relaxation energies.

The self-trapping energy is defined relative to the bottom of the band of free polaron states, which in turn depends on the crystal under study. Thus, whilst E_{st} may be used as a criterion for self-trapping in any particular crystal, it does not provide a good basis for comparison with other crystals or for mixed crystals. Such a comparison needs energy cycles to calculate the appropriate defect formation energy. This formation energy can also be compared with the crystal ionisation and excitation thresholds, and so is linked to the effectiveness of the polaron or exciton production.

3.3.6 Small-polaron formation energy: Energy cycles

Although the general idea is straightforward, energy cycles differ from case to case. Moreover, there can be several cycles for any particular case (see, e.g., Cade *et al.* (1984), N Itoh *et al.* (1989a), Shluger *et al.* (1993)). Consider the V_k-type self-trapped hole in alkali-metal halides or alkaline-earth fluorides. In all these systems, the self-trapped hole is sensibly regarded as a halogen molecular ion X_2^- occupying two anion lattice sites (see Stoneham (1975), Tasker & Stoneham (1977)).

The energy required to form a V_k centre may be considered as the sum of energies for several distinct steps. First is the energy required to remove the two halogen lattice ions (X^-) separately to infinity, which is the vacancy energy E_v. Secondly, there is the energy $D(R)$ for the formation of X_2^- (gaseous, at chosen spacing R) + e^- from two X^- anions. The internuclear separation R for X_2^- may be taken as R_e for free X_2^- although other values of R do not change the final result. Thirdly comes the energy associated with introducing X_2^- into the prepared (perfect) vacancy site, which is the interstitial energy E_I. Finally, there is the energy arising from relaxation of ions in region I, region II, and of course, within the V_k centre, the relaxation or lattice energy E_R. It is necessary to consider the electron released in the second step, and this is usually placed at the bottom of the conduction band of the crystal, defined by the crystal electron affinity χ. The V_k centre formation energy may be written in the form:

$$E_D = E_v + E_I + E_R + [A(X) - \chi] - D(R),$$

where $A(X)$ is the electron affinity of the halide X. Values of the electron affinity for crystals are not well known; those quoted range from 0.1 to 1.5 eV (Poole *et al.* 1975). The result is that the defect energy has a consistent reference point within a given crystal, but comparisons of different energies involving different crystals do not have identical reference points.

Alternative energy cycles might be useful, e.g. the uncertaincy in χ can be eliminated if the electron is removed from the conduction band and placed on an 'electron trap', i.e. it reduces a metal ion doped into the crystal for this purpose. Typically, for V_k centre formation, Pb^{2+}, Tl^+, or Ag^+ cations are employed as electron traps. The energy scale (zero) involving the electron trap corresponds more to what is involved in practice in self-trapped hole formation, and in that sense, this zero point may be more satisfying. However it has serious technical drawbacks, especially if the Mott–Littleton technique is employed for calculations. These concern the pair potentials between the host lattice ions and impurities in different charge states and the accuracy of the relaxation energy calculation for the impurity states. However, the location of the electron on a trap preserves the crystal neutrality and permits comparison with uncharged defects on a common ground. It enables us to use the periodic boundary conditions in the calculations of charged polaronic states.

One simple estimate of the energy E_G of a hole at the top of the valence band of an undistorted crystal assumes that the centre of the valence band corresponds to the Madelung potential at the anion, so that:

$$E_G = M/a_0 - (1/2)\, E_{val},$$

where M is the Madelung constant, a_0 is the lattice constant, and E_{val} is the valence bandwidth. The corresponding approximate condition for self-trapping becomes:

$$-E_D - M/a_0 + (1/2)\, E_{val} \leq 0,$$

an equation which has been widely used in previous calculations (e.g. Cade *et al.*, 1984; Monnier *et al.* 1977; Norgett & Stoneham 1973).

3.3.7 *Specific properties of the self-trapped exciton (STX) state*

The structure and characteristics of a self-trapped exciton involve both the basic issues of self-trapping, and the interaction and correlated behaviour of the electron and the hole. The conditions for exciton self-trapping have been studied phenomenologically and reviewed by Ueta *et al.* (1986) and Rashba (1982). The theory predicts that an exciton can self-trap even if neither an electron nor a hole can do so alone. Another striking prediction is that, in certain circumstances, the parity of the relative motion of an electron and a hole may be broken (Ueta *et al.* 1986). In particular, this may happen when a hole has a very large effective mass and its interaction with an electron consists of the long-range Coulomb attraction and short-range strong repulsion. These predictions of the phenomenological theory are supported by the results of quantum-mechanical calculations.

Starting from the first calculations by Wood (1966) and Stoneham (1974), successful attempts have been made to use for self-trapped excitons the methods developed for calculations of point defects (Stoneham 1974; Shluger *et al.* 1992). Such an approach is only possible when the exciton is well localised in a small region of the crystal host. Even then there are clear distinctions between the ground or excited state of the localised point defect and the self-trapped exciton.

An exciton is an excited state of an otherwise perfect (but deformable) lattice. A key characteristic of the self-trapped exciton is its luminescence: self-trapped excitons are short-lived luminescent states of excited crystals, with properties very different from those for the crystal ground state (see for reviews Stoneham (1979a), Luschik & Luschik (1989), N Itoh (1982), Williams & Song (1990) and references therein). The initial states for the luminescence are treated as approximately stationary, and correspond to the minima of the adiabatic potential of the excited crystal. However, the methods and basis sets which are developed for the calculation of the properties of the crystal ground state are not well suited for its excited state. Moreover, usual adiabatic calculations are meaningful only in the close vicinity of these minima, for which atomic structures are considered as models of self-trapped excitons. The dynamics of the processes leading to these states requires further concepts and complexities.

The electron and the hole overlap strongly, but become localised in the lattice for different reasons. In alkali halides and alkaline-earth fluorides, in effect, the hole is self-trapped and localised in the form of a quasimolecule, whereas the electron is localised mainly by its Coulomb interaction with the hole. In crystalline quartz the holes do not self-trap, while the self-trapping of excitons gives rise to the luminescence with a very large Stokes shift. According to experimental ESR data, the electron of the self-trapped exciton in these crystals is less localised than the hole. Further, the self-trapping of the exciton in many crystals is accompanied by its decay producing primary Frenkel defects (Itoh & Tanimura 1990). The degree of localisation of the electronic component of the nearest-neighbour defect pair depends sharply on the relative position of the hole component. Since the radius of the electron localisation is large and changes significantly in the calculation of the adiabatic potential for the exciton localisation and decomposition, special efforts are needed in order make the basis set flexible enough and to take into account the resulting changes in polarisation and other energies.

Another issue arises with the wavefunction of the self-trapped exciton. Experimentally, both singlet and triplet states of the self-trapped exciton in the alkali-metal halides have been detected or inferred. In crystalline quartz, only the triplet state of the self-trapped exciton has been observed so far. Most theoretical studies concentrate on the triplet excited state alone, solely for technical reasons. In these wide-gap insulators, the crystal triplet state is always an excited state. For the lowest multiplicity, it may be treated in the single-determinant unrestricted Hartree–Fock approximation. This approach has been adopted in all many-electron calculations made so far. It is much more time consuming to treat the singlet excited state using a many-determinant approximation for the wavefunction of the crystal. The first calculations for the singlet state of the self-trapped exciton in NaCl were those of Stoneham (1974). Later, similar calculations were made for several alkali-metal halides using the generalised valence bond (GVB) technique (Song & Baetzold 1992).

3.4 Mobilities and charge transport in non-metals

3.4.1 Experimental data on mobilities

The mobilities of carriers in non-metals have several key characteristics. In semiconductors like Si or Si/Ge, where the electron–phonon coupling is weak, the mobility is high (175 000 $cm^2/(V\ s)$ has been quoted at low temperatures), falling with rising temperature. Even in KBr, if zone-refined, the electron mobility at low temperatures can be as high as 130 000 $cm^2/(V\ s)$. In SrF_2, where the hole is strongly coupled to lattice distortion, there are two thermally-activated processes, one causing the V_k centre to reorient (e.g. from [100] to [010]), the other leading to motion along a single row (e.g. [100]). In an oxide like UO_2, both electron and hole are self-trapped; in addition, there is trapping of these carriers by defects, such as interstitial oxygen, which introduces extra temperature-dependent processes. In oxides like MgO, Al_2O_3 or SiO_2, the free carriers are either just self-trapped or not self-trapped. If the self-trapping energy is very small, then the stable configuration may be a small polaron, but the transport may be primarily by the population of large polarons in thermal equilibrium. Disorder can have many effects. The simplest effect is trapping at relatively isolated sites. A second effect is to require percolation in non-stoichiometric systems, since there will be only certain regions in which relatively free carrier motion is possible. This can be modelled by effective medium theories as well as percolation models. One extreme case is Anderson localisation (Anderson 1958), in which there are no delocalised states, a situation very different from the well-known case of trapping (the term Anderson localisation is often misused; see Section 8.2.2 of Hayes & Stoneham (1985) for a fuller discussion).

3.4.2 *Small polarons and large polarons: Ideas about motion*

The mobility of a carrier provides some of the information to distinguish between large-polaron motion and small-polaron motion (Fig. 3.4). In a perfect (or, at least, trap-free) crystal, large polarons must have a mean free path at least equal to the atomic spacing. If the large polarons have thermal

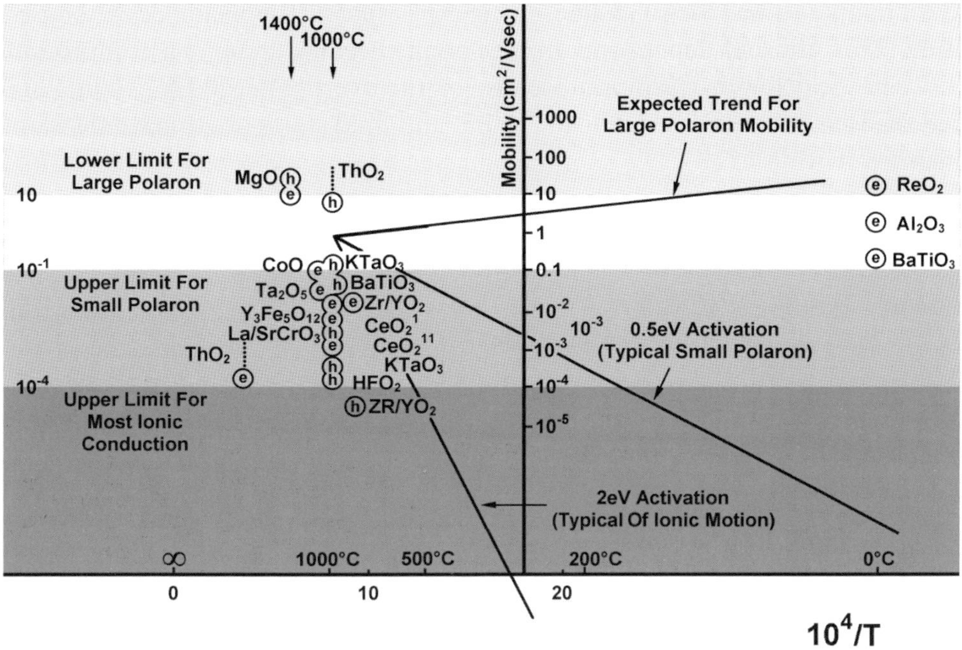

Fig. 3.4. Mobilities of carriers in oxides. The limiting cases require: (a) that the mean free path exceed the interatomic spacing for the large-polaron regime, and (b) that the incoherent jumps in the small-polaron regime do not occur more often than the highest phonon frequency. The effects of trapping are not shown.

energies, then their mobilities must be at least of order 1–10 $cm^2/(V\,s)$. For large polarons, the mobility falls as temperature rises, since scattering by phonons increases. Small polarons, on the other hand, have the fastest incoherent jump rate between sites of the order of the fastest phonon frequency. This implies a mobility less than 0.1–1 $cm^2/(V\,s)$. The mobility rises as the temperature rises, since small polarons move by thermally-activated incoherent jumps. Strictly, the temperature dependence should deviate from Arrhenius behaviour: the apparent activation energy will fall as the temperature falls, and there should be a power-law dependence at very low temperatures (see, e.g., Norgett & Stoneham (1973)).

Both of these arguments are simplistic; nevertheless, there is a division for mobilities of order 1 $cm^2/(V\,s)$. Trapping is one of the complications, since it will slow the carrier motion, and will involve thermally-activated release from traps. There are various checks, such as the Seebeck effect, which enable one to distinguish trapping from self-trapping. There are, of course, other transport mechanisms. In amorphous materials, variable range hopping can occur at lower temperatures, with a characteristic $\exp[-(\theta/T)^{1/4}]$ dependence on temperature. This unusual dependence comes about because there is a compromise between short jumps (favoured because of the overlap needed for tunnelling) and the longer jumps needed if the initial and final states are to have the same energy to within kT.

3.4.3 Self-trapped excitons versus self-trapped holes: Exciton bandwidths

For free carriers, the valence band (hole) and conduction band (electron) widths lead to a measure of the kinetic energy required to localise the carrier in a single cell. The balance between this kinetic energy term and the other contributions favouring localisation is what decides whether large or small polarons will have the lowest energy. For excitons, there is a similar balance, but the corresponding bandwidth is much smaller, since it is determined by an interionic exchange energy, rather than an overlap-like term.

Free excitons emit sharp luminescence near to the absorption edge, and close to the first exciton absorption band. In some crystals, like alkali iodides, recombination luminescence due to both free and self-trapped excitons is observed, indicating that free and self-trapped excitons can coexist. The intensity of the free-exciton luminescence is much higher after excitation to the first exciton peak than when electron–hole pairs are generated. It decreases as temperature increases (Nishimura 1983), as expected if there is an energy barrier against exciton self-trapping. The free-exciton luminescence in RbI has been shown to arise from the triplet state, not from the singlet state, using magneto-optical circular dichroism (Tanimura & Itoh 1992). It is likely that the combination of exciton–phonon interaction and the large iodine spin–orbit coupling converts singlet free excitons to triplet free excitons. The motion of self-trapped excitons has been measured through its interaction with impurities. Free-exciton luminescence perturbed by the K^+ impurity in NaI falls in intensity as the temperature rises from liquid helium temperature to 40 K, below which the free-exciton luminescence is not quenched, suggesting coherent migration of free excitons (Nishimura 1989). On the other hand, the hopping migration of self-trapped excitons is seen in NaCl (Tanimura & Itoh 1981) and in NaI (Popp & Murray 1972) with an activation energy of 0.15 eV (NaCl) and 0.18 eV (NaI).

3.4.4 Classical diffusion of ions and other over-the-barrier processes

In classical diffusion, the jump process is one which occurs on an adiabatic surface in a system which is usually close to equilibrium. The diffusion constant

D which describes incoherent jumps may be related to the jump rate Γ for an atom (here meaning the number of jumps per unit time to one of the N neighbouring sites); the N jumps are assumed equivalent in our simple present discussion, with jump distance a. Again, taking only these simple cases, the relationship has the form:

$$D = (Na^2/3)\,\Gamma.$$

For practical purposes, we note that an atom making one jump every lattice vibration period will have a diffusion constant of order 10^{-4} cm^2/s (10^{-8} m^2/s), similar to diffusion constants observed for liquids, and a mobility of order 1 (cm/s)/(V/cm) (10^{-4} m^2/(V s). Both these numbers are approximate upper limits for ionic motion, but are frequently approached for the electronic analogue, small-polaron motion, in which self-trapped carriers also undergo an incoherent sequence of jumps. We remark that these bounds apply strictly to D, the diffusion rate itself, rather than to the prefactor D_0. The extra factor $\exp(\alpha)$ can raise D_0 by more than an order of magnitude.

In the simplest cases, one can use the known D_0 and jump distance a to deduce a jump probability Γ. What value is expected? For classical diffusion, there are three main approaches. One is molecular dynamics and, when it is practical, this approach has many advantages. The other two approaches are *Rice–Slater dynamical theory* (Rice 1958; Slater 1959) and *reaction rate theory (Vineyard theory)* (Vineyard 1957). These we now describe, since they have somewhat different underlying physics, and their ranges of applicability are different.

Rice–Slater dynamical theory, and related first-passage theories

Suppose diffusion takes place by motion along some reaction coordinate Q, in which it passes through a saddle-point value Q_s. Suppose further that Q is dynamically independent of the other degrees of freedom (see Section 3.2 for a fuller discussion). How often will one find the system with $Q > Q_s$ and $dQ/dt > 0$, i.e. beyond the barrier and moving towards the final site? Within certain working assumptions, e.g. that the energy surface is harmonic on both sides of the saddle point, this can be related to the 'first-passage' problem. With some generalisation, the result is an expression for the jump rate of the form:

$$\Gamma = \Gamma_0 \exp\left(-E/\langle\epsilon\rangle\right),$$

in which $\langle\epsilon\rangle$ is simply proportional to the energy associated with the reaction coordinate, the proportionality constant being such that its value would be kT for thermal equilibrium.

Γ_0 will be some average of the vibrational frequencies for the (defect) lattice. This has two corollaries. First, any 'first-passage' theory will yield a jump rate bounded by the largest lattice frequency. Indeed, it is likely (but not strictly essential) that the frequency prefactor will have this same upper bound. *These upper bounds will apply in any cases for which it is necessary to describe the atomic dynamics of the route by which the saddle point is reached.* The second corollary is that there is an implicit assumption of negligible energy transfer between degrees of freedom. This is reasonable for the molecular systems for which the theory was first devised. The third corollary concerns the expected temperature dependence for enhanced diffusion. Suppose that some extra, non-thermal, energy can be made available, increasing $\langle \epsilon \rangle$ from kT to $kT + \delta\epsilon$. Then we might expect the original Arrhenius expression to become the *non*-Arrhenius form:

$$\Gamma = \Gamma_0 \exp[-E/(kT + \delta\epsilon)].$$

Reaction rate theory (Vineyard theory) and related thermodynamic theories

The original ideas of Marcelin, Hirschenfelder, Wigner, and others concentrate on a system in thermodynamic equilibrium. For a system of classical particles in thermal equilibrium, the probability of some particular combination of momenta and positions is defined (with assumed boundary conditions, such as constant volume). How the system reached that state is not asked. When one talks of 'ground' and 'saddle-point' states, there is no assumption about how the system moved from one to the other. Instead, there are certain general assumptions, like ergodicity (so, for example, there is no selection rule preventing the system reaching the saddle point from an assumed initial state). It is also implicit that the rate of energy transfer between degrees of freedom is not a limiting step, a very different assumption from that implicit in the dyanamical theories. These ideas yield a rate which, in general terms, involves three factors. The first factor is a population factor, essentially the probability that the system is close to the saddle point (more strictly, close to the saddle hypersurface in the phase space of the system). The second factor is a flux factor, describing the motion near to the saddle point. The third factor is an efficiency factor, which allows for the transient motions which cross the barrier without being countable as jumps; this third factor is likely to differ from unity by only a few per cent.

The practical implementation of reaction rate theory was by Vineyard (1957), who assumed locally harmonic surfaces. In the initial (and equivalent final) states, the vibration frequencies of the normal modes are ν_i; at the saddlepoint, the real frequencies are ν_j', and there is at least one imaginary frequency, corresponding to motion in the direction of the reaction coordinate. The key frequency which enters into the jump rate is Γ, given by:

$$\Gamma = \prod_{i=1}^{N} \nu_i \Big/ \prod_{j=1}^{N-1} \nu_j'$$

Note that the expression involves products of factors from each mode separately. This means that if one mode is highly anharmonic, it can indeed be handled in a special way. This was necessary in applications of Vineyard theory to cation diffusion in MgO. It is such anharmonicity which tends to give large prefactors (as observed) and not usually a small value of ν_j'. Small values of ν_j' often point to anharmonicity, of course, since the region of configuration space over which the energy surface is flat is rarely very extensive.

The reaction rate theory can be exploited for a variety of activated processes. These include non-radiative decay from state C of Fig. 3.1 via the cross-over X where, in effect, the system goes over a barrier from a lower state. This description is exploited by Struck & Fonger (1991) and others (see also Section 3.6).

3.4.5 *Diffusion of self-trapped carriers*

As pointed out by Flynn & Stoneham (1970), in many cases the wavefunctions of the localised polaron states may be treated as sensibly-exact eigenfunctions of the total Hamiltonian. They bear the same relation to the exact eigenstates that Wannier functions bear to Bloch functions. Note too that for the mean energy of the band of the localised states both Bloch and Wannier representations have exactly the same value. Just as in the construction of Wannier functions, the localised functions of different sites may be orthogonalised. Nevertheless they are mixed by the exact crystal Hamiltonian, and the transitions induced between different polaron configurations may be related to the polaron hopping motion through the crystal.

This approach has been used, for instance, by Holstein (1959). In the latter work the electronic overlap term of the total Hamiltonian, which is responsible for the polaron jumps between sites, was treated as a small perturbation. In zeroth order, i.e. in the absence of this term, the polaron may be treated as permanently localised at a given site. Within the static approach, which concerns the microscopic characteristics of the small polaron rather than the dynamics of its diffusion, this state is treated as a approximately stationary.

3.5 Non-radiative transitions I: Cooling transitions

We now discuss some of the ways in which equilibrium can be reestablished after excitation. Equilibrium can be disturbed in several ways. First, a change of electronic state (such as a transition between two bound states of an impurity) will cause changes in the forces on the atoms or ions near to that impurity. Switching on these forces will drive an alteration in the mean position about which the nearby ions move, and also gives energy to motion in some reaction coordinate. The evolution to a new equilibrium is discussed in Section 3.5.1. The same ideas apply to subthreshold collisions. Secondly, we recognise that the enhanced local motion just described can aid a diffusion jump, in which an atom changes site. This situation, which allows us to relate cooling to thermal diffusion, we discuss in Section 3.5.2. Thirdly, when a carrier is excited into the conduction band, the excited electron will lose kinetic energy until it has been thermalised or, in some cases, captured. In the same way, excitation from the valence band will generate a hole, which may also thermalise. The dynamics of this thermalisation process is discussed in Section 3.5.3. Fourthly, during the cooling process, there may be a transition from one adiabatic energy surface to another. This is important in deciding whether luminescence will be observed or not (Section 3.5.4). Related to this electronic cooling process are the cascade processes whereby carriers captured into the closely-spaced, shallow, higher-energy states of a defect either decay to the ground state or are thermally ionised.

3.5.1 Cooling of atomic motion

In a perfectly harmonic solid, the modes are dynamically independent. The energy of excitation in a single normal mode is not redistributed over other degrees of freedom. However, electronic or vibrational excitation usually leads to forces on atoms which drive a reaction coordinate, not a normal mode. When an F centre is excited from a 1s state to a 2p state, there are strong radial forces on the cations neighbouring the anion vacancy, and the radial motions of these ions (together with some suitable motions of more distant ions) represent a reaction coordinate, not a single mode. The distinction is important. Motion in this reaction coordinate involves modes of many frequencies, and the dephasing of these contributions is a significant effect. This is one of the many situations in which the traditional one-frequency models must be generalised. Such generalisations are readily done by analysing molecular dynamics.

In some cases, vibrational energy transfer can be observable. Thus, there is a 1000-fold increase in relaxation rate for the stretching vibration of CN^- at a halogen site on going from an alkali halide host to an Ag halide host (Happek *et al.* 1994); this difference is the effect of a high-frequency local mode in the case of the silver halides. Another example concerns the dynamics of H or D in water (Gale *et al.* 1999) or hydroxyl. Bonn *et al.* (1996) discussed the vibrational relaxation of the O–D mode for surface-bonded hydroxyls with various H-bondings. There is a transfer of energy to other modes, and especially to the OD...X hydrogen bond vibration. We shall not usually need to consider this sort of energy redistribution. A third example (Kremerman *et al.* 1997) is the non-radiative decay of excited Cr^{4+} ions in forsterite, where there is energy transfer from the electronically-excited ion to local modes; these local modes then lose their energy to other modes in times of a few picoseconds.

There are, of course, many examples in which the phonon bottleneck has been seen for the low-energy phonons taking part in the direct process of spin–lattice relaxation (van Vleck 1941; Stoneham 1975, Chapter 14). In such cases, the degree of excitation of some group of phonons remains out of equilibrium with the bulk of the phonon modes, evolving to the lattice temperature only by energy transfer to modes of other frequencies or through coupling to another thermal reservoir.

It is helpful to begin with some simple analytic arguments. First, suppose the reaction coordinate is nearly a normal mode Q. If it were exactly a normal mode in a perfectly harmonic solid, there would be no energy transfers to other modes. Real crystals are always anharmonic, so terms in the Hamiltonian like AQQ_bQ_b' will transfer energy to other modes Q_b, Q_b' which constitute a heat bath. However, even in a perfectly harmonic crystal, we would expect terms to redistribute energy, simply because there is a difference between the Hamiltonian which would make the reaction coordinate a normal mode and the actual Hamiltonian. This difference h is small in some sense, since we have assumed the reaction coordinate is close to a normal mode; h will have the form BQQ_b''. The important terms in both the anharmonic and harmonic perturbations are linear in Q. A simple argument based on the usual golden rule for transition probabilities then shows (Hayes & Stoneham 1985; Stoneham 1977, 1981b) that the rate at which energy is transferred from the reaction coordinate to other modes is proportional to the degree of excitation of the reaction coordinate. In essence, the key transition matrix element includes a factor $\langle n+1|h|n\rangle$, and the rate is proportional to square modulus of this, i.e. proportional to $n+1$. This result, which we shall use later, only applies when the damping is relatively small, such that the vibrational quantum number changes every few vibrations. The key result is that the rate of loss of energy is proportional to the energy of the oscillator. This result is also predicted by a phenomenological argument in which a harmonic oscillator is damped by a friction term proportional to velocity.

Another simple description follows from the ideas of dephasing (Masri & Stoneham 1984). Suppose the modes which contribute to the reaction coordinate lie within the frequency range ω_1 to ω_2. Then vibrations with these two extreme frequencies which are initially in phase will be completely out of phase after a time $2\pi/(\omega_1 - \omega_2)$, and one can easily rewrite this to say after how many periods at the average frequency will this dephasing happen. A natural measure of dephasing for a continuous distribution is the ratio of the dispersion to the root mean square frequency $(\langle\omega^2\rangle - \langle\omega\rangle^2)^{1/2}/\langle\omega^2\rangle^{1/2}$. For a Debye model, this is about 1.03. Masri & Stoneham found this simple approach agreed well with molecular dynamics using interatomic potentials (Fig. 3.5). Their results suggested that any coherent vibrational motion in the reaction coordinate would usually be damped rapidly, thus reducing the efficiency of the recombination-enhanced processes which we discuss later (Chapter 7).

Later studies (Mainwood & Stoneham 1994; Stoneham & McKinnon 1998) used self-consistent molecular dynamics to examine vibrational behaviour following core exciton creation (Fig. 3.6). What is important is the appreciation that molecular dynamics can be used to predict the *lineshape function*. This function, in addition to determining optical lineshapes (as its name implies) is a response function, a measure of how readily the lattice takes up energy in response to the sudden imposition of a force, as when electronic excitation occurs. It is therefore also a critical function in non-radiative transitions.

The availability of self-consistent molecular dynamics means that cases can be examined which have significantly complex behaviour. An example is the excitation of trans-polyacetylene or similar conducting polymers, where exciton formation or soliton creation are possible, and where the subsequent evolution is very complicated. The dimerisation pattern is particularly important: for the Nth C atom in the conjugated polymer chain, the dimerisation measures the difference in bond length for its bonds with the $(N+1)$th C and the $(N-1)$th C; there is also a factor $(-1)^N$ to make interpretation easier. The polymer can be modified by excitation in ways which affect the topology; for example, a soliton will have dimerisation of different sign on opposite sides. Figure 3.7 shows the evolution in one case, which illustrates that it is still possible to model systems where it is necessary to go beyond simple non-self-consistent, harmonic theories. There is a wealth of experimental information for conducting polymers. In polydiacetylene, for instance, Pham *et al.* (1995) show that C=C double bonds and C≡C triple bonds are involved in dephasing and population dynamics during transient structural changes just after excitation.

Experiment, when available, supports the general picture described. Nisoli *et al.* (1996) reported subpicosecond dynamics for the KBr F centre. They observed coherent dynamics, which lead to subpicosecond oscillations in the

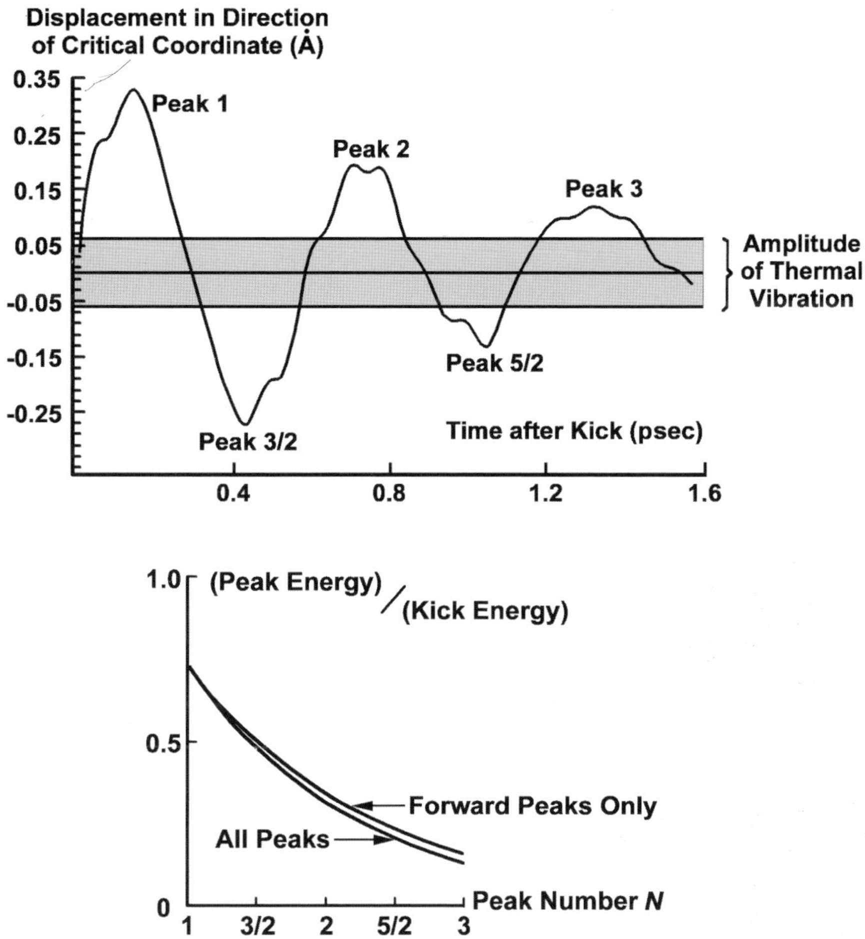

Fig. 3.5. Damping of vibrations in a reaction coordinate. These results (Masri & Stoneham 1984) are for the interstitial motion in a bcc crystal, and mimic recombination-enhanced diffusion (Chapter 7). The damping can be understood as dephasing of the modes of different frequencies which contribute to the reaction coordinate. In the upper figure, the damped motion gradually decreases into the thermal background. The lower figure shows the roughly exponential decay of the heights of the forward peaks (peaks 1, 2, . . .) and all (including backward peaks 1/2, 3/2, . . .) with time (peak number).

sample transmissivity. There were oscillations at 3.2 THz (period 0.31 ps) and at 3.5 THz (period 0.29 ps), frequencies which agree well with Raman data, although not with continuous wave measurements. The authors believed what was seen corresponds only to the A_{1g} oscillations. They deduced that the oscillations were due to off-diagonal matrix elements of the density matrix between consecutive excited and ground states; there is no equilibrium counterpart.

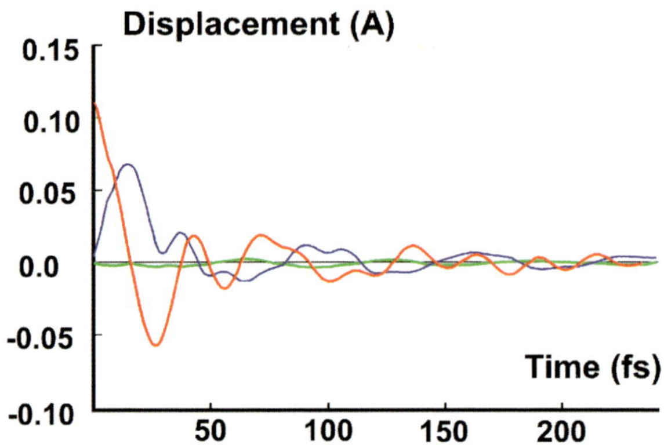

Fig. 3.6. Damping of vibrations following core excitation in diamond, from self-consistent molecular dynamics (after Mainwood & Stoneham 1994). The motion is of the excited atom along the 111 axis.

From the temperature dependence of the luminescence and absorption line-widths, they concluded that, for the ground state the force constant is 9.47 eV/A, the frequency is 3.2 THz, and the effective mass is 6 times the cation mass. For the excited state, the force constant is marginally less, 9.39 eV/A, but the effective mass is reduced to 4.5 cation masses.

Complex behaviour is the rule, rather than the exception, for biological systems. A fascinating example is that of electron transfer in a genetically-modified bacterial reaction centre following optical excitation with an 80 fs laser pulse, sufficiently short that its energy width spans the absorption band (Vos *et al.* 1993). Monitoring of the emission shows features which might be found in many systems. First, it is clear that there are at least two modes involved (77 cm^{-1}, 15 cm^{-1}), and that these are modes of the excited-state system, not the ground-state system. The relaxation time for electron transfer process is about 11 cycles of the higher-frequency mode or over 2 cycles of the lower-frequency vibration. Secondly, there is some coherent nuclear motion in the primary electron transfer process, contrary to the usual view that dephasing is complete before transfer.

Another interesting example of vibrational energy relaxation and the associated structural dynamics is found in the heme proteins (R J D Miller 1991). We emphasise once more that there is no need to have translational (or any other) symmetry to have normal modes. For the heme proteins, one can distinguish between the *intramolecular* heat bath (vibrations mainly within the heme chromophore) and the *intermolecular* heat bath, with which the heme is linked by van der Waals interactions to the rest of the protein, and hence to the water. In these systems, anharmonic interactions may well be important, in addition

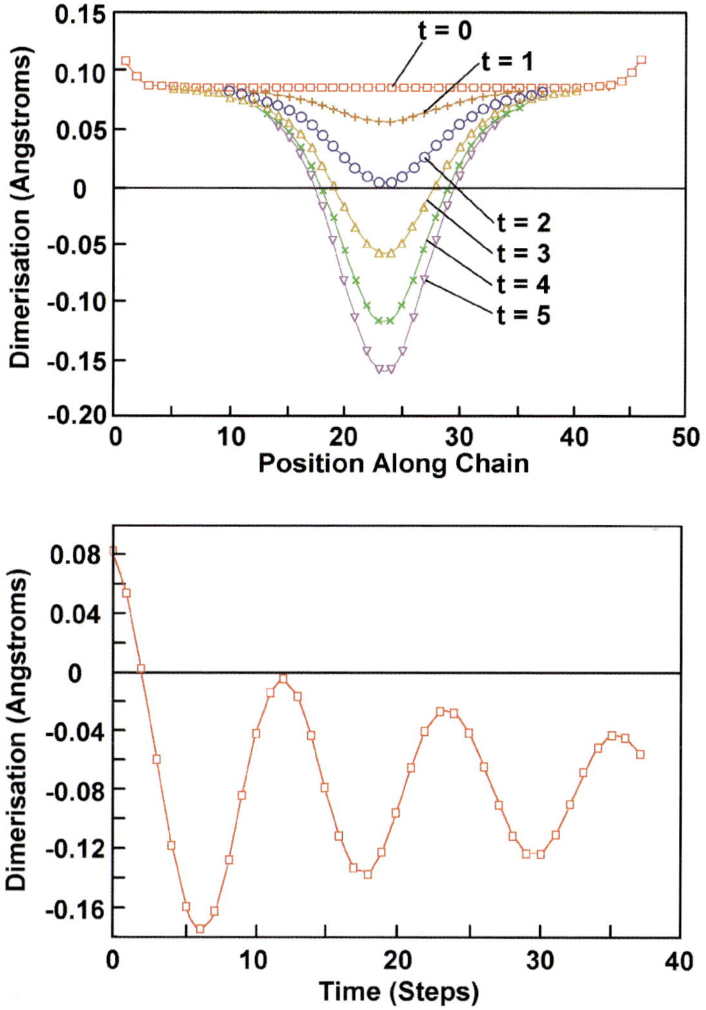

Fig. 3.7. Evolution of trans-polyacetylene following the injection of an electron–hole pair (from Wallace *et al.* 1991). The dimerisation corresponds to a soliton–antisoliton pair, plus a local charge-neutral oscillation (breather) at the centre of the chain. The breather motion is weakly damped, not unlike the damping shown in Fig. 3.5.

to any of the harmonic processes already discussed in relation to dephasing. The transfer of energy from the intramolecular heat bath to the intermolecular heat bath may be rate-limiting in some cases, which is an example of a phonon bottleneck. The energy transfer may well be thermal diffusion of heat, i.e. transfer of energy in space, as opposed to the transfer of energy from modes of one frequency band to another, which can occur in other systems.

The studies of redistribution of vibrational energy through anharmonic terms are very varied. One example is the work of Schofield *et al.* (1996). In their computational study of non-linear anharmonic oscillators, they showed that the non-linear coupling implied a spectrum of state survival probabilities, and that this implied fluctuation in these probabilities. Two limiting cases emerged. For the *strong coupling limit,* there were small fluctuations, and the average survival probability followed semiclassical diffusion, with a linear dependence of the rate of energy flow on the coupling strength. In *weak coupling,* there were large fluctuations, and the average survival probability followed a $1/t$ power-law decay; there were strong beats in the probabilities, and quantum interference.

3.5.2 Transitions from one energy surface to another

We have mentioned weakly-damped and strongly-damped systems. A simple distinction in terms of the configuration coordinate diagram would regard a strongly-damped system as one which lost energy at a few vibrational quanta per vibrational cycle, whereas a weakly-coupled system would vibrate for several cycles between losses of one vibrational energy quantum. These distinctions can be made more precise in specific cases. The differences have significant consequences, for example, in predictions of whether or not a particular defect will luminesce after excitation.

We take up again the specific model of the F centre outlined in Section 3.2.3 (Fig. 3.1). After excitation (transition AB), if the damping were strong (a model once used by Seitz (1940)) the system would evolve directly to C, and would only be likely (subject to certain qualifications about tunnelling) to make a non-radiative transition if the lower energy surface intersected the upper energy surface between B and C. This leads to several contradictions with experiment. However, if the damping is weak (a view taken by Dexter (1956), Dexter, Klick & Russell (1956) and Bartram & Stoneham (1975)), another picture emerges. A crucial issue is whether or not the cross-over X between the upper and lower configuration coordinate curves lies at an energy above that of B or below it. There is a simple and effective rule: luminescence will occur efficiently if B lies below X; if B lies above X, there will be little or no luminescence at zero temperature.

The rule can be understood as follows (Stoneham 1977). Near the cross-over (E close to E_x) the vibronic states involve strong admixtures of both ground and excited electronic states. At low temperatures, these cross-over states will be involved in the dominant non-radiative deexcitation route from the excited electronic state to the ground state. If the cross-over has energy well above B, these cross-over states will not be encountered in the evolution after excitation; the system can decay without interruption to C, and luminesce. If the cross-over lies below B, the important factor is the branching ratio, the fraction of cases that the system ends in the ground electronic state or excited electronic state after passing through a cross-over state. An important point is that whether or not X lies above B can be decided from optical absorption data, subject to some modest simplifications. The critical parameter (Bartram & Stoneham 1975; Hayes & Stoneham 1985) is Λ, which is the ratio of the relaxation energy $S\hbar\omega$ to the optical absorption energy. If Λ is less than 0.25, luminescence is expected; F centres in many halides and oxides are in this category. For values above 0.5 (LiI), no luminescence is expected. Values in between (NaI, NaBr, LiCl, LiBr, LiF; NaCl is marginal, with a Λ value of 0.26) need a slightly more sophisticated argument, but luminescence is expected to be weak. The several obvious tests of the ideas work very well: predictions of luminescence are in essentially full accord with what is seen. Moreover, changing the excitation energy by tuning a laser through the cross-over region does indeed show the expected dependence of luminescence energy on laser energy (Hirai & Wakita 1983).

Turning to the weak luminescence for $0.25<\Lambda<0.5$, one expects that if $N_{CD}{}^{rad}$ is the number of radiative recoveries C to D, if $N_{xg}{}^{non\text{-}rad}$ is the number of non-radiative recoveries, and if F_{cx} = fraction of cases leaving the cross-over states and cooling down the vibrationally excited states of the upper curve (excited electronic state) and F_{cg} = fraction of cases leaving the cross-over states and cooling down the vibrationally excited states of the lower curve (ground electronic state), then

$$N_{CD}{}^{rad}/N_{xg}{}^{non\text{-}rad} = F_{cx}/F_{cg}.$$

But we know that the cross-over states will have comparable components of both electronic states. Further, we know that the transition probabilities are proportional to the degree of excitation (Section 3.5.1). We surmise therefore that the ratio is also given by $N_{CD}{}^{rad}/N_{xg}{}^{non\text{-}rad} = (E_X - E_C)/(E_x - E_A)$. This leads to predictions (Fig. 3.8) of luminescent efficiencies which are in respectable accord with experiment, including the effects of applied hydrostatic pressure (Asami *et al.* 1986). It is clear that there are other processes as well (Bartram & Stoneham 1983; Luty 1983).

One ambitious idea is that reactions might be tuned by selective vibrational excitation. This would have parallels with molecular reactions, where such tuning is achievable. Thus Tannor (1994) have considered reactions of the sort $X + HCN \rightarrow HX + CN$, in which the vibrational state of the HCN can be critical. The issue here is the possible vibrational state control of the branching ratio. Several different types of behaviour are seen. If X = H or O, the CN is a spectator, and remains vibrationally cold. If X = Cl, there is evidence for an HClCN intermediate and for a statistical (chaotic) process where the CN ends with a vibrational state independent of initial excitation.

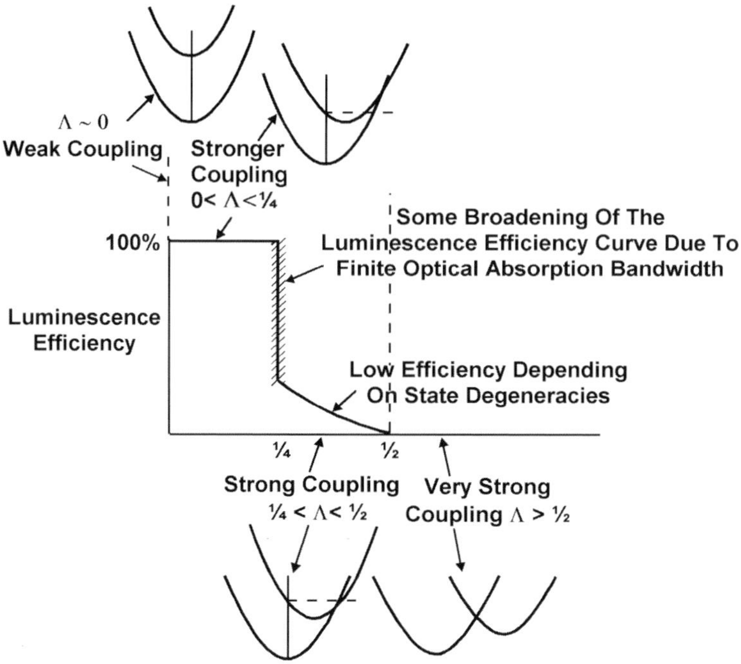

Fig. 3.8. Luminescent efficiencies predicted for F centres or similar two-level systems (Bartram & Stoneham 1983). Real F centres follow these predictions quite well, although there are certainly further complications, such as the consequences of the other excited states (including ionised states) and finite concentrations.

3.5.3 Cooling of electronic excitation: Free carrier states

Suppose an electron is excited into the conduction band of an ionic crystal, above the band minimum. How will the electron lose energy? The standard answer is that it will lose kinetic energy mainly by emitting longitudinal optic (LO) phonons. These have a finite energy, so that energy loss will be in discrete units of the order of 0.05–0.1 eV. If this process were the only fast process, the kinetic energy would fall until it was too small for an LO phonon to be emitted, possibly still leaving the electron at energies significantly more than kT above the bottom of the conduction band (we suppose that the electron does not self-trap). Weaker interactions, such as with acoustic modes via the deformation potential, would then enable equilibrium to be established. Discussions of energy loss to optic and to acoustic modes are given, e.g., by Kittel (1964 p 136 *et seq.*, p 148). Such behaviour is consistent with the substantial amount of information surveyed by Harper *et al.* (1973), especially that which indicates oscillations in photoconductivity and luminescence corresponding to the quantised emission of LO phonons.

The idea that the interactions with acoustic modes are weak, and so can be ignored, is certainly false in certain critical cases. In particular, the energy losses of electrons injected into thin films of oxide on silicon can only be understood if acoustic mode interactions are treated seriously (DiMaria 1991). Even in Si, which is non-polar, electrons excited to 1.4 eV above the gap have a relaxation time as fast as 240 fs (Sjodin *et al.* 1998), although this time is longer because of screening when the electron density rises above $5 \times 10^{21}/cm^3$.

The simplest description is incomplete in several respects. Obviously, photon emission is possible, although it is often far too slow to be important. Capture into traps is another possibility. There are selection rules which we have not mentioned. But, perhaps more important is our assumption that the energy loss is slow, with the average time between emission long compared with the period of the LO phonons.

This view has been challenged by Daguzan *et al.* (1995; see also Guizard *et al.* 1996, Martin *et al.* 1997). Daguzan *et al.* (1995) looked at two distinct models for the time dependence of an electron gas injected into a wide-gap (Frohlich) insulator. The parameters they chose were those for quartz. The first model used a semiclassical Boltzmann equation. The results were broadly in line with the simple picture. The momentum distribution of the gas became isotropic after 5 fs (Daguzan *et al.* 1995, Fig. 1), collapsing after 10 fs as energy loss took place to other modes. After 20 fs, almost all the electrons had lost several phonons of energy. Significant phonon energy-related structure persisted during the decay (Daguzan *et al.* 1995, Fig. 2). Almost all the electron energy was lost (the electrons were thermalised) after 40 fs, which is comparable with the phonon periods.

The second approach chose a quantum (density matrix) method, with truncation. This indicated very rapid damping (Daguzan *et al.* 1995, Fig. 3), although thermal equilibrium was not reached for a very long time, if at all (Daguzan *et al.* 1995, Fig. 4). The energy loss rate was greater than in the semiclassical case. There are no obvious experimental data to confirm or dispute the rapid energy loss predicted. The result certainly implies caution in using the simple picture of well-defined, separate, LO phonon energy loss processes.

Experimentally, there is some information, but it is incomplete. The situation for metals is clearest, both as regards physics (where the key parameters are the plasmon frequency and the electron collision time) and consistency of behaviour. The oscillations expected for weak coupling, with the period and amplitude of the LO phonon energy, are seen not only in the III–Vs, but also in CuCl and CuBr crystallites in thin films (Nakayama *et al.* 1997; we discuss other quantum dot systems below). A phonon bottleneck is seen in $GaAs/Ga_{1-x}Al_xAs$ quantum wells for which subband separations were less than the LO phonon energy (Murdin *et al.* 1997). Above about 35 K, LO phonon emission determines the equilibration; below this temperature, acoustic phonons dominate. For oxides, in experiments on commercial and laboratory samples of alumina, magnesia, and titania (Warman *et al.* 1991), effective mobilities of carriers were obtained from the microwave conductivity (26.5–38 GHz) observed after pulses of 3 MeV electrons. For alumina, the mobility was close to limit of measurement. For magnesia, the mobility was much larger than for small polarons. There was rapid localisation in virgin materal; the lifetime reached nanoseconds after larger doses of a few kilogreys, because traps were suppressed. For titania, the mobilities were within the small-polaron range. The effective mobility decreased with decreasing particle size, apparently due to rapid localisation at surface traps (there were consistent effects with Pt on the surface, iso-propynol covering, and with Cr^{3+} bulk doping).

Cooling of 'free' carriers in quantum dots

Photoluminescence (PL) and photo-luminescence excitation (PLE) spectra were obtained for self-organised (In/Ga)As quantum dots (Steer *et al.* 1996) and show two deexcitation mechanisms in relaxing from excited state to ground state. In the electronic excited state, a non-resonant mechanism operates (but apparently is based on a selection process which singles out those dots with an energy level spacing close to an integer multiple of the LO frequency) and a resonant process operates in the ground state which involves the emission of one or more LO phonons with well-defined energies. In another paper on similar (In/Ga)As dot systems (Heitz *et al.* 1997), inelastic intradot phonon scatter is the major effect, with up to four LO phonons involved. Carrier capture (e by h?) is a cascade process, with initial capture within a few picoseconds. Multiphonon relaxation of the first excited hole states involves three LO phonons and takes 40 ps. The $|001\rangle$ hole state 'presents a relaxation bottleneck which determines the ground state population after non-resonant excitation.' For smaller dots, the intradot carrier relaxation is faster than either the radiative (>1 ns$=1000$ ps) or non-radiative (100 ps) recombination; this explains the phonon bottleneck effect seen in the PL spectra.

In quantum dots of CdSe, electron–hole pairs with kinetic energies larger than the LO phonon energy relax rapidly (Woggon *et al.* 1996). Relaxation occurs within 500 fs; there is no bottleneck. There is no relationship seen between the precise energy of excitation and the LO phonon energy.

In other circumstances, the precise value of a vibrational mode energy can be critical. An example is GaAs with H, D, T implants, where there is a non-monotonic isotope effect (Steeples *et al.* 1980). What appears to be happening is that recombination-enhanced diffusion is suppressed when the bandgap is equal to a multiple of the local mode energy plus only a very small additional energy. Another example concerns energy transfer to SiH surface vibrations in porous silicon nanocrystals at low temperatures (Letant & Fishman 1998).

3.5.4 Cooling of electronic excitation: Capture and cooling of bound carrier states

The capture of a carrier can occur with the emission of a photon, with the emission of a phonon, or by transferring energy to another carrier, as in the various Auger processes. For each of these mechanisms, we may derive a cross-section σ_{capt}. What is measured in many experiments is a rate coefficient, which is an average (a thermal average in equilibrium) $\langle v\sigma_{capt}(v)\rangle$. Cross-sections depend on temperature and on the charge state of the trap (Table 3.2). If there is Coulomb attraction, as in the capture of an electron by an ionised donor, the cross-section can be large. At low temperatures, the cross-section for Ge:Sb^+ can be as large as 10^4 Å^2. Centres which have no long-range Coulomb attraction often have cross-sections in the range 0.1–10 Å^2, similar to atomic dimensions. If the long-range Coulomb interaction is repulsive, the cross-section can be very small, perhaps 10^{-8}–10^{-5} Å^2, although values as high as those for neutral centres are sometimes found. These Coulomb factors are discussed by Bonch-Bruevich & Landsberg (1968), by Pässler (1977), and by Ridley (1982) (see also Stoneham (1981b)).

At finite temperatures, a carrier may gain energy (absorb phonons) as well as lose it during the cooling process. For free carriers at low temperatures, this is unimportant. However, when capture is involved, whether at an impurity, or a defect, or a suitable fluctuation in structure in an amorphous solid or liquid, an increase in energy may mean loss from the trap, i.e. a qualitative change in carrier state. For this reason, it is common to define two quantities for each of the shallower states of a trap. There is a cross-section σ_j of capture from a free state into bound state j. The capture probability is defined in the usual way, and depends on the initial energy E of the carrier. The *sticking probability* P_j is the probability that an electron captured into state j will enter the ground state before escaping from the trap. The sticking probability has been calculated for donors in semiconductors like Si and Ge, and some of the main results are shown in Fig. 14.5 of Stoneham (1975). In these systems, there are many states, separated one from the other by less than the maximum phonon energy, and a sequence of steps (a 'cascade') must be considered (Fig. 3.9). The several sets of results differ substantially, but agree that the sticking probability for binding energy E_j falls off rapidly as a function of E_j/kT and of E_j/E, as one would have

Table 3.2. *Capture mechanisms for carriers (after Stoneham (1975)).*

Mechanism	Radiative	Auger	Phonon
Typical cross-section (Å^2).	10^{-4}–10^{-5}	10^{-1} typical; varies widely	10^{-2}–10^{4}
Conditions for dominance	Depends greatly on details; favoured by large energies	High carrier and defect concentrations	Low defect concentrations
Temperature dependence	Small dependence for allowed transitions	Small dependence	Power law or Arrhenius dependence
Concentration dependence	Little direct effect	Linear or faster with concentration until screening[a] is important	Little direct effect

Note: [a] Dynamic screening is usually small; static screening can be large (Burt 1981)

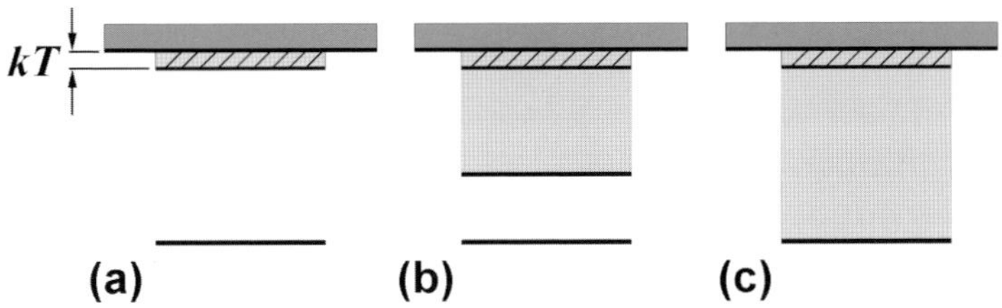

Fig. 3.9. Cascade processes. Three different regimes are indicated. The upper shaded region comprises continuum states. The lower shaded region indicates localised states which are dense in energy (spacings less than kT or phonon energies) among which a cascade can occur. The hatched region indicates kT and the typical phonon energies. In (a), multiphonon processes are essential. The bottom line represents the ground state. In (c), cascade processes alone are needed. (b) shows an intermediate situation.

expected. We remark that the acoustic phonons become significant when the carrier kinetic energy falls below mv_s^2, where v_s is the velocity of sound and m the carrier mass (Abakumov & Yassievich 1976).

For deep traps, the initial capture event may be into a shallow state, and a key process will be the transition from the shallow state to the deep ground state. Clearly, thermal ionisation from the shallow state defines one timescale, and the deexcitation process defines another (Rees *et al.* 1980). We shall discuss the non-radiative transitions in Section 3.6, but remark here that radiative transitions become more probable as the energy E increases (roughly as E^2 for constant oscillator strength), whereas non-radiative transitions become less probable (often decreasing exponentially with E). The approaches to free-bound transitions of Sumi (1980a,b) and of Henry & Lang (1977) are reviewed by Stoneham (1981b).

3.6 Non-radiative transitions II: Absolute rates

3.6.1 Kinetics and dynamics

In discussing non-radiative transitions, it is helpful to separate kinetics from dynamics. Rate equations are a typical tool for kinetics. Equations are constructed involving the populations of various states and transition probabilities between them, possibly with empirical damping terms. Kinetics is concerned with the different types of behaviour which can result, and hence with the ratios of the transition probabilities (branching ratios). Dynamics is concerned with the absolute magnitudes of the rates.

In kinetics, one may be concerned with choosing parameters so as to end in a particular state (cf. the analysis of self-trapped exciton kinetics by Song &

Stoneham (1976)). One may be aiming at control of the competition between radiative and non-radiative processes (Struck & Fonger 1991; Bartram & Stoneham 1975). We shall not discuss kinetics in detail, although a few points need to be mentioned. First, quantum behaviour can mean that there is interference between different processes. Secondly, it is tempting to assume relations between processes which make simple thermodynamic assumptions which, in turn, imply the existence of a well-defined temperature. It is often the case that either there is no such temperature, or that subsystems have different temperatures (e.g. electrons and nuclei, or spins and nuclei, or one group of phonons and the main thermal vibrational reservoir). Thirdly, some systems are special: they may be chaotic, or non-ergodic, or otherwise need a different treatment. There is an analysis of the limitations of the usual rate equations in Section 14.5 of Stoneham (1975).

In dynamics, the central question is how one can calculate how fast a process will occur. Often, this reduces to whether or not non-radiative transitions overwhelm radiative ones. Sometimes the question reduces to relative rates and branching ratios. We shall be concerned with two main classes of non-radiative transition: those which transfer energy to vibrations (multiphonon processes), and those which transfer energy to other carriers (Auger processes).

The dynamics of systems with electron–lattice coupling encompasses a rich variety of behaviour. This is reflected in the review of Jortner (1997).

3.6.2 Multiphonon non-radiative transitions

Lattice vibrations have two distinct roles. One role is as an energy sink: the electronic energy released is transferred to vibrational modes; these modes subsequently exchange energy to achieve thermal equilibrium. The modes involved here are the so-called *accepting modes*. For each of these modes, the mean displacement in the initial state ($Q_{\mathrm{i}\alpha 0}$ for mode α) and that in the final electronic state ($Q_{\mathrm{f}\alpha 0}$) will differ. The modes with the larger differences $Q_{\mathrm{i}\alpha 0} - Q_{\mathrm{f}\alpha 0}$ will take up most of the energy. A second role is inducing the transition itself. It is the non-adiabaticity associated with the *promoting modes* which provides the matrix element causing the transition.

Sometimes promoting modes and accepting modes are easily distinguished by symmetry. For the 2p to 1s transition of an F centre, it is the $T_{1\mathrm{u}}$ modes with the symmetry of an electric field which drive the transition, and it is primarily the symmetric modes of $A_{1\mathrm{g}}$ symmetry which take up the energy as the lattice readjusts to the different radial charge densities of the 2p and 1s states. This simple situation is not universal. The general situation is that modes may contribute in both roles. Nor, when making accurate calculations, can we ignore the fact that the modes themselves will not be the same combinations of ionic displacements in the initial and final states.

The calculation of the matrix element M_{if} inducing the transition continues to cause problems for serious calculations. The issues are discussed by several authors (Englman 1979; Stoneham 1981b; Stoneham 1990; Bartram & Stoneham 1985) and involve two major concerns. The first arises because the electronic matrix element depends strongly on the accepting mode displacement. Here, there are routes which look promising (Bartram & Stoneham 1985; Stoneham 1990), but which have still to be tested in detail. Secondly, we would wish to use the best available electronic structure codes (whether based on a density functional approach or a Hartree–Fock approach). However, these normally provide valence states which depend adiabatically on the ion cores. The proper inclusion of the cores introduces awkward issues of accuracy.

It might have been hoped that self-consistent molecular dynamics would, somehow, make non-adiabatic transition probabilities as easy as predictions of energy surfaces. This is not so. There is a fundamental problem: in these approaches, the nuclei are treated classically, and only the (valence) electrons are treated quantally. The exchange of energy between these two systems cannot be obtained by simple means. Fortunately, for many of the results we shall need, a simple estimate of the matrix elements suffices (examples from Huang & Rhys (1950) and Henry & Lang (1977) are given in a common form in Stoneham (1981b)). Sometimes a bound on the rate is sufficient. Occasionally, the familiar Landau–Zener theory is applicable, at least in spirit, although this is rare in solids, as opposed to molecules, since the large number of degrees of freedom means the underlying assumptions fail.

In addition to the transition matrix element, or its equivalent, the second factor in non-radiative rates is the lattice response: how readily will the accepting modes take up energy when there is a sudden perturbation? This response is conveniently expressed in terms of a *lineshape function* $G(\omega)$ or its Fourier transform, the *characteristic function* $g(t)$. These functions appear in expressions for optical cross-sections (see Stoneham (1975, Section 10.7)), where $G(\omega)$ is the generalisation of what would be a simple delta function in the absence of electron lattice coupling. We shall not discuss the calculations here (see, e.g., Stoneham (1975, 1981b) and references therein) but simply give some of the useful limits. These will be expressed in terms of p, where $p\hbar\omega$ is E_0, the energy to be taken up by the accepting mode, and S, the Huang–Rhys factor, such that $S\hbar\omega$ is the relaxation energy (S_0 is the value of S at absolute zero temperature). In most cases we shall be interested in the *statistical limit* $p \gg 1$.

Expressions holding at $T=0$.

$$G(T=0)=(2\pi\hbar\omega E_0)^{1/2}\exp(-S_0)\exp\{-p[\ln(p/S_0)-1]\}.$$

In this expression, the second exponential can be rewritten, replacing the logarithmic term $p\ln(p/S_0)$ by $\gamma E_0/\hbar\omega$, where γ is $\ln(p/S_0)$ and varies only slowly with the various system parameters. This factor in $G(T=0)$ is now proportional to $\exp(-\gamma E_0/\hbar\omega)$, which is a special case of the *energy gap law*: the rate falls off exponentially with the amount of energy to be taken up by the lattice. In contrast, the rate of a radiative transition increases with energy because of the density of states factor. The other exponential factor, $\exp(-S_0)$ can be regarded as a 'phonon dressing' of the transition matrix element M_{if}, which takes an effective value $M_{if}\exp(-S_0/2)$; at high temperatures, the corresponding factor is absorbed into other parts of $G(\omega)$, so that care must be taken to use the various forms consistently.

Expressions for weak coupling ($S_0 \ll 1$)

$$G(T)=G(T=0)[1+n(T)]^p.$$

Here n is the phonon occupation number; at high temperatures, it will lead to a T^p dependence, rather than an Arrhenius form. We can also obtain a further form of the energy gap law in a relatively general way. Suppose the rates of a p-phonon process and a $(p-1)$ phonon process are related by

$$W_p=\beta W_{p-1}$$

where β is a small factor. Suppose further that the lowest-order process will always dominate, i.e. the largest value of $\hbar\omega$ (or the smallest value of p) dominates. Then, provided p is not too small, the major process has $p=E_0/\hbar\omega_{max}$ and a rate $W_0\beta^p$, which can be rewritten in the form

$$W=W_0\exp(-\gamma'' E_0/\hbar\omega_{max}),$$

with $\gamma''=-\ln(\beta)$; since β is small, γ'' is positive.

Expressions for strong coupling ($S_0 \gg 1$)

$$G(T)=(1/\hbar\omega)\exp[E_0/2kT-S_0(2n+1)]\,I_{-p}\{2S_0[n(n+1]^{1/2}\}$$

where $I_N(x)$ is a Bessel function of order N. For excitation to a state of higher energy, E_0 changes sign, so the ratio of deexcitation probability to that for excitation has the form expected for detailed balance, namely $\exp(E_0/kT)$. There is no simple energy gap law, nor is there a simple 'dressing' of the transition matrix element. However, in suitable (and quite general) limits, one finds the simpler and better-known form:

$$G(T) = [2\pi E_M kT^*]^{-1/2} \exp(-E_A/kT^*)$$

where $kT^* = (\hbar\omega/2)\coth(\hbar\omega/2kT)$, which tends to kT at high temperatures. Ignoring the modest temperature dependence of the prefactor, the activation energy E_A is $E_A = (E_0 - E_M)^2/4E_M$, just the energy of the cross-over in the configuration coordinate diagram (Fig. 3.1). This limiting case has a natural and intuitive appearance, and coincides with expressions assumed by many early workers.

There are a few useful generalisations of these expressions (see, e.g., Stoneham (1981b)). However, the most useful developments are ones exploiting molecular dynamics, since this approach is not restricted to a one-frequency model.

Bounds on rates

We can use this final Arrhenius expression to guide us to likely upper limits on non-radiative transition rates. These are necessary, in that naive use of formulae can predict physically unacceptable rates. The difficulty can be understood by considering optical transitions, for which there is a fairly clear division between the electronic system to be excited or deexcited, the electromagnetic field, and the interaction between the field and the electronic system. There is rarely any difficulty in deciding between which states the transitions occur, since the energy transfer takes place between distinct subsystems. Further, the *f*-sum rule can often put significant bounds on transition probabilities.

Non-radiative transitions, however, occur in a closed system. Problems arise if the states are chosen carelessly (see also Englman (1979, Section 8.3)). As a trivial example, suppose one has a time-independent Hamiltonian H_0 with time-independent eigenstates ψ_0; likewise, let ψ be the eigenstate of time-independent Hamiltonian $H+h$. If the system is in state ψ at time $t=0$, it will simply stay in that state. If the system is in state ψ_0, however, the term h will cause the system to change with time, possibly quite rapidly. This points to a need for consistency and accuracy, often beyond what might be thought convenient. What, then, can be said about rates?

The Arrhenius expression suggests that here, as in many other processes (like thermal diffusion), a *lattice condition* must be satisfied. If Q is the reaction coordinate (here that of the accepting mode), then Q has either to exceed some threshold value (as in standard classical 'over-the barrier' transitions), or to achieve some critical value to some specified accuracy (often determined by a tunnel probability). The important point is that not more than one transition can occur each time the lattice condition is met. Even if the electronic matrix element is so large than many such transitions might be deemed possible, the net effect of all these transitions (forward and backward) is at most one transition. Thus the overall transition rate is the product of (number of times per second that the lattice condition is satisfied) × (efficiency factor, with value between 0 and 1). Putting the efficiency factor at its maximum value of unity, the maximum rate is given by the first factor. This (again, with possible exceptions, but still in rather general circumstances) is bounded by the largest vibrational frequency. Essentially the same result obtains when the limit is either dephasing or energy relaxation towards some final state.

For Jahn–Teller systems, one must remember that there is a multivalued energy surface. In simple cases, the rate for Jahn–Teller reorientation can be modelled in the same way as for single-valued energy surfaces, but other situations do occur.

3.7 Non-radiative transitions III: Localisation processes and their rates

3.7.1 Routes to the self-trapped state

Suppose we have a system for which there is a small-polaron state and also a large-polaron state, these states being separated by an adiabatic potential barrier. The phenomenological theory (Emin & Holstein 1976) shows that this is possible in three dimensions for sensible values of both long-range and short-range polaron couplings. If the long-range component is too strong, the barrier may be suppressed. Now suppose the carrier (electron, hole, muon, or even an exciton) is produced in a free state. For self-trapping to occur, the system may have to pass through a potential barrier (see Fig. 3.3). The height of this barrier depends on the nature (one-centre, two-centre, etc.) of the self-trapped state and on the degree of localisation in the final stages of this transition.

Route 1: Free state → nucleation → localised state

If the localisation occurs at low temperatures, after the polaron has thermalised, the system should tunnel from the bottom of the band of free states (or from the large-polaron state). To calculate the tunnelling rate, one needs to know the wavefunction of the localised state (the so-called *nucleation state* (Sumi 1984)) at the tunnelling energy. The nucleation state has appreciable lattice distortion, but is less localised than the final self-trapped state. The system then has to relax further, beyond the distortion associated with the nucleation state. The self-trapping rate is the rate at which the most probable nucleation state is formed. In some cases, the initial nucleation state can relax into a transient metastable localised state, subsequently evolving into a final stable state. Model calculations suggest there may be cases for which the adiabatic picture fails. The localisation rate is then determined by the rate of lattice relaxation (Wagner & Koengeter 1988).

Route 2: Free state→non-radiative transition to localised state

The transition from the free-polaron state to the localised state is a non-radiative transition. As such, the rate can often be factorised into an electron matrix element of the coupling which induces the non-radiative transition and a vibrational factor which determines how readily the lattice will give or take up energy. The latter, which we may refer to as the *lineshape function,* is itself determined by the configurational changes between the initial and the final states. Both short-range structural modifications and the long-range polarisation associated with optical modes are involved. These appreciable configurational changes can happen in crystals and in polar liquids because of short-range structural fluctuations.

Route 3: Free state + pre-existing trap→localised state

Although the arguments have been given for crystals, they are more general. The ideas apply also to excess electron localisation in water or other polar solutions (Hart & Anbar 1970), and even for inert fluid hosts like liquid helium and neon (Jortner *et al.* 1965). However, in liquids or amorphous solids, pre-existing disorder can cause localisation, possibly even prior to any relaxation induced by the carrier. This has been shown in quantum molecular dynamics. A study (Motakabbir & Rossky 1989) of the electronic states accessible to an excess electron in pure liquid water showed that there were many pre-existing fluctuations capable of trapping an electron. These traps were shallow; solvent configurations which could act as deep traps were rare in the pure liquid. Clearly, even shallow pre-existing fluctuations could be effective in nucleating a strongly-localised state. Pre-existing potential wells would offer a route to localisation which avoided the barrier. This process could occur in parallel with other mechanisms involving tunnelling into a nucleation state or localising adiabatically, as simulated in (Rossky & Schnitke 1988). It is hard to determine from experiment which channel dominates.

The calculation of the self-trapping rate for the adiabatic barrier scenario has been studied by several authors (see, e.g., Rashba (1982), Nasu & Toyozawa (1981), Sumi (1984), Ioselevich & Rashba (1985)). The most complete work relevant to real systems is that of Sumi (1984), who used a phenomenological approach to self-trapping of excitons in alkali halides. Sumi considered both one- and two-centre nucleation states, with different degrees of localisation of the exciton wavefunction; the scale lengths of these states are of the order of the lattice paramenter. Using non-radiative transition theory, he calculated the tunnelling rate between the free state and the different nucleation states as a function of temperature and of the parameters of the Hamiltonian. One key parameter is the localisation energy, B, the increase of the exciton kinetic energy on localisation into the nucleation state. A second key parameter is the relaxation energy, S, the energy gained by lattice relaxation and polarisation in this state. Both B and S depend strongly on the degree of localisation of the exciton. The relaxation energy S must exceed the localisation energy B if there is to be self-trapping. At low temperatures, the tunnelling rate from the free state to the nucleation state is given approximately in terms of the gamma function by:

$$W_0 = 4p^2 \nu B^2 e^{-S} S^{S-B} \Gamma(S - B + 1)$$

for S greater than or equal to B, and with S and B given in units of the phonon energy (Sumi 1984). The characteristic frequency of those lattice vibrations which govern relaxation of the system is ν. Sumi's theory gave a reasonable qualitative account of the self-trapping of excitons in alkali iodides, where the potential barrier between the free and self-trapped exciton states is known experimentally. In particular the theory predicts that self-trapping of excitons in KI and RbI is likely to take place through a one-centre nucleation state. This is an interesting idea, and such a one-centre nucleation state might have been observed for the self-trapping of holes (rather than excitons) in certain alkali halides. Certainly a broad band is seen just after excitation, but the nature of the state involved remains to be proved. However, there are complications (Thoma *et al.* 1997) which leave this interpretation doubtful.

Free holes in alkali halides self-trap into the well-known two-centred V_k configuration. It is natural to assume that they reach that state via a two-centre nucleation state from which further relaxation occurs. However, femtosecond spectroscopic experiments suggest this is not so in KI, RbI (Iwai *et al.* 1994, 1996). The experiments use two-photon excitation, and observe the transient absorption spectrum in the time domain from 0.3 ps to 20 ps. The energy of about 8 eV produces a free electron–hole pair in the bulk of these crystals. The initial intense absorption spectrum, seen from about 0.3 ps, differs strongly from the V_k centre spectrum, which appears only after 3 ps, and rises until 10 ps. Similar transient spectra are seen in RbI. In KBr, a broad featureless optical absorption band is observed just after the pulse; this band transforms into the absorption spectrum of the V_k centre after about 10 ps. The behaviour of the transient optical absorption is the same in both pure KI and in those doped with the electron-trapping impurity (NO_2^-), which suggests that the band be attributed to a metastable state of the hole in these crystals.

3.7.2 Quantum molecular dynamics approaches

In quantum molecular dynamics, the electrons are treated by quantum methods; the electrons normally follow the nuclear motion adiabatically, and the nuclear motion itself usually follows classical dynamics. The dynamics of electron excitation, localisation and migration can be simulated. This can be done using a quantum path-integral molecular dynamics techniques. The approach has been applied successfully to excess electron localisation and diffusion in molten ionic salts (Selloni *et al.* 1987), liquid water (Rossky & Schnitke 1988), small polar clusters (Landman *et al.* 1987) and alkali halide clusters (Landman *et al.* 1985). The path-integral technique stems from the ideas of Feynman (1955; see also Feynman & Hibbs 1965) and his early approach to the polaron problem. Feynman's ideas exploit the equivalence between the quantum system and related (but different) classical system. These ideas were taken up by many workers and developed into powerful simulation techniques (Chandler & Wolynes 1981; De Raedt *et al.* 1984; Chandler 1984; Parrinello & Rahman 1984). It is this quantum-classical 'isomorphism' that leads to a simple representation of quantum particles in computer simulation.

An alternative approach is to use a direct description of the electron in terms of wavefunctions. This needs efficient methods for evaluating the Born–Oppenheimer, or adiabatic, electronic states associated with a given molecular configuration. Such methods have been developed for an excess electron in polar media (Kosloff 1988; Leforestier *et al.* 1991). The excess electron–molecule interaction is usually treated using pseudopotentials (see for example Barnett *et al.* 1988, Schnitke & Rossky 1987). The efficiency of the simulation of the propagation of the wavefunction in time is based on the algorithms using only simple multiplicative operations and repeated fast Fourier transforms. A different approach, mainly applied to polarons, solitons and excitons in conducting polymers, uses semiempirical molecular orbital methods (Wallace *et al.* 1991). In these approaches, the forces on the atoms due to the electrons can be calculated efficiently at each timestep of the molecular dynamics.

3.7.3 Solvation of an electron in water

An electron, injected into water, rapidly polarises its environment in a process resembling self-trapping. The process appears to involve two stages. The first occurs in 110–240 fs after injection, when a 'wet' electron is formed. The observed time-resolved femtosecond spectra at this early stage, well before the spectrum of the hydrated electron has emerged (Migus *et al.* 1987), show a diffuse infrared spectrum: the electron has evolved to a prelocalised state. Loosely speaking, the wet electron can be considered as an electron in an excited state in the potential well created by the polarisation of the nearby waters. Even from the earliest experiments (Rentzepis *et al.* 1973) it was realised that the relaxation process leading to initial infrared absorption was far shorter than the dielectric relaxation time of water (7.4 ps at 30 °C).

The evolution of the solvated electron absorption band (on a timescale of about 4 ps) shows solvent relaxation around the localised state initially formed. The wet electron makes a non-radiative transition to become the solvated electron, the ground state of the system. This process has been simulated in the Born–Oppenheimer approximation (Motakabbir & Rossky 1988; Rossky & Schnitke 1988); a new electronic state is evaluated by solving the time-independent one-electron problem for the new solvent configuration for each molecular dynamic timestep. The results appear to be reasonable, although limited quantitatively because of simplifications in the electron–water and water–water interactions (see also the discussions in Del Buono *et al.* (1992)).

The solvated electron has a characteristic absorption at 1.7 eV. Simple theories which omit the excited state (those based on dielectric theory with inner and outer shells of water molecules, or indeed more advanced theories which follow a single adiabatic energy surface) fail to account for the time dependence: the route through the excited state needs to be treated explicitly.

A puzzling second feature is the relative lack of an isotope effect when light water (based on H) is replaced by heavy water (based on D). There is only a very modest effect in either the rate of solvation or the energies of the solvated form. The lack of an effect on the energies can be rationalised partly by realising (Hughes 1966; Stoneham 1975, Section 8.2) that relaxation energies are isotope-independent in a harmonic host lattice; water is, of course, a liquid, but it is not too surprising that some features carry over from one case to the other. The lack of isotope effect in the rate has a more subtle origin, and leads to important issues in non-radiative transitions (Neria & Nitzan 1993; Schwarz *et al.* 1996). Some of these issues go beyond our present needs, but others are of importance to us.

3.7.4 Frozen Gaussian methods

The approach to non-radiative transitions which we have defined above emphasises the parallels with optical transitions. This can be both appropriate and convenient. However, we are clearly regarding the lattice vibrational motion as classical, except that we use quantal statistics, or evaluate simple matrix elements using an assumed linear coupling. But treating the lattice as quantal involves further features, notably the existence of multiple histories. In these multiple histories, the system might make several transitions from one electronic state to the other, with the nuclear motion appropriate to the current electronic state at each stage. These alternative histories can interfere with each other. In special cases, we can use intuition to guide us, but this can be suspect when we have more complicated situations. Yet we still have to find an approach which is practical. This is what Neria & Nitzan (1993) and Schwarz *et al.* (1996) have done for the solvation problem, by exploiting a different way of describing non-radiative transition rates, and using the so-called frozen Gaussian technique of Heller (1975, 1981, 1991). The important feature of the frozen Gaussian method is that classical nuclear trajectories can be exploited in a treatment which incorporates key quantum aspects.

The approach of Neria & Nitzan and of Schwarz *et al.* is to write the non-radiative transition rate in terms of a correlation function $C(t)$. This correlation function involves two factors. The first factor is a time-correlation component involving $\mathbf{F}\cdot\mathbf{v}$ for the initial energy surface at time zero and $\mathbf{F}\cdot\mathbf{v}$ for the second energy surface at time t; $\mathbf{F}$ is the force matrix element between the two states, and is effectively the non-adiabatic coupling term. It is the projection of this force on the nuclear velocities $\mathbf{v}$ which is important. Here, and in several other places (such as Chapter 7), we should note that the *direction* of a coupling vector (here $\mathbf{F}$) is important, not just its magnitude. The second factor in $C(t)$ is a function $J(t)$ which modulates the $\mathbf{F}\cdot\mathbf{v}$ time correlation. $J(t)$ is the time-dependent overlap of the nuclear wavefunctions propogated along the two different energy surfaces, starting from a time $t=0$ for which the nuclear coordinates were the same. Again, we can see that the nuclear overlaps are an important component, as in our earlier, simpler, discussions.The function $J(t)$ decays with time and, at least in the case of electron solvation, decays faster than the correlation of $\mathbf{F}\cdot\mathbf{v}$, the non-adiabatic matrix elements. The characteristic decay time of $J(t)$ is the *quantum decoherence time.* This characteristic time can be calculated by methods involving frozen Gaussians (Schwartz *et al.* 1996), and it is the isotopic dependence of $J(t)$ which leads to the relatively modest dependence of the rate of solvation on the water isotopic composition.

Chapter Four
Self-trapping

In this chapter, we shall analyse the way self-trapping occurs in real systems. The first issue is whether or not self-trapping is observed in a given material. We have discussed the criteria for self-trapping in Section 3.3, but experiment is the decisive factor. Table 4.1 shows the occurrence of self-trapping of electrons, holes, and excitons in a number of insulators. The absence of self-trapping of electrons and holes can be determined by measuring their mobilities; the absence of self-trapping of excitons is demonstrated by observing free-exciton luminescence. When the interaction of free charge carriers and of excitons with defects or impurities indicates long-range motion, this can imply the absence of self-trapping. Once self-trapping is proven, the atomic and electronic structures of the self-trapped states are needed. The relationship between the self-trapped state and the host atomic and electronic structure is important, especially for self-trapped excitons, since exciton self-trapping provides the energy localisation needed for local modification of the lattice structure and defect formation.

Although self-trapping is induced as a consequence of coupling with the lattice, existing defects play a role in self-trapping in two ways: extrinsic self-trapping and defect-perturbed self-trapping. Extrinsic self-trapping occurs by the interaction with the lattice and defects in materials in which the interaction with the lattice is not sufficient to induce intrinsic self-trapping, while defect-perturbed self-trapping occurs by the interaction between intrinsic self-trapping and defects (Fig. 4.1). Trapping by defects without self-trapping does not involve large lattice distortion.

4.1 Self-trapped carriers in halides

Detection of self-trapping

As discussed in Section 3.4, the mobility of charge carriers is correlated with trapping or with self-trapping. The mobility is extremely small for carriers

Table 4.1. *Presence and absence of self-trapping in major materials. The character of the self-trapped state is indicated. For the oxides, the situation is not always clear, and is discussed in the text.*

Material	Electrons	Holes	Excitons
Diamond, Si, Ge	No	No	No
III–V compounds	No	No	No
II–VI compounds	No	No	No
NaCl structure alkali halides	No	Yes (X_2^-)	Yes
CsCl structure alkali halides	No	Yes (X_2^-)	Yes
Alkaline-earth fluorides	No	Yes (F_2^-)	Yes
Ba fluorohalides BaFX	No	X_2^- off plane	Yes
Pb halides	$PbCl_2$ (Pb_2^{3+})	$PbCl_2(Cl_2^-)$	Yes?
Cu halides	No	No	No
Ag halides	No	AgCl(Ag^{2+})	AgCl
Tl halides	No	No	No
Closed-shell oxides	No	Marginal, but even weak disorder will localise	SiO_2 (h on O, e on Si). Others marginal

which are self-trapped, less than 0.1 cm²/(V s), and increases with rising temperature. This behaviour is seen in the alkaline-earth halides. Carriers which are not self-trapped (large polarons) have large mobilities, greater than 10 cm²/(V s), and become less mobile as phonon scatter increases with rising temperature. For example, the hole mobilities of Tl halides (Kawai *et al.* 1971) and Cu halides are of order 10^2 cm²/(V s), in the range expected for large polarons. It can be hard to distinguish self-trapping from trapping by mobility alone, because of the possible effects of defects or impurities. Electron paramagnetic resonance gives the most convincing evidence for the self-trapping of charged carriers, since it will normally show if there are any impurities or other defects present. Self-trapped electrons and holes often have characteristic optical absorption bands. It can be hard to identify these bands unambiguously unless they correlate with EPR spectra.

Band structures of halides

For the alkali halides (Kunz 1982a) and alkaline-earth fluorides (Albert *et al.* 1977 for CaF_2), the valence-band wavefunctions are dominated by the p orbitals of the halogen ions. Pressure can modify the orbitals involved in the free exciton. Tsujimoto *et al.* (1996) show that, under pressure (to 1.3 GPa), the lowest free exciton state changes from the s-like (Γ_8^-, Γ_6^+) state to the d-like (Γ_8^-, Γ_8^+) state; this, in turn, affects the self-trapped exciton. For the silver

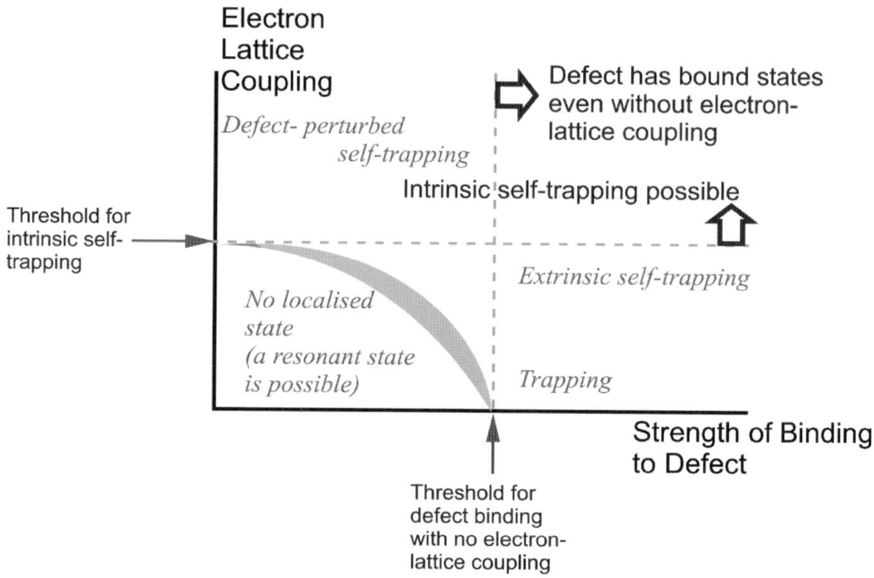

Fig. 4.1. Regimes for localised states, as a function of defect coupling strength and self-trapping coupling strength. *Self-trapping* occurs when the lattice relaxation energy exceeds the localisation energy. *Trapping* need not involve relaxation, e.g. bound excitons in semiconductors are trapped, not self-trapped. *Extrinsic self-trapping* means that both defect and self-trapping interactions are needed to localise a carrier or exciton, as for the trapping of a hole by Ge in α-quartz. *Defect-perturbed self-trapping* includes cases like self-trapped excitons in alkali halides doped with a monovalent impurity: self-trapping occurs anyway, but the configuration is modified in detail.

halides (Kunz 1982b) and alkali silver halides (Edamatsu *et al.* 1989), thallium halides (Van Dyke & Samara 1975; Schreiber & Schafer 1984), and copper halides (Song 1967), there is a significant metal orbital contribution to the states at the top of the valence band.

4.1.1 Self-trapped electrons

High electron mobilities are observed in most halides and are characteristic of electrons which are not self-trapped (Section 3.4). The reason is that the conduction bands are wide (large transfer energy), so that there is a large energy cost to localise an electron in one unit cell. Self-trapping of electrons has been seen in only one halide ($PbCl_2$), for which, unusually, holes are more mobile than electrons.

$PbCl_2$

Nistor *et al.* (1993) observed an EPR spectrum in $PbCl_2$ produced at 60 K after optical excitation with photons above the bandgap energy. The g value was smaller than the free-electron value, suggesting that the signal is due to an electron trap. The hyperfine structure shows that the electron is shared equally by two Pb nuclei, and can be described as a weakly-perturbed Pb_2^{3+} molecular ion, oriented along the direction of the nearest-neighbour Pb ions in the perfect lattice. The spectrum has been attributed to self-trapped electrons, probably with the electron localised in a covalent σ_g-type orbital between two neighbouring Pb^{2+} ions. The centres are stable below 130 K and annealed completely above 210 K. Electron self-trapping in $PbCl_2$ is consistent with the positive sign of the Hall constant (Verwey & Westerlink 1969).

4.1.2 Self-trapped holes

AgBr

Cyclotron resonance (Hodby *et al.* 1974) confirms that the holes in AgBr are not self-trapped. The hole mobility shows an anomalous decrease with temperature. Toyozawa & Sumi (1974) suggest this is because the free hole is metastable, with a potential barrier against self-trapping. The mobility of the holes in AgBr is as high as 3×10^4 cm^2/(V s), comparable to that of electrons in the same system (Komiyama *et al.* 1979). Such mobilities for AgBr are three orders of magnitude greater than the hole mobility in AgCl (Hanson 1962), for which self-trapping of holes is established.

AgCl

The self-trapped hole in AgCl has been identified in paramagnetic resonance (Hohne & Stasiw 1968; see also von der Osten 1984, Hamilton 1988). The hole can localise on the Ag^+ ion because the Ag d orbitals contribute to the states near the top of the valence band, and the localised hole forms an Ag^{2+} ion (Fig. 4.2(a)). The electronic degeneracy of the ($4d^9$) Ag^{2+} ion in octahedral symmetry leads to a tetragonal Jahn–Teller distortion of the lattice, i.e. self-trapping is driven by both Coulomb and Jahn–Teller interactions. The simulation of self-trapping by Heifets & Shluger (1992) suggests that 67% of the hole is localised on the d orbital of the silver ion and the remaining 33% is localised on the p orbitals of the four chlorine ions distorted toward the localised hole. They predict a relaxation energy E_R for a free hole is 0.9 eV and a self-trapping energy E_{ST} of 0.1 eV. This value agrees with the experimental thermal depth 0.12 eV of the self-trapped hole for thermal migration (Laredo *et al.* 1983) provided that transport occurs by excitation from self-trapped small-polaron states to large-polaron states; this is consistent with the mobility of about 30 cm^2/(V s) observed, which would be very high for small-polaron hopping.

Kao *et al.* (1992) measured the migration of the self-trapped hole from 10 to 30 K, and found a rate which varied as $T^{1.3}$, becoming thermally activated

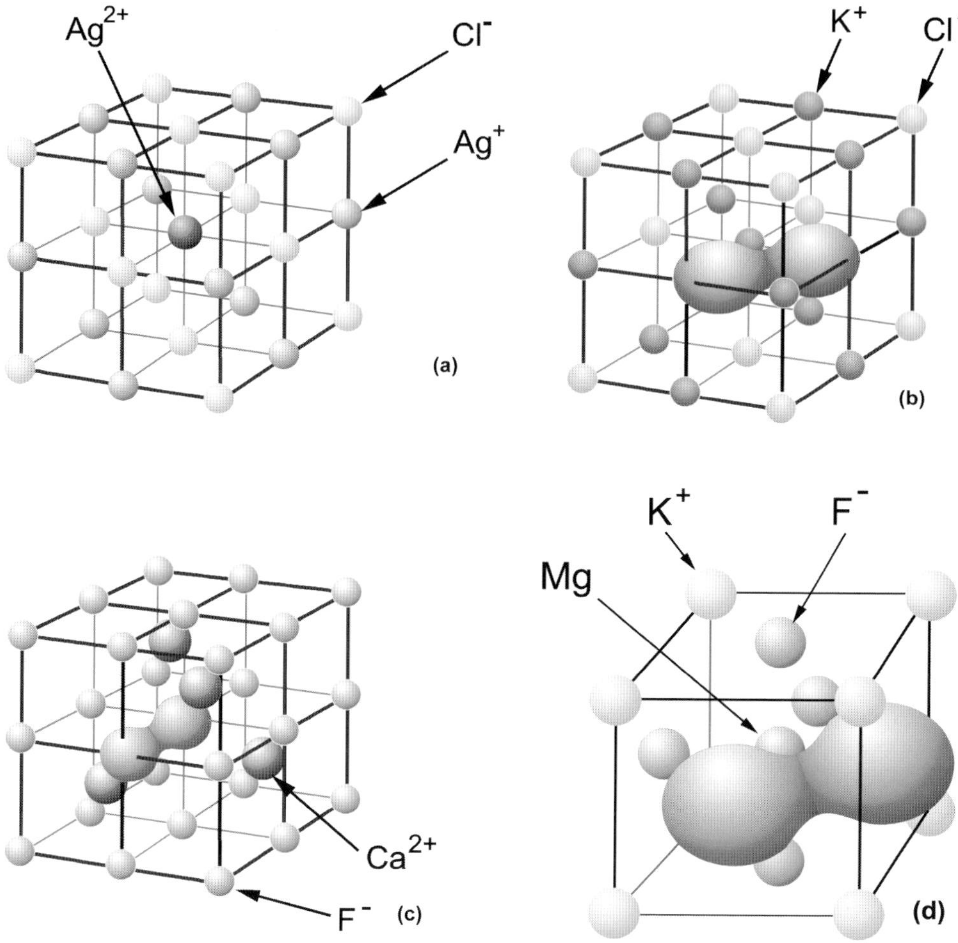

Fig. 4.2. Atomic structure of the self-trapped hole in halides: (a) AgCl; (b) alkali halides of NaCl structure; (c) CaF_2; and (d) $KMgF_3$.

above 30 K. The broad optical absorption bands at 1.2 eV and 1.7 eV observed by Marquardt *et al.* (1971) and Grigorjeva *et al.* (1989) are attributed to the self-trapped hole in AgCl. Grigorjeva *et al.* (1989) also found luminescence at 2.4 eV from recombination with an electron.

The self-trapped hole in AgCl is also observed trapped by a cation vacancy. There are two types of centre. The hole on the nearest-neighbour Ag survives to 110 K; that at the next nearest-neighbour site survives only to 70 K (Kao *et al.* 1992).

Alkali halides

The stable self-trapped holes are the so-called V_k centres (halogen molecular ions), whose structure is shown in Fig. 4.2(b). This halogen atom at a single

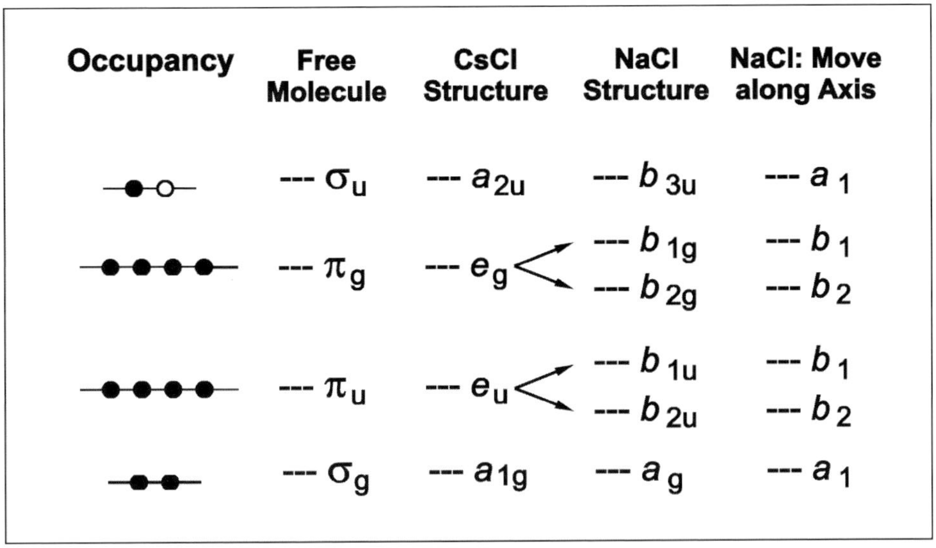

Fig. 4.3. Electronic energy levels of a halogen molecular ion (X_2^-), in free space and in crystalline environments.

lattice site may be metastable. The structure of the V_k centre was established by electron paramagnetic resonance (EPR) studies, and the key parameters are given by Schoemaker (1973).

Since the hyperfine interactions show that the hole is localised on two halogen ions, the V_k centre is often called a two-centre self-trapped hole, and its electronic structure can be described in terms of the molecular orbitals of a free X_2^- ion (X denotes a halogen atom). This 'molecule in a crystal' picture has been analysed in detail by Tasker & Stoneham (1977) and by Cade *et al.* (1984, 1986). The energy levels of the free ion X_2^- and of the molecular ion in several crystalline configurations are illustrated in Fig. 4.3. In alkali halides with the NaCl structure, the V_k centre has D_{2h} symmetry, reducing to C_{2V} as the X_2^- molecular ion is displaced along its symmetry axis. In the CsCl structure, the centre has D_{4h} symmetry. V_k centres in ammonium halides have very similar structures. The H centre, an interstitial halogen atom, can also be described sensibly as an X_2^- halogen molecular ion, and has properties similar to those of the V_k centre.

There is abundant information on the optical and EPR properties of the V_k centre. It is reviewed by Kabler (1972), Stoneham (1975), Hayes & Stoneham (1985) and Song & Williams (1993). To a good approximation, the EPR and optical properties can be explained in terms of the electronic structure of the X_2^- molecular ion with a spacing close to that in free space, although certain details are influenced by the ion being embedded in the crystal lattice. There are two prominent optical bands: one in the ultraviolet due to the σ_g–σ_u

transition, the other in the infrared, due to the π_g–σ_u transition. The fact that the spacing is close to the free-space value suggests that there is weak short-range interatomic repulsion along the V_k centre's axis. This is in line with the early analysis of V_k centre hopping in the alkaline-earth fluorides, where Norgett & Stoneham (1973) found that the short-range repulsion between the X_2^- and its X^- axial neighbours had to be reduced to yield the observed relative proportions of 90° and 180° jumps. In essence, there is open space for translational motion: the adiabatic potential energy surface near the stable configuration of the V_k centre is soft. We shall see later that this influences the atomic structure of the self-trapped exciton.

Magnetic circular dichroism (MCD) signals are expected when there is a change in orbital angular momentum. Thus MCD is expected for the π_g–σ_u infrared band, but not for the σ_g–σ_u ultraviolet band of X_2^-, as observed by Spaeth *et al.* (1994) for the H centre in KBr and KI. However, Spaeth *et al.* did observe MCD for the ultraviolet band of the V_k centre. This they attributed to a ${}^2\Pi_u$– ${}^2\Sigma_u$ admixture from translational displacement of the halogen molecular ion. Such an admixture does not occur for the H centre, for which the row of halogens is more close packed. The close packing should mean that, for a given crystal, the ultraviolet band of the H centre should be higher than that of the V_k centre. This is usually the case. The reason for exceptions, such as KI, has been suggested as configuration admixtures with higher-energy host states (Shluger *et al.* 1995).

Other halides

The self-trapped hole in alkaline-earth fluorides is also an F_2^- molecular ion formed by the two nearest fluorine ions in the lattice. The V_k centres in CaF_2, SrF_2, and BaF_2 are oriented along a ⟨100⟩ direction (Fig. 4.2(c)) (Hayes & Stoneham 1974; Stoneham 1975); in MgF_2 the orientation is close to a ⟨110⟩ direction (Ueda 1976). The V_k centre in $SrCl_2$ (Hayes *et al.* 1973; Repzka *et al.* 1973) appears to have the same structure as that in alkaline-earth fluorides of fluorite structure.

Nistor *et al.* (1998) observed an EPR spectrum due to hole centres below 80 K in undoped $PbCl_2$, which they ascribed to the self-trapped hole. The spectrum is not seen in Tl-doped $PbCl_2$, where holes are trapped by Tl^+ to form $(PbTl)^{2+}$ centres. Although the atomic structure of the self-trapped hole has not been analysed in detail, it appears that both electrons and holes are self-trapped in $PbCl_2$ below 80 K.

In the perovskite-structured $KMgF_3$, as in the alkali halides (Fig. 4.2(d)), the self-trapped hole is an F_2^- molecular ion, oriented along a [110] direction. Lewis *et al.* (1973) have shown the activation energy for 90° jumps (0.26 eV) to be marginally less than for 60° jumps (0.29 eV). The alkaline fluorohalide structure BaXY comprises planes in the sequence X/Ba/Y/Y/Ba/X, with two types of halogen row. Thus, in the SrFCl lattice, there is a [110] F^- row and a [100] Cl^- row. It is possible to form a F_2^- halogen molecular ion along the [110] direction, and a Cl_2^- molecular ion along either a [100] direction or a [111], out-of-plane, direction. Such out-of-plane Br_2^- self-trapped holes have been seen in BaFCl (Yuste *et al.* 1975). Theory (Baetzold 1987) indicates that another configuration has lower energy in SrFCl and BaFBr.

In potassium and rubidium silver iodides (Awano *et al.* 1989), the self-trapped holes are I^0 atoms, loosely bound to K^+ or Rb^+. The formation of an I_2^- molecular ion, as in the alkali iodides, is not seen in these silver halides, partly because the iodines are not all equivalent. Holes localise on I^-, rather than Ag^+ as in AgCl, because of the small contribution of Ag 4d orbitals to the valence band (Edamatsu *et al.* 1983): the upper valence bands are composed of mainly iodine 5p and silver 4d orbitals and the lower valence bands are composed of alkali p orbitals.

4.1.3 Relaxation processes of self-trapped holes

One-centre versus two-centre localisation

We have noted two types of self-trapped hole: one-centre, as in AgCl, and two-centre, as in the V_k centre. In the one-centre case, lattice polarisation by the localised positive charge and the Jahn–Teller distortion combine to stabilise the small polaron. The localisation process is believed to be tunnelling from a large-polaron state to the small-polaron state or nucleation state (see Section 3.3).

For the two-centre configuration, which is stabilised in part by bond formation, the question arises as to whether the final two-centre configuration is reached directly by tunnelling from the large-polaron state, or whether it is a final state reached through an initial metastable one-centre configuration. The removal of an electron from the top of the valence band in alkali halides produces X^0; this results in lattice polarisation. The X^0 atom at an alkali halide lattice point can lower its energy further by bond formation with a neighbouring of X^- ion. Even if the one-centre self-trapped hole is a metastable state, it should be able to collapse to the V_k centre in a short time. The arguments of Sumi (1984) for the self-trapping of excitons in alkali halides should apply, so that tunnelling from the one-centre to the two-centre configuration will depend on the strengths of the site-diagonal and off-diagonal phonon couplings.

There have been time-resolved optical absorption measurements of the V_k

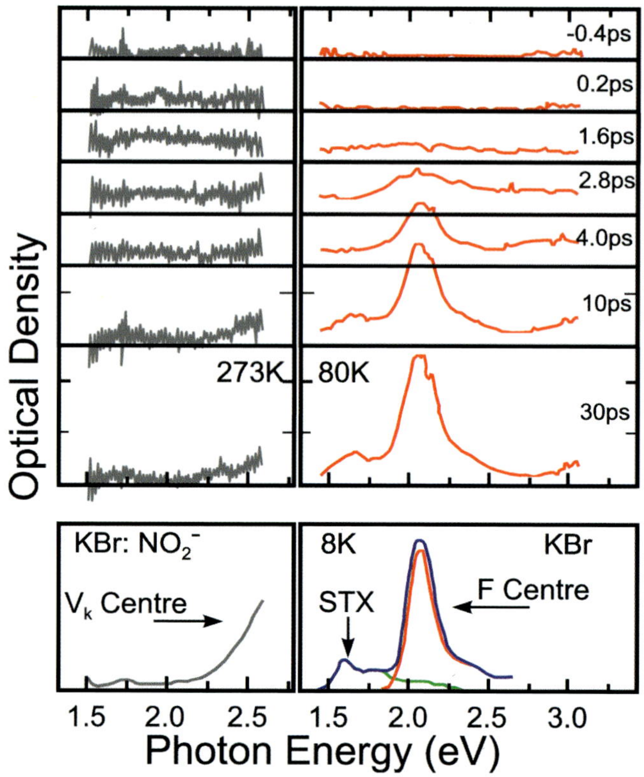

Fig. 4.4. Time-resolved spectra of self-trapped hole evolution after formation of electron–hole pairs in KBr : NO_2^- The electrons are trapped by NO_2^-, and the observed spectra represent the hole relaxation (Iwai *et al.* 1994).

centre evolution following the creation of electron–hole pairs by 100 fs laser pulses. The result for KI is shown in Fig. 4.4. These experiments have been done in KI and RbI (Iwai *et al.* 1994) and in KBr and RbBr, both doped with NO_2^- electron traps (Shibata *et al.* 1994; Sugiyama *et al.* 1996). Whilst some doubts remain, these studies appear to show three common features. First, the hole relaxes to form a V_k centre after about 10 ps. Secondly, a broad band (the B band, 0.5–2.5 eV) emerges about 0.1 ps after the initiation of the laser pulse. Thirdly, this broad band disappears as the V_k centre develops. In KI and RbI doped with NO_2^-, there are further transient bands, sharper than the B band, which are seen before the V_k centre is established. Their origin is not yet known. Sharper bands seen in KBr were shown to arise from two-photon absorption (Thoma *et al.* 1997). The broad transient band could be attributed to the σ_g–σ_u transition of the halogen molecular ion during vibrational cooling, i.e. to the

Table 4.2. *Relaxation energies (in electron volts) of the one- and two-centre self-trapped hole in alkali halides.*

System	One-centre self-trapped hole relaxation energy	Two-centre self-trapped hole relaxation energy	Reference
NaCl		1.82 eV	Tolpygo & Shaterman 1987
		0.4 eV below one-centre self-trapped hole	Puchin *et al.* 1996
KCl	1.1		Shluger, unpublished
		1.53	Tolpygo & Shaterman 1987
KI	1.3	1.6	Shluger, unpublished

early bonding of the halogen molecular ion and subsequent transfer of vibrational energy from the halogen–halogen stretch to other modes.

The relaxation energy for self-trapped holes

The major contribution to the relaxation energy arises from the polarisation energy (Gilbert 1966; Stoneham 1975). The polarisation energy drives charge localisation onto a minimum number of sites, as can be seen from the simple argument in Section 3.3. Although the one-centre self-trapped hole should have C_{4V} symmetry from the Jahn–Teller distortion, the relaxation energy from this term is modest, a few tenths of an electron volt in an alkali halide. A significant energy comes from specific rebonding to form a two-centre molecular ion. The bonding energy between X^- and X^0 in free space is about 1.5 eV for Cl_2^-. It is not yet clear that one-centre self-trapping of a hole is stable or even metastable; in any case, it should relax rapidly to the V_k centre. Predicted relaxation energies are given in Table 4.2.

Motion of self-trapped holes

V_k centres are seen to reorient or to disappear above certain characteristic temperatures. The reorientation and decay processes are both the result of motion which is best described as an incoherent hopping motion (Stoneham 1975). In the NaCl structure, these jumps result in reorientations of V_k centres by 60°, 90° and 180°. In the CsCl structure, the 90° jumps reorient the centre and the 180° jumps lead to long-range motion to traps, which provide a decay process. Experiment indicates that 60° reorientation dominates in KCl and KI (Keller & Murray 1966). In Cs halides, the 180° translational motion occurs at temperatures lower than the 90° reorientation process (Pilloud & Jaccard 1975). Thus, motion along a $\langle 100 \rangle$ linear row occurs before three-dimensional motion starts.

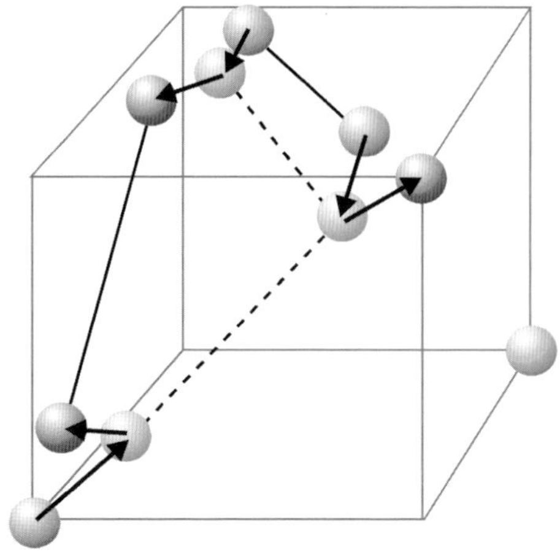

Fig. 4.5. Intermediate states of migration of the V_k centre in alkali halides. Only the halogen ions are shown explicitly. The hole starts on a diagonal of the top face of the cube, and ends on a diagonal of the face on the left. This shows a 60° jump.

In the fluorite structure (see Hayes & Stoneham (1974), p. 244) for a review of CaF_2, SrF_2, BaF_2 and $SrCl_2$), the 180° jump occurs more readily than the 90° jump; this, in turn, has implications for the interatomic potentials for the F—$F^{-1/2}$ interactions (Norgett & Stoneham 1973). In CaF_2, only the 180° jumps are seen prior to disappearance. The apparent activation energies (Beaumont *et al.* 1970) are 0.19 eV (90 K) and 0.31 eV (120 K), in broad agreement with later work (Huisinga *et al.* 1997) giving 0.30 ± 0.06 eV (490–620 K). An increase with increasing temperature is expected from small-polaron theory, as discussed for this system by Norgett & Stoneham.

The jumps leading to thermal reorientation depend strongly on the jump angle; in simple models (Flynn 1972) the activation energy increases monotonically with the distance moved by the charge. In NaCl-structure alkali halides, thermal reorientation of the V_k centre occurs primarily through 60° jumps. However, the reorientation induced by the ${}^2\Sigma_u^+ - {}^2\Sigma_g^+$ optical transition has no preference for a particular change in angle (Murray & Bethers 1969). The ${}^2\Sigma_g^+$ excited state is not a bonding state of the X_2^- molecular ion, but the thermal disorientation appears to occur before complete dissociation of the X_2^- molecular ion. Cluster calculations (Shluger *et al.* 1992) suggest a triatomic intermediate state is involved, as shown in Fig. 4.5. The configuration is somewhat similar to a hypothetical one-centre self-trapped hole, and its energy is higher that the one-centre form by only 0.1 eV. It seems likely that the one-centre self-trapped hole relaxes through this saddle-point configuration, and that it is from this configuration that relaxation branches into the different channels.

4.1.4 Extrinsic and perturbed self-trapped holes

Extrinsic self-trapping of holes

As defined earlier, extrinsic self-trapping is induced by an impurity or defect where self-trapping does not occur in the perfect crystal. We will only use the term when the extrinsic self-trapped state is in a state like that of intrinsic self-trapped holes in similar crystals; we shall not use the term for the trapping of conventional self-trapped holes by defects, e.g. of V_k centres by impurity alkali ions in alkali halides. Extrinsic self-trapped *excitons* have been seen in TlCl crystals doped with TlBr (see Section 4.3). We know of no cases of extrinsic self-trapped holes, so defined, in halides. If extrinsic self-trapping were to be observed in halides, it should occur in Cu or Tl halides, in which holes are not self-trapped.

Perturbed self-trapped holes

Self-trapped holes can be perturbed by defects and impurities. At the temperatures for which the self-trapped holes are mobile, they can move to these defects or impurities, forming perturbed self-trapped holes. A self-trapped hole stabilised by an alkali impurity of a smaller radius is called a V_{kA} centre. The structure of the V_{kA} centres is similar to that of the V_k centres, with slight displacements of their centre of the mass, due to the mismatch of the impurity alkali ions (Delbecq *et al.* 1969; Schoemaker 1970). Self-trapped holes stabilised by larger alkalis are not reported, presumably because the relative positions favoured by the elastic interaction are more distant, and the signals harder to resolve.

Two classes of structure are seen for self-trapped holes stabilised by a halogen impurity in alkali halides: those oriented along a $\langle 110 \rangle$ axis, and those oriented along a $\langle 111 \rangle$ axis. Those with a $\langle 110 \rangle$ orientation include $BrCl^-$ in KBr:Cl (Delbecq *et al.* 1971, 1974) and ICl^- ions in KI:Cl (Wilkins & Gabriel 1963; Goldberg & Meistrich 1968). Those molecular ions with a $\langle 111 \rangle$ orientation include FCl^-, FBr^- and FI^- in KCl, KBr and KI doped with F^-, respectively (Schoemaker 1966). The FCl^- ion formed in KF:Cl is oriented along a $\langle 110 \rangle$ direction (Delbecq *et al.* 1974), while that in KCl:F is oriented along a $\langle 111 \rangle$ direction.

The orientation of the halogen molecular ion of a self-trapped hole involves a balance between bond formation, polarisation, and short-range repulsive forces (Cade *et al.* 1984). When there are impurity halogens Y, whether or not there will be XY^- hole centres or only X_2^- ions will depend on the relative electronegativities (Cade *et al.* 1984). If XY^- molecular ions are formed, the distribution of the hole over the two ions will be unequal. When the hole is equally distributed, the Madelung potential favours the halogen molecular ion taking a ⟨110⟩ orientation. However, if the hole is unequally distributed, the ion with higher negative charge is favoured to be at the anion site. In such a case, the orientation of the halogen molecular ion no longer needs to be along ⟨110⟩, and Pauli repulsion is more likely to favour a ⟨111⟩ orientation. The importance of short-range repulsion depends on the ion sizes. For a small impurity ion, e.g. KCl:F, the short-range forces are less critical, and the ⟨110⟩ configuration remains. For larger impurity ions, KF:Cl (again a FCl^- centre) the repulsions are more significant, and the molecular ion has a ⟨111⟩ orientation. Such issues also arise when we discuss self-trapped excitons.

4.2 Self-trapped carriers in oxides

We may consider two main classes of oxides: those for which there is a closed-shell cation (e.g. MgO, Al_2O_3; we shall regard SiO_2 as being in this category) and those for which the cation can exist readily in several different charge states (e.g. CoO, UO_2, and insulating phases of the oxides considered for high-temperature superconductivity).

Self-trapped carriers are commonly formed in the second class of oxides, since they can be thought of as a change of charge state of one of the cations: the hole in UO_2 corresponds to a U^{5+} ion, the electron corresponds to a U^{3+} ion. In some cases, the change of charge state simply changes the number of electrons in a partly-filled shell, e.g. the 3d shell for first series transition metal ions. However, this description can be naive: in many cases, the hole charge is distributed so as to involve oxygens as well as the cations, e.g. NiO:Li (Kuiper *et al.* 1989).

In oxides with closed-shell cations, the self-trapping of holes has not been observed directly, although theory predicts that self-trapped holes are stable by very small energies in several oxides (MgO, SiO_2). In such cases, charge transport may be via delocalised large-polaron excited states. These systems also show extrinsic self-trapping, e.g. in Ge-doped SiO_2 (Hayes & Jenkin 1986). If the oxide is amorphised, the site-to-site variations can lead to self-trapping, as for amorphous SiO_2 (Griscom 1989, 1992a).

Hayes & Jenkin (1986) showed that Ge-trapped holes can be created in Ge-doped quartz, indicating that the holes in pure quartz are very mobile and probably not self-trapped. They observed the EPR spectra of the Ge-trapped hole in quartz, finding g values very similar to those for the relaxed hole in amorphous SiO_2. This suggests that the configuration of the relaxed hole is governed not by the local disorder before hole relaxation, but the distortion induced by hole localisation. If so, the Ge impurity yields a site for hole localisation, and

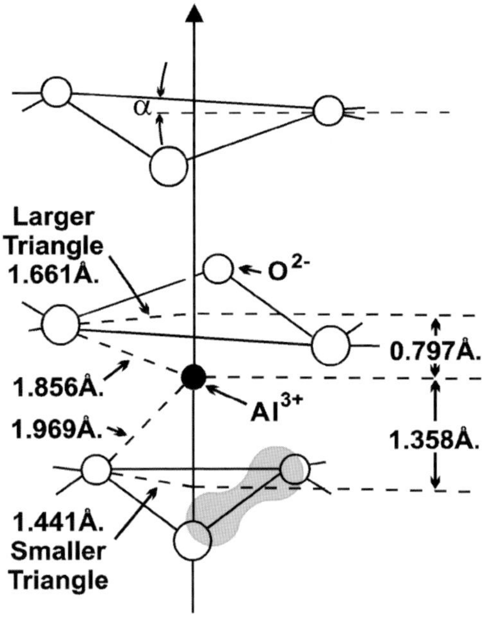

Fig. 4.6. Predicted self-trapped hole in Al_2O_3 (after Jacobs & Kotomin 1992).

the relaxation after the localisation is governed by the electron–lattice interaction. This is the phenomenon of extrinsic self-trapping defined earlier: the self-trapped hole in crystalline and amorphous SiO_2 appears to show extrinsic self-trapping.

In Fig. 4.6 we show predictions for the configurations which result from a constrained relaxation with holes localised at chosen lattice sites in several oxides. We cannot assume that predictions for these 'relaxed holes' are the configurations of self-trapped holes. For a hole localised on oxygen in MgO (Shluger *et al.* 1991c) the distortion is primarily a symmetrical breathing motion of the neighbours; the O^- also moves slightly off-centre, perhaps from a Jahn–Teller instability. A more stable configuration for a relaxed hole in Al_2O_3 (Jacobs & Kotomin 1992; see also Jacobs *et al.* 1992) shares the hole between two oxygen atoms; the oxygen–oxygen distance is reduced, and the cations are displaced (Fig. 4.6). The relaxation of the oxygen network is ascribed to the Jahn–Teller effects. The lattice relaxation energy has been obtained to be 3.7 eV. We note that the relaxation energies of the holes in MgO and Al_2O_3 oxides are large: it is the large bandwidths (large transfer energies, and hence large energies needed to localise a carrier on a site) which inhibit self-trapping. However, experimental studies by Kirm *et al.* (1999) conclude that electrons and holes do not self-trap, but that a self-trapped exciton can form, and emits at 7.6 eV.

For MgO, there is no evidence for self-trapped holes from spin resonance. If self-trapping occurs, excitation to a large-polaron form is easy. For other oxides, there is less evidence. EPR could give unambiguous proof of the existence of self-trapped holes. Evidence against self-trapping, although not so certain, might come from the observation of impurity-trapped holes in a concentration almost equal to the total number of holes created. Kulis *et al.* (1991) have seen a slow inertial increase in thermally-stimulated recombination following a step-like increase in the temperature of X-irradiated crystals. This delayed recombination is observed for alkali halides and alumina, but not observed for MgO. This suggests (cf. the calculations of Kotomin *et al.* (1989)) a delay caused by the slow motion of self-trapped holes in these materials. Thus the experimental results of Kulis *et al.*, imply that the hole is self-trapped in Al_2O_3. Nevertheless, we should reserve judgement until more direct evidence is available.

In amorphous SiO_2, EPR data (Griscom 1989) show that the hole is localised on a 2p non-bonding orbital of an oxygen atom bridging two Si atoms. The localisation is accompanied by a symmetrical distortion of these two Si atoms. In crystalline silicas, there is no direct evidence for self-trapping: there is certainly rapid hole motion. One possible interpretation is that the lowest hole state is indeed self-trapped, but that the excitation energy to mobile large-polaron states is small (Zhang *et al.* 1994). The several calculations (Edwards 1993; Zhang *et al.* 1994; Shluger & Stefanovich 1990) agree that self-trapping is only just achieved, if achieved at all.

Predictions for hole self-trapping in SiO_2 with the β-cristobalite structure (Shluger & Stefanovich 1990) conclude that the self-trapping of a hole is unlikely. Edwards concludes that the hole should be self-trapped in quartz. Zhang *et al.* looked at a number of crystalline forms, and found that self-trapping was marginal, becoming easier in oxides with higher densities. Stoneham (1991) pointed out that self-trapping will be easier close to the Si/oxide interface, because of the image interaction and the higher polarisability of the silicon.

4.3 Self-trapped excitons in halides

Halides are the materials in which self-trapped exciton behaviour is best documented. As early as 1964, Kabler suggested that the intrinsic luminescence following ionising radiation of alkali halides was due to a self-trapped exciton formed by recombination of an electron with a self-trapped hole. Just as there are two types of self-trapped holes, so there are two types of self-trapped exciton. The first type is an electron trapped by a self-trapped hole localised on the cation sublattice; the second type consists of an electron and a halogen molecular ion (X_2^-). The presence of the bound electron means that the structures of self-trapped excitons are often significantly different from those of the self-trapped holes. The coupling of both electron and hole to lattice ion displacements leads to a variety of self-trapped exciton structures, and these are the main topic of this chapter. Readers may refer to several reviews (N Itoh 1982; Lushchik 1982; R T Williams & Song 1990; Song & Williams 1993) for further details of the historical development and of the properties for specific materials.

The self-trapped excitons are characterised by their recombination luminescence, with its large Stokes shift, by optical absorption to higher excited states, and by EPR tagged by the luminescence. Optically-detected EPR (ODEPR) and electron–nuclear double resonance (ODENDOR) are especially important. Results from EPR tagged with one of the recombination luminescence bands (ODEPR: Marrone *et al.* 1973; Wasiela *et al.* 1973; Call *et al.* 1975; ODENDOR: Block *et al.* 1978) show that the self-trapped exciton in both alkali halides and alkaline-earth fluorides consists of an electron bound to a halogen molecular ion. The precise structure of the molecular ion and its neighbours has still to be determined. Results from ODEPR are also available for AgCl (Hayes *et al.* 1977).

The EPR parameters of the self-trapped excitons can usefully be compared with those for other centres involving a halogen molecular ion, namely the V_k and H centres. Likewise, optical absorption spectra for self-trapped excitons have been observed which are similar to those of V_k and H centres (see, e.g., Song & Williams (1993), or Stoneham (1975)). The luminescence energy for electron–hole recombination of self-trapped excitons has a value significantly less than the optical absorption energy for exciton formation. Optical absorption and luminescence measurements are evidence for formation of self-trapped excitons when defects or impurities can be ruled out.

Relaxed self-trapped excitons in halides can adopt one of several configurations. One has an electron trapped by a self-trapped hole; this we shall write [STH:e]. Another has both electron and hole effectively self-trapped, [S_e:S_h]; since the electron does not self-trap in isolation, S_e is essentially an F centre. Often there are several different [S_e:S_h] configurations. Schematic diagrams of the configurations are illustrated in Fig. 4.7. In AgCl, the self-trapped excitons are of the [STH:e] type. In alkali halides, the singlet self-trapped exciton is of [STH:e] form, whereas the triplet self-trapped exciton is mostly of the [S_e:S_h] type. In the alkaline earth fluorides, only self-trapped excitons of [S_e:S_h] type have been found; these are shown in Fig. 4.7.

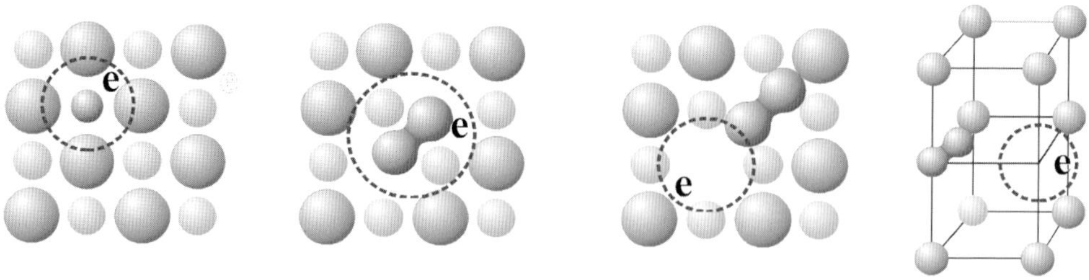

Fig. 4.7. Self-trapped excitons in halides. From the left, these show AgCl, an [STH : e] system, an [S_e : S_h] system, and CaF_2. Two cases correspond to an electron trapped by a self-trapped hole. In the other cases, the hole state changes substantially.

4.3.1 AgCl

In AgCl, both excitons and holes are self-trapped; in AgBr, neither excitons nor holes are self-trapped. Self-trapped excitons are observed in mixed $AgBr_{1-x}Cl_x$. ODEPR (Hayes *et al.* 1977) shows that the triplet self-trapped exciton in AgCl is of the [STH:e] type: an electron trapped by a Ag^{2+} ion on a cation lattice site, with D_{4h} symmetry. A transient optical absorption band associated with the self-trapped exciton is observed at 1.2 eV (Marquardt *et al.* 1971), the same energy as the transition observed for the self-trapped hole. There is a 2.5 eV luminescence from electron–hole recombination (Marquardt *et al.* 1971), and this is used in ODEPR.

The EPR signal of the AgCl self-trapped exciton is very similar to that of the corresponding self-trapped hole, except that the hyperfine parameter is smaller by a factor of 2, and so is too small to be resolved. The hole wavefunction from electron–nuclear double resonance (ENDOR) studies (Bennebroek *et al.* 1996) is similar to that for the self-trapped hole. The intraionic Ag^{2+} optical absorption band is almost exactly the same for the self-trapped hole and for the self-trapped exciton. The electron is relatively weakly bound; its wavefunction radius is about of 1.5 nm, as expected from the high AgCl dielectric constant. The exchange energy is small, approximately 5 cm^{-1} (Donckers *et al.* 1992). The negative zero-field splitting (Yoshioka *et al.* 1985; Donckers *et al.* 1992) is probably due to spin–spin interaction. The recombination of electrons and holes created by band-to-band transitions is complex, and the luminescence does not show a simple exponential decay (Marquardt *et al.* 1971). ODEPR tagged by recombination luminescence detects signals from self-trapped holes and from trapped electrons, as well as from self-trapped excitons (Marchetti & Tinti 1981). The electron, localised weakly by the self-trapped hole, is also affected by the defects which form shallow traps, as observed by Sakuragi & Kanzaki (1977).

Kanzaki *et al.* (1971) measured the luminescence of $AgCl_xBr_{1-x}$. They showed that the broad luminescence bands with large Stokes shifts are observed for $x<0.4$ (Br-rich), whereas sharp free-exciton luminescence is seen for $x>0.4$ (Cl-rich). ODEPR from hole localisation on an Ag^{2+} site has been detected (Marchetti & Tinti 1981; Grigorjeva *et al.* 1995). It is not yet clear whether excitons or holes are self-trapped at higher Br^- coordinations. A search for a transition from a self-trapped exciton state to a free-exciton state under applied pressure failed (Yokohama & Kobayashi 1985). The luminescence band due to self-trapped excitons disappeared, but was replaced by another broad band, not by free-exciton luminescence. Apparently the conversion of excitons from self-trapped to free excitons is local in nature.

4.3.2 Alkali halides with the NaCl structure

The atomic configurations of self-trapped excitons in alkali halides are still not clear, despite extensive studies. Self-trapped excitons in alkali halides comprise an electron and an X_2^- molecular ion. Early suggestions that the [STH:e] or [V_k:e] configurations were unstable (N Itoh & Saidoh 1973; Toyozawa 1974) were developed only much later, after Leung *et al.* (1985) suggested that the self-trapped exciton has an off-centre [S_e:S_h] configuration. It now appears that there are several possible configurations depending on the spin multiplicity and material.

The luminescence emitted in alkali halides includes contributions from singlet and triplet initial states. The singlet transitions are usually in the ultraviolet, whereas the triplet transitions are normally in the visible or the infrared. The current view is that the large energy difference between singlet and triplet luminescence is not due to the exchange interaction, and is probably not due to transitions from different orbital states, but results from the difference in the configuration. Measurements of excitation spectra show that only the triplet luminescence is seen after excitation at the exciton peak; band-to-band excitation induces both singlet and triplet luminescence (Nishimura 1983).

Three types of configuration for self-trapped excitons

Kan'no *et al.* (1990) surveyed the luminescence spectra of all alkali halides with the NaCl structure. Fig. 4.8 shows the main features of the luminescence bands. The singlet luminescence bands (unshaded) decay in nanoseconds; the triplet luminescence bands decay in microseconds to milliseconds. Evidently, the luminescence bands can be divided into three groups, designated Types I, II and III.

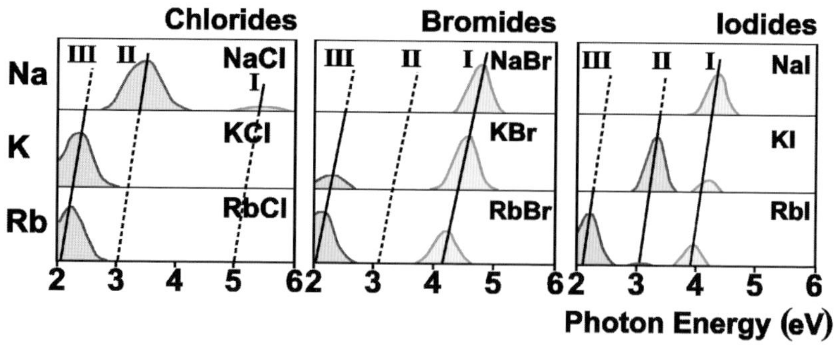

Fig. 4.8. Luminescence spectra of alkali halides of the NaCl structure, showing spectra of Types I, II and III (after Kan'no *et al.* 1990).

The three types of luminescence band suggest that there should be three types of relaxed exciton state, depending on the system and on the spin multiplicity.

Type I Singlet excitons relax to a configuration giving rise to Type I luminescence. We call this configuration a Type I self-trapped exciton. The triplet self-trapped excitons of NaBr and NaI are also of Type I. There are no observations of Type I self-trapped excitons in KCl and RbCl.

Type II Type II self-trapped excitons are limited to the triplet self-trapped excitons in NaCl, KI and RbI; RbI also shows Type III behaviour.

Type III Triplet excitons in most alkali halides (but not NaCl and KI) relax to the Type III configuration. Both Type II and Type III luminescence is seen for the triplet self-trapped excitons in RbI.

Empirical rules for the optical absorption energy

The optical transition energy E_T of a defect or impurity often shows systematic behaviour from one host to another. The best known empirical rule describing such behaviour is the Mollwo–Ivey relation: if a is the lattice parameter, then $E_T = Aa^{-n}$. The rule holds for the optical absorption energies of F centres and M centres (Fowler 1966), and can be given theoretical support in terms of the point-ion and more general models (Stoneham 1975). Tanimura & Itoh (1992) have shown that the optical absorption due to the electron transitions of Type II and Type III self-trapped excitons follow the Ivey–Mollwo relation (see Fig. 4.9). There is a clear difference between Type II and Type III excitons. This suggests that, for each material, there are two distinct types of potential well in which the electron is trapped; these wells are similar in all these materials. The optical transition energy for Type I self-trapped excitons is much smaller than those of Types II and III, suggesting a much shallower potential well.

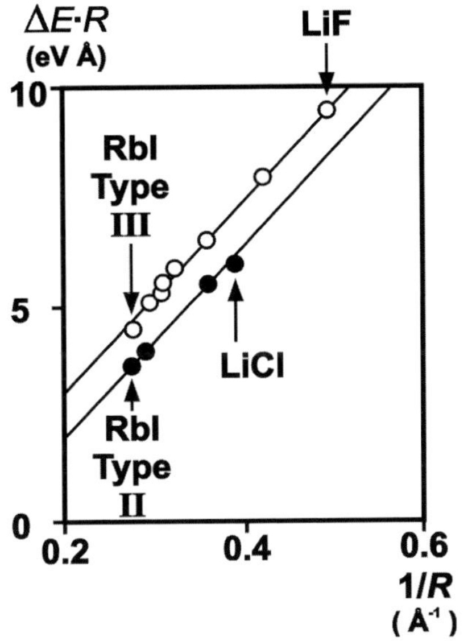

Fig. 4.9. Ivey–Mollwo plots of the optical absorption energies of the self-trapped excitons in alkali halides of the NaCl structure. The points on the upper line include LiF, KCl, RbCl, KBr, RbBr and RbI (Type III). Those on the lower line include LiCl, NaCl, KI and RbI (Type II) (after Tanimura & Itoh 1992).

Empirical rule for the luminescence energy

Kan'no *et al.* (1990) found an empirical relation for the luminscence energies of self-trapped excitons in alkali halides in terms of the *Stokes shift* and the *Klick–Rabin parameter S/D*, where S is the space between two nearest halogen ions in the perfect lattice and D is the diameter of a halogen atom. The rule shows that the plots follow three different straight lines for Type I, II and III exciton luminescence bands. Now S/D does not have any direct relationship to the optical transition; it was first introduced by Rabin & Klick (1960) to correlate the defect yields of alkali halides to the compatibility of an interstitial halogen atom in the lattice (see Chapter 5). We organise the data in another way. In Fig. 4.10, we plot the Stokes shift divided by $m\omega_0^2$ (essentially the force constant of the configuration coordinate diagram, so the ratio is a measure of the square of the change in coordinate on relaxation) as a function of R_+/R_-. Here m is the reduced alkali and halogen mass, ω_0 is the LO phonon energy , and R_+ and R_- are the radii of the alkali and halogen ions. We find again that the data fall onto three groups, representing Type I, II and III self-trapped excitons.

The empirical rules for absorption and emission group the self-trapped excitons in alkali halides in a consistent way. They suggest that there are three

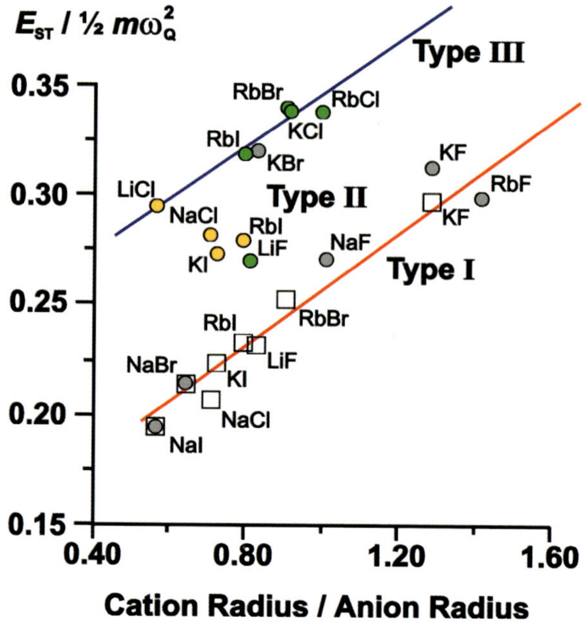

Fig. 4.10. Luminescence energies of the self-trapped excitons of alkali halides with the NaCl structure, showing scaling. The ordinate is essentially the square of a configuration coordinate displacement, Q^2, namely the Stokes shift divided by $2m\omega_Q{}^2$, where m is the reduced alkali and halogen mass and ω_Q is the LO phonon energy and the abscissa is the radius of the alkali ion divided by that of the halogen ion.

different configurations for these self-trapped excitons, each having a typical but different atomic structure. The absorption energies for Type II and Type III excitons are close to the corresponding F centre transition energies and follow the Mollwo–Ivey scaling with nearest-neighbour distance. Such scalings are often found for centres including electrons trapped at anion vacancies. Here, the scaling is consistent with an electron trapped in the potential well of a perturbed anion vacancy, as in the $[S_e:S_h]$ off-centre model for these Type II and Type III excitons. The absorption energies by themselves are inconclusive, and must be supplemented by other data. For instance, theory for the on-centre $[V_k:e]$ in NaCl (Stoneham 1974) actually agrees marginally better with experiment than that for the $F+V_k$ model expected from the Type II classification. It may be too simple to assume that the lowest singlet and triplet adiabatic surfaces alone are the source of the observed luminescence.

We can express the empirical rule in another way. In a single-frequency model, the Stokes shift is proportional to $(1/2)m\omega^2Q^2$, where m is the mass, ω the frequency and Q the displacement of the minima for the configuration coordinate. We have also an empirical equation:

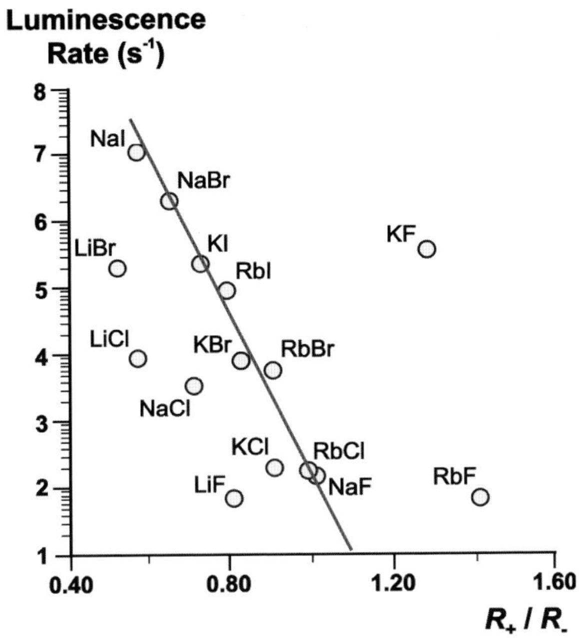

Fig. 4.11. Relation between the luminescence rate of triplet self-trapped excitons and the ratio of cation radius to anion radius, R_+/R_-.

$$E_{ST} = (1/2)\, m\, Q^2 \omega_Q^{\ 2}\, A(R_+/R_-),$$

where ω_Q is the LO phonon frequency and A is a linear function of the ratio of ionic radii, R_+/R_-. Together, these imply that $Q^2\omega^2/\omega_Q^{\ 2}$ increases linearly with R_+/R_-. Two components are involved in the factor A: the softening of the average force constant for the displacement of a halogen atom compared to that of a halogen ion and the magnitude of the displacement. Both factors increases clearly as R_+/R_- increases.

Empirical rule for the optical transition lifetime

A simple empirical relation holds between the luminescence rates of triplet self-trapped excitons and a/R_- (and hence to R_+/R_-), as shown in Fig. 4.11. It is clear that lifetime τ may be written:

$$\tau/\tau_0 = e^{\xi a/R_-},$$

where a is the lattice constant and τ_0 and ξ are constants. Since the recombination of the self-trapped excitons of the $[S_e{:}S_h]$ type involves charge transfer, the a/R_- dependence appears to relate to the range of the charge transfer process.

Table 4.3. *Theoretical and experimental optical absorption energies for various transitions. Experimental values are boldface. Type I [V_k:e] cases and predictions are in italics.*

Transition if Type I Transition if Type II, III	$a_{1g}\rightarrow b_{3u}$ $a_1\rightarrow a_2$	$a_{1g}\rightarrow b_{2u}$ $a_1\rightarrow b_2$	$a_{1g}\rightarrow b_{1u}$ $a_1\rightarrow b_1$
NaCl (Type I)	*0.7–1.1*[(a)] *1.49*[(b)]	*2.3–2.4*[(a)] *1.95*[(b)]	*2.3–2.5*[(a)] *1.8*[(b)]
NaCl ([S_e:S_h])		2.6[(f)]	
NaCl (Type II)	**2.00**[(e)]	**1.95**[(e)]	**2.13**[(e)]
NaBr (Type I)	**<0.2**[(c)]		
NaI (Type I)	**<0.2**[(c)]		
KCl (Type I)	*0.7–1.1*[(a)] *1.36*[(d)]	*2.1–2.2*[(a)] *1.9*[(d)]	*2.1–2.2*[(a)] *2.2*[(d)]
KCl ([S_e:S_h])		2.0[(f)]	
KCl (Type III)	**2.10**[(e)]	**1.87**[(e)]	**2.19**[(e)]
KBr (Type III)	**1.70**[(e)]		**1.77**[(e)]
RbCl (Type III)	**1.89**[(e)]	**1.70**[(e)]	**1.99**[(e)]

Notes:
References are: [(a)] Song *et al.* 1975; [(b)] Stoneham 1974; [(c)] Hirota *et al.* 1991; [(d)] N Itoh *et al.* 1977; [(e)] Tanimura 1989; [(f)] Puchin *et al.* 1993.

Properties of Type I self-trapped excitons

The small Stokes shift and low optical absorption energies of Type I self-trapped excitons suggest a relatively small lattice distortion. In Table 4.3, we compare the experimental energies of the self-trapped excitons with the theoretical values predicted for the [V_k:e] configuration. The good agreement implies that the configurations of Type I self-trapped excitons are close to [V_k:e] configurations. Predictions for the $a_{1g}\rightarrow b_{3u}$ transition are much smaller than those for other transitions. In fact, the electron optical absorption transition appears in the infrared for Type I self-trapped excitons, as seen for NaBr and NaI (Edamatsu *et al.* 1993) because of the small Stokes shifts. The information is not sufficient to say whether or not there is a slight off-centring because of the instability mentioned above. The work of Puchin *et al.* (1996) included configuration interaction in predictions of the configuration of the self-trapped excitons. They concluded that there is a local minimum near the [V_k:e] configuration. The electronic structure of a many-electron system with a flat energy surface close to a symmetrical configuration is a challenging problem.

Properties of Type II and III self-trapped excitons

Type II and III self-trapped excitons are regarded as having the [S_e:S_h] configuration, although their structures are not yet clear. The electron-transition energies follow the Mollwo–Ivey plots, suggesting that the potential well for the electron has nearly the same shape for all alkali halides. Several calculations have been made to elucidate the structures. Leung *et al.* (1985) found that the adiabatic potential energy surface of these self-trapped excitons has a minimum at a $[S_e{:}S_h]$ configuration. Later work exploited *ab initio* Hartree–Fock methods (Baetzold & Song 1991; Shluger *et al.* 1991; Puchin *et al.* 1993; Song & Baetzold 1992). These calculations all conclude that the potential energy surface near the minimum is extremely flat, so it is difficult to define an equilibrium geometry accurately. Table 4.3 summarises the predictions of optical transition energies obtained by *ab initio* Hartree–Fock calculations for the $[S_e{:}S_h]$ configuration. The off-centre displacement of the halogen molecular ion may vary little from one alkali halide to another; one of the plausible configurations locates the halogen atom of the halogen molecular ion between two neighbouring alkali ions.

The assignment of Type II and III self-trapped excitons to $[S_e{:}S_h]$ configurations is supported by the results of Raman spectroscopy. Suzuki *et al.* (1994) measured the Raman spectra of triplet self-trapped excitons in several alkali halides in resonance with the electron and hole transitions. The NaCl Raman spectrum in resonance with the electron transition is very similar to that for the F centre. The Raman spectra in resonance with the hole transition represents the stretching vibration of the halogen molecular ion. The frequencies of stretching vibrations of self-trapped excitons in NaCl and RbI agree with those of the H centres in the same crystal. This suggests that the hole part of the self-trapped exciton closely resembles the H centre. The differences between the stretching vibrations of the halogen molecular ions constituting the V_k and H centres have been studied very carefully for almost all alkali halides with NaCl structures (Suzuki *et al.* 1993). The systematic differences are attributed to the differences in the crystal potential on each atom of the halogen molecular ion. In NaBr, the fact that the Raman frequency of the self-trapped exciton and that of the V_k centre are essentially the same is consistent with the assignment of the NaBr Type I self-trapped exciton to the $[V_k{:}e]$ configuration.

ODEPR and ODENDOR studies

EPR and ENDOR data for Type III self-trapped excitons do not fully specify their atomic structure. Some of the results suggest that the Type III self-trapped exciton is an F–H pair. First, the zero-field splitting parameters in alkali fluorides are positive (Block & Wasiela 1979), whereas those of the alkaline-earth fluorides are negative (see Section 4.3.3). According to Leung *et al.* (1985), the zero-field splitting of the self-trapped excitons in fluorides is due to the spin–spin interaction, and takes positive values when the axis of the F_2^- molecular ion and the spin–spin axis are reasonably close. Secondly, for KBr, the superhyperfine structure from ODENDOR experiments (Block *et al.* 1978) shows a strong hyperfine interaction with two K^+ ions and one Br^- ion. These indicate a nearest F–H pair model of the self-trapped exciton.

Relaxation pathways to Type I, II and III configurations

As noted, the Type I configuration is taken mainly by *singlet-state* self-trapped excitons, whereas Type II and III configurations are taken by the *triplet* self-trapped excitons, apart from NaBr and NaI. There is a dependence on the principal quantum number, n, of the exciton created in absorption. Measurements of the excitation spectra for the singlet and triplet self-trapped excitons show that the triplet state is induced by creation of the $n=1$ free exciton, and that this state does not lead to singlet exciton luminescence. The Type I emission bands in NaBr and NaI include *fast luminescence* and *slow luminescence* (Matsumoto *et al.* 1992). The fast luminescence has been ascribed to the emission from the singlet state of a Type I configuration, and is not emitted on the formation of the $n=1$ free exciton (Matsumoto *et al.* 1995a). Thus $n=1$ free excitons and $n=2$ free excitons appear to relax through completely different pathways.

Femtosecond time-resolved studies of the relaxation of electron–hole pairs in KBr and RbBr (Shibata *et al.* 1994; Sugiyama *et al.* 1996) and KCl and RbCl (Fujiwara *et al.* 1997) show the formation of self-trapped holes; these holes are gradually converted to self-trapped excitons and to F–H pairs. It seems likely that the electron has little effect on the relaxation of the hole part of an $n=2$ exciton and the electron–hole pair. There are two decay channels: a fast generation of F–H pairs within a few picoseconds, and the formation of self-trapped excitons and F–H pairs with a delay of about 10 ps. The 10 ps delay is about the same as the vibrational cooling time for halogen molecular ions. On the other hand, the $n=1$ free exciton is converted to the triplet state very rapidly, and its relaxation appears to be very strongly influenced by the electron–hole correlation, although no femtosecond measurements have been made.

The $n=2$ excitons and electron–hole pairs lead both to singlet and triplet luminescence, with approximate relative efficiencies of 1:3. Most of the singlet luminescence occurs from the Type I configuration, whereas most of the triplet luminescence comes from the Type II or III configuration, except for NaBr and NaI. From this information, one can construct the adiabatic potential energy diagram shown in Fig. 4.12. The $n=2$ excitons and electron–hole pairs presumably relax first to a Type I configuration. For the Type I self-trapped exciton, the radiative lifetime for singlet luminescence is shorter than the lifetime for jumping over the potential barrier to a Type II or Type III configuration, so singlet luminescence will be observed from the Type I configuration. On the other hand, triplet excitons dominate for Type II and III con-

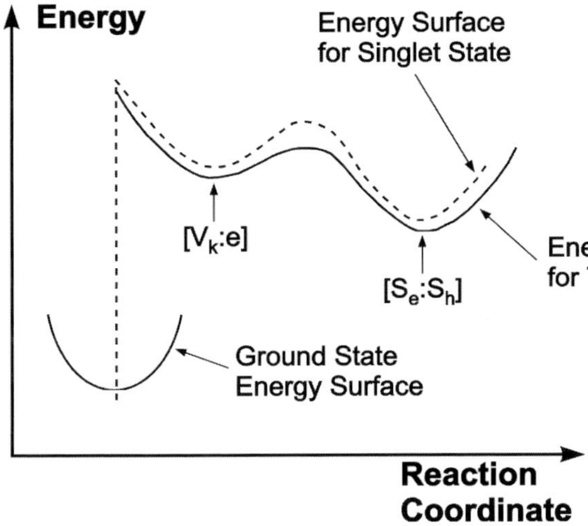

Fig. 4.12. A schematic adiabatic potential diagram indicating several configuration types of self-trapped excitons in alkali halides.

figurations since, for them, the radiative lifetime is longer than the lifetime the transition from the Type I configuration. The relaxation of the $n=1$ excitons may bypass the Type I configuration in many alkali halides. The reason for Type I triplet luminesecence in NaBr and NaI could be that there is a potential barrier which prevents jumps from the Type I configuration. This barrier is probably associated with the large halogen-to-alkali ion radius ratio.

Although the major fraction of Type I self-trapped exciton luminescence is singlet, and most of the Type II or Type III self-trapped exciton luminescence is triplet, a small fraction of the other multiplicity is observed. R T Williams *et al.* (1991) observe that the Type II and Type III self-trapped excitons in RbI exhibit a fast luminescence which can be ascribed to singlet emission. They show that the self-trapped exciton lifetimes of singlet states increase in the order Type I, Type II, Type III, which would fit with increasing electron–hole separation. Matsumoto *et al.* (1992, 1993) show the Type I self-trapped exciton luminescence of NaCl, KBr, RbBr, KI and RbI includes a slower component which can be ascribed to triplet luminescence; moreover, the self-trapped exciton luminescence of NaBr and NaI, normally predominantly triplet, includes some singlet luminescence. The experimental singlet–triplet splittings for Types I, II, III are all of order 0.1 eV, consistent with singlet and triplet energy surfaces separated by an energy of that order.

The exchange energy for the Type III configuration in KCl has been evaluated to be 15 meV from the temperature dependence of the triplet luminescence (Song & Stoneham 1976; Purdy *et al.* 1977). Singlet–triplet splittings in alkali

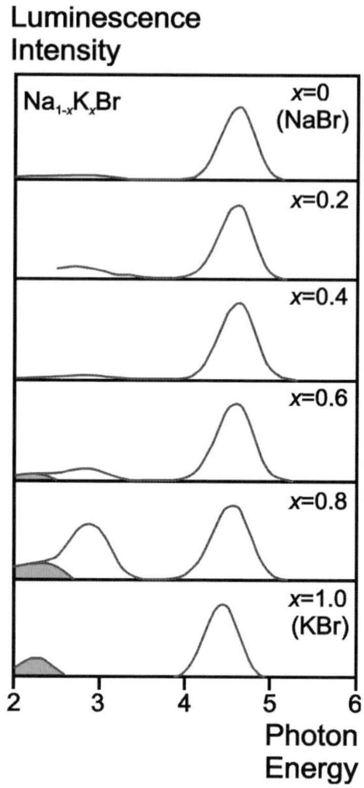

Fig. 4.13. Luminescence spectra of $Na_{1-x}K_xBr$ at 10 K following stimulation by 7.7 eV light, for various values of composition x. The band at 4–5 eV is Type I, while the shaded band for $x > 0.6$ is Type III (after Kan'no *et al.* 1990).

halide excitons are discussed by Song *et al.* (1981). That for the Type I configuration in KCl has been evaluated by Song *et al.* (1981) to be 14 meV, and much the same for other alkali halides. Matsumoto *et al.* (1995a) suggest that the exchange energy will be larger in states intermediate between Type I and Type III configurations, because of larger electron–hole overlap.

Hetero-alkali mixed crystals: (Na/K)Br

The classification into Types I, II, III also holds for peak optical transition energies in mixed crystals. The peak energy of the luminescence of each type shifts as the composition changes. Fig. 4.13 shows an example of such a shift for $Na_{1-x}K_xBr$ (Kan'no *et al.* 1990). The triplet luminescence band of NaBr is at 4.65 eV; as the composition moves towards KBr, there is a gradual shift to lower energy, approaching the singlet luminescence band of KBr at 4.40 eV. The decay at intermediate energies has fast and slow components. The fast component becomes more important as the K content increases, substantiating the

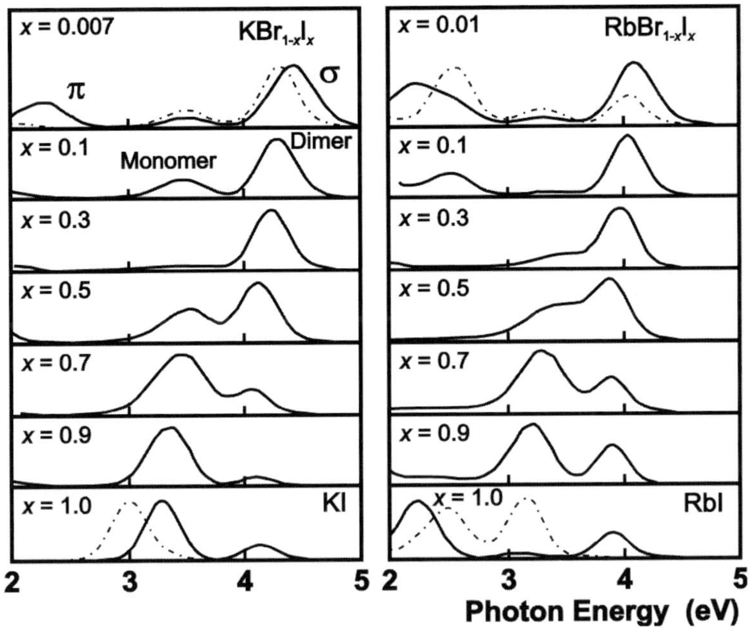

Fig. 4.14. Luminescence spectra of $KBr_{1-x}I_x$ and $RbBr_{1-x}I_x$ (at 7 K following 7.7 eV excitation (Hayashi *et al.* 1992)) for various values of x. The chain curves for the smallest values of x follow excitation of I^- absorption; the dotted curves for $x=1$ are spectra following stimulation in the $n=1$ exciton band (after Hirabayashi *et al.* 1992).

assignment of the 4.65 eV band in NaBr and 4.40 eV band in KBr to Type I. The triplet luminscence in KBr at 2.28 eV is observed only for $x>0.6$, indicating that the relaxation to this configuration in the triplet manifold occurs only after most of the Na^+ in NaBr has been replaced by K^+. A luminescence band seen at intermediate concentrations is ascribed to the self-trapped exciton perturbed by Na impurity in KBr.

Hetero-halogen mixed crystals: K(Cl/Br), K(Br/I) and Rb(Br/I)

The luminescence bands in hetero-halogen mixed crystals show similar, but not identical, behaviour. The complication is that the heavy halogen impurity can form a hetero-halogen molecular ion in some cases (Cade *et al.* 1984). Results are available for $KCl_{1-x}Br_x$ (Tanaka *et al.* 1990), $KBr_{1-x}I_x$ and $RbBr_{1-x}I_x$ (Hayashi *et al.* 1992) (Fig. 4.14). As the concentration of heavier halogen Z in $MX_{1-x}Z_x$ increases, the Type I bands of MX are replaced by luminescence bands from the recombination of an electron with a hole localised on two impurity halogens, Z_2^-. The peak energies of the luminescence of these bands changes continuously, and approaches the Type I luminescence energies of the crystal with the heavier halogen, MZ. The Type II and Type III luminescence bands of the crystal MX with the lighter halogen disappear when the concentration of the heavier halogen rises above about 10%. New bands with a larger Stokes shift emerge when the heavier halogen concentration rises above about 50%. From the dependence on light halogen concentration, these bands are ascribed to recombination of electrons with holes at mixed halogen molecular ions, XZ^-. These bands approach the Type II or Type III bands of MZ as the lighter halogens X are replaced by the heavier species Z. An ODMR study of the Type II BrI^- perturbed self-trapped exciton has been reported by Shirai & Kan'no (1998).

Edamatsu *et al.* (1993) have shown that homonuclear self-trapped excitons in $KCl_{1-x}I_x$ have an optical absorption energy which is as small as that for the Type I self-trapped exciton in NaBr, indicating on-centre structure. This is what one would expect for a large I_2^- molecular ion. The hetero-halogen molecular ions have a natural asymmetry, however, and one might expect off-centre character. A large Stokes shift is seen. Two luminescence bands have been found for heteronuclear self-trapped excitons in $KCl_{1-x}I_x$ and $RbCl_{1-x}I_x$. The bands are attributed to Type II and Type III self-trapped excitons (Kan'no *et al.* 1990).

4.3.3 Other halides

Alkaline-earth fluorides

In alkaline-earth fluorides, unlike alkali halides, only a single luminescence band arises from photoexcitation at the exciton peaks. The luminescence band, and the two major peaks of the absorption band from the luminescent state, have a fast decay component with a lifetime of 20 ns and much slower components (R T Williams *et al.* 1976). The fast component, which is thought to arise from the singlet state, is much weaker than the slow component. ODEPR measurements tagged by the slow component in CaF_2, SrF_2 and BaF_2 show that the luminescence comes from triplet states, with the hole distributed asymmetrically in an F_2^- molecular ion oriented along a ⟨111⟩ direction (Call *et al.* 1975; Block & Wasiela 1979). Adair *et al.* (1985) showed that the three $[S_e{:}S_h]$ type configurations shown in Fig. 4.15 are nearly degenerate. The optical absorption bands in the ultraviolet have been ascribed to hole transitions, and those in the visible to electron transitions. The negative zero-field splitting parameter D, attributed to the interaction between the electron and hole spins because of small spin–orbit interaction of F ions, indicates that the axis of the F_2^- molecular ion deviates from the spin–spin direction (Leung *et al.* 1985) (Section 4.3.2). Only for configurations (b) and (c) of Fig. 4.15 is D negative. It appears that the on-centre-type self-trapped exciton (Type I in alkali halides) has no local energy minimum in alkaline-earth fluorides. The luminescence comes from both singlet and triplet states; the energy separation is small

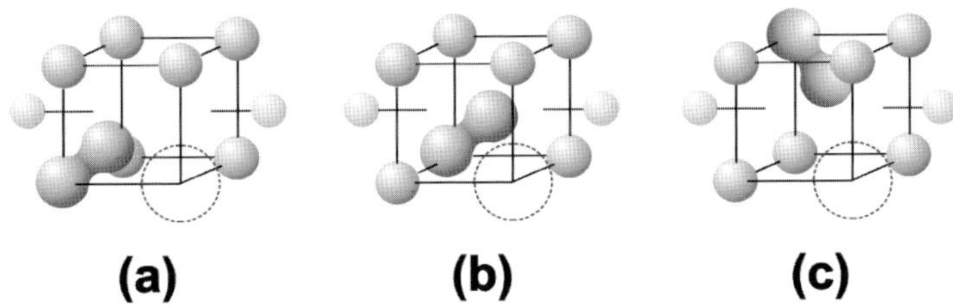

Fig. 4.15. Lattice configurations of three possible configurations for self-trapped excitons in alkaline-earth fluorides. Each structure has a different relative position of the V_k centre and F centre components.

because of the small exchange energy. The difference in the configurations of self-trapped excitons in alkali halides and in alkaline-earth fluorides is certainly due in part to the open fluorite structure, since fluorine can move off-site at relatively low cost in energy.

There is some evidence for several nearly-degenerate configurations of self-trapped excitons in CaF_2 and SrF_2. First, the slow decay of the luminescence does not follow a single exponential curve (R T Williams *et al.* 1976). Secondly, the optical absorption band due to transitions to higher excited states in CaF_2 can be divided into at least two components through their differences in decay time (Tanimura *et al.* 1989). Thirdly, the optical absorption into higher excited states of self-trapped excitons in SrF_2 can be divided into two components, such that each part can be partially bleached. The bleaching of each component enhances the other (Eshita *et al.* 1984), implying that the triplet self-trapped excitons consist of at least two components which can be inter-converted by photoexcitation. Photoexcitation of the hole component of the self-trapped exciton in CaF_2 stimulates F centre production (Tanimura *et al.* 1989).

The luminescence and transient absorption of CaF_2, SrF_2 and BaF_2 (Lindner *et al.* 1999) following two-photon excitation by femtosecond pulses of 6 eV photons show several effects. First, there are unidentified long-lived excited states which can be accessed and controlled by excitation with 3 eV photons. The 3 eV energy is reasonably close to the electron excitation energy of the self-trapped excitons in CaF_2, and to the hole excitation in BaF_2. Secondly, there is a 3 eV absorption which emerges with a single time constant (780 fs for CaF_2, less than 300 fs for the Sr and Ba cases) which may be the characteristic time for electron trapping to form the F centre part of the F-H-like self-trapped exciton.

CsCl-structure caesium halides

The self-trapped excitons have D_{4h} syymetry. In CsCl and CsBr, only the π band has been observed (A Lushchik *et al.* 1994), whereas both π and σ bands have

Table 4.4. *Excitons in caesium and ammonium halides.*

Host crystal	Exciton in absorption	Exciton luminescence	Character	Lifetime
CsCl	7.85 eV	2.82 eV	π	
CsBr	6.83 eV	3.55 eV	π	15 μs
CsI	5.30 eV	4.1 eV	σ	2 ns
		3.7 eV	π	1.8 μs
NH_4Cl	7.5 eV	4.91 eV	σ	5–10 ns
NH_4Br	6.4 eV	4.18 eV	π	1.4 μs
NH_4I	5.3 eV	3.74 eV		290 ns

been seen in CsI (Pooley & Runciman 1970; Falco *et al.* 1980) (see Table 4.4). The Stokes shift decreases as the halogen radius increases, as for the alkali halides with the NaCl structure. For CsI, as for some other alkali halides, the singlet and triplet states seem to have adiabatic energy surfaces which are nearly parallel, with a small energy separation; each surface has two energy minima, with a small barrier in between. Nishimura *et al.* (1995) found that the σ luminescence peak at 4.1 eV, which decays in 2 ns at low temperatures, is replaced by a band at 3.7 eV, which decays with time constants 0.6 μs and 20 μs. The intensity of the 0.6 μs component falls for temperatures above 20 K. As the temperature rises above 20 K, the 20 μs component of the 3.7 eV band is partly replaced by the 4.1 eV band, both components having the same decay time. It seems that, for the singlet adiabatic surface, excitons in the minimum which gives the 4.1 eV peak can overcome the barrier between it and the state giving the 3.7 eV peak in 2 ns. The minimum associated with the 4.1 eV transition is ascribed to the on-centre configuration, and that responsible for the 3.7 eV minimum to an off-centre configuration. The 0.6 μs component is ascribed to a singlet initial state, and the 20 μs component to a triplet state.

Below 20 K, the singlet–triplet separation is practically quenched; the barrier height between the 4.1 eV and 3.7 eV minima is much larger than the energy difference between the minima. The energy difference between the on- and off-centre configurations is small for CsI, less than for NaCl-structure halides. For CsCl and CsBr, an explanation of the lack of a σ component might be a low potential barrier between the on- and off-centre configurations, but time-resolved spectroscopy is needed to confirm this. Calculations by Fu *et al.* (1999) suggest the (100) off-centre displacement of the triplet self-trapped excitons in Cs halides is larger for smaller anions.

Ammonium halides

The ammonium halides have the CsCl structure, apart from any orientational structures of the NH_4 groups, and the self-trapped excitons thus have D_{4h} sym-

Table 4.5. *Exciton bands in alkaline-earth fluorohalides.*

System	Exciton absorption peak (eV) (Nicklaus 1979)	Luminescence peak (eV) (Ohnishi *et al.* 1994, 1996)
SrFCl	8.74, 8.86	
BaFCl		5.36, 3.37
BaFBr	7.64, 8.15	5.06, 4.10

metry again. Only one luminescence band is seen in NH_4Cl, NH_4Br and NH_4I, as seen in Table 4.4. In NH_4Cl, the luminescence is σ polarised and has a nanosecond decay time (M Itoh 1988), typical of recombination from a singlet state. The band in NH_4Br shows signs of recombination from a triplet state; that in NH_4I is of unknown character. The Stokes shifts are smaller than for the π polarised luminescence of Cs halides. Song & Williams (1993) suggest the self-trapped excitons in ammonium halides are on-centre; whether a singlet or a triplet is seen will depend on the magnitude of the exchange energy relative to thermal energies.

Alkaline-earth fluorohalides

Alkaline-earth fluorohalides are materials with a high stability used for two-dimensional X-ray image plates. BaFCl and BaFBr each have two luminescence bands which is definitely intrinsic (Table 4.5; Ohnishi *et al.* 1994, 1996). The higher absorption bands have been ascribed to the recombination of on-centre self-trapped excitons, and the lower energy bands to off-centre self-trapped excitons. The excitation spectra for the luminescence of BaFCl and BaFBr have peaks at the Cl and Br excitons, respectively. This excludes the possibility of F_2^- being involved in these self-trapped excitons. Baetzold & Song (1993) have calculated the Stokes shift of the self-trapped exciton luminescence for several configurations, and show that off-centre self-trapped excitons have energies lower than on-centre ones for any of these configurations. The largest Stokes shift is predicted for the self-trapped exciton with Cl_2^- along a [100] direction. No comparison has been made so far for which configuration has the lowest energy.

$KMgF_3$ and related perovskites ($LiBaF_3$, $KZnF_3$, $NaMgF_3$, $RbMgF_3$)

The self-trapping of excitons in $KMgF_3$ is well established by ODMR (Hayes *et al.* 1975). The luminescence band is at 3.9 eV. Alcala *et al.* (1975) showed a correlation between the polarisation of V_k centres and the luminescence emitted as they recombined with electrons. The recombination of an electron with a V_k centre gives luminescence at 2.89 eV (Koumvakalis & Sibley 1976).

It is likely that the self-trapped excitons in these materials have a [V_k:e] configuration.

MgF_2, AlF_3

The luminescence and transient optical absorption induced by electronic excitation of the rutile-structure MgF_2 is similar to that for the fluorites, except that the optical absorption is single-peaked and the decay is single-exponential over three decades. Plots of the ultraviolet and visible absorption bands versus lattice parameter for the MF_2 systems suggest that the electron- and hole-excitation bands overlap (R T Williams *et al.* 1977). The single exponential decay suggests one single deep energy minimum which, in view of the large Stokes shift, is probably of the [S_e: S_h] type. For MgF_2 at least, the self-trapped exciton formed from a free exciton does not lead to F centre production (Tanimura & Itoh 1991). The F centre production process which underlies the damage mechanism involves an excited state of the self-trapped exciton, which can be accessed from the recombination of a free electron and a self-trapped hole. The doping of MgF_2 with rare-earth ions has two effects on F centre production under γ irradiation (Nuritidinov *et al.* 1999): the rare earths provide other recombination channels, reducing F centre production, but random strains appear to enhance production. MgF_2 and AlF_3 show the same nanolithography behaviour as halides with self-trapped excitons (Broers 1988). Crystalline AlF_3 is far more resistant to electron beam damage than the thermally-evaporated amorphous form (Chen 1999).

$La_{1-x}Ce_xF_3$

Cerium-doped lanthanum fluoride is a scintillator and has been discussed as a candidate for use in the proton collider at CERN (Blasse 1994). CeF_3 has a bandgap of 10.4 eV. The luminescence of the La/Ce fluorides (Moses *et al.* 1994) shows a broad peak from about 2.75 to 5 eV, centred around 3.6 eV, with an excitation spectrum peaked at the band edge. This is attributed to a self-trapped exciton, the suggestion being that it is an electron trapped by a V_k centre. If so, there is a very large Stokes' shift, consistent with the large line-width. No defect production was reported by Moses *et al.*, since damage was not looked for in their experiments. Clearly, the effectiveness of these crystals as scintillators in a radiation environment would demand resistance to radiation.

Lead halides

The exciton peaks of $PbCl_2$ and $PbBr_2$ are in the same energy range as the $6s^2$ to 6s6p transition of Pb^{2+} (Liidja & Plekhanov 1973), suggesting a large contribution of the Pb^{2+} orbitals in the conduction band. The luminescence bands seen after excitation at the exciton peak of $PbCl_2$ and $PbBr_2$ are observed at 3.79 and 2.73 eV, with a Stokes' shift of 20% of exciton peak energy for $PbCl_2$. The bands are attributed to the recombination of a self-trapped located on Pb^{2+}. Band-to-band excitation of both $PbCl_2$ and $PbBr_2$ generates a luminescence band at 2.62 eV, with larger Stokes' shifts. The intensity of this luminescence band decreases as the temperature increases above 140 K, at which the self-trapped electron becomes unstable (Kitaura & Nakagawa 1996). This latter type of luminescence band is suggested to arise from the recombination of the self-trapped electron with a hole. In contrast to alkali halides, the first type of luminescence may be due to the recombination of on-centre self-trapped excitons, while the second type may be due to the recombination of off-centre self-trapped excitons. The off-centre self-trapped exciton might be describable as a self-trapped electron (essentially Pb_2^{3+}) with a hole on a halogen molecular ion (X_2^-).

Heavy metal fluoride glasses

Amorphous systems are discussed further in Chapter 8. Here we remark that the heavy metal fluoride glasses based on Zr or Hf show hole species analogous to those seen in crystals (Griscom & Friebele 1991; Griscom 1993). Species detected include F_2^-, Br_2^-, FCl^-, FBr^-, and F_3^{2-}. In another fluoride glass (basically $AlF_3/YF_3/RF_2$, with R an alkaline earth) no intrinsic defect production was found using a 6.42 eV ArF excimer (Hosono *et al.* 1999a). Intrinsic here means associated with Al, R, Y or F. The component molar percentages were Al(35%), Ca(26%), Sr(14%), Ba(9%), Mg(7%), Y(5%)). However, defects were found associated with oxygen (possibly a hole centre absorption at about 5 eV) and with P (conceivably a P-related E′ centre absorbing at 6.9 eV). We remark that several O-related hole centres have been identified in the X-ray storage phosphor BaFBr (see, e.g., Schweitzer & Spaeth (1999)).

4.4 Self-trapped excitons in oxides

4.4.1 Self-trapped excitons in oxides with closed-shell cations

Criteria for self-trapped excitons

Whether or not excitons are self-trapped in oxides with closed-shell cations is controversial. Partly this is because excitons have relatively short lifetimes, and it is possible that the state of lowest energy is not reached for kinetic reasons. The emission and luminescence spectra alone do not determine the presence and absence of self-trapped excitons unambiguously. Even the information derived from ODEPR studies is insufficient to be sure of the atomic and electronic structures of the self-trapped excitons. Some guidelines for determining whether excitons are self-trapped or not are these:

(1) Free excitons have a relatively long lifetime. We might assume that self-trapping does not happen when resonant free-exciton luminescence is seen at the photon energy of the exciton peak, or separated from it by a phonon energies. However, self-trapping might merely be inhibited, or self-trapped excitons might be formed and coexist with free excitons. There might be a delay time before free excitons convert to self-trapped excitons, as for alkali iodides (Section 3.4).
(2) One indicator of exciton self-trapping is the appearance of a broad luminescence band whose excitation spectrum either matches the exciton absorption peak or has a cutoff at the band edge. In materials which lack a clear exciton absorption peak, determination of the band edge is not feasible. In them, especially if electrons and holes are highly mobile, the luminescence band could arise from electron–hole pair recombination at a defect or impurity centre.
(3) When a triplet ODEPR signal is seen, tagged by the broad luminescence band emitted following excitation at the exciton peak or at the bandgap energy, this indicates the presence of a triplet state which may be that of a self-trapped exciton. Further guidance as to whether or not the signals are from a self-trapped exciton can come from observations of fine and hyperfine structure.

Band structures: Optical absorption in closed-shell oxides

The oxides which we shall discuss range from highly ionic (e.g. ZnO and MgO) to less ionic, such as SiO_2. Much of the information on their electronic structures comes from theory, since there is far less experimental information (essentially none from techniques like cyclotron resonance, for instance) than for the semiconductors.

ZnO is ionic (Liang & Yoffe, 1968), with a direct bandgap. The band structure is similar to that of alkali halides: the top of the valence band consists of the anion (oxygen) p orbitals and the bottom of the conduction band consists of cation (Zn) s orbitals. The contribution of the d orbitals is greater in MgO and CaO. MgO has a direct bandgap (Pantelides *et al.* 1974), whereas CaO has an indirect bandgap, since the d orbitals lower the conduction band minimum at the X point (Seth & Chaney 1975). In MgO, and also in alumina, the valence bands have separate regions, one essentially an O 2p part, the other mainly Mg 3s. The top of the valence band in α quartz comprises non-bonding oxygen orbitals (Chelikowsky & Schlüter 1977). The Si–O bonding orbitals appear about 5.5 eV below the top of the valence band, and Si–O antibonding orbitals form the conduction band. Al_2O_3 is also an indirect bandgap material.

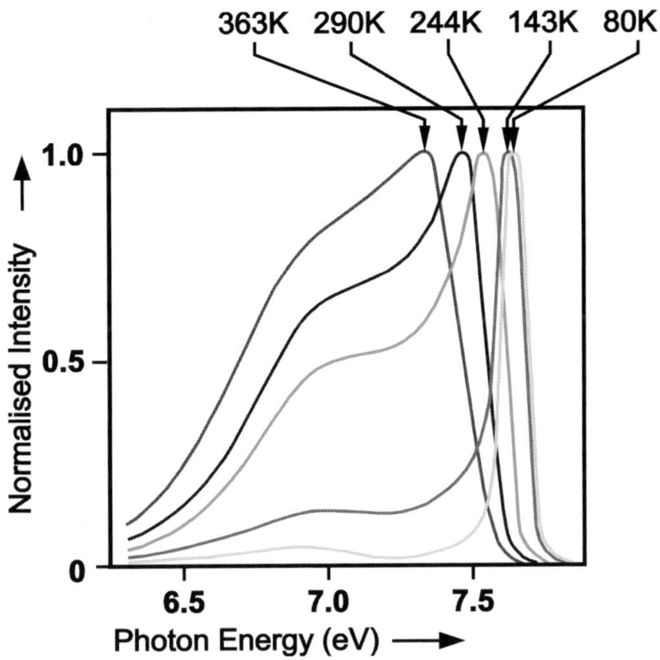

Fig. 4.16. The cathodoluminescence spectra of MgO at various temperatures (after Valbis *et al.* (1975) and Rachko & Valbis (1979)). The maxima fit the approximate formula $E_{max} = 7.65\ \text{eV} - 0.184(T/300\ \text{K})^2$.

Again, the top of the valence band consists of the O non-bonding orbitals (Ciraci & Batra 1983). Other oxides with closed-shell cations are expected to have band structures similar to those mentioned of comparable ionicity.

Fundamental optical absorption

Of the oxides which we discuss, the exciton absorption peak can been seen only in BeO (Roessler *et al.* 1969), ZnO, and MgO (Roessler & Walker 1967). For other oxides, the optical absorption spectra do not have a resonance structure, but show a slow increase toward a plateau. The absorption edges of complex oxides are determined by the Urbach analysis (Section 1.4).

Free-exciton luminescence

Free-exciton luminescence has been seen in ZnO (Nikitenko & Tereshchenko 1979), BeO (Ivanov *et al.* 1988), and MgO (Valbis *et al.* 1975; Rachko & Valbis 1979). In BeO and MgO, both a sharp luminescence band and a broad luminescence tail are observed at liquid helium temperatures. The sharp band in MgO becomes weaker and the broad tail stronger at higher temperatures, as shown in Fig. 4.16. It has been suggested that the sharp band is the free-exciton luminescence, and the broad tail luminescence from a self-trapped exciton,

Table 4.6. *Exciton data for closed shell oxides.*

Material	Exciton in absorption	Luminescence energy	Remarks
MgO	6.9 eV	6.9 eV	Free-exciton luminescence
Al_2O_3	8.95 eV	7.5 eV (20 K)	20 ns (80 K) lifetime luminescence and excitation spectra
$YAlO_3$ (YAP)	8.0 eV	5.9 eV	Optically detected magnetic resonance
$Y_3Al_5O_{12}$ (YAG)	7.0 eV	4.9 eV	
$Y_3Ga_5O_{12}$ (YGG)	6.0 eV	4.1 eV	
$Lu_3Al_5O_{12}$ (LuAG)	7.3 eV	5.1 eV	
Y_2O_3		3.45 eV	Optically detected magnetic resonance
Sc_2O_3		5.55 eV	Optically detected magnetic resonance

with Stokes' shift 0.75 eV. This interpretation would imply a potential barrier for self-trapping in MgO.

Broad luminescence bands ascribed to the self-trapped exciton

Apart from BeO, ZnO, and MgO, all the oxides shown in Table 4.6 emit only broad luminescence bands. The characteristics of these luminescence bands (Murk *et al.* 1994, 1995; Murk 1997) are these. First, the excitation spectra are close in energy to the exciton peak determined from the merging point of the Urbach tail (Fig. 4.17). Secondly, the luminescence is fairly strong, suggesting it might be intrinsic. Thirdly, the Stokes' shift is larger in the less-ionic (more covalent) oxides. Murk (1997) also found luminescence bands with higher Stokes shifts and ascribed them to another form of self-trapped exciton. However, the excitation spectra of these luminescence bands do not always coincide with the absorption edge. Alternatively, these bands might be ascribed to self-trapped excitons perturbed by defects or impurities.

Relaxation energy of excitons

The relaxation energy of excitons in MgO (Shluger *et al.*, 1991c) has been estimated as 0.3 eV, and that in Li_2O as 0.7 eV (Shluger & Itoh 1990). In these highly ionic oxides, the self-trapping distortion involves only Jahn–Teller and

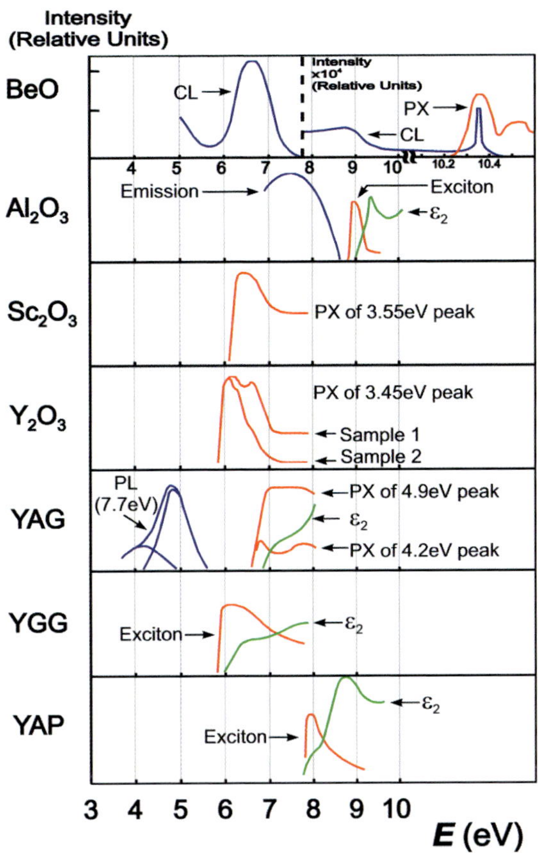

Fig. 4.17. Typical optical absorption and luminescence spectra and the excitation spectra of the luminescence in oxides. PX(E eV) (red line) indicates the excitation spectrum of the band at E eV; CL and PL (blue lines) indicate cathodoluminescence and photoluminescence, respectively; ϵ_2 (green line) corresponds essentially to the absorption spectrum.

breathing modes, with little off-centre displacement of the oxygen on which the hole is localised. Theoretical estimates of the relaxation energy of an exciton in SiO_2 are of order 2 eV for beta crystoballite (Shluger 1988; Shluger & Stefanovich 1990) and 3.8 eV or 2.6 eV (depending on the cluster used) for quartz (Fisher *et al.* 1990a,b). The experimental Stokes' shift is about 5.3 eV for quartz (see below). The self-trapping distortion around the exciton involves a substantial displacement of the oxygen on which the hole is mainly localised.

A comparison of these systems suggests that the relaxation energy increases with increasing covalency. This is partly because of the ways that bonding patterns change on excitation, and partly because the directed bonds in more covalent crystals lead to more open structures. The exciton–phonon coupling is also

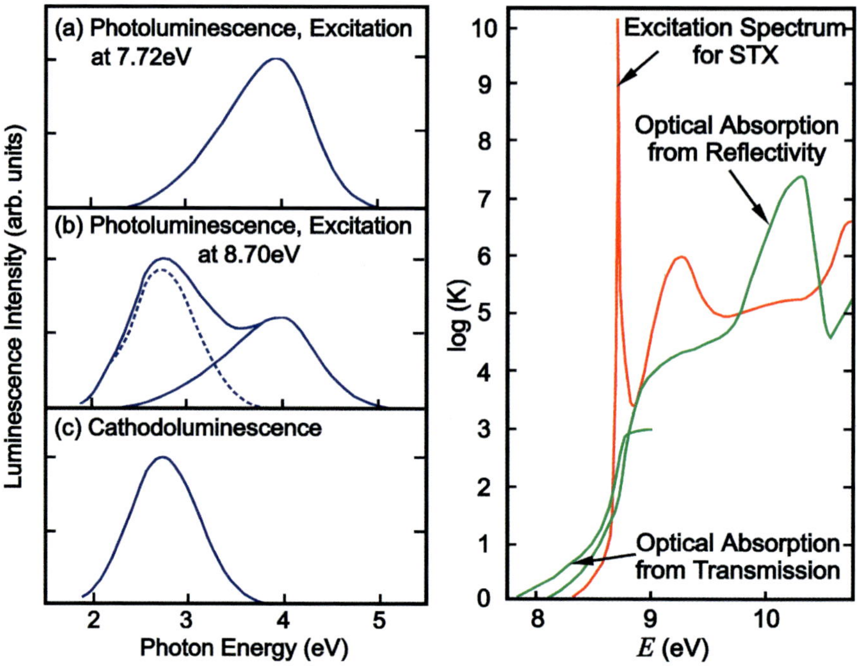

Fig. 4.18. The optical absorption spectrum, the photo and cathode luminescence spectra and the excitation spectra for the intrinsic luminescence in SiO_2.

weaker in ionic oxides than in alkali halides. One reason that the interaction is strong in the halides is that a localised hole can form a halogen molecular ion, X_2^-, with a substantial binding energy. The corresponding oxide species would be O_2^{3-} in a fully ionic oxide, and there is a strong Coulomb repulsion.

Self-trapped excitons in SiO_2: Experiments

The self-trapped exciton in alpha-quartz has been studied in most detail among the oxides with closed-shell cations. ODEPR work by Hayes *et al.* (1984b) showed clearly that the initial state of the 2.8 eV luminescence (Griscom 1979; Trukhin 1979, 1982) was a triplet. The excitation spectrum for the 2.8 eV peak starts to increase near the absorption edge (Trukhin 1980; C Itoh *et al.* 1989), and a sharp threshold has been found near the band edge at 8.1 eV, indicating a Stokes' shift as large as 5.3 eV. Figure 4.18 shows the optical absorption spectrum, the photo and cathode luminescence spectra and the excitation spectrum of quartz. The large Stokes' shift of about 5 eV agrees with theory, and indicates a large self-trapping distortion. Observations of the macroscopic volume expansion due to exciton self-trapping (Tanimura *et al.* 1983) show that the volume change per self-trapped exciton is of the order of the volume per

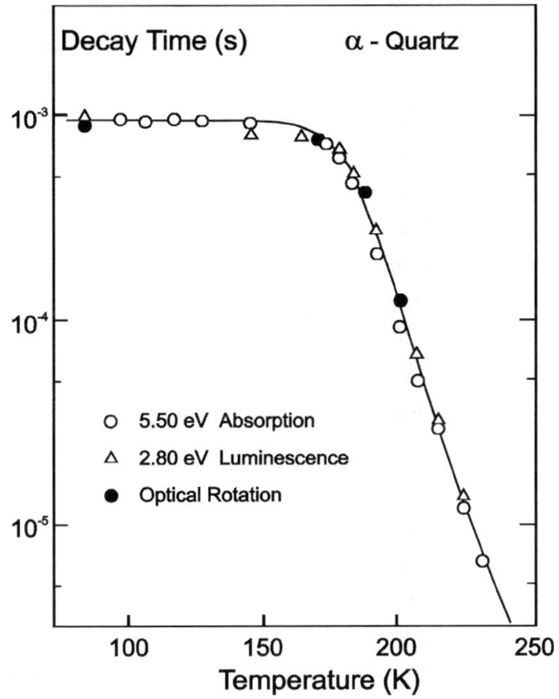

Fig. 4.19. The temperature dependence of the lifetime of the self-trapped excitons in α quartz as obtained by measurements of the decays of the intrinsic luminescence, of the volume expansion, and of the optical absorption band at 5.0 eV due to optical excitation from the lowest state of the self-trapped excitons to higher state.

molecular unit in the lattice. Figure 4.19 shows the observed temperature dependence of the lifetime of self-trapped excitons, as obtained from the decay of the intrinsic luminescence, the volume expansion as detected by optical rotation, and the optical absorption band at 5.0 eV from the self-trapped exciton to higher excited states.

Atomic and electronic structures of self-trapped excitons in SiO_2

Detailed modelling of the self-trapped excitons in crystalline SiO_2 (Fisher *et al.* 1990a; Shluger 1988; Shluger & Stefanovich 1990) shows that quantum-chemical methods predict correctly many of the observed features, including lattice relaxation. The results support a picture of the self-trapped exciton as a hole primarily localised on an oxygen, this oxygen being displaced substantially from its perfect lattice site, and an electron primarily localised on a neighbouring silicon (Fig. 4.20). In parallel with similar ideas for As_2Se_3, the presence of a hole in an oxygen non-bonding orbital weakens the Si–O bond to localise the electron and hole on the dangling bonds of Si and O, respectively.

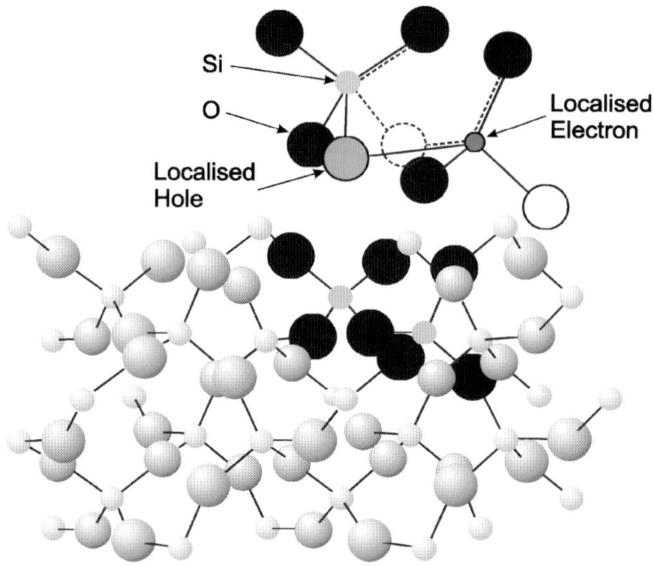

Fig. 4.20. The self-trapped exciton in α-quartz (after Fisher *et al.* (1990b)). The hole is primarily localised on the oxygen, which is displaced, as shown in the inset; the electron is mainly localised on a neighbouring silicon.

We now outline the predictions of one of these models (Fisher *et al.* 1990a,b), which used Hartree–Fock methods for atoms in an embedded cluster which was relaxed to equilibrium. The cluster corresponded to alpha-quartz. Similar conclusions for cubic crystoballite were reached by Shluger (1988) and Shluger & Stefanovich (1990), using semiempirical methods. Fisher *et al.* made predictions for the ground state and a number of spin-triplet $2p^5 3s$ excited states from the $2p^6$ ground stable for two clusters, one centred on Si, the other centred on O. A significant number of comparisons with experiment follow.

First, the observed *Stokes' shift* (the difference between the absorption energy of 8.7 eV and the emission energy of 2.8 eV) is 69% of the perfect crystal bandgap. Theory gives 74% (Si-centred cluster) and 89% (O-centred cluster). Secondly, the *radiative lifetime* observed (of order 1 ms) is typical of a spin-forbidden transition; indeed, this value would be given by many models of the self-trapped exciton. Thirdly, magnetic resonance experiments measure the components of the *fine-structure tensor.* The cluster models (both Si-centred and O-centred) predict values in reasonable agreement with those measured. Both clusters correctly give the observed principal axis of the fine-structure tensor, which is primarily determined by the anisotropic occupancy of the p orbitals on the displaced oxygen. Fourthly, the *polarisation of the blue luminescence* (C Itoh *et al.* 1988) provides a strong test, being +0.48 to +0.65 polarised parallel to the *c* axis. Theory predicts a polarisation of +0.44, and shows that this results from the transition being primarily the charge transfer of an electron from an Si to the oxygen. Fifthly, the *transient volume change* can be predicted from the virial of the defect forces. The predictions are of the order of 1–2 molecular volumes, in line with the observed value of about 1 molecular volume.

Approximate estimates of excitation from the long-lived triplet excited state give both electron excitations (4.9 eV and 7.3 eV for the Si- and O-centred clusters, respectively) and hole excitations (6.1 eV and 5.7 eV for the Si- and O-centred clusters, respectively). These energies are similar to the observed values of 4.2 eV and 5.2 eV, though it is not possible to decide whether these are electron or hole excitations.

Self-trapped excitons in Al_2O_3

It has been suggested that the broad luminescence band in Al_2O_3 with a Stokes' shift of 1.5 eV arises from a self-trapped exciton (Kulis *et al.* 1986; Murk *et al.* 1995; Kirm *et al.* 1999). The quantum yield is high, and the excitation spectrum matches the absorption edge (Fig. 4.20). Studies of free-electron lifetimes for a photon energy of 2 eV (Daguzan *et al.* 1994) were unable to detect the lifetime, since the photon energy is larger than the relaxation energy; it would be of interest to investigate the lifetime of the free electrons with a photon energy smaller than the relaxation energy. Although the evidence for self-trapping is incomplete, the experimental results on excitation spectra and the trends with ionicity support it.

Free and self-trapped excitons in MgO

Even though free-exciton luminescence is observed in MgO, self-trapping of excitons is likely, according to Rachko & Valbis (1979) (Fig. 4.16). The free-exciton luminescence is observed only below 80 K; broad luminescence bands with a Stokes' shift of 0.6 eV appear above this temperature. This suggests that the self-trapped state of an exciton in MgO is stable, but that there is a potential barrier for self-trapping. If the broad luminescence band arises from the self-trapped exciton, the localisation energy of an exciton (essentially half the exciton bandwidth, which is, of course, not the same as the valence bandwidth) should be smaller than the relaxation energy, approximately a half of the Stokes' shift. The long lifetime of the free electrons measured with 2 eV photons (Daguzan *et al.* 1994) can also be reconciled with the shallow self-trapped state of excitons.

Self-trapping of excitons in more complex materials

For more complex structures, there are other issues. Suppose we know that, for the aluminosilicate with the smallest unit cell (perhaps corresponding to

$(SiO_2)_n(Al_2O_3)_m$ with $n = m = 1$; let us suppose that there is a structure for which all Si are equivalent, and all Al are equivalent), self-trapping is marginal. The energy to localise the exciton or carrier in one unit cell (or perhaps in one silica unit) is similar to the energy gain from lattice distortion. What happens when we go to a more complex structure with the same components, e.g. to a zeolite? The relaxation energy may not change much: the force constants and polarisabilities will often have values depending little on n or m, and the forces driving the displacements will often be much the same as for the simpler structure. However, the complex structure will have many inequivalent Si sites and many inequivalent Al sites. The energy differences for carriers on these sites will favour certain of them, making localisation easier. Usually the band structure will split into narrower bands. Overall, self-trapping will be easier. These arguments apply equally to amorphous forms compared with crystalline forms, and indeed holes are observed to self-trap in a-SiO_2, whereas the situation is marginal at most in crystalline forms (X Zhang *et al.* 1994).

Self-trapped excitons are well established in mica, a silicate mineral with a layered structure composed of linked SiO_4 tetrahedra (M Itoh 1995). A luminescence band at 5.46 eV is observed in synthetic fluor-phlogopite crystals following excitation above the fundamental absorption edge and in a peak assigned to an exciton. Two decay components have been observed, with decay times of 1.8 ns and 0.5 μs.

Another class of complex systems includes layered systems. If the carrier is confined to a less-polarisable layer, then proximity to a more-polarisable layer will increase the energy gain on localisation; this can be regarded as a contribution of the classical image interaction. Such extra stabilisation may occur in the Si/SiO_2 system. In layered systems, the order of the atomic planes can be chosen to give electric fields across a layer. In such cases, the field will tend to separate electron and hole in an exciton, and this will also modify self-trapping.

Other complex structures involve ions which can take more than one valence, especially transition-metal cations. These we discuss separately.

4.4.2 Self-trapped excitons of oxides with open-shell cations

When there are open-shell cations, questions of definition arise. If we take (say) CoO and excite the Co^{2+} to a higher crystal field state, this might be described as a self-trapped hole, namely Co^{3+}, which traps an electron in the upper state. But is this an exciton? What about the charge-transfer transitions which are so often the basis of colour, and where again an excitonic description is possible?

Our view is pragmatic. We shall consider these systems whenever there is a materials modification process involving the excited state, and we shall use parallels with the closed-shell cation systems so long as they are useful. Thus, self-trapped carriers (electrons or holes) appear to be common, and hold many analogies with halides; self-trapped excitons show less similarity, and we shall not force the comparison. When there are open-shell cations, non-stoichiometry is common, and high defect densities are to be expected (see, e.g., Fig. 3.10 of Hayes & Stoneham (1985)). Defect-associated trapping will occur, even if self-trapping is strictly absent.

4.5 Self-trapped excitons in crystalline semiconductors

The self-trapping of excitons is neither reported nor expected in the well-known III–V or Group IV semiconductors, either in zinc-blende or wurzite structures. Self-trapping is observed in chain and layered structures, namely in Se, As_2S_3 and As_2Se_3, in both crystalline and amorphous forms. This stems partly from differences in the phonon coupling of three-dimensional and low-dimensional excitons (Section 3.3). The relation between the self-trapping of excitons and photo-induced processes of amorphous materials is treated in Chapter 8.

Band structures

S, Se, and Te have six valence electrons per atom. In the helical chain structures, two of these electrons form a bond, and the other four occupy non-bonding orbitals (Harrison 1989). The top of the valence band consists of non-bonding p orbitals (Joannopoulos *et al.* 1975), just as in SiO_2. Transport within chains dominates, and interchain bonding can be regarded as a perturbation. As_2S_3 and As_2Se_3 form similar linear chains, but adjacent chains are connected by bonds to form a layer structures shown in Fig. 4.21. The valence bands of As_2S_3 and As_2Se_3 consist of antibonding orbitals of S and Se, respectively. The top of the valence band of elemental Se consists almost entirely of non-bonding Se orbitals; the bonding orbitals are lower in energy (Althaus *et al.* 1978). Se has an indirect bandgap of 1.86 eV. There is a 0.72 eV luminescence, with absorption at 0.6–1.6 eV from the initial state of the luminescence. Most chalcogenides have indirect bandgaps, with direct bandgaps a few electron volts larger. For As_2Se_3, which has an indirect bandgap of 2.15 eV, there is luminescence at 1.2 eV; ODEPR has been done on the luminescing state with optical absorption at 0.4–2.0 eV.

Are excitons self-trapped?

Self-trapping is suggested in those semiconductors showing broad intrinsic luminescence bands with large Stokes' shifts. The excitation spectra of the

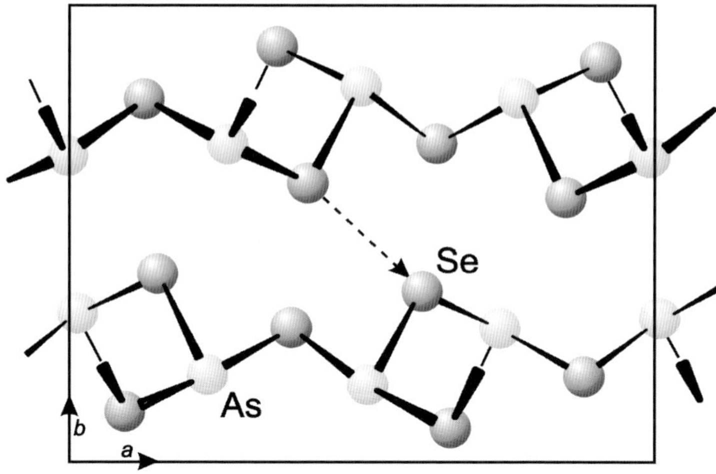

Fig. 4.21. Atomic structure of As_2Se_3. In the exciton, an electron is transferred (as shown by the arrow) leaving a hole in an antibonding orbital, which enhances bonding.

luminescence bands match the optical absorption spectra near the band edge; this is shown for Se in Fig. 4.22. The excitation spectrum shows peaks at both the indirect exciton and the direct exciton peak. That for the indirect peak resembles that for SiO_2 (Fig. 4.18). The luminescence is almost certainly intrinsic, partly because the peaks of the excitation spectrum match those for band edge absorption, and partly because the quantum yield is close to unity (Ristein & Weiser 1986). Applying an electric field reduces the luminescence yield of As_2Se_3 (Ristein & Weiser 1986) and of Se (Lundt & Weiser 1983), and enhances the photoconductivity. For a given electric field, the enhancement of photoconductivity increases with temperature, as expected for field ionisation of self-trapped excitons (cf. Section 2.6)

Studies of EPR tagged to the luminescence bands of As_2Se_3 show that the initial state of the luminescence is a triplet state. The centre responsible is highly localised and anisotropic, indeed almost uniaxial. Its principal axis lies along a newly-formed As–Se axis after relaxation. Optical absorption spectra have been found from the initial state of the luminescence for Se (Chen & Kastner 1986) and for As_2Se_3 (Robins & Kastner 1984). These experimental results are all analogous to those for SiO_2, and support the view that excitons are self-trapped in Se and in As_2Se_3.

Atomic structure of self-trapped excitons

A proposed atomic structure of the self-trapped excitons in As_2Se_3 has been given by Ristein & Weiser (1988; see Fig. 4.21) and Robins & Kastner (1987).

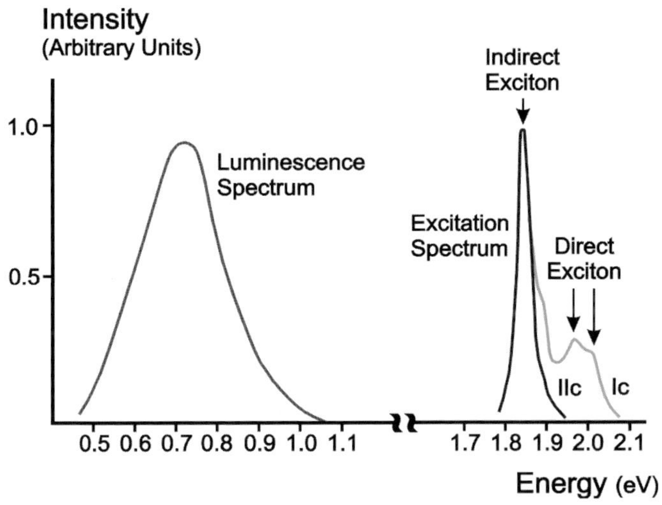

Fig. 4.22. The optical absorption spectrum and the excitation and emission spectra of the intrinsic luminescence of Se (Lundt & Weiser 1983).

First, hole localisation strengthens the bonding between two neighbouring Se atoms (dotted arrow in the figure), since the top of the valence band consists of antibonding orbitals. This reduces the Se–Se distance, but also makes an As–Se distance longer, which lowers the energy of the antibonding orbitals which form the bottom of the conduction band. Thus, the electron of the self-trapped exciton is localised on the antibonding orbital and the hole is localised on the antibonding orbital, displaced into the forbidden gap by the distortion. EPR data show that the principal axis deviates only by 10° from the axis of the As–Se bond, and that the electron interacts with one As nucleus. The large ratio of the axial and off-axial zero-field splitting parameters, $D/2E$, for the self-trapped excitons in As_2Se_3 (15) and in quartz (7.0) indicates that the structure is almost exactly axial.

Localisation of more than one hole (multiholes)

There are two cases when we need to consider the localisation at the same position of more than one carrier of the same sign (multiholes or multielectrons). In the first case, the multicharge is localised sufficiently long (of order 0.1 ps) for an atom to be displaced by Coulomb repulsion. In the second case, the additional lattice distortion from the extra carrier is sufficient to overcome the Coulomb repulsion energy.

Localisation leading to ion displacement was first suggested by Varley (1954) to explain defect formation in the bulk of alkali halides. His idea was that two holes h^+ might localise on a halogen ion X^-, giving a halogen cation $[X^-,h^+h^+]$, or X^+. Coulomb forces will eject the positively-charged ion from the site which favours anions. Later, Feibelman & Knotek (1978) used a distinct but similar mechanism to explain ion emission from surfaces of insulating materials and from metal–adsorbate systems. As we shall see, there is good experimental evidence for these surface processes (Chapter 9), but there is no conclusive evidence for a Varley mechanism for the bulk.

The localisation of two holes by virtue of the lattice relaxation is called negative U (Anderson 1975), which appears to be reasonably common in semiconductors. How it works can be seen by a simple example (Hayes & Stoneham 1985, p 401). If there are defect forces F acting on a lattice, and if the restoring elastic force constant is K, then there will be a relaxation energy $F^2/2K$. Suppose we have a defect which could have 0, 1, or 2 electrons, and the defect forces F in these cases are 0, f, $2f$, respectively. We shall ignore changes in the force constant K, although these may reinforce the effect in real systems. The relaxation energies in these three occupancies are 0, $f^2/2K$, $(2f)^2/2K$, respectively. This means that these relaxation energies favour charge disproportionation, for there is energy released in the transition from two defects each with one electron (relaxation energy $2f^2/2K$) to one defect with no trapped electrons and the other with two electrons (relaxation energy $0+4f^2/2K$). This model description is a reasonable first description of the vacancy V in Si and its dependence on the number of electrons associated with the t_2 states constructed from the vacancy dangling bonds. The V^+ state is metastable, so two V^+ centres are higher in energy than V^0 and V^{2+} (Baraff & Schluter 1978). In Chapter 9, we shall discuss the possibility that the localisation of two holes in ionic oxides on surfaces can cause atomic emission.

Localisation of multiexcitons

When more than one excitation takes place, it can be possible to localise the excitation energy from the several excitations in the same place. Multiple excitations are more likely to give displaced atoms, for two main reasons. First, there is more energy available. Secondly, the excitation will affect interatomic forces locally, and may make displacements easier. There are several ways to generate such localised excitons, such as photoexcitation of the exciton band perturbed by self-trapped exciton (Section 5.2.1), or by electronic excitation by particles with high stopping power. Increasing the concentration of excitons by intense photoexcitation may not cause condensation resulting in defect formation, unless the exciton radius is sufficiently small that localisation of two holes is effective. The creation of dense excitons by energetic heavy ions, able to ionise all molecules along a track, is discussed in Section 11.6. Clearly, strong electron–lattice coupling is necessary for defect production. Biexcitons, multiple bound excitons, and even an exciton gas, are well documented in semiconductors, yet they do not seem to generate damage.

Summary

Halide crystals with closed-shell cations (NaCl, CaF_2) Self-trapped holes (STH) give charge localisation, usually as the halogen molecular ion X_2^-; the electron is usually not self-trapped. The self-trapped exciton (STX) provides energy localisation. Usually X_2^- plays a critical role, even when the STX is not of the form [V_k:e]. The STX can show incipient defect formation: its off-centre form is evolving towards an F–H pair. The STX can be classified as Types I (all singlets; triplet NaBr, NaI), II (triplet NaCl, KI, RbI), or III (most other triplets, including RbI).

Amorphous halides and halide glasses Charge and energy localisation occur as in their crystalline counterparts. The halogen molecular ion X_2^- is seen in solution and in glasses. Presumably electrons can be trapped at special sites.

Halide crystals in which the cation can change charge state (AgCl, $PbCl_2$) Charge localisation: the STH can be on the cation (AgCl), and the self-trapped electron (STE) is possible ($PbCl_2$).

Amorphous halide or halide glass in which the cation can change its charge state Very little information is available; presumably behaviour is similar to the crystalline form.

Oxide crystals with closed shell cations (MgO, ZnO, Al_2O_3, usually SiO_2) STH, STE either do not exist at perfect bulk sites, or have such low stability that the charge is mobile and moves via a non-self-trapped state. Charge localisation is associated with defects or impurities. Energy localisation can occur: the STX is unambiguous in quartz and other silicas. The STX structure suggests incipient defect formation (O moves towards a configuration of an O vacancy and O interstitial). In Al_2O_3 and in less ionic oxides, STX are probably the cause when luminescence is observed with a large Stokes' shift.

Oxide glasses with closed-shell cations, like borosilicate or aluminosilicate glasses STH, STE can be stabilised at special sites; STX provides energy localisation.

Oxide crystals with cations which can have several charge states ($Fe_{1-x}O$ and other transition metal oxides) STH, STE can both exist. The STX corresponds to charge transfer. Behaviour will depend strongly on the degree of non-stoichiometry.

Oxide glasses which include transition metals Charge localisation as STE, STH simply corresponds to charge state changes.

Semiconductor crystals (III–V, Group IV) with relatively wide bands and small gaps No STE, STH, or STX are reported in Group IV, III–V, II–VI semiconductors, but exciton self-trapping (STX) is found for Se and chalcogenides.

Amorphous semiconductors STE or STH may be limited to special sites.

Chapter Five

Local lattice modification by electronic excitation of halides

In Chapter 4, we discussed the relaxation of excitons and of electron–hole pairs to form self-trapped excitons. In halides, these relaxations may involve asymetrical atomic displacements, including the $[S_e, S_h]$ self-trapped exciton. In Chapter 5, we discuss how these asymmetric displacements relate to the asymmetric displacements which lead to defect formation. In the most important case, the displacement creates a pair of neutral defects on the halogen sublattice. This defect pair comprises a halogen atom vacancy (F centre) and a halogen atom interstitial (H centre). The F–H pair is sometimes referred to as a Frenkel pair, but this is misleading, since the term is rightly used for a defect pair comprising a halogen ion vacancy (α centre, X^- missing) and a halogen ion interstitial (I centre, X_i^- rather than X_i^0, differing in charge, site, and symmetry from the H centre).

Almost all of the discussion in this chapter relates to the alkali halides and alkaline-earth fluorides, in which excitons can be self-trapped. The arguments would apply in similar ways to many other halides, but not necessarily to those for which self-trapping does not occur (like the thallous and cuprous halides). Since the self-trapped exciton has a substantial latent energy, only a small energy is needed to transform a self-trapped exciton into an interstitial-vacancy pair on the halogen sublattice. The exciton–phonon coupling which causes exciton self-trapping also assists the creation of defect pairs. Likewise, a closely related process leads to defect formation following the recombination of an electron–hole pair to an exciton. We shall describe the defect formation process associated with the relaxation of excitons [eh] or of correlated electron–hole pairs [e+h] as the *excitonic mechanism.* This excitonic mechanism is not the only process of defect formation observed. Other mechanisms are observed following high-density excitation, core excitation, and defect-related charge separation. We shall concentrate on the microscopic description of these processes in alkali halides. These models are often of use for understanding similar processes in other materials.

The idea that defects are produced from electron–hole pairs or from excitons was first suggested by Pooley (1965). He argued that energy emitted in the non-radiative recombination of an exciton could be converted to energy for defect production (essentially the local heating or phonon kick mechanism, described in Chapter 7). Aside from problems underlying the local heating mechanism, the main difficulty of the Pooley mechanism, as first suggested, is that it predicts the generation of α–I pairs. However, most of the defect pairs generated by ionising radiation at low temperatures are neutral F–H pairs, which may be transient or stable. The transformation of an exciton into a defect pair is now considered to occur by the system following an adiabatic potential energy surface for the excited state (a local excitation mechanism, as described in Chapter 7).

The creation of defects in halides originating from electron–hole pairs is well known; reviews include those by Sonder & Sibley (1972), N Itoh (1982), Hayes & Stoneham (1985), N Itoh & Tanimura (1990) and K S Song & Williams (1993). The process can be very efficient, and is often as efficient as the formation of self-trapped excitons. At low temperatures, the primary defects created by the excitonic mechanism are F–H pairs, which may be transient or stable. Most of the defect pairs at low temperatures are transient; stable α–I pairs are more numerous than F–H pairs.

Table 5.1 summarises the yields of stable and transient pairs produced by ionising radiation and some of the other significant features of defect formation in halides. At temperatures too high for H centres to be stable, the interstitials are stabilised by forming complexes; this allows the formation of the F centres.

5.1 Excitonic mechanisms for defect formation

5.1.1 Adiabatic potential energy surfaces and relaxation channels

In certain alkali halides (those with large Rabin–Klick parameters, as defined in Section 4.3.2), the symmetric $[V_k:e]$ configuration is unstable against translation of the halogen molecular ion (X_2^-) along its molecular axis. In alkaline-earth fluorides, the analogous instability also involves rotation. The self-trapped exciton then adopts an off-centre configuration. This we might regard as the nearest-neighbour pair of an interstitial halogen atom and a halogen atom vacancy. At the least, it is a first stage in the production of a separated interstitial and vacancy pair. Figure 5.1 shows the configurations of defect pairs as the interstitial halogen atom is displaced along its molecular axis. The adiabatic energy surface has an energy barrier between the different possible spacings, for bond switching is involved. We

Table 5.1. *Defect formation in NaCl-structure alkali halides by ionising radiation.*

The rows are in order of increasing Rabin–Klick parameter. Note that the yields follow almost exactly the same order (LiF is marginally out of place, RbCl more so). There is some information for other halides, e.g. yield and luminescence do not anticorrelate for CaF_2.

Substance	Rabin–Klick parameter	Yield per 1 keV (T for transient defects; stable, otherwise)	Activation energy (eV) for defect formation	Do defect yield and luminescence anticorrelate?
LiI	0.05	0.9×10^{-3}		Yes
LiBr	0.13			Yes
NaI	0.18	0.65×10^{-4}		Yes
LiCl	0.18	1.9×10^{-3}		Yes
NaBr	0.28	2.7×10^{-4}		Yes
KI	0.33	1.4×10^{-4}	0.14[c]	Yes
NaCl	0.36	0.64×10^{-2}	0.07[d]	Yes
LiF	0.39	0.395		Yes
RbI	0.40	0.7×10^{-4}		Yes
KBr	0.45	0.32 (T: 10–50[a])	0.03[e]	No
RbBr	0.56	0.27		No
KCl	0.60	0.45 (T: 10[b])	0.07[f]	No
NaF	0.70	0.22		No
RbCl	0.71	0.04	0.06	No
KF	1.07	0.15		No
RbF	1.22	0.42		No

Notes:
[a] Kondoh *et al.* 1972; [b] From the data of R T Williams *et al.* (1978), assuming 20 eV per electron–hole pair; [c] Guillot *et al.* (1975); [d] Sonder (1975); [e] Hughes *et al.* (1967); [f] R T Williams *et al.* (1978).

shall discuss later the heights of the potential barriers for defect formation and recombination. The energy per defect pair increases as their separation increases, partly due to the elastic interaction between the components of the defect pair (Hayes & Stoneham 1985). Those defect pairs with separations less than some critical distance are unstable, and will show geminate recombination (Section 2.6).

Adiabatic potential energy surfaces

The energy surfaces which we must consider are these. First, we will have the adiabatic energy of the ground-state configuration as the ions are displaced (this is also the recombined state). Secondly, we will have a number of excited

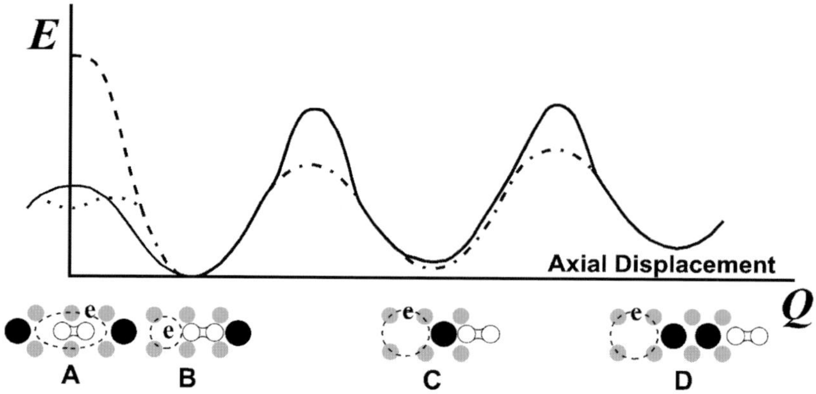

Fig. 5.1. Atomic configurations of the self-trapped excitons (A is the on-centre [V_k:e] configuration, which is either metastable or unstable here; B corresponds to Type II or III excitons) or defect pairs (C shows next-nearest-neighbour pairs and D shows second-neighbour pairs) in alkali halides. Also shown is a schematic configuration coordinate diagram for the transformation of a self-trapped exciton to an F centre–H centre pair.

states. The lowest excited adiabatic energy surface includes that corresponding to the free exciton, that corresponding to the self-trapped exciton, and that corresponding to a separated defect pair. The self-trapped exciton can, of course, occur in several forms, depending on spin state, electronic orbital state, and on whether or not it can exist in a one-centre metastable form. Figure 5.2 shows a possible set of energy surfaces. In the figure, the defect pair configuration is not luminescent, and can be regarded as a metastable state. The energy surfaces intersect. Switching of the adiabatic potential surface for the ground and lowest excited states is predicted by several calculations for alkali halides (K S Song & Baetzold 1992, Puchin *et al.* 1993). The energy surfaces of the singlet and triplet excited states are separated by about 0.1 eV.

Possible relaxation channels

There are at least two channels for relaxation from a free exciton to a defect pair. The first is direct conversion (the *direct channel*), bypassing at least the lowest states of the self-trapped exciton; the second is *conversion via the ground state of self-trapped exciton* (the self-trapped exciton, or STX, channel). The STX channel itself has two alternatives. For the *dynamic STX channel*, the system transfers from one adiabatic surface to another at the cross-over energy, when the two sets of vibronic states are strongly mixed. This is the case expected for relatively slow vibrational cooling (the underdamped case). For the *thermal STX channel*, vibrational cooling is very rapid (the overdamped

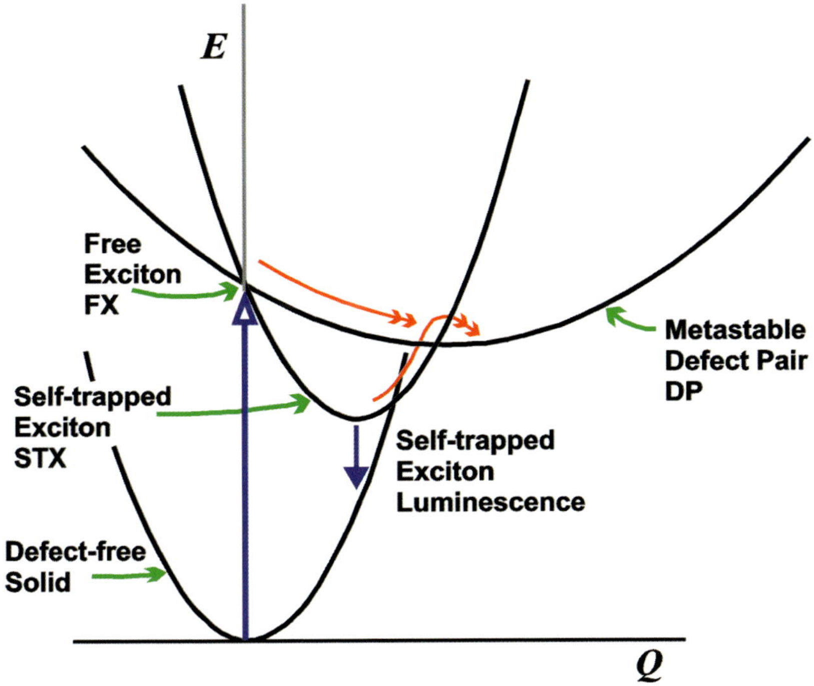

Fig. 5.2. Schematic configuration coordinate diagram for the free exciton (FX), self-trapped exciton (STX) and the defect pair (DP). As shown, the defect pair configuration is not luminescent, and can be regarded as a metastable state.

case), and the self-trapped exciton reaches its lowest vibronic states for its initial adiabatic surface, and there is then thermal activation over a barrier to the other surface.

Direct channel

A direct channel, bypassing the lowest states of the self-trapped exciton, has been suggested by N Itoh & Saidoh (1973) and Kabler (1975). This direct channel is favoured for three main reasons. First, relaxation to the defect pair via the STX channel needs bond switching. Secondly, the creation of a self-trapped exciton converts a substantial part of the electronic excitation energy to stretching vibrations of the X_2^- molecular ion; this part of the energy is no longer available for the self-trapped exciton–defect pair conversion. Thirdly, there is no energy barrier to transforming from a hole excited state of the [V_k:e] configuration into a nearest-neighbour defect pair (N Itoh *et al.* 1977). As N Itoh & Saidoh (1973) have suggested, dynamic relaxation from the antibonding hole excited states of the X_2^- can lead to large interstitial-vacancy separations.

The dynamic STX channel

R T Williams *et al.* (1986) have suggested that it is the dynamic STX channel which leads to defect formation at low temperatures. The energy differences (Section 4.3) between the [V_k:e] configuration (A in Fig. 5.1) and the Type II or III configurations (B in Fig. 5.1) can drive the X_2^- molecular ion away from the vacancy site. For this to happen, a non-radiative transition to the lowest state must occur at a configuration for which the energy exceeds the barrier height. Level crossing does not occur at the [V_k:e] configuration. We discuss this problem in Section 5.1.3.

Differences between the direct and dynamic STX channels

Clearly, the direct channel and the dynamic STX channels have different relaxation paths. Relaxation through the direct channel is more rapid than the dynamic STX channel. The dynamic STX channel follows the adiabatic potential energy surface of the lowest exciton state with the hole at the bonding σ orbital of the halogen molecular ion. Obviously, the defect pair state could be generated much faster on an adiabatic potential energy surface with an excited hole, as suggested previously (N Itoh & Saidoh 1973; N Itoh *et al.* 1977). The lowest hole-excited state of the self-trapped exciton has the hole occupying the π_g orbital of the halogen molecular ion. There is no potential barrier to relaxation from this state to a defect pair (N Itoh *et al.* 1977). Further, the halogen molecular ion is only weakly bonded in this state. The motion of the halogen molecular ion along the molecular axis is essentially a replacement collision sequence of a halogen atom, accompanied by a hole tunnelling through π bonds. However, the motion of the halogen molecular ion in the ground state is accompanied by rebonding neighbouring halogen ions. Given this difference in the nature of relaxation, F–H pairs formed by the direct channel are likely to have a larger separation than those formed through the dynamic STX channel.

The branching ratio between the direct and dynamic STX channels depends on when an electron is trapped by the hole during relaxation. If the electron is trapped by a hole at low temperatures, after the hole has begun to be converted into a halogen molecular ion, the dynamic STX channel results. On the other hand, if the electron is trapped before relaxation starts, forming a one-centre self-trapped exciton, the hole has equal probabilities of adopting a σ- or π-bonding state. The one-centre hole will have an instability for the same reason as the on-centre self-trapped exciton, so relaxation will proceed via the direct channel with probability 2/3. The hole transition to the ground state will take place after the halogen molecular ion has lost its kinetic energy by emitting phonons.

Thermal STX channel

Thermally-activated jumps over the barrier between the self-trapped exciton and nearest-neighbour defect pairs are another channel. Further thermally-activated jumps, leading to more distant pairs, are feasible at higher temperatures, separating the interstitial from the vacancy. The defect yield through this thermal STX channel should increase with increasing temperature, as in geminate recombination (Section 2.6).

5.1.2 Experimental evidence for three channels for defect pair formation in alkali halides

Subpicosecond measurements

Subpicosecond spectroscopy monitors the growth of defect pairs after the initial generation of electron–hole pairs. Figure 5.3 shows the emergence of F centres and self-trapped excitons following the two-photon excitation of KBr (Shibata *et al.* 1994). It indicates the occurrence of both direct and dynamic STX channels. Clearly, F centre concentration grows initally rapidly, within a picosecond, and then more gradually. The slow component parallels the growth of the self-trapped excitons, discussed in Section 4.3. The rapid growth has no parallel for the self-trapped excitons, and must be ascribed a direct channel, bypassing the self-trapped exciton. Similar results have been found for RbBr (Sugiyama *et al.* 1996) and KCl (Fujiwara *et al.* 1997). The direct channel and the dynamic STX channel have been seen for KCl, KBr, and RbBr, but not for KI.

F centre growth during oscillation on the lowest adiabatic energy surface

The slower growth of F centres, in parallel with self-trapped excitons, arises from non-radiative transitions to the lowest state. It can be ascribed to defect formation by the dynamic STX channel. Tokizaki *et al.* (1991) (see also Makimura *et al.* (1994)) found growth of the F band associated with oscillatory motion on the lowest adiabatic potential energy surface of the self-trapped exciton. They started from the ground state of triplet self-trapped excitons in NaCl. These self-trapped excitons were excited optically to higher states. The subsequent formation of F centres and the recovery of the ground-state triplet self-trapped excitons were monitored simultaneously. There were no signs of the rapid growth associated with the direct channel. What was seen were oscillations between two potential minima on the lowest adiabatic potential energy surface of the self-trapped exciton, and the growth of the F centres occurred during this oscillation. Evidently the F centres were generated dynamically during oscillation on the lowest adiabatic potential energy surface. The lowest

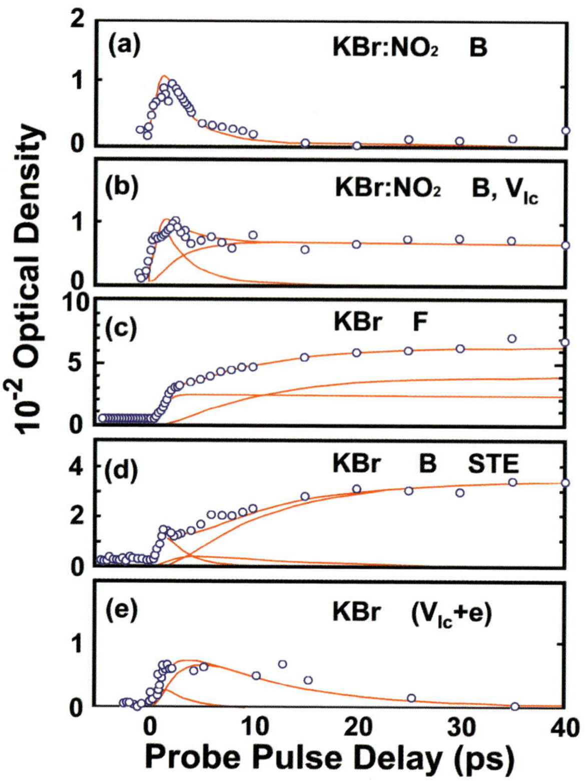

Fig. 5.3. The growth curves of the F centres and self-trapped excitons induced by two-photon excitation of KBr crystal at 4.2 K. Also shown are results for NO_2^--doped KBr, in which electrons are trapped by NO_2^-. Neither self-trapped excitons nor F centres are found (after Shibata *et al.* 1994).

electronic state is reached within 1 ps, followed by vibrational cooling, which takes about 10 ps.

Hirai *et al.* (1987) showed that there are fast and slow components in the growth of the F centres in KI due to a picosecond laser pulse at room temperature. The slow component arises from the thermal STX mechanism. The temperature dependence of the growth rate and the magnitude of the slow component agree with those of Type II luminescence. The origin of the fast component (growth time of less than 20 ps at room temperature) is less clear; it may be also be a thermal STX mechanism. The magnitude of this fast component increases almost linearly with temperature above 100 K. The yield is small at low temperatures for KI, for which the *S*/*D* parameter is small, but increases with temperature. There are several possible reasons for the increase. R T Williams *et al.* (1978) suggested that more energy is available for the off-centre relaxation because of an earlier onset of off-centre relaxation as the

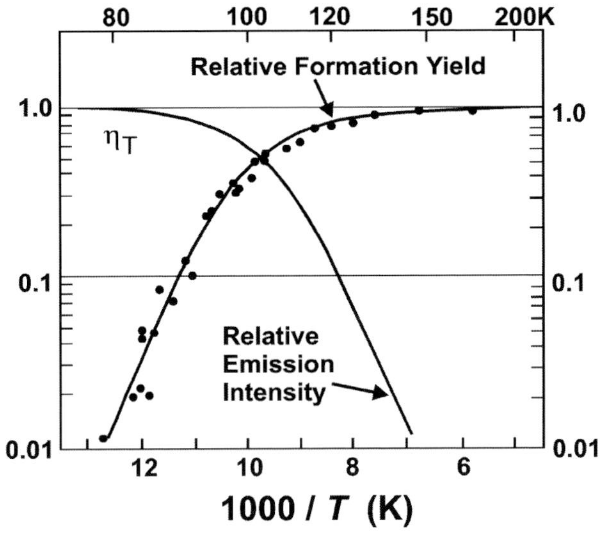

Fig. 5.4. Temperature dependencies of the yield of the triplet self-trapped excitons and defect pairs in KI (after Karasawa & Hirai 1975).

halogen molecular ion is formed. Anharmonicity is also likely to have an important effect on the rate of the transfer.

Thermal conversion from self-trapped exciton to defect pair

The yield of the defect pairs is thermally activated, increasing as the temperature rises (see N Itoh & Tanimura (1990) for a review). Table 5.1 lists activation energies obtained from the Arrhenius plots of the yield. The yields of both stable defects (Pooley & Runciman 1970; Sonder & Sibley 1972; Sonder 1975) and transient defects (R T Williams *et al.*, 1978) increase with temperature. For certain of the alkali halides (those with a small S/D ratio) the yield of defect pairs increases at that temperature at which the yield of the triplet luminescence decreases, as shown in Fig. 5.4. This anticorrelation shows that the triplet luminescent state is depopulated as defect pairs are created. In these materials, defect pair formation competes with radiative and non-radiative recombination. In other alkali halides, non-radiative recombination precedes defect pair formation.

5.1.3 Branching between the relaxation channels from exciton to defect pair

Branching between the direct and dynamic STX channels

As we have argued, the [V_k:e] configuration can be unstable against asymmetric displacement of the X_2^- molecular ion. Similarly, a free exciton (essentially an excited halogen ion, [X^0:e]) may be unstable, leading to the direct channel of defect formation. It is likely that electron–hole pairs branch primarily to the direct channel if an electron is trapped by a hole early in the process of hole self-trapping. Otherwise, the self-trapped exciton channel dominates.

Branching in excited hole states The excited hole states of the X_2^- molecular ion are antibonding (Section 4.1), so hole excitation of the self-trapped exciton should lead to the [X^0:e] configuration. Thus the direct channel should be activated by hole excitation. Enhanced formation of defect pairs has been observed following hole excitation of self-trapped excitons in CaF_2 (Tanimura *et al.* 1989).

Temperature dependence For KBr (Sugiyama *et al.* 1996; Fujiwara *et al.* 1997), it has been shown that the branching ratio to the direct channel decreases with increasing temperature, whereas the total yield from both direct and dynamic channels is constant. This they interpreted through an X_2^- vibrational relaxation rate which increases with temperature. The range of the replacement collision sequence may also influence the defect yield through the direct channel. It is known (e.g. Thompson 1969; Agranovich & Kirsanov 1971) that the range of focused collision sequences can be reduced significantly by thermal vibrations. If the direct process cannot provide a large enough separation of the interstitial and vacancy through focused collisions, the pair will collapse to the self-trapped exciton.

Branching at electron excited states of the self-trapped excitons

Experimental survey Subpicosecond pulse excitation of self-trapped excitons in alkali halides appears to create defect pairs by a mechanism which is neither the dynamic STX channel, nor the direct channel. Excitation of the electron component of [X_2^-:e] does not dissociate the X_2^- molecular ion. This is seen from detailed measurements using polarised light to excite the electronic components of self-trapped excitons in alkali halides with large Klick–Rabin parameters (Tanimura & Itoh 1984; Mizuno *et al.* 1986). The results, summarised in Table 5.2 (Tanimura 1989), give the branching ratio after electron excitation to each symmetry-split 2p orbital of the self-trapped exciton.

What emerges is this. Excitation to the highest excited state, with the electron in the a_2 orbital, leads to the singlet self-trapped exciton in KBr. Excitation in KCl and RbCl, for which no singlet self-trapped exciton is observed, leads to recombination to the ground state instead. Thus excitation to the higher excited state leads either to singlet luminescence or to the ground state. The non-radiative transition from the highest excited state to the lowest state appears to occur rapidly, as observed for NaCl (Makimura *et al.* 1994).

Table 5.2. *Quantum yields for the conversion of 2p electron excited states of triplet self-trapped excitons to defect pairs and to singlet self-trapped excitons in alkali halides.*

Material	Initial excited state	Relative yield of ejection from triplet self-trapped excitons	Fractional yield, defect pairs	Fractional yield, singlet self-trapped excitons	Fractional yield, ground state
NaCl	b_1	1.0	Details uncertain	Uncertain	<0.01
	b_2	1.0	Defect pairs	in detail	<0.01
	a_2	0.8	dominate from b_1		<0.01
KCl	b_1	1.0	0.46		0.54
	b_2	1.0	0.46		0.54
	a_2	0.36	0.16		0.20
RbCl	b_1	1.0	0.61		0.39
	b_2	0.42	0.26		0.16
	a_2	0.48	0.21		0.20
KBr	b_1	1.0	0.85	0.15	<0.01
	b_2	1.0	0.69	0.31	<0.01
	a_2	1.0	0.52	0.48	<0.01

Excitation to the lower electron excited states, with the electron in the b_1 or b_2 orbitals (for notation, see Section 4.3) favours the conversion to defect pairs. Self-trapped excitons only activate the STX defect channel when the electron is excited to the low-lying b_1 or b_2 levels.

Interpretation Putting together the theoretical predictions and experimental observations of optical transition energies for self-trapped excitons in the [V_k:e] and off-centre configurations (Section 4.3.2), one derives Fig. 5.5. This shows the configuration coordinate diagram for the ground state, the lowest and next higher excited states of self-trapped excitons in alkali halides with large Rabin–Klick parameters. The figure exploits the following information. First, experiment shows the a_{1g}–b_{3u} level separation for the [V_k:e] configuration to be smaller than the a_1–a_2 level separation at the off-centre configuration by about 1.5 eV. Secondly, theory indicates that the a_{1g}–b_{1u} and a_{1g}–b_{2u} splittings of the [V_k:e] configuration are similar to the a_1–b_1 and a_1–b_2 level splittings for the off-centre configuration. Thirdly, theory suggests an energy lowering of the X_2^- molecular ion on moving off-centre from the [V_k:e] configuration. Fourthly, experiment (Tanimura 1989) shows that the widths of the optical absorption bands for excitation to the b_1 and b_2 levels are narrower than that to the a_2 level.

Figure 5.5 suggests that evolution from the [V_k:e] geometry to the off-centre configuration lowers the energy for the [a_2:b_1] and [a_2:b_2] states, but has the

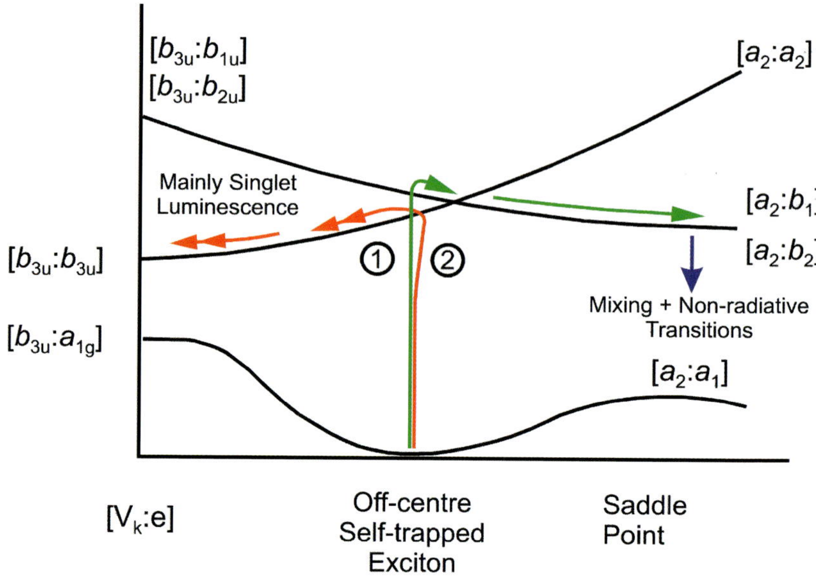

Fig. 5.5. Schematic configuration coordinate diagram for self-trapped excitons in alkali halides with either [V_k:e] or off-centre configurations.

opposite effect on the [a_2:a_2] state. When excited to one of the former two states, defect formation may occur after lattice relaxation to the saddle-point configuration. A rapid non-radiative transition could occur by bond-switching of the X_2 molecular ion at the saddle point configuration. When excited to the [a_2:a_2] state, the halogen molecular ion is pushed in a direction *opposite* to that for which the defect formation occurs, unless there is some means of transfer to the other states at the crossing point. If there is relaxation to the [V_k:e] configuration, the non-radiative transition plus intersystem crossing will lead either to singlet self-trapped excitons, or to the non-radiative transitions to the ground state which we discuss later in this section.

Higher excited states of the self-trapped excitons Excitation spectra for the conversion of self-trapped excitons to defect pairs for NaCl, KBr, and RbBr are shown in Fig. 5.6 (Tanimura 1989). The conversion yield is extremely small for excitation to the higher electron excited states of the off-centre self-trapped excitons. This result is consistent with the observation (Hall *et al.* 1976) that the recombination of an electron with a V_k centre does not create defect pairs, but does give rise to singlet and triplet luminescence.

Excitation of the lowest triplet self-trapped exciton leads to singlet luminescence and also to recovery to the lowest triplet state. The dynamic STX channel appears only in rather special circumstances, e.g. the recombination of an electron with a hot (electronically and usually vibrationally excited) V_k centre.

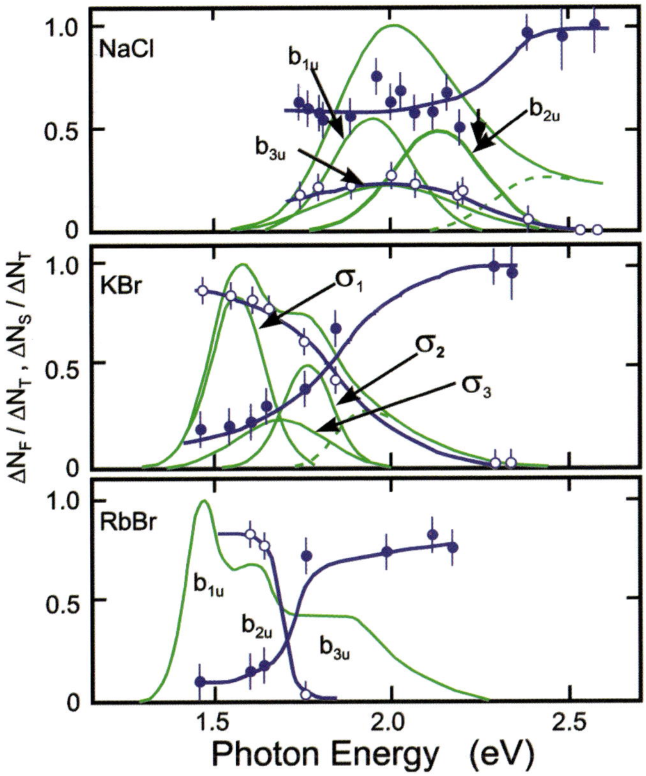

Fig. 5.6. Photo-induced conversion of lowest triplet self-trapped excitons at 4.2 K: yields of singlet self-trapped excitons and defect pairs in NaCl, KBr and RbBr. The green lines show the total STX optical absorption and the components to each sub-level. Open circles give the defect yield per triplet STX eliminated. Full circles give the singlet luminescence per triplet STX eliminated (after Tanimura 1989).

Further studies are needed of the electronic structure at the critical saddle-point configuration, including hole excited states, electron excited states and their configuration interactions.

Defect formation and spin multiplicity

STX channel and spin multiplicity A number of experiments suggest that defect pairs may be produced in the triplet manifold of an exciton through STX channels. First, the yields of defect pairs and of triplet luminescence anticorrelate in certain alkali halides (Fig. 5.4). In the alkali halides which show this anticorrelation, the singlet luminescence decays at a lower temperature than does the triplet luminescence (Pooley & Runciman 1970) without an increase in defect yield at this temperature. It would seem that defects are created only in the triplet manifold. Secondly, the photo-induced conversion of triplet self-trapped exci-

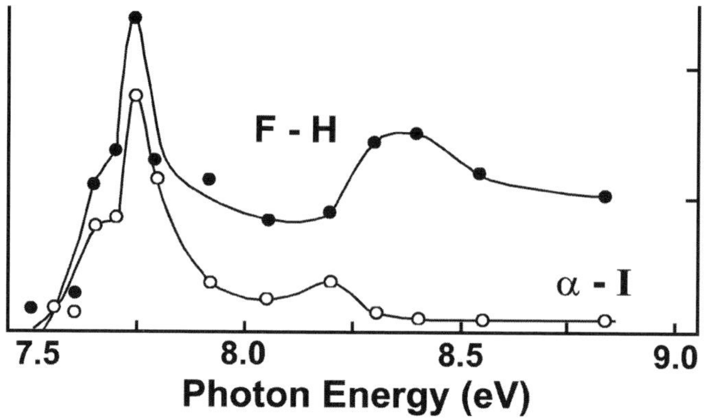

Fig. 5.7. Excitation spectrum for the generation of stable F–H pairs and α–I pairs in KCl at 4.2 K.

tons to defect pairs anticorrelates with photo-induced conversions to the singlet self-trapped excitons (Fig. 5.6). The defect yield is extremely low at those photon energies for which a significant fraction of triplet excitons is converted to singlet self-trapped excitons, and vice versa. Singlet self-trapped excitons convert to defect pairs only with low yield. The low defect yield can be ascribed to rapid non-radiative recombination to the ground state in the singlet manifold.

Excitation spectrum for the STX channel: Dynamic and thermal At low temperatures, only the direct and dynamic STX channels are effective in producing stable defect pairs. Excitation spectra for KCl at 4.2 K are shown in Fig. 5.7 (A Ch Lushchik & Lushchik 1986). The defect yield is clearly highest at the first exciton peak, where the triplet exciton yield also has a peak. The finite yield of the triplet self-trapped excitons and defect pairs above the bandgap energy is consistent with the result that ionising radiation induces defect pairs at low temperatures. Similar results have been obtained for KI at 100 K and RbBr at 190 K (Ch B Lushchik 1982). Thus it seems that both dynamic and thermal STX channels are effective in defect formation after the generation of excitons or electron-hole pairs. Figure 5.8 summarises the relation between the type of excitation and the channel of the relaxation.

The α–I pairs in KCl (Ch B Lushchik *et al.* 1982) and CsBr (Ch B Lushchik *et al.* 1981) are produced only by excitation in the exciton energy range, not in the energy range for electron–hole generation. This implies that some of the stable F–H pairs are converted to α–I pairs, even though the F–H pairs are more stable. We remark that excitons can react with defect pairs to produce modified products. Kirin *et al.* (1998) find that synchrotron radiation in the 6–25 eV range can cause reactions such as $F+H+[e,h] \rightarrow F+V_k+I$.

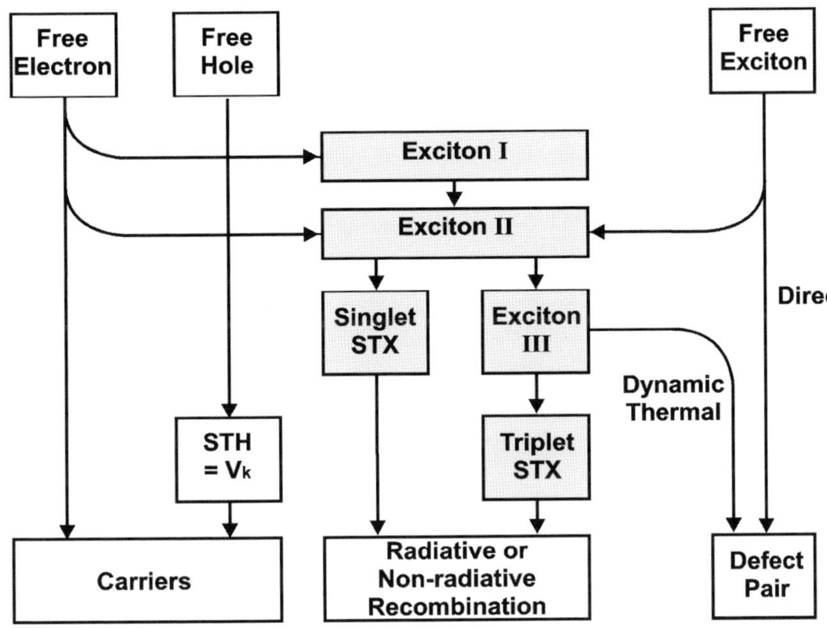

Fig. 5.8. A schematic diagram of the relaxation of an exciton and of an electron–hole pair in alkali halides. The direct channel is effective only from an exciton or from a hole with a trapped electron early in its relaxation. Excitons I, II, III describe the evolution of the exciton: exciton I corresponds to an electron and an on-centre hole; for exciton II, the hole relaxation proceeds towards the V_k geometry and to the relaxed form III.

5.1.4 Thermal conversion from self-trapped exciton to defect pair

Defect formation and non-radiative transitions

The effectiveness of thermal conversion depends on the relative anion and cation radii. In alkali halides with small Klick–Rabin parameters, evidence for thermal conversion of self-trapped excitons in their lowest triplet state to defect pairs comes from the anticorrelation between triplet luminescence and defect pair formation. There is no such anticorrelation for alkali halides with large Klick–Rabin parameters. N Itoh *et al.* (1986) ascribe this to a non-radiative transition from the lowest state of the self-trapped exciton to the ground state, as schematically shown in Fig. 5.9. Two channels compete here: non-radiative recombination to the ground state, and defect formation. The non-radiative rate may be expressed as:

$$\nu_{NR} = \nu_{NR0} \exp(-E_{NR}/kT)$$

and the rate of defect formation as:

$$\nu_D = \nu_{D0} \exp(-E_D/kT).$$

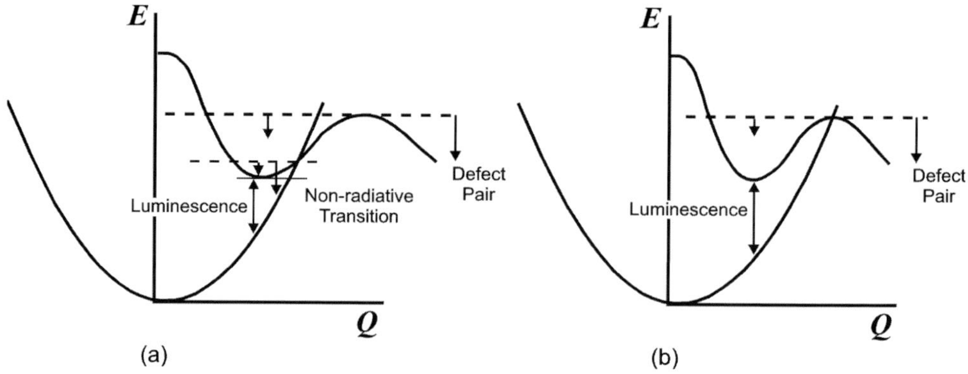

Fig. 5.9. Non-radiative quenching of the self-trapped exciton luminescence shown schematically for: (a) no anticorrelation between the temperature dependences of the defect and luminescence yields and (b) anticorrelation.

Since the non-radiative transition involves both a spin conversion and a level crossing, we may assume $\nu_{NR0} \ll \nu_{D0}$; defect formation only involves the surmounting of a potential barrier. Thus, at very high temperatures, where the prefactor is crucial, defect production would be the dominant one of these two processes. At low temperatures, it is the exponential Arrhenius factor which dominates. Anticorrelation is expected only if the two channels have similar rates for some reasonable temperature, below which non-radiative transitions can dominate. This is possible only if $\nu_{D0} \gg \nu_{NR0}$. Anticorrelation is expected if E_D and E_{NR} are similar in magnitude.

Non-radiative recombination and defect formation in KCl

A quantitative picture can be given for defect formation by the thermal process competing with non-radiative recombination. K S Song & Stoneham (1976) analysed experimental data for KCl and obtained as a quenching rate

$$1/\tau_Q = 5 \times 10^8\ [\mathrm{s}^{-1}]\ \exp\,(-0.016\ [\mathrm{eV}]/kT).$$

On the other hand, R T Williams *et al.* (1978) measured the temperature dependence of the F centre yield 47 ps after the generation of electron–hole pairs. The yield per electron–hole pair increases with increasing temperature above about 100 K with an activation energy of 0.075 eV, reaching unity at high temperatures. If the prefactor is $10^{13}\ \mathrm{s}^{-1}$, defect formation will dominate over non-radiative transitions above 68 K. If so, the thermal STX channel is governed by the liberation of the interstitial halogen atom from the vacancy site. The activation energy for migration of the liberated H centre is lower than the energy for the liberation (Table 5.1). The activation energies for the formation

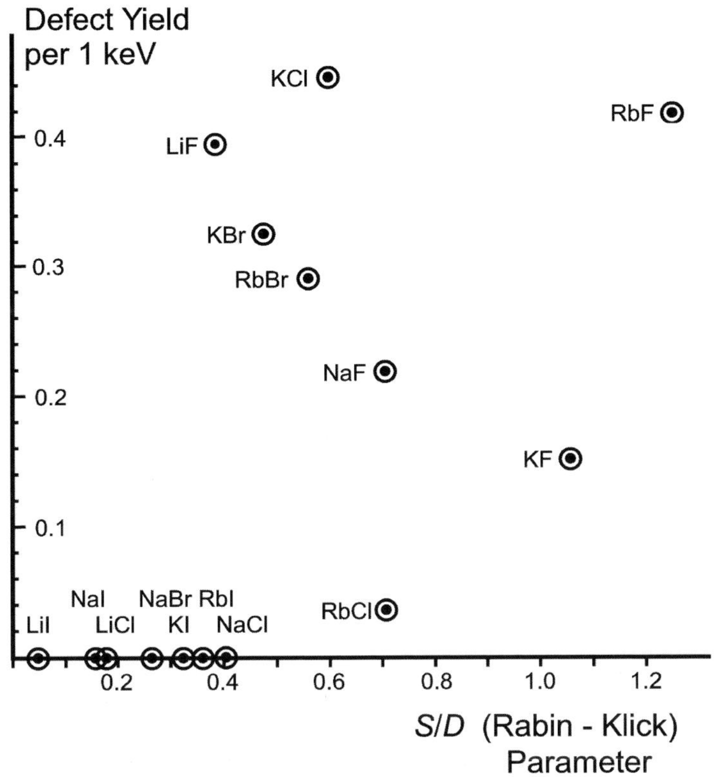

Fig. 5.10. Yields of F–H pairs which have avoided geminate recombination, as a function of Rabin–Klick parameter.

of stable and transient defects have been shown to be the same. However, we note that the very small activation energies may correspond to non-Arrhenius (power-law) behaviour (Chapter 3)

In Section 5.1.2, we argued that the yield of stable defects at low temperatures, where thermal STX mechanism is not effective, can be related to the Rabin–Klick parameter *S*/*D*. The yields of F–H pairs which have avoided geminate recombination are extremely small for alkali halides with $S/D<0.5$, but increases rapidly as *S*/*D* increases (Fig. 5.10; Rabin & Klick 1960; Townsend 1973). Although the fraction of geminate recombination, about 90% in KCl, has not been measured for all alkali halides, the dependence of the yield shown in Fig. 5.10 probably indicates the trends of the direct and dynamic STX channels. As the Rabin–Klick parameter *S*/*D* increases, the halogen molecular ion has more space for jumping to the direction of the molecular axis, making the direct and dynamic STX channels more effective.

Table 5.3 summarises the main features of the three channels of defect

Table 5.3. *Summary of defect formation channels. The main details of the three channels of defect formation identified clearly in alkali halides.*

	Direct channel	Dynamical STX channel	Thermal STX channel
Relaxation path	Direct conversion from a free exciton to a defect pair	Dynamical conversion on the lowest adiabatic potential energy surface through the self-trapped exciton	Thermal conversion from a self-trapped exciton to a defect pair
Origin of the instability	$[X^0{:}e]$	$[V_k{:}e]$	
Timescale	Within 1 ps	A few picoseconds	Slow
Excitation spectrum	Largest at the $n=1$ free-exciton excitation peak, but appreciable at higher energies	Only from electron–hole pairs at low temperatures; from all excitation ranges at higher temperatures	All excitation ranges
Temperature	Only at low temperatures	Wide temperature range	Only above about 100 K
Properties of defects	Close defect pairs	Stable defect pairs	Stable defect pairs, contributing to the component which increases with rise in temperature

formation in the alkali halides. We also remark that excitons can react with defect pairs to produce modified products. Kirin *et al.* (1998) found that synchrotron radiation in the 6–25 eV range can cause reactions such as F + H + [e, h] → F + V_k + I.

5.1.5 Other materials in which the excitonic mechanism is effective

Conditions for the excitonic mechanism

The excitonic mechanism is not effective in all materials. If it is to operate, several conditions must be fulfilled. First, excitons must self-trap. Secondly, the energy (E_X) of an exciton should be higher than the energy (E_{DP}) of a defect pair. Thirdly, the potential barrier between the self-trapped exciton and the defect pair should be small, substantially lower than E_X. The first condition, self-trapping, was discussed in Chapter 3. Here we consider the other two conditions.

Experimental proof of defect generation comes from studies using band-to-band excitation or ionising radiation. Care is needed to draw a distinction between *defect generation* and *defect conversion*, such as the change of a defect charge state, as in the trapping of an electron by a vacancy to form an F centre. One useful observation is saturation of particular radiation products under irradiation. Measurements of volume expansion (Sakaguchi & Suita 1952; Balzer *et al.* 1968) and the identification of interstitials (Kanzig & Woodruff 1958) can provide direct evidence for defect generation, as for alkali halides. The conversion of self-trapped excitons to defect pairs in a picosecond time range is evidence for the dynamic STX process; a long timescale indicates thermal STX process.

Table 5.4 compares the bandgap energy and the energy for Frenkel pair formation in materials for which the exciton is self-trapped. Obviously, the condition $E_X > E_{DP}$ is satisfied for alkaline-earth fluorides. However, E_X is unlikely to exceed E_{DP} for oxides.

The third condition, that of a low potential barrier, is not straightforward, and is different for the dynamic STX and thermal STX mechanisms. The dynamic STX mechanism requires only $E_X > E_{DP}$, so long as the cooling transitions are not too rapid (cf. Chapter 3 and our discussion of the Dexter–Klick–Russell mechanism). However, there will be dephasing of the atomic motion to the STX configuration, and this damping means that a certain amount of energy is not available for passing over the potential barrier. It seems that this third condition (relating to the barrier) for the dynamic STX and direct mechanisms is not satisfied for some of the alkali halides with small *S*/*D* values, although defects can still be generated by the thermal STX mechanism.

Table 5.4. *Bandgap energies compared with defect formation energies.*

Host crystal	Defect formation energy (eV) Theory	Experiment	Bandgap (eV)
Alkali halides (Schottky defects)			
LiF	2.5		14
KF	2.37	2.64–2.72	10.5
NaCl	2.22	2.2–2.75	9
KCl	2.5	2.30–2.59	8.5
KBr	2.27	2.39–2.53	7.5
KI	2.18	1.60–2.21	6
Alkaline-earth halides (Frenkel defects)			
CaF_2	2.7		12.2
SrF_2	2.3		11.44
BaF_2	1.9		10.59

Alkaline-earth and other fluorides

The excitonic mechanism has been proved effective in a few materials which are not alkali halides. In alkaline-earth fluorides, the generation of H centres (interstitial halogen atom centres) is detected by EPR (Hall *et al.* 1969; Hayes *et al.* 1974), and suggests strongly that the excitonic mechanism is effective. About 100 self-trapped excitons are needed to yield one F–H pair (N Itoh & Tanimura 1986), which is of much the same order of magnitude as the yield for stable F–H pairs in alkali halides. Subpicosecond time-resolved spectroscopy data are needed for F–H pair generation, as well as the temperature dependence of the yield, to decide whether the dynamic or the thermal STX mechanism operates. In CaF_2 at low temperature, the major products are transient F–H pairs (Tanimura *et al.* 1989) and a few types of self-trapped exciton, and the yield of stable F–H pairs is small (lower by about three orders of magnitude). These results may reflect the fact that the self-trapped excitons are not oriented along a close-packed halogen row in these materials. The thermal STX mechanism appears to occur in CaF_2 and MgF_2; there is an anticorrelation between self-trapped exciton luminescence and F centre creation in MgF_2 (R T Williams *et al.* 1977) but the self-trapped exciton luminescence is quenched at temperatures lower than F centre growth in CaF_2 (Williams *et al.* 1977).

Evidence for CaF_2 (Tanimura *et al.* 1989) and MgF_2 (Tanimura & Itoh 1991) indicates that F centres are formed more readily when the hole component of the self-trapped exciton is excited. In both cases, an electron pulse by itself creates very few F centres; likewise, excitation of the electron creates a transient state which does not seem to lead to F centre production. The inefficiency of defect production is an advantage for possible lithography optics (CaF_2) and for windows for ultraviolet sources (MgF_2).

The evidence for excitonic processes is less clear in other halides, although such processes seem likely. In $BaF_{1.1}Br_{0.9}$ (Assmann *et al.* 1998), X-irradiation yields F centres at the Br sites and F_2^- molecules at fluorine sites, apparently from the displacement of F_{Br} antisites. In MgF_2 and AlF_3, the electron nanolithography process suggests an easy damage mechanism. Other systems for which there are at least hints of excitonic mechanisms include $KMgF_3$, the alkali silver halides, and $LiYF_4$. However, no intrinsic defects appear to be formed under ArF excimer irradiation in Al/Y/alkaline-earth fluoride glasses (Hosono *et al.* 1998). Defects associated with impurities, like oxygen, are more readily formed.

5.2 Defect formation by other mechanisms

5.2.1 Defect formation from interacting excitons

Processes involving two excitons

We have noted several cases where two electrons [ee] or two holes [hh] can be localised by local lattice deformation when such localisation was not possible with a single electron or a single hole (the so-called negative-U situations). Given the existence of biexcitons in semiconductors, and analogues like the hydrogen molecule, it should be no surprise that there are situations where two excitons 2[eh] can be localised together, aided by local lattice deformation. Again, not surprisingly, the distortion will be different from that which induces localisation of a single exciton [eh]. The relaxation of a second exciton under the influence of a pre-existing self-trapped exciton may generate qualitatively different forces than those from the initial self-trapped exciton. For example, there is a possible negative-U instability: 2[eh] may collapse to form 2[ee] and 2[hh], which will drive a different overall relaxation from that found for a single exciton [eh]. Relaxation and Auger recombination will compete, and the Auger process may convert 2[eh] to some excited state of the single exciton $[eh]^*$.

High density excitation

The relaxation of two interacting excitons can be studied following dense electronic excitation. As discussed in Chapter 11, the effects of dense electronic excitation can be seen in the tracks of energetic heavy ions and also under laser irradiation. The observation of the relaxation of an exciton perturbed by a self-trapped exciton is of value for understanding of microscopic processes involved.

Relaxation of an exciton perturbed by a self-trapped exciton

R T Williams & Kabler (1972) identified the band of an exciton perturbed by a self-trapped exciton in RbI. Tanimura & Itoh (1992) studied the effects of excitation on this band at 4.2 K. They found singlet luminescence was produced with a yield of 0.38 and defect pairs with a yield of 0.04, the yield being expressed as a fraction of self-trapped excitons destroyed. The yield of defect pairs was larger by a factor of 500 than the F centre yield following ionising radiation of RbI at 4.2 K (see Table 5.1). Evidently the presence of a self-trapped exciton close to a newly-created exciton enhances the creation of the defect pairs.

The singlet luminescence observed shows that 2[eh] can collapse to an excited exciton [eh] by an Auger process. However, the singlet luminescence and defect production yields add to less than unity, showing that a substantial fraction (0.58) of excitons created in the proximity of self-trapped excitons is lost via different channels, probably by non-radiative processes. This loss may well involve the migration to the surface of free excitons generated by Auger transitions. Thus, Auger transitions dominate for 2[eh]. Only a small fraction of the excitation energy (albeit much larger than for [eh]) is converted to defect pairs. The enhanced defect yield for RbI under laser irradiation at 4.2 K (Tanimura & Itoh 1988b) is probably due to the interaction of an exciton with a free exciton.

5.2.2 *Defect generation by two-hole localisation*

The Varley mechanism and related mechanisms

Varley (1954) suggested that defect pairs were generated in alkali halides by the localisation of two valence band holes [2h] following an Auger transition to core holes. Two holes [2h] localised on an anion site convert a halogen ion (X^-) to X^+, and there is a Coulomb instability which ejects the halogen to an interstitial site. Klick (1960) suggested a somewhat similar model, postulating that two neighbouring neutral halogen atoms X^0 along a track of an energetic electron could initiate a collision sequence and force a halogen ion into an interstitial site. Both these mechanisms assume that [2h] cause an asymmetric displacement, but they differ from the negative-U situation discussed in Sections 3.3 and 5.1, for these two-hole localisation processes occur much faster than lattice relaxation. They occur within about 10^{-15} s after core excitation for the Varley mechanism, for the Klick mechanism, or for successive excitation by energetic electrons. The localisation is temporary, and needs last only for a lattice relaxation time, 10^{-13} s. Localisation for this time is feasible if the on-site two-hole localisation energy is larger than the width of the valence band (Cini 1978; Sawatzky 1977).

A distinct but similar mechanism based on anion charge switching has been suggested for the desorption of surface ions (Feibelman & Knotek 1978). The [2h]-induced surface emission process is often referred to as the Feibelman–Knotek mechanism, which we discuss in Sections 9.2 and 9.3.

Experimental observations

The Varley mechanism implies a change in F centre yield when the incident X-ray energy crosses the anion core excitation energy. This was observed by Sharma & Smoluchowski (1965). However, later work on other alkali halides showed a change in yield, even on crossing cation core excitation energies (Brown *et al.* 1987). Experimental work at 20 K shows no difference in the F centre yield for X-ray energies below and above the K edges of Cl in KCl and Br in KBr (Kondo *et al.* 1993) and Br in NaBr (Balanzat *et al.* 1996).

The problem with these observations is that the defect pairs are generated from electron–hole pairs and excitons by X-rays below and above core-excitation energies. Any change in the defect yield by the excitonic mechanism on crossing the core-excitation energy has to be distinguished from the core effect. The yield change on crossing core-excitation energies could arise from the change in the local density of excitation, since there is a large local deposition of energy by the Auger electrons generated on filling a core hole, and the energy distribution of Auger electrons is altered (Brown *et al.* 1987). In NaBr, for which there is no complication of defect formation by the excitonic mechanism, the lack of a core effect suggests strongly stable defects are not formed from [2h] in bulk alkali halides. Nor does theory favour defect formation in the bulk by [2h], although halogen ions may be emitted from surfaces (N Itoh *et al.* 1989b). Ion emission by [2h] processes at surfaces is discussed in Section 9.3.

5.2.3 The photographic process in silver halides

Description of general features

The photographic process of silver halides has been known for over a century, and the basis of a major industry for many years. Even in 1839, Herschel was assessing four different possible explanations (Herschel 1839; see also James 1987). Yet its microscopic understanding is still incomplete. What is seen is the growth of silver colloids after prolonged irradiation with photons above the bandgap energy (Hayes & Stoneham 1985; Hamilton 1988). Colloids are observed both in bulk crystal specimens at dislocations and grain boundaries, and in small crystallites ranging from 0.05–2 μm. The absorption of a few photons in small crystallites creates latent images, and then these crystallites can be reduced to silver by a photographic developer. The formation of the latent image, a cluster of silver atoms, is local lattice modification in the sense of this book.

Comparison of colloid formation in alkali and silver halides

Colloid growth in alkali halides (Section 2.6) and silver halides shows both common features and differences. In both cases, colloid formation is due to photolytic decomposition of the material, with halogen emission. However, the colloids in the bulk of silver halides are formed only along dislocations and grain boundaries (Mitchell 1957), whereas colloids in alkali halides can nucleate and grow even at perfect lattice sites (see Sections 2.6.2 and 5.3.2). This difference appears to involve two factors. First, there are differences in the natures of self-trapped excitons in Ag halides and in alkali halides. In AgBr, excitons do not self-trap, and the self-trapped exciton in AgCl is of the [Ag^{2+}:e]-type, whereas the self-trapped exciton in alkali halides is of the [V_k:e]-type. The [V_k:e] self-trapped excitons can convert to defect pairs in the halogen sublattice, leading to colloid formation in alkali halides as halogen vacancies coagulate. This mechanism cannot be effective in silver halides. Secondly, two H centres in KCl can punch out a divacancy, leading to a Cl_2 molecule in a (KCl) divacancy, and to K and Cl ions which move to dislocations. It is not clear that this process has an analogy in the silver halides.

Mott–Gurney theory

The underlying mechanism can be understood following the ideas of Mott & Gurney (1948). They suggested that the optically-produced electron [e] migrates and is trapped at an electron trap T at a grain boundary or near the surface. The negative charge attracts one of the interstitial Ag^+ ions, present in thermal equilibrium, which moves to the trapped electron. This is the first step in a sequence in which electrons and interstitials are trapped, building up the silver cluster which constitutes the latent image. The Mott–Gurney theory has been modified and extended in many ways, but the basic idea of a combination of ionic and electronic processes is accepted.

One problem for the Mott–Gurney description is the behaviour of holes. The recombination of [Te] with a hole competes with the growth of the silver cluster. Possibly the hole is trapped, and somehow converted to a halogen atom (Hamilton 1988). However, the traps are not identified, and the mechanism for conversion of a trapped hole to a halogen molecule is not known. Moreover, the process of hole transport is very different in AgCl and AgBr (Section 4.1). The hole is self-trapped (as Jahn–Teller distorted Ag^{2+}) at low temperatures in AgCl, but not in AgBr; the hole mobility at room temperature in AgCl is 10^{-2} $cm^2/(V\ s)$ (Muller *et al.* 1970) and 1 $cm^2/(V\ s)$ in AgBr (Ahrenkiel & van Heyningen 1966). Yet there is no significant difference in photographic sensitivity.

Absolute measurements of the photographic sensitivity for photographic films show that the most efficient films need eight to twelve photons to be converted to silver ions for development (Hamilton 1988). This emphasises that latent image formation is a very efficient process. In Section 9.2, we deal with the photo-induced surface processes and their relevance to the photographic sensitivity.

The reaction of the growth of colloids may be described by the alternate capture of electrons and Ag^+ ions. The processes might be these:

$$T + e \rightarrow T^-,$$

$$T^- + Ag^+ \rightarrow T{:}Ag^0,$$

$$T{:}Ag^0 + e \rightarrow T^-{:}Ag^0,$$

$$T^-{:}Ag^0 + Ag^+ \rightarrow T^-{:}Ag_2^0.$$

Baetzold's (1973, 1975) calculations suggest that (starting from Ag^+) possible reactions at perfect AgBr surfaces might be the addition of either two electrons then two Ag_i^+ interstitials alternately, or of one electron and one interstitial Ag_i^+ added in sequence. It is assumed that carriers and interstitial ions move independently, but motion of Ag atoms cannot be ruled out.

The reactions may differ at the surface. One possible reaction sequence might start with surface photolysis to release a Br^0 atom and create an F centre. Hole capture could then lead to an (AgBr) divacancy and an Ag atom. The role of the surface is certainly partly to provide space for the Ag atom; various other defects, like dislocations, and possibly twin planes, can fulfil this role. The Ag atom, in whatever state, appears to be stable for only about 1 s. Larger clusters are more stable. Stable deep impurity traps cause loss of efficiency by competing with cluster formation.

The important factors are the mobile ionic defects, Ag_i^+, and electrons. The relatively high stability of Ag_2^0 is also significant. Photo-produced holes are not involved to any real degree, although any electron–hole recombination will reduce efficiency. Many holes are trapped at iodine sites, which act as non-radiative recombination centres at room temperature, but as luminescing centres at low temperatures.

Latent image formation

A modern emulsion consists of a suspension in gelatin of small crystallites of AgBr or AgBr:I with dimensions varying from a few hundredths of a micron to several microns. Growth of these microcrystals is carefully controlled, both to avoid structural defects like dislocations, which may trap photoelectrons in the bulk of the crystal, and to encourage growth of silver specks only on the surface of the grains.

Exposure of photographic films to light produces latent images, which catalyse the chemical reduction of the crystallite by reducing agents known as developers. The grains containing 2–3 silver atoms are reduced to silver, while the other grains are unaffected. Since the number of photons absorbed by a grain is approximately 10 for the latent image to be produced and a typical crystallite contains perhaps 10^9 atoms, a gain (the number of silver atoms produced per photon) of 10^8 results. The formation of the latent image is accompanied with halogen evaporation from the surface (Kanzaki 1984). The emitted halogen is presumed to react with gelatin, which is known to be an efficient halogen absorber. This is important: recombination occurs in photochromic glasses, where the halogen cannot escape. Halogen removal is one of the limiting factors in photography (Hamilton 1988; Baetzold 1997).

The formation of the latent image is aided by sensitisers. One role of the sensitisers may be to create a locally defective region to provide space for the Ag atom. Work by Kanzaki & Tadakuma (1994, 1997) on luminescence associated with sensitisers shows that the sensitisers involve two types of trap: electron traps consisting of sulphur dimers, and hole traps. Analysis by Ramsden (1984) suggests that two Ag atoms (Ag_2) suffice to initiate development in sulphur–gold sensitised AgBr, and that a four-atom cluster Ag_4 is needed for sulphur-sensitised AgBr. The electron trap is found to be deep, so as to keep an electron long enough for it to be neutralised by an Ag_i^+. The sensitisation is effective only in the presence of both electron and hole traps. Apparently, latent image formation involves the Mott–Gurney process, in so far as the hole traps are effective. The continuing growth of silver colloids requires the emission of halogen atoms or molecules from the surface. We discuss this problem in Chapter 9, along with other surface phenomena. We remark that there is a surface space charge potential, typically −0.1 to −0.3 V (depending on surface and measurement method), and this must play a role either directly or through the defect concentrations which are the source of the potential (Section 1.3.3; also Hayes & Stoneham (1985) section 7.5). The higher defect concentrations mean that the ionic conductivity in crystallites may be up to two orders of magnitude higher than in a bulk crystal.

Table 5.5. *Measured enthalpy (h_f) and entropy (s_f) of formation of Frenkel pairs and enthalpies of migration for vacancies (h_v) and interstitials (h_i) in AgCl and AgBr. No allowance has been made for defect interactions in analysing the data.*

	hf (eV)	Sf/k	hv (eV)	hi (eV)
AgCl	1.7	9	0.4	0.15
AgBr	1.2	7	0.35	0.15

Hamilton (1988) noted a further factor specific to surfaces. Kink sites at surfaces always have a long-range Coulomb field which cannot be saturated by capture of ions or electrons. This can be seen from an argument due to Seitz (1954). Suppose that a kink site at a ledge ends with a Br^- ion and has a charge $-q|e|$. If an Ag^+ ion is captured the kink charge, by symmetry, will be $+q|e|$. Since the extra charge introduced is $+|e|$ we have $-q|e|+|e|=+q|e|$, giving $q=1/2$. The one-half integral kink charge encourages alternate capture of e^- and Ag^+. The second point is that, even though only shallow intrinsic electron traps have been observed in AgBr, deep traps are necessary since the photographic process occurs at room temperature. One possibility is that the shallow traps are merely metastable, separated by a potential barrier from a deep stable state.

Photolysis and formation of colloids in AgCl

Optical absorption in the fundamental band of AgCl (416 nm) at 300 K produces colloidal silver particles. The photochemical reaction involves several steps, probably: (i) electron and hole production, (ii) the generation of silver atoms ($Ag^+ + e \rightarrow Ag^0$ or $Ag^+ \rightarrow Ag^0 + h$), and (iii) the aggregation of silver atoms. Colloids form more readily in AgCl than KCl because of the ready mobility of silver ions (Table 5.5). The precise structure of the hole centres produced at room temperature is uncertain. In the silver halides the colloidal particles tend to separate out along dislocations, providing an early example of dislocation decoration.

The colloids have a broad optical absorption band, independent of temperature; its peak varies between 450 and 500 nm. The position of the band can be understood qualitatively from simple dielectric theory. Regard the colloid as a sphere of radius R and dielectric constant ϵ_c, in a host continuum of dielectric constant ϵ_H. In a long-wavelength electric field E_0, the sphere has an electric dipole moment μ given by (Coulson 1956)

$$\mu = E_0 R^3 (\epsilon_c - \epsilon_H)/(\epsilon_c + 2\epsilon_H).$$

Clearly $\mu/E_0 = -\alpha$ is the polarisability. If there are N colloidal particles per unit volume, the overall dielectric constant (for $NR^3 << 1$) is given by

$$\epsilon_{eff} = \epsilon_H + 4\pi N\alpha.$$

Optical absorption is proportional to $\mathrm{Im}[\epsilon_{eff}(\omega)]$. For the host insulator, ϵ_H is real at frequencies less than the bandgap. The metal colloid has, however, a complex dielectric constant $\epsilon_c = \epsilon_c' - i\epsilon_c''$, with

$$\epsilon_c' = 1 - (\omega_p/\omega)^2,$$

where ω_p is the plasma frequency. The imaginary part of ϵ_{eff} contains a denominator $(\epsilon_c' + 2\epsilon_H)^2 + (\epsilon''^2)$, which has a resonance when $\epsilon' + 2\epsilon_H = 0$. Simple manipulation shows that the resonance frequency, for which optical absorption is a maximum, is given by

$$\omega = \omega_p / (1 + 2\epsilon_H)^{1/2}.$$

The peak of the colloid band is at the plasma frequency of the colloid metal, modified by the host insulator in a shape-dependent way. For a full discussion of this simple picture and more general models, see A E Hughes & Jain (1979).

Other photographic systems

The silver halides give a negative picture, i.e. dark where they have been illuminated. This is characteristic of crystals where an electron–cation reaction produces metal atoms; other examples include mercury halides and the oxalates of iron and palladium. Lead halides also exhibit a photographic process, but based on an electronic process with self-trapped electrons and mobile holes and an ionic process with rapid anion vacancy motion. In the lead halides, therefore, one gets a positive image, with the illuminated regions light below 200 K (Kerssen *et al.* 1973), although negative images appear at higher temperatures (Verwey 1970). Similar systems include thallium halides and metal layers on As_2S_3 (Jaenicke 1977). Metal precipitation is also found in transparent organic ionic crystals with layer structures following X- or γ-irradiation at 77 K (Yoshinari *et al.* 1985). After warming to room temperature, these $(C_nH_{2n+1}NH_3)_2MCl_4$ crystals (with M = Mn or Cd) show a metallic lustre consistent with chlorine loss and metal precipitation.

5.2.4 Photochromic and photosensitive glasses

Silver halide photochromic glasses are examples of a range of materials that undergo a reversible colour change upon optical irradiation. If we assume irradiation with light of wavelength λ_1, at t_1, the absorption coefficient at λ_2 will rise linearly at first, then saturating at a time of order t_2. Then an equilibrium is reached in which the rates of the forward and reverse reactions are equal. On removal of the exciting light at t_2, the original situation is gradually restored.

By a suitable heat treatment, one can prepare an oxide glass containing a suspension of silver-halide-enriched particles containing traces of copper in the form Cu^+. These glasses are transparent because the suspended non-metallic particles are too small (about 10 nm) to scatter light appreciably. However, the glass becomes dark on exposure to ultraviolet light due to the formation of silver colloids and of Cu^{2+}, both of which absorb in the visible over a wide region. In the absence of ultraviolet light the transparency of the glass is restored by visible light that is absorbed by the colloids and Cu^{2+}, releasing free carriers. Such glasses have commercial applications, for example, as sunglasses.

These effects are specific examples of more general solid-state photochromic phenomena. Wide bandgap solids may be coloured by absorption bands associated with defects or impurities. The absorption spectrum (and hence the colour) can be changed reversibly by optical bleaching, through reversible changes in oxidation states. Information storage and retrieval therefore becomes possible (Kiss 1970). Another application for charge transfer transitions is avoiding colouration in glasses for viewing chambers in which radioactive materials are handled. Gamma radiation causes Compton scattering, which moves electrons away from the hot side of the glass, and this can cause changes of charge state which lead to colouration. Certain impurities, notably Ce, prevent colouration, since both the key two-charge states (3+, 4+) absorb only light outside the visible range.

In photoreactive glasses based on Cu-doped Ag halides, ultraviolet light ionises the Cu; the electron is captured by the Ag^+, and the Ag atoms aggregate. When the light ceases, the Cu regain their electrons and the Ag reverts to halide. In photosensitive glasses (Trotter 1991; Trotter & Smith 1984; Stookey *et al.* 1978) containing Ag, the electrons come from donors like Ce, which lose electrons to traps of various sorts. Heat treatment releases electrons, which are captured by Ag and aggregate to form specks. However, there are parallel reactions in which crystals of NaF or of Li metasilicate are formed, and this creates a more permanent opaque region. The colour can be controlled through the ultraviolet intensity, which affects the sizes and shapes of the needles of Ag which form.

5.2.5 *Creation of defect pairs in the cation sublattice*

Can defect pairs in cation sublattice be created as a result of exciton relaxation? Ch Lushchik *et al.* (1969) (see also Ch Lushchik *et al.* (1981)) have suggested a

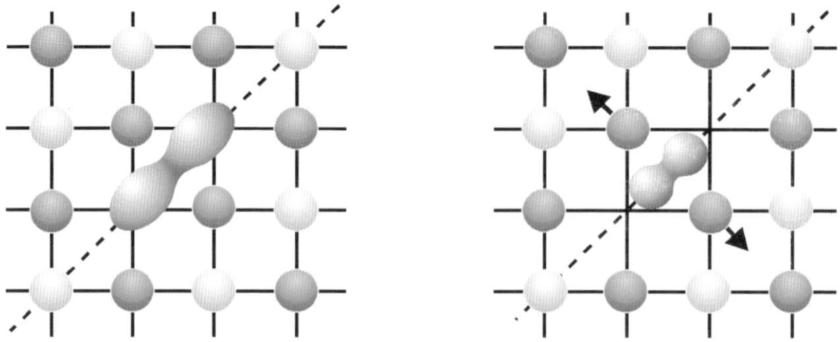

Fig. 5.11. A schematic model for a possible relaxation of an excited self-trapped exciton to a cation Frenkel pair.

dynamic model. In particular, their discussion favours either some process based on an anion exciton (not discussed in detail), or a reaction between two H centres leading to a halogen molecular ion in a divacancy, plus an anion and a cation interstitial. The processes are feasible, in that the exciton energy is larger than the defect formation energy. As we see later, there are several reasons for believing that irradiation at higher temperatures should generated defects on the cation sublattice as well as in the anion sublattice. However, there seems to be no direct experimental evidence for the dynamical production of cation defect pairs, even though thermal defect creation on the cation sublattice does play a role in defect coagulation.

One possible mechanism (Stoneham 1981, unpublished, Fig. 5.11) involving an anion exciton simply recognises that a change of electronic state of the halogen molecular ion X_2^-, especially from an antibonding excited hole state to a bonding state, will lead to a sudden change of X–X interatomic distance. This will give an impulse to the close-packed halogen row normal to the X–X axis. In certain cases, this impulse might suffice to create defects. Certainly there is evidence from resonance Raman experiments (Suzuki *et al.* 1993) that there is cation involvement in V_k and H centre vibrational frequencies. At surfaces, cation defect formation is possible, but has a different mechanism (Puchin *et al.* 1995).

5.3 Defects created by ionising radiation

We now turn to the properties of the defect pairs produced by electronic excitation. At temperatures where thermal conversion is effective, H centres are not stable, but form complexes. F centres survive and are stable even in this temperature range.

5.3.1 Defect pairs created at low temperatures

Transient and stable defects

Defect formation at low temperatures is appreciable only in alkali halides with large Klick–Rabin parameters. A substantial fraction of the F–H pairs formed at low temperatures is annihilated, showing an exponential decay (Y Kondo *et al.* (1972); see also a review by N Itoh & Tanimura (1990)). Those that survive include F–H pairs, which are present in a concentration about an order of magnitude smaller than that of the α–I (halogen ion vacancy plus interstitial halogen ion) pairs. When the temperature is raised after irradiation, a large fraction of the F–H (Behr *et al.* 1967a; Saidoh & Itoh 1973) and α–I (N Itoh *et al.* 1965; Behr *et al.* 1967b) pairs anneal by first-order reactions (see also N Itoh (1972)). The exponential annihilation and first-order annealing indicate geminate recombination (Section 2.6). This implies that most of the defect pairs formed by the direct and dynamic STX channels are close pairs.

Separation of dynamic defect pairs

The separation of stable F–H pairs in KBr formed by irradiation at 4.2 K has been determined (Spaeth *et al.* 1994) from the magnetic circular dichroism of the optical absorption (MCDA). The technique uses the fact that the MCDA signal is altered by the change in spin polarisation. Conversion of the spin of one of the two interacting defects causes a change in the MCDA signal of the other by cross-relaxation. This shows the vacancy and interstitial to be separated by four lattice spaces along the $\langle 110 \rangle$ direction. The theory (Kotomin & Kuzokov 1996) of the correlated annealing of close F–H pairs in KBr concludes that first-order annealing should be seen only for separations less than the fourth nearest neighbours, in agreement with experiment. The transient F–H pairs are presumably those with a smaller separation.

α–I pairs

The stable defect pairs include a substantial fraction of α–I pairs. Why is this, since the primary products from halogen sublattice excited states are F–H pairs? Two mechanisms are feasible for α–I pair creation. The first is local heating or non-radiative recombination followed by transformation to a defect pair (Pooley 1966). The second is electron tunnelling from a close F–H pair to form an α–I pair (Ch B Lushchik *et al.* 1982; Shluger *et al.* 1982). For distant pairs, an F–H pair is more favourable energetically than an α–I pair. It appears that there are specific sites at which the tunnelling is feasible (Shluger & Kotomin 1982). We note that the yield of those stable F–H pairs which anneal via a second order reaction is extremely small; it is only about 10% of the transient pairs. The α–I pairs in KCl (Ch B Lushchik *et al.* 1982) and in CsBr (Ch Lushchik *et al.* 1981) are produced only by excitation in the exciton energy range, but not in the energy range for electron-hole generation.

Recombination of defect pairs

The F–H pair energy appears to be higher than the energy of Type II and III self-trapped excitons, whereas the pair energy seems to be lower than that of Type I self-trapped excitons. Tanimura & Okada (1980) observed that thermal recombination of stable Frenkel pairs at 28 K due to H centre migration induces Type III luminescence in KBr with an efficiency of 0.03, but no Type I luminescence is induced. Similar observations have been made for NaCl and KI (Tanimura & Okada 1980) and KCl (Tanimura *et al.* 1974; Purdy & Murray 1975). Thus the recombination of an H centre with an F centre can create a self-trapped exciton of Type II or III. Alvarez Rivas (1980) suggested that F–H centre recombination gives rise to some of the higher-temperature thermoluminescence bands when H centres, released from interstitial clusters, annihilate F centres to cause luminescence.

For iodides irradiated at liquid helium temperatures, Berzina (1992, 1995) found that excitation at the F-band maximum led to all types of self-trapped exciton luminescence. Excitation of the F centre electron in a close F–H pair appears to produce an electron-excited, self-trapped exciton.

Defect pairs in other halides

Optical spectroscopic measurements of alkaline-earth fluorides for times longer than nanoseconds show there are transient and stable F-H pairs, just as in alkali halides. Transient F–H pairs in CaF_2 have a yield of about 1/10 of that of self-trapped excitons, while that of stable F–H pairs is about 1/1000, i.e. about 1% of the transient F–H pairs. The optical absorption bands of CaF_2 in which there are F–H pairs have a band slightly blue shifted from the stable F band, lying between it and the electron transition band of the self-trapped excitons. Given the presence of several self-trapped exciton configurations, there seems to be a variety of unstable defect pair configurations in CaF_2 and SrF_2. Electron excitation of the self-trapped excitons converts them partly to transient F–H pairs and partly to self-trapped excitons with other configurations. The thermal STX mechanism appears to be effective in MgF_2, in which self-trapped excitons appears to take only a single form. The creation of self-trapped excitons dominates below 190 K, whereas transient F–H pairs dominate above 190 K. Excitation of the hole of the self-trapped exciton in CaF_2 and MgF_2 creates F–H pairs. This leads to the high-density excitation effects discussed in Chapter 11.

Table 5.6. *Centres (excluding the F band) produced by ionising radiation at higher temperatures, where isolated H centres are mobile.*

Material	Optical absorption band (eV)	Raman frequency (cm^{-1})
KCl	5.2[(a)]	
KBr	4.4[(a)]	175[(d)]
KI	3.8, 4.4[(b)]	113[(c)]
RbI	3.5, 4.3[(c)]	113[(c)]

Notes:
[(a)] Faraday & Compton (1965); [(b)] Allen *et al.* (1985); [(c)] Allen & Comins (1992); [(d)] Taurel *et al.* (1983).

5.3.2 *Stabilisation of interstitials*

The primary defects are mobile above characteristic temperatures. At these temperatures, stable defects will survive when there are processes which prevent annihilation of defect pairs. Such processes include: (1) the formation of small defect clusters, (2) the formation of larger defect aggregates, and (3) the trapping of defects by impurities.

In the halides, H centres are mobile at temperatures above liquid nitrogen temperature. Stabilising the interstitial governs the defect yield above liquid nitrogen temperatures (Hayes & Stoneham 1985). F centres are stable to about 600 K, and are converted to metal colloids only at high temperatures. Usually, dislocation loops are a sink for interstitials, but the perfect dislocation loops which are observed require equal numbers of interstitials from both sublattices, even though the primary defects are created only on the anion sublattice. The formation of Frenkel pairs on the cation sublattice should also be accompanied by the formation of dislocation loops. Voids should be also stoichiometric: voids and dislocation loops are complementary, in the sense that the perfect lattice can be restored by recombination. Processes are needed which generate defects on the cation sublattice as well as the primary defects on the anion sublattice.

Formation of interstitial clusters

Table 5.6 lists the prominent optical absorption bands due to the interstitials which are produced slightly above the temperatures at which H centres become unstable. These bands have been given different names, depending on the host (N Itoh 1972), but are called the H_2 band here. These bands can have concentrations much higher than those of nominal impurities and defects, and are considered intrinsic (Faraday & Compton (1965); reviewed by Sonder & Sibley

Table 5.7. *Stretching vibration frequencies for X_2^- ions in alkali halides (Suzuki et al. 1993).*

Material	V_k centre	H centre	LO frequency
NaCl	228, 258	361	260
KCl	241	329	211
RbCl	244	318	171
NaBr		225, 241	201
KBr	144	182, 209	163
RbBr		195	129
NaI			170
KI		139, 174	141
RbI		138	106

(1972) and N Itoh (1972)). The table also lists vibrational frequencies for the H_2 bands. We remark that, in solution, the stretching vibrations of X_3^- ions are observed to be at 175 cm^{-1} (X=Br) and 113 cm^{-1} (X=I) (Maki & Forneris 1966).

Stretching vibration frequencies for X_2^- ions in alkali halides

Suzuki *et al.* (1993) compared the stretching vibrational frequencies of V_k and H centres (Table 5.7 also shows the longitudinal optic frequencies of the host crystals). They interpreted the two values for H centres as an interaction between the X–X stretch and the longitudinal optic (LO) modes of the bulk crystal. Values are in cm^{-1}.

H_2 centres

Dimer formation is the most plausible mechanism of stabilisation, and this is supported by the quadratic relation observed between the heights of the H_2 and H bands in KCl and KBr (N Itoh & Saidoh 1969). The defects which give rise to the H_2 bands are not paramagnetic. The H_2 centres in KBr (Taurel *et al.* 1983), KI (Allen *et al.* 1985) and RbI (Allen & Comins 1992) show a Raman peak at a frequency matching the stretching vibration frequency of the corresponding tri-halogen molecular ions (X_3^-) in solution. It is likely that the H_2 centre consists of an X_3^- molecular ion. However, the precise structure of the H_2 centre is not yet clear. Some of the possible configurations of the di-H centres in alkali halides are shown in Fig. 5.12. The configuration shown in the right-hand figure needs the formation of a cation vacancy, as predicted by Catlow *et al.* (1980)

These models explain the observation that prolonged irradiation leads to F centre formation and to dislocation loops (Hobbs *et al.* 1973). No Raman signal

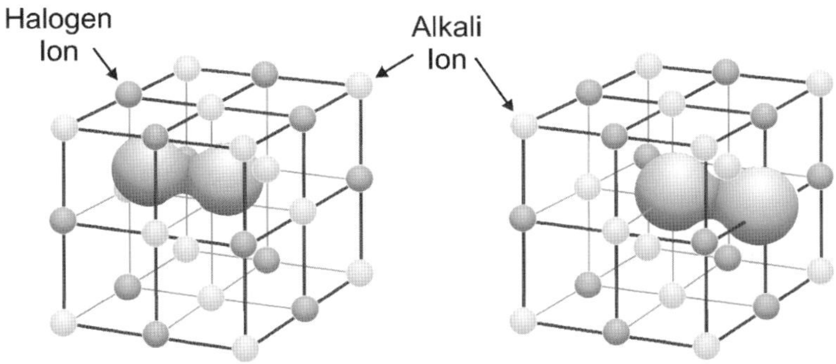

Fig. 5.12. Possible configurations of di-H centres: (left) an X_2 molecule occupying an interstitial position; (right) an X_2 molecule occupying a cation–anion vacancy pair. The combination of the X_2 with an X^- ion forms an X_3^- molecular ion.

has been seen associated with the X_2 molecule, except at higher temperatures, even though configuration (a) is predicted to be energetically favourable among those di-interstitial centres not associated with defects (Catlow *et al.* 1980).

Interstitial stabilisation by impurities

An H centre can be stabilised by a monovalent alkali impurity of smaller radius to form an H_A centre. This H_A centre has a structure similar to that of the original H centre, although the H_A centre symmetry is lowered by the impurity. The hole in the X_2^- molecular ion is equally populated for H_A(Na) in potasium halides, which has a ⟨110⟩ molecular axis (Delbecq *et al.* 1969), but not for H_A(Li), for which the molecular axis is nearly parallel to ⟨111⟩ (Schoemaker & Kolopus 1970). H centres in chlorides and bromides are stabilised by F^- impurities, forming ⟨111⟩-oriented heteromolecular ClF^- or BrF^- ions (Wilkins & Gabriel 1963). Fluorine has a high electron affinity, so the F^- ion carries a larger negative charge and lies closer to the anion site; this favours a ⟨111⟩ orientation. The H centre in KBr does not seem to be stabilised by a Cl ion: F–H and α–I pairs are created by ionising radiation of KBr doped with Cl^- (Taiirov 1983). The trapping of an electron by a $ClBr^-$ V_k centre in KCl generates an F–H pair (Keller & Pattern 1969). Similarly, formation of an exciton localised by an I^- ion in KCl and KBr generates an F–H pair, the same as is produced in pure crystals; this is detected by optical absorption (Arimoto *et al.* 1984). Hirai (1990) gave a review of the H centre stabilisation by anion impurities.

H_D centres (interstitials associated with divalent impurities)

EPR studies (Van Puymbroeck & Schoemaker 1981) reveal several types of interstitial centre associated with divalent impurities (M^{2+}) in KCl. All are created by ionising radiation at 77 K. These centres, known as H_D centres, include Cl_2 stabilised by an M^{2+}–V_C complex (V_C being a cation vacancy), and involve either a single interstitial Cl^0 or two interstitial Cl^0, and Cl_3^{2-} stabilised by an M^{2+}–V_C dimer. The formation of the centre incorporating two interstitials needs two interstitial–vacancy pairs in the halogen sublattice. Optical absorption studies show two bands: one saturates at relatively low irradiation doses; the other increases as the square of the intensity of the first one (Hoshi *et al.* 1975). This indicates defect-initiated di-interstitial formation.

General view of cluster formation

Summarising the experimental and theoretical information, we may draw the following view of defect formation by ionising radiation. At liquid nitrogen temperatures, the dominant interstitial centres are di-H centres, which consist of an X_3^- molecular ion occupying either an anion site or a cation–anion site. In the latter case, the two interacting H centres need the formation of a Frenkel pair in the cation sublattice. Since an alkali ion and a halogen ion go to form a dislocation loop, two F centres and a halogen molecule occupying a cation–anion site remain (Hobbs 1973). Interstitial aggregation at high temperatures should be accompanied by the formation of cation Frenkel pairs. Raman observations (Allen & Comins 1992) show halogen molecules occupying a cation–anion site form a cluster $(X_2)_m$ or, by combining with a halogen ion on a lattice site, X_n^-, where n is an odd number.

Clustering of vacancies

As the concentration of F centres rises, aggregates of F centres are formed, including F_2 (M) centres, F_3 (R) centres, and F_4 (N) centres. These centres are formed more rapidly at higher temperatures because of the faster diffusion of F centres (Luty, 1968). Colloids (effectively aggregates of many F centres) are formed at high temperatures for high doses of ionising radiation. (Compton 1957; A E Hughes 1983) and neutron irradiation (den Hartog *et al.* 1993). The heavy irradiation of alkali halides at high temperatures produces colloids and dislocation loops, whereas, in oxides, the irradiation products at high temperatures are generally voids or vacancy clusters and oxygen bubbles (Clinard & Hobbs 1986).

Summary

Halide crystals with closed-shell cations ($NaCl$, CaF_2) There are three channels for defect formation: direct; dynamic **STX**; thermal **STX**. Which dominates depends on the system, the mode of excitation, and the temperature. Large Rabin–Klick ratios favour direct or dynamic channels. For NaCl, the dynamic channel is favoured over the direct channel. The defects produced are normally neutral anion defects (F and H centres). Even at low temperatures, most of these defects are unstable. Note that there must be a mechanism to prevent recombination in order to generate stable pairs (e.g. a collision sequence) so it is important to know what happens to the interstitial. Stable pairs and other defects are created inefficiently, e.g. charged anion defects (α–I pair) or cation defects.

Amorphous halides and halide glasses Presumably these behave like the crystalline forms, except that one expects F + H recombination likely because there is no mechanism to inhibit recombination.

Halide crystals in which the cation can change charge state ($AgCl$, $PbCl_2$) The normal photographic process yields a negative image, and is based on mobile h, mobile I.

Oxide crystals with closed shell cations (MgO, ZnO, Al_2O_3, usually SiO_2) and oxide glasses with closed-shell cations (borosilicate or aluminosilicate glasses) There are only inefficient processes (if any) to create O vacancies and interstitials.

Oxide crystals with cations which can have several charge states ($Fe_{1-x}O$ and other transition metal oxides); Oxide glasses which include transition metals; Semiconductor crystals (III–V, Group IV) with relatively wide bands and small gaps No defect formation processes established. Enhanced diffusion occurs instead (Chapter 7).

Amorphous semiconductors Defect pairs are produced with small yield. The defects comprise an electron centre and a hole centre (Chapter 8).

Chapter Six

Local lattice modification by electronic excitation of crystalline insulating oxides

6.1 Basic phenomena

6.1.1 Oxides and halides: Basics and similarities

Our discussion of halides showed the central importance of self-trapped excitons and of the processes which compete with self-trapping. When we turn to oxides, we have to ask whether the same classes of process occur. Moreover, we cannot confine ourselves to ideal single crystals, but should assess the situation for oxide ceramics, since these polygrannular materials are typical of functional materials in technology. In this chapter we emphasise materials modification in oxides due to excitation, rather than knock-ons, and attempt to understand the mechanisms. For many technological materials, the mechanisms are not always clear. Most probably, there are further complications, such as diffusion, chemical reactions, and changes of phase.

Excitons are self-trapped in some oxides. Yet there seem to be no cases of defect formation as a consequence of a single bandgap electronic excitation, even when the exciton is self-trapped. Why is this? One possibility is simply that there is a potential barrier between the self-trapped exciton and the defect pair. If this potential barrier could be overcome or bypassed, electronic mechanisms for defect formation would be possible in oxides. If this were the explanation, then the differences between halides and oxides would be differences of degree only. However, there are several other common differences between oxides and halides, and these may be important.

First, the thermal formation energies of defect pairs in oxides are comparable with the bandgap for many oxides, whereas formation energies are significantly less in halides (Table 6.1). In this sense, oxides are closer to semiconductors (Chapter 7) than halides.

Secondly, many important oxides include cations which can exist in several valence states, like the oxides of transition metals. In these systems, electronic

Table 6.1. *Comparisons of key energies for oxides. Results in electron volts (Hayes & Stoneham 1985; Harding 1990). For displacement energies from momentum transfer collisions, the first number is for cations, the second for anions.*

System	Schottky defect formation energy	Bandgap	Displacement energy
MgO	7.5 (theory) 5.7 (expt)	7.7	52/54
ZnO	2.5 (theory) 3.4 (expt)	3.3	30–60/60–120
SiO_2	7–8 (theory)	9	Excitonic mechanisms overshadow knock-ons

disorder can occur without ionic defects or disorder. This is exploited in a number of applications, such as the prevention of colouration in glasses exposed to high gamma fluxes. Oxygen can exist in several possible charge states (especially neutral, O^- and O^{2-}); these are seen normally in systems in which the cation does not change charge state readily, like alumina or magnesia.

Thirdly, the self-trapping of free carriers is marginal in a number of oxides. It is clear that if holes self-trap in oxides like magnesia, alumina, or silica, they do so with a very small self-trapping energy, and most charge transport is as large polarons. This variety of possible charge states leads to a different range of possible reactions for oxides. Luminescence data (de Haart *et al.* 1985) suggest a systematic trend towards self-trapping on titanate groups along the sequence TiO_2 (in which self-trapping is less favoured), $SrTiO_3$, $K_{1.8}Mg_{0.9}Ti_{7.1}O_{16}$, $MgTi_2O_5$, $MgTiO_3$, $La_2Ti_2O_7$ (self-trapping most strongly favoured). De Haart *et al.* suggested that, in $SrTiO_3$ optical absorption can occur by $O \rightarrow Ti$ charge transfer, and that exciton self-trapping on the titanate group appears to occur below 35 K.

Fourthly, many oxide crystals have a sufficient dislocation density that pre-existing line defects or planar defects offer significant competition to the processes which occur in halides. These may contribute to the radiation-induced electrical degradation discussed in Section 6.3.2. Moreover, oxides form amorphous systems (including glasses) more readily than halides.

6.1.2 *Self-trapping in oxides*

The crystalline oxides in which there are believed to be self-trapped excitons may include Li_2O (Liu *et al.* 1996), BeO (Grobunov *et al.* 1990; Kislov *et al.*

1991), quartz (α-SiO_2), Al_2O_3 (Arutyunyan *et al.* 1994; Georgiev & Itoh 1990; P W M Jacobs & Kotomin 1992), $YAlO_3$, $Y_3Al_5O_{12}$, and Sc_2O_3. The spectroscopic evidence is outlined in Chapter 4. With a very few possible exceptions, it is safe to assume that energy localisation only occurs at some pre-existing defect (including a surface) in closed-shell oxides.

Arutyunyan *et al.* (1994) use synchrotron radiation, and observe the formation of O vacancies in various charge states, Al interstitial/O vacancy centres, and Al interstitial dislocation clusters. They suggest that excitons are involved. In MgO, Al_2O_3, and quartz, there is evidence of transient defect production (N Itoh *et al.* 1991), but it is not yet clear whether the transient defects result from excitation alone or are associated with pre-existing defects. Apart from the possible case of Al_2O_3, amorphous SiO_2 (Chapter 8) is the only oxide for which electronic excitation creates Frenkel pairs. For a-SiO_2, multiple excitation is crucial in defect generation. Frenkel pairs are only generated by excitation on irradiation by high-density light ion beams (Hosono *et al.* 1998). At low densities, only a Si dangling bond and an O dangling bond are created.

6.1.3 Charge transfer and colour

Transition metals are the traditional source of colour in gemstones (Mattson & Rossman 1987; Nassau 1983; G Smith & Strens 1976; Burns 1981; Blasse 1981) and in glasses and ceramic glazes (Midwinter 1979; Nassau 1983). The colours result from three major types of transition: (i) crystal field transitions, in which the transition is internal to the specific ionic species; (ii) charge transfer, where an electron is transferred from one ion to another, e.g. Ce^{3+} transferring an electron to a Ce^{4+} ion; (iii) colloids, where the colour arises from the change in dielectric constant asssociated with small particles (e.g. of Au or of Pb). These mechanisms are additional to the colours induced by radiation damage centres. Transfer from a metal ion like Fe or V or Cr to an oxygen neighbour is also commonly observed in silicate minerals. Intervalence transitions, such as electron exchange between Fe^{2+} and Fe^{3+}, were recognised by Weyl (1951).

Tables 6.2(a), 6.2(b) list data for glasses, for which the most systematic studies have been done. Many of the features will hold for crystalline forms as well.

Charge state control can be achieved by modifying the oxygen activity. The differences between oxidised states (higher positive charges) and reduced states (lower positive charges) can be exploited to reduce colour. Adding both Mn^{3+} and Fe^{2+} leads to uniform absorption over the spectrum, the purple Mn colour complementing the greenish Fe. This approach removes colour at the expense

Table 6.2(a). *Colour changes observed by eye in heavily-doped glasses (Midwinter 1979).*

Species	Cr	Mn	Fe	Co	Ni	Cu
Charge state and colour, oxidised form	6+ green	3+ purple	3+ yellow to brown (pale yellow-green)	2+ purple to blue	2+ brown to green	2+ blue
Charge state and colour, reduced form	3+ yellow (green)		2+ green to blue			

Table 6.2(b). *Colours of glasses (Nassau 1983).*

Number of 3d electrons	Species	Charge state	Colour in sodalime silicate glasses	Colour in other glasses
1	Ti	3+	Violet purple	
	V	4+		Blue
2	V	3+	Yellow-green	
3	Cr	3+	Green, yellow	Yellow-orange
4	Cr	2+		Blue
	Mn	3+	Purple	
5	Mn	2+	Colourless	Yellow, brown
	Fe	3+	Pale yellow (yellow to brown)	Colourless
6	Fe	2+	Blue-green	Yellow-pink
	Co	3+		Yellow
7	Co	2+	Violet-blue	Pink, red
8	Ni	2+	Yellow-brown	Purple
9	Cu	2+	Blue-green	

of increasing absorption. The Mn^{4+} can also be beneficial in reducing absorption too, since there are charge transfer transitions which yield colourless products, such as $Mn^{4+} + 2\,Fe^{2+} \rightarrow Mn^{2+} + 2\,Fe^{3+}$. Control of colour or of refractive index is the basis of optical storage and optical holography. Lithium niobate is a common choice of material, with doping by Fe and Mn. With the two traps, the material can be made either absorptive or transparent by a suitable light source; for example, red light could write and blue light erase the colour. Such a memory is relatively non-volatile.

Charge transfer can involve intrinsic defects as well as impurities. The anion and cation vacancies in MgO, for instance, can occur in several charge states. The photoconversion of vacancy centres in MgO and charge transfers involving H impurity species have been noted by Gonzalez *et al.* (1999), for example.

The main changes proposed are ones involving oxygen vacancies (F^+, F^0), cation vacancies (V^-, V^0) with charge-compensating impurities, and H ions at anion sites (H^-, H^{2-}).

6.1.4 Non-linear processes and negative U

Unlike alkali halides, the modification of oxides by electronic excitation cannot be achieved by conventional photoexcitation. High intensities are needed, or dense electronic excitation, for instance laser beams, strong electron beams or energetic heavy ions with a very high stopping power. This suggests that the processes of materials modification require non-linear electronic excitation processes, involving more than one simultaneous electronic excitation step. One factor is the energy available for defect creation, since Table 6.1 suggests one electron–hole pair has barely enough energy to create a defect. As argued in Sections 5.2.1 and 5.2.2, such multiple excitation can be achieved by excitation in the proximity of a self-trapped exciton or by simultaneous excitation of nearest neighbours. There are thus two questions: is it likely that two carriers can localise on the same site (the negative-U situation) in oxides? And will such a situation be unstable and lead to defect formation?

If two holes are localised on one oxide ion following multiple excitation, the result would be a neutral oxygen atom. This would be a negative-U system, which is favoured by large polarisation energies and by small ionisation energies. In principle, the cation could show negative-U behaviour. Stoneham & Sangster (1983) examined the situation for a number of defects and impurities in halides and oxides. They found very few cases in which charge disproportionation (negative-U behaviour) would take place spontaneously. If a neutral atom, O^0, is formed, it will not influenced by the Madelung potential, so it will not be driven off site (Varley mechanism) but it could react with a neighbouring oxide ion to form a peroxide ion. There are several oxygen-based ions observed at surfaces, and these seem to be stable in a number of cases. So one can see routes to charge localisation, but the routes to damage are less obvious. It may be that pre-existing defects are an important ingredient. The calculations of Shluger & Itoh (1990) suggest that, in Li_2O, two excitons should relax to form a peroxy ion, O_2^{2-}, with two electrons transferred to close to a cation site. Free surfaces (Chapter 9) allow other options. Shluger *et al.* (1991c) predict that O^{2-} at the MgO surface may transform into a loosely bound O^0 and a nearby O_2^{2-}. Even though there is no direct evidence about the states which lead to emission, there are good indications of changes due to multiple excitation which make emission easier. The evidence is strongest at high intensities, where laser-induced sputtering is widely demonstrated (see Chapter 11).

6.1.5 Amorphisation

Calling a material 'amorphous' has several meanings (Chapter 1; see also Stoneham (1987b)); often it indicates only that diffraction rings, rather than spots, are seen. Amorphisation has as many meanings. Ways to amorphise a crystalline material, such as an oxides, are varied. They include rapid thermal quenching methods, growth methods (the oxide on silicon is amorphous), and methods involving radiation; crystalline Se films can even be amorphised when placed under stress (Chapter 8). The radiation methods can include high-density electronic excitation, such as irradiation with intense electron beams or with heavy ions with energies of order of megaelectron volts per nucleon; they may also include elastic encounters, irradiation with ion beams in the kilo- to mega-electron volt range or with neutrons. Many oxides form glasses. Some do so from the melt; others do so on rapid cooling; still others amorphise under electron irradiation (Section 6.2.2).

The nature of amorphisation, and whether or not it occurs, depends on the material and the means of amorphisation. Elastic encounters amorphise the widest range of materials, whereas amorphisation by energetic heavy ions appears to be limited to materials in which excitons are self-trapped. Amorphisation induced by electron beams occurs only in a few materials. Thermal amorphisation occurs mainly in systems close to standard glass-formers. Most of the newer oxide glasses formed under rapid solidification (Revcolevschi & Livage 1981) involve one large cation (usually a rare earth) and an oxide of Nb, Ta, Ti, Al, Ga, or Fe.

The reason for the differences from one material to another can be obvious: amorphisation requires reorganisation of bonds to occur in some way. The elastic encounters by ion beams induce knock-ons and hot zones which give energy to the target ions and excite electrons (create excitons) so as to make reconstruction relatively easy. Such bond reorganisation is possible in almost any material, although we remark that the excitonic fluids and dense electron–hole excitation in low bandgap semiconductors like Ge do not generate damage. Under electronic excitation, there is energy localisation when excitons can self-trap; if self-trapping is not possible, the only sites at which energy can be localised are defect sites. Energetic heavy ions create dense electron–hole pairs or excitons in a short time whereas, under an electron beam, excitons are generated slowly and reach an equilibrium concentration.

The possibility of bond reorganisation involves topological considerations as well as issues of critical energy localisation. The change in topology may be a way to create barriers which inhibit the reverse recystallisation reaction. Amorphisation is only one of the various possible changes of phase which can be induced by excitation. Some of the changes result from charge transfer among ions which can have several possible charge states. Others result from composition changes, such as changes in oxygen stoichiometry. Still other changes are possible in oxides with many distinct forms, like SiO_2, where alternative amorphous forms can be identified. Phase changes in oxides are often believed to occur by nucleation and growth. The critical size of a nucleus will include energies associated with the interface between the phases, as well as the bulk molar energies and strain energies for the two phases. Changes of condition under intense excitation are discussed in Chapter 11.

6.2 Effects induced under electron beam excitation

Generally speaking, irradiation of crystalline oxides with electron beams does not give any appreciable radiation damage, unless the incident energy is so high to give an energy to an atom in the solid higher than the threshold energy of displacement. This is because there is no self-trapping in many oxides and even if there is self-trapping no defect formation follows. However, there are a few exceptions observed by intense electron irradiation: amorphisation in a limited oxides such as SiO_2, and drilling holes. Hole-drilling, which is seen for a number of oxides, appears to involve surface effects and will be treated in Chapter 11.

6.2.1 Damage and degradation

Displacement of atoms by elastic encounters

The anion and cation sublattices of oxides can have distinctly different threshold energies for defect generation by knock-ons. The threshold energies are obtained by looking at the rise of some feature as the energy transfer is changed. Thus, for the anion sublattice, the generation of F^0 and F^+ centres is common. For the cation sublattice, the threshold is determined from electron microscope observations of defects involving metallic atoms, such as the formation of aluminum precipitates for Al_2O_3 (Pells & Phillips 1979). We emphasise that there is considerable room for uncertainty. Threshold energies might be *underestimated* if only electrons are moved (as in charge transfer) or if there are pre-existing defects, including dislocations. However, values for UO_2 and for the oxide superconductor YBCO are probably genuinely small, since there are natural low-energy processes. The oxide superconductors can also be damaged very extensively (see Stoneham & Smith (1991)); effects on superconductivity are seen at about 0.01 displacements per atom (dpa), and 0.05 dpa destroys superconducting behaviour (amorphisation requires 1–2 dpa). Equally, threshold

Table 6.3. *Knock-on threshold energies (eV) in oxides. Data are from various sources, including Kotomin & Popov (1998), Hayes & Stoneham (1985), p 299, and Stoneham & Smith (1991). Values are given in order of increasing energy.*

Type of oxide	Material	Cation	Anion
M_2O	Li_2O		43
MO (6-fold coordinated)	MgO	37, 52, 60, 64	30, 53, 55, 60
	CaO	50	50, 58
	CaO:Mg	65	33
MO (4-fold coordinated)	BeO		76
	ZnO	57	57, 60–120
M_2O_3	Al_2O_3	18	76
MO_2	UO_2	40	20
Complex oxides with more than one cation	$LiNbO_3$		53
	$KNbO_3$		64
	$BaTiO_3$		60
	$MgAl_2O_4$		59 (77 K) 130 (300 K)
	$Bi_{12}SiO_{20}$ (BSO)		49
	YBCO		20

energies can be *overestimated* if there is annealing (which should show in the temperature dependence, e.g. $MgAl_2O_4$, unless there is recombination-enhanced diffusion) or if the primary defect gives no clear signal. It is clear from Table 6.3 that displacement energies are in the 50–70 eV range for almost all oxides.

Formation of colloids in Li_2O

The evolution of Li colloids under irradiation with 1 MeV electron beams has been observed in polycrystalline Li_2O (Vajda & Beuneu 1996) and single crystal Li_2O (Beuneu & Vajda 1996). For the polycrystalline form, irradiation at liquid helium temperatures creates only F^+ centres, which are detected by EPR. The yield of the F^+ centres is in accord with that expected for knock-ons with a 43 eV threshold energy. Irradiation at temperatures above 150 K produces EPR signals with insignificant hyperfine structure; the signals are ascribed to F-aggregates. Irradiation at temperatures above 200 K creates Li colloids, with an EPR signal with the free electron *g* value 2.0023. The number of the spins in the Li colloids created at room temperature, where Li colloids are only the radiation products, is about 1/10 of the number of F^+ centres created at liquid helium temperatures for the same electron fluence. Yet colloids are not formed by heating the specimens after F^+ centres have been created by irradiation at liquid helium temperatures. Only at high temperatures do F^+ centres collapse to form F-aggregate centres or collids. Beuneu *et al.* (1997) have also shown that the evolution of microstructure occurs only using electron beams above 250 keV, corrresponding to a threshold energy near 43 eV. Evidently, F^+ centres are formed in knock-ons. The oxgyen interstitials form some configuration which appears to become unstable around 400 °C, where the F^+ centres and colloids anneal.

The loss of a substantial fraction of F^+ centres and the formation of colloids at room temperature suggest that the F^+ centres in Li_2O are mobile under electron irradiation at that temperature. The radiation-enhanced diffusion is probably due to electronic excitation, since room-temperature neutron irradiation (Noda *et al.* 1981) or O^+ bombardment (Noda *et al.* 1985) create only F^+ centres and a few colloids. The mechanism of this radiation-enhanced diffusion is not known; one possibility is that the trapped electron of the F^+ centre is excited, and this leads to a modified energy surface permitting faster diffusion. We discuss enhanced diffusion in Chapter 7. Since oxygen vacancies appear to occupy the same site in all charge states (F^0, F^+, and oxygen vacancy) we can rule out the Bougoin–Corbett (Bougoin & Corbett (1978); see also Section 7.4.3) mechanism in this case.

The creation of colloids alters both the real and imaginary parts of the dielectric constant of Li_2O. This effect on the dielectric constant is seen only above 200 K, where the formation of colloids is significant, and both dielectric constant and the colloids anneal at the same temperature. In single crystals of Li_2O, two types of colloids have been seen (Beuneu & Vajda 1996; Beuneu *et al.* 1997): one gives a sharp EPR line with the free-electron g value, the other a broad EPR line. For the colloid giving the broad line, the g value and its temperature dependence follow the Dyson form (Feher & Kip 1955; Dyson 1955). It is suggested that the sharp line, observed also in polycrystalline specimens, arises from colloids around 1 μm in size, whereas the broad line comes from colloids ten times larger. Optical microscopy reveals colloids as large as 20 μm. The smaller colloids are seen only in electron microscopy, and are found to be anisotropic. This is consistent with the magnitude of the change in dielectric constant, which can be explained only by assuming anisotropic colloids.

6.2.2 Amorphisation by electron beams

Amorphisation by intense electron beams has been observed in SiO_2 and $NaAlSi_3O_8$ crystals (Hobbs 1994). The electrons which amorphise these materials have too low an energy to induce knock-ons, so electronic excitation, not elastic encounters, must be responsible. At room temperature, the dose for full amorphisation by electrons from an electron microscope is 2×10^{25} electrons/cm^2 for 100 keV electrons (Pascucci & Hobbs 1980). The energy dependence of this amorphisation dose is in accordance with the energy dependence of excitation cross-section. The fluence for amorphisation corresponds to 10^8 eV/atom. At about 1/100 of the fluence for amorphisation, centres of strain are nucleated with concentrations of 10^{15}–10^{16} cm^{-3}. These nuclei are three-dimensional inclusions; they are probably amorphous, and have a different density from their surroundings. There is a wide range of amorphous silicas, discussed further in Chapter 8. The inclusions grow with increasing fluence, and finally collapse together as the whole specimen becomes amorphised. It appears amorphisation proceeds by depositing electronic excitation energy at the interfaces between the crystalline part and amorphous parts which evolve from the nuclei. Excitation elsewhere is dissipated as heat, which is why the amorphisation process is not especially efficient. In Si, 2 MeV electrons, which have an energy sufficient to cause displacement damage, cause amorphisation at a dose which corresponds to about 5.0 displacements per atom (Takeda & Yamasaki 1999).

The atomic process of amorphisation at the interfaces is not yet clear. Obviously, electronic excitation (and hence electron–hole pairs) is involved. It is less clear whether excitons are important, or whether multiple excitations are needed. Amorphisation by electron beams can be contrasted with that induced by energetic heavy ions (Chapter 11), even though both are caused by electronic excitation. Heavy ions cause dense electronic excitation within 10 fs, leading to spontaneous non-nucleated amorphisation, whereas amorphisation under intense electron irradiation needs a nucleation step. It is not known what these nuclei are; probably they involve defects, either point defects or extended defects such as dislocations, grain boundaries.

6.2.3 *Transient defects*

There is evidence that transient defects are formed in oxides under electron irradiation (for $SrTiO_3$, Tanimura *et al.* (1984); for MgO, Al_2O_3, and SiO_2, N Itoh *et al.* (1991)). These transient defects are important for two reasons. First, oxides have been proposed for use as insulators in radiation environments, and the dielectric loss from these transient defects could be important. N Itoh *et al.* show that the transient defects could indeed make a substantial contribution to the high-frequency dielectric loss factor. Secondly, the combination of transient defects and energetic electrons suggests links to the more general phenomenon of electrical breakdown, where mechanisms (as opposed to phenomenology) are still emerging.

The experimental situation is that there are transient volume changes under pulsed electron beams. N Itoh *et al.* (1991) used 10 ns pulses of 2 MeV electrons; the current was 1 A/cm^2, and there were typically 10^{21} electron–hole pairs/pulse. The distinctive feature was a volume change $\Delta V/V$ per eV energy absorbed of 0.12 (α-quartz), 0.05 (MgO), and 0.03 (alumina). In all cases, the characteristic decay time was of order 1 ms at 80 K. Only in quartz could the volume change be associated with a luminescence; for quartz, this permitted an estimate of the volume per transient defect.

6.3 Electrical breakdown and related phenomena

Electrical breakdown is a commonplace phenomenon. It occurs in insulators which are electrically stressed. In all cases of which we are aware, breakdown involves the permanent creation of defects or damage. It is distinct from switching and reversible changes in resistivity, which may simply be due to charge trapping. For engineering purposes, breakdown is characterised partly by extremal statistics (such as Weibull statistics) and partly by some averaged quantity (averaged macroscopically or mesoscopically) such as the applied electric field, or the electrical energy density, or some mix of electrical and elastic strain energies. It seems likely that the role of the elastic energy is the reduction of the energy of some defect process, just as applied pressures or uniaxial stresses change activation energies, since the maximum average elastic energy per ion is very small. Despite the common nature of electrical breakdown, and despite a wealth of empirical information, there is no general agreement as to the atomistic processes involved.

We shall argue that breakdown involves *energy localisation* or *charge localisation*, exactly as in photolytic or radiation-induced defect production processes. This is potentially a powerful hypothesis (Stoneham 1995, 1998). It links the femtosecond spectroscopies, the fundamental atomistic theories, and a critical performance-limiting property in insulator performance. The hypothesis allows us to exploit what we know about excitation-induced processes in quartz to understand the electrical breakdown of ultrathin microelectronic oxide insulators. Needless to say, there are cases where other classes of process prove more important. For example, in MgO, the defect processes mainly concern dislocations, although they require charge or energy localisation associated with the dislocation. Still other cases, like organic insulators, show unexpected parallels in the effects of excitation. The way that cross-linking or scission proceeds and the nature of the products (including gaseous products) of electron or gamma damage are usually the same as those in the electrical degradation processes, both ultimately causing carburisation.

The acceleration of carriers in the conduction band can also lead to electrical breakdown (O'Dwyer 1964), although this need not involve permanent damage through defect formation. In semiconductors, for instance, the onset of impact ionisation can lead to electron multiplication (see, e.g., K Seeger (1973)). This may or may not lead to irreversible damage of the material; the damage itself may be the result of heating.

6.3.1 Metal–insulator transitions in oxide films

In quite a few systems, a thin oxide film may 'form', and be switchable without permanent damage (breakdown) between low- and high-resistance states (Dearnaley *et al.* 1970). These systems usually show differential negative resistance. Differential negative resistance is itself of two main types (N-type and S-type), according to the shape of their $I(V)$ curves. Dearnaley *et al.* classified models for this behaviour into several groups. First, there are models loosely associated with electron excitation and with Joule heating. These include the Gunn effect (N-type; electrons populate band structure regions with different mobilities), electric-field-dependent capture (N-type), impact ionisation (S-type) and superheating (S-type). Another possibility which merits consideration is simply the excitation of small polarons to metastable large-polaron states. Secondly, there are mechanisms involving special space charge distributions. These include the models of Hickmott (1964; N-type, where the field suppresses an impurity band), Simmons & Verderber (1967; N-type, involving injection of anode ions into the dielectric and charge redistribution over a resulting depletion layer) and Mott's theory for chalcogenides (Mott 1967; Mott & Davis 1979). Thirdly, there are mechanisms which involve ionic motion (Barriac *et al.* (1966), also involving ion injection; S-type) or atomic rearrangements. Some correspond to conventional phase transitions. Thus vanadium oxides have long been known to show a metal–insulator transition (Cope & Penn 1968). The charge-ordered colossal magneto-resistance oxides like $Pr_{0.7}Ca_{0.3}MnO_3$ are antiferromagnetic insulators below a characteristic temperature, and ferromagnetic metals just above this temperature (Fiebig *et al.* 1998; Keimer 1998; Miyano *et al.* 1997).

The systems considered in most detail by Dearnaley *et al.* involved primarily SiO or alumina insulators; many halides appear to show forming as well. Whether or not forming occurs depends on the anode metal. For SiO, Simmons & Verderber found easy forming for Pd, Ir, Pt, Au, Ag, Cu, C, Si, Ni, Co, Fe, and Sn, whereas anodes made from Zn, Cr, In, Mn, Be, Al, and Mg did not form. Hickmott found that alumina formed with Au, Ag, Cu, and Sn anodes, formed with difficulty for Co, Pb, Bi, and Al anodes, and did not form with an Mg anode. Clearly, noble metals are good for forming, reactive metals ineffective.

Dearnaley *et al.* developed a model based on defect formation within the insulator. The defects postulated were sufficiently close to constitute some sort of conducting filament (Dearnaley *et al.* 1970; Rakshani *et al.* 1976; Ray & Hogarth 1990). The conducting filaments are rendered ineffective at large currents, but can be reformed as switching continues. The behaviour is observed to be voltage-controlled, rather than field-controlled, and this can be understood if the filaments propagate from asperities which cause local field enhancement. These several ideas led to a consistent quantitative phenomenological model. Filaments are predicted to be a few nanometres across, with a density of perhaps 10^6–10^7/cm^2.

Analogous ideas involving filaments are found in later studies of silicon oxide electrical breakdown, and also in the metal–insulator transition in giant magneto-resistance oxides. In the colossal magneto-resistance oxides, two behaviours are seen. In one, $Pr_{0.7}Ca_{0.3}MnO_3$ (Fiebig *et al.* 1998; Keimer 1998; Miyano *et al.* 1997), optical excitation plus an electric field are needed to create low-resistance current filaments. Once established, light is not needed, but the electric field must remain. There is a suggestion of avalanche behaviour, which is one way to create filaments. Simple heating is ruled out. There are several unusual features. Thus there is a threshold laser power needed, but more power is needed for larger photons (for corresponding geometries, 60% more power is needed for 1.6 eV photons than for 0.6 eV photons). There are two components in the photocurrent: fast photocurrent injection, and a slow process driven by the applied potential. The second type of behaviour is seen for $Nd_{0.5}Sr_{0.5}O_3$ (Mori 1998). The change in resistance does not need an applied voltage to sustain it. The process can be reversed by a room-temperature anneal. The explanation of these phenomena is far from certain. In many cases, conduction in the colossal magneto-resistance oxides can be regarded as a combination of small-polaron behaviour (in which the Jahn–Teller coupling of the $3d^4$ Mn^{3+} ions is an important feature) and spin-dependent control of hopping which provides a link to the applied field. One can envisage many ways to give rise to a change in conduction, some perhaps initiated at defect sites. Possibilities might include the generation of metastable large polarons, or simply charge transfer which reduces the effectiveness of the Jahn–Teller interaction. Kiryukin *et al.* (1999) reported X-ray-induced structural transitions involving Mn^{3+}/Mn^{4+} charge ordering in $La_{0.875}Sr_{0.125}MnO_3$. Similar behaviour has been seen for $Pr_{1-x}Ca_xMnO_3$.

6.3.2 Electrical breakdown in simple ceramic oxides, like MgO and alumina

For MgO, the experimental evidence connects electrical degradation to dislocation production. In a typical MgO crystal, the dislocation density is of the order 10^5 cm/cm^3. This can be measured directly in electron microscopy, or monitored through the effects of dislocations on spin resonance linewidths (Stoneham 1966, 1969). Under electrical stress, immediately before breakdown, the dislocation densities in MgO are typically 10^{10} cm/cm^3 (Zong *et al.* 1996; Dr Yok Chen, November 1996, private communication; see also Sonder *et al.* 1978; Narayan *et al.* 1978). We remark that this density corresponds to a typical spacing of 0.1 μm (100 nm), which is similar to the 10–100 nm dimensions of terraces seen on the [100] surface of MgO by atomic force microscopy (Abriou *et al.* 1996). This similarity does not necessarily mean that new dislocations are nucleated, although this may happen to some degree. It is perhaps more likely that there is the climb of charged dislocations under the applied field, the charge arising from injection. The dislocation mobility needed for this process to occur is not excessive, provided charge densities can be high. It is known that quite high dislocation charges can occur (Whitworth 1974; Heinrich *et al.* 1996). Similar processes may occur in the poling of polycrystalline $BaTiO_3$, where dislocation densities rise to about 10^9 cm/cm^3 (Balzar *et al.* 1999). These large dislocation densities have implications for accelerated ageing tests, which may not be representative of in-service degradation.

In a separate study (Tsang *et al.* 1984), several significant features were identified. Breakdown did not have a thermal mechanism, since an AC field did not cause breakdown. Nor was breakdown due to polarisation, as the field tended towards a constant value as the current increased. What was proposed was a double-injection model, involving drift of ionic species or vacancies. The current increased from an increase in hole injection, which was enhanced by the migration of negatively-charged species to the anode. For MgO, typically at 1473 K and a field of 1500 V/cm which was reversed at chosen intervals, the time-dependent current had the form:

$$I = I_{01}\exp(-t/\tau_1) + I_{02}\exp(-t/\tau_2).$$

The characteristic times were τ_1 of order 150 s and τ_2 of order 2300 s (this latter time could be as large as about 10000 s under somewhat different conditions). The characteristic time τ_2 had the same activation energy (2.3 eV for MgO:Ni with 4000 ppm Ni) as the short-circuit current. This time constant varies with field as $\exp(-\beta E)$. We remark that the Mg vacancy in MgO (which is negatively charged) has an activation energy of about 2.3 eV (Sempolinsky & Kingery 1980; Yang & Flynn 1994, 1996; Sangster & Stoneham 1984). The relatively short breakdown times imply that, if it is ionic diffusion which is the controlling process, the range of ionic motion is short.

The electrode materials do matter, as in most cases of switching or breakdown. In the work of Tsang *et al.*, Ir speeded breakdown relative to Pt and Rh. There were slight effects on the τ_2 activation energy, which was 2.9 eV for Ir, 2.7 eV for Rh and 2.6 eV for Pt. These changes presumably reflect different barriers for injection. There was also an effect of impurities: H, Co, Cu slowed breakdown, whereas Cr, V, Ni, Fe speeded degradation. Such dopant effects could be the result of many separate effects, some electrical (because the impurities can take several different charge states) and some mechanical (through effects on dislocation motion). It is no surprise that apparently contradictory results are found for somewhat similar systems; thus Miller (1981) found that Mn, Ti, and Cr additions improved breakdown characteristics for alumina.

6.3.3 Breakdown in the oxide on silicon

One major reason for silicon's dominance in microelectronics is its oxide, which both passivates the surface and is a remarkably good insulator. The performance needed from the oxide has become more challenging as miniaturisation demands thinner and thinner layers. We shall relate experimental data for breakdown in the amorphous oxide to defect production by electronic excitation in quartz and vitreous silicas. Breakdown implies damage, with chemical rebonding, or defect production. Such damage indicates both energy localisation and an energy release mechanism so as to generate damage. In our view, the breakdown process is due in part to electron–hole recombination, leading to atomic displacements by some mechanisms very similar to the damage mechanisms in photolysis or electron beam damage. This should provide an opportunity to exploit such basic studies in the optimisation of breakdown resistance.

The usual breakdown criterion is expressed as a critical electric field or as an electrical energy per unit volume. These phenomenological criteria are useful, but can easily be misleading. Why is a critical field likely to mislead? A certain minimum energy is needed to create a defect, probably a few electron volts, say 5 eV, similar to bond energies. If we consider a thin oxide film, of thickness d, and apply 1 V across it, there will be a field of 10^6 V/cm for a film of thickness 10 nm. But, however thin the oxide is, an electron will never gain more than 1 eV in free transit. However efficiently the electron can use its energy, it will never have the postulated 5 eV energy needed to cause the basic displacement required to initiate damage. In other words, there will be a critical voltage in some cases, not a critical field. This can be deduced from the work of DiMaria (1996), who found that, for very thin oxides, a minimum voltage is appropriate, not a minimum field. Is the stored energy/unit volume a better criterion? Let us ignore the lack of a mechanism by which this energy might be used. If this is electrical energy, taking an electrical field typical of breakdown, the maximum energy stored is very small, of order 0.5–5 meV/atom. It is interesting that the maximum elastic energy which can be stored in a solid before plastic deformation is also of order 0.5–5 meV/atom. These energies should be compared with the 5 eV or so needed to create a defect, typical of bond energies. Without energy localisation, damage is very hard to create, and would certainly need some special site or circumstance.

Charge localisation in the amorphous oxide on silicon should be easy. Holes do self-trap in vitreous silica, and there is a further stabilisation of carrier self-trapping because Si has a high dielectric constant (and hence polarisability). This affects the mode of oxidation (Stoneham & Sofield 1998) and encourages self-trapping of carriers immediately adjacent to the Si/oxide interface. Such self-trapping may be associated with some of the electrical noise associated with electron transfer between the silicon and the oxide closest to it (Kirton & Uren 1989). We may need to recall that some of charge-related phenomena are ionic with hydrogen ion motion, or with a forming process (presumably involving ionic defects) under applied electric fields. This forming process appears to be a precursor to breakdown (Dearnaley *et al.* 1970). The dissipation of contact-electrified charge seen by atomic force microscopy (Morita *et al.* 1994) may well involve ionic motion. These experiments on thin Si oxide show charge asymmetry (negative charges are deposited more easily; initially positive charges dissipate faster, but after 100 s the negative charges dissipate faster) so at least two species are involved (perhaps one related to oxygen, one to hydrogen). The charges are small, typically 1 per 10000–100000 sites.

There is a wealth of empirical information on breakown statistics, excellently summarised by Hori (1997). The first important point is that, empirically (Wolters 1981), there exists a *charge to breakdown* Q_{pc}: the number of holes injected before breakdown is broadly independent of the current or voltage applied. The charge referred to is integrated over the hole current injected. This applies only to the so-called Mode C Intrinsic Breakdown regime, i.e. to failures at above fields E of 10 MV/cm, possibly due to local thinning or asperities. Thus one must exclude Mode A (failures for E less than 1 MV/cm, attributed to pinholes or particulates) and Mode B (failures at intermediate fields, perhaps due to oxide or carbide precipitates or stacking faults in the Si).

The role of this charge can be made clearer by relating it to the Fowler–Nordheim current J_G, given approximately by the expression:

$$J_G = J_{G0} \exp(-B/E)$$

(strictly, J_{G0} is proportional to E^2, but this is a much slower dependence and can be ignored). The hole current J_p is proportional to J_G, the proportionality constant being the hole generation coefficient α, which also depends on the applied field E:

$$\alpha = \alpha_0 \exp(-H/E).$$

In the simplest case of a constant applied field, we would have $Q_p = J_p\, t$. The time to breakdown is then given by $t_{BD} = Q_{pc}/J_p$, or, bringing the terms together:

$$t_{BD} = [Q_{pc}/\alpha_0\, J_{G0}] \exp[(B+H)/E].$$

Typical values are $B = 80$ MV/cm, $H = 240$ MV/cm, $B+H = 320$ MV/cm. Another useful way to describe the data is to introduce a *field acceleration factor* defined by

$$\beta = -\mathrm{d}\,[\log_{10}(t_{BD})]/\mathrm{d}E,$$

which can be obtained easily from the expression above as $\beta = (B+H)/2.3E^2$.

Since the charge to breakdown has no explicit dependence on the oxide volume, we should presume that some sort of nucleation or initiation step is involved. This could involve particular interface sites or particular bulk sites. Hori noted that the data are not consistent with electron trapping leading to a field which builds up to a critical value. Instead, he argued that hole trapping near the cathode enhances the Fowler–Nordheim current, and that this leads to a field sufficient for breakdown.

As it stands, Hori's phenomenology seems right, but the description of the microscopic processes needs modification. For the thinnest oxides, there cannot be a critical *field*, but there must be a critical *voltage* or similar, since a critical energy must be supplied to create a permanent defect or bond reorganisation (Stoneham & Sofield 1997). This is seen experimentally (DiMaria 1996); indeed, Hori's Fig. 4.34 seems to show a breakdown field which rises as the oxide gets thinner. However, there is clearly dependence in detail on the samples used: Hori's map of breakdown mechanisms (2.19) does not agree quantitatively with DiMaria's data, for instance.

We can make initial comparisons of breakdown with the spectroscopic studies in terms of exciton models by noting that typical values of Q_{pc} are 0.1–1 C/cm^2. This corresponds to about 1000–10000 electrons per molecular unit of the Si/SiO_2 interface. For defect processes due to electronic excitation, we find as examples (N Itoh *et al.* 1998):

100 excitons are needed per molecular unit removed from the free surface of SiO_2;
1000 excitons are needed to create one E′ centre in a-SiO_2;
1000–10 000 holes plus associated electrons are needed to reach Q_{pc}.

The efficiencies are credible for an electronically-induced initiation process, and in line with standard defect mechanisms. This view has support from two other pieces of information. First, scanning tunnelling microscope (STM) work (Jeffery *et al.* 1998) shows that the mechanical stress of an STM tip affects the breakdown behaviour of an oxide film. One likely explanation is that stress lowers the defect formation energies. Secondly, Hori's data show (Hori's Fig. 4.35) that t_{BD} gets shorter as T is raised, again in line with other defect phenomena.

The microscopic picture of breakdown as a defect production process cannot be the whole picture. It is necessary to ask how a conductive path extends across the oxide layer. This raises questions at the mesoscopic scale. Here, the picture which emerges is of filamentary conduction (see also the discussion above of switching). One description is based on the assumption that defects are created in the oxide, partly randomly, partly in response to the modification of the electric field distribution by the defects already formed. A Monte Carlo approach (Despas *et al.* 1996; Degraeve *et al.* 1996) leads to failure statistics which are in good agreement with experiment for assumed defect radii of 0.45 nm, a sensible value.

Hori's data also point to some more complicated phenomena and issues. First, there is a dependence of the median t_{BD} on the sign of the gate bias (Hori's Fig. 4.35). The values of t_{BD} are lower when the Si/oxide layer (which is where the hole traps tend to be) is the anode. Secondly, the hole charge to breakdown charge Q_{pc} rises as the oxide gets thicker (Hori's Fig. 4.34). Thirdly, t_{BD} has different dependences on oxide thickness, depending on the nature of the electrode used on top of the oxide (Hori's Fig. 4.37). The performance is better for p-Si than for Al; indeed, Al shows a pronounced field dependence of t_{BD}. Hori suggested that there is a reaction between Al and the oxide, which makes the effective thickness smaller. There is an obvious check, not yet made, in that a consistent value should be found, independent of the total thickness, for the effective reduction of thickness.

We may compare this behaviour with electrical breakdown in polymers. Polymers have negative electron affinity, and some structures have a possible instability against loss of H from the excited state. In the presence of energetic electrons (whether from irradiation or from injection from an electrode) there will be changes in cross-linking or from scission. The network structure will alter, and even transient changes will affect electron trapping, vulnerability to electron damage, attack by interstitial oxygen, and restructuring to allow open spaces/voids. Once damage has been initiated, there are reliable ways to predict growth of degradation (Champion & Dodd 1995, 1996; Champion *et al.* 1996). In organics, such as polymers and diamond-like carbons, one major consequence at high doses is the loss of H and the creation of relatively-conducting, possibly graphitic, tracks. For 2 MeV Ar^+ ions, low doses of $10^{10}/cm^2$ show initial beam-induced structural effects. At doses of $10^{11}/cm^2$, scission or cross-linking becomes evident; at $10^{12}/cm^2$, the role of oxygen becomes important; at doses above about $10^{12}/cm^2$, carbonisation occurs. This carbonisation is associated with an insulator–metal transition as the random carbonaceous tracks overlap to form a connected network (Wasserman 1986).

6.3.4 Radiation-induced electrical degradation

The electrical conductivity of some oxides may be enhanced irreversibly under irradiation at high temperatures. The phenomenon, which is both observed by many workers and disputed by others, is known as RIED (radiation-induced electrical degradation). Most work has been done on Al_2O_3 and to a lesser extent on MgO; a small amount of work suggests there is no effect in vitreous silica. Studies of RIED have been concerned primarily with whether or not the phenomenon posed problems for insulators in fusion technology. Present indications are that the problem is not severe. Our concern in this section is whether or not there is a photolytic process.

The main features of the phenomena can be summarised as follows. First, RIED is observed primarily under electron irradiation, although it has been observed under reactor irradiation as well (Shikama *et al.* 1992). Secondly, RIED is observed at temperatures between 450 and 800 K (Hodgson 1994). Thirdly, an applied electric field, even a field as small as 1 keV/cm, is necessary for RIED to be induced. The fields are so small that it is hard to believe their effect is direct. What may be happening (Stoneham 1994) is some mechanism of field amplification, for example through an analogue of the photoconductive instabilities (Basun *et al.* 1984a,b,c; we are indebted to Professor A A Kaplyanskii for the comment that these instabilities, like RIED, have been observed only in alumina). Fourthly, RIED due to electron beams requires electron energies sufficient to generate knock-ons. This shows that both elastic encounters and electronic excitation are necessary for RIED (Hodgson 1991). Impurities in virgin crystals can be a substitute for defects created by irradiation (B L Zhang & Raghavachari 1995, 1997). In fact, RIED due to neutron irradiation depends on the purity of the sample (Shikama *et al.* 1998). Fifthly, RIED is accompanied by what appears to be colloid formation (Hodgson 1992), and by the generation of dislocations (Zong *et al.* 1994, 1995). Colloid formation is not necessarily accompanied by RIED, but RIED is enhanced when surfaces are involved (Shikama & Pells 1983).

Evidently RIED is not caused by heavy damage from elastic encounters, but is caused by electronic excitation. Indeed, ion irradiation can generate a large number of defects, but does not necessarily generate RIED (Pells 1991; Kesternich *et al.* 1993, 1995). There is an incubation period for RIED; the electrical conductivity starts to increase after irradiation with a certain dose. This suggests some materials modification has to build up before enhancement of electrical conductivity becomes observable. The modification is related to the presence of an electric field: reversing the electric field reduces electrical conductivity previously enhanced by RIED before reversing the electric field (Zong *et al.* 1996). Apparently the products of ionising radiation under an electric field can be eliminated after reversing the electric field.

It is natural to compare RIED and dielectric breakdown. Electrical breakdown is often a rapid irreversible increase in conductivity of some part of an insulator. The irreversibility stems from defect formation or perhaps the formation of a new phase, i.e. a permanent change of properties, rather than just redistribution of space charge. RIED is closer to what is sometimes called 'soft' breakdown: the insulator continues to insulate, but its performance degrades. The degree of degradation observed, whilst worrying, is modest, and does not rule out alumina for fusion insulator applications (Shikama *et al.* 1999).

What is the origin of the conductivity increase?

Irreversible changes in conductivity imply atomic defects or reconstructions, rather than simply the redistribution of electronic charge over traps. Certainly there appear to be atomic defects involved. Hodgson (1992) has suggested that the presence of metal colloid was the cause of conductivity change. Zong *et al.* (1996) suggested that dislocations generated are the cause of the conductivity change. They proposed that carriers injected from electrodes are trapped by dislocations and become mobile under the applied electric field. The last two models offer explanations of the polarity effects, in that reversing the field supplies carriers of the opposite charge (or removes carriers of the initial sign) and might immobilise the oxygen vacancies or dislocations, respectively. An alternative view of the conductivity increase is that it may result from a stoichiometry change (the development of metallic clusters or oxidation) near the electrodes, as a result of a defect-related photolytic process. Metallic clusters are indeed observed to develop in the bulk. Reversing the field would then reduce the number of metallic clusters and hence reduce the conductivity. It is also possible that photolytic processes might generate dislocations, partly through the effects of the large volume expansion which accompanies the generation of a self-trapped exciton (Chapter 3).

There is one further possibility, namely that the effect is not what it seems. This has been argued forcibly by Kesternich (1999) following studies of sapphire and polycrystalline alumina. He concluded that there is no permanent volume term after surface effects have been eliminated, and that the radiation-induced conduction actually decreases with dose. There is a transient increase in conduction after the beam has been switched off. Kesternich suggested that the earlier results were associated with surface leakage currents and with undetected cracks.

The situation is unclear at present. It seems unlikely that there is a major photolytic phenomenon in defect-free sapphire. Nor is there evidence that self-trapping or similar energy localisation is involved. Yet there may well be redistribution of charge of an unconventional sort, and this might involve intrinsic defects (like dislocations or grain boundaries) as well as impurities.

Summary

Oxide crystals with closed shell cations (MgO, ZnO, Al_2O_3, usually SiO_2) Charge localisation (STH, STE) either does not exist at perfect bulk sites or has such low stability that charge is mobile and moves in a non-self-trapped state. Energy localisation (STX) is unambiguous in quartz and other silicas, and is very plausible in other oxides (such as BeO, YAG).

Oxide glasses with closed-shell cations, like borosilicate or aluminosilicate glasses STH, STE can be stabilised at special sites. Oxygen vacancies and interstitials can be formed with low efficiency. The STX in fused quartz shows incipient defect formation (an O anion moves, tending towards an O vacancy and O interstitial). Defect formation can occur: from (1) two excitons formed consecutively, and also from (2) defect excitation (multiple processes).

Oxide crystals and glasses with cations which can have several charge states ($Fe_{1-x}O$ and other transition metal oxides) STH, STE can both exist as altered charge states of ions; the STX corresponds to charge transfer. If defects are formed (as opposed to alterations in charge states), it is probably with low efficiency or associated with impurities or inclusions.

Discussions are given later of effects at high energy densities (Chapter 11) , effects in amorphous materials (Chapter 8), surface phenomena (Chapter 9) and interface phenomena (Chapter 10).

Chapter Seven

Local lattice modification of semiconductors by electronic excitation

7.1 General comparisons: Switching between states and motion

For wide-bandgap ionic solids, electron–hole recombination energies can exceed defect formation energies. In semiconductors, the gap is smaller, and usually significantly less than the energy needed to create a defect. Defect production is predicted to occur only in special cases, as in porous silicon (Caldas *et al.* 1997) and perhaps in other nanocrystals. Most wide-gap ionic solids have strong electron–phonon coupling, so that self-trapping of excitons and certain carriers is common. In semiconductors, the lower electron–phonon coupling makes self-trapping unlikely. Even excitonic complexes (e.g. Kuvalovskii *et al.* 1986) are relatively ineffective, since the electrons and holes tend to recombine one at a time. The role of electronic excitation takes a different form in semiconductors from that observed in wide-gap ionic solids. In semiconductors, it is processes involving impurities, or pre-existing defects such as dislocations, which are affected. The major consequences of excitation are enhanced processes like diffusion, or the production of metastable, non-equilibrium states of defects. The ways in which defect processes are influenced by excitations are often identified only after frustrations. A common example might be that devices survive for long times in storage, but degrade rapidly when used because of dislocation climb.

Discussions of materials modification imply modification controlled in space or time, or with selected defect states. Control in space is common, e.g. lithography and photography. Control in time is reasonably common, and includes processes which occur in operation. Control of the type or state of one critical species is less common. Control of the defect state can include charge transfer, which allows control of optical absorption, or may involve the manipulation of a defect between its ground state and a metastable state. In this chapter, we analyse some of the ways to influence rates of processes of selected species, especially diffusion of chosen species.

The defect processes we shall discuss are primarily diffusion processes. Diffusion processes generally have maximum impact at the mesoscopic scale. Thus we shall not be repeating our discussion of defect formation, except to remark that excitation leading to defect production is primarily an issue for wide-bandgap materials. Only when bandgaps exceed formation energies is the production process efficient. Effects on defect formation are the exception (e.g., Emtsev *et al.* (1997) suggested that gamma irradiation of p-Si can affect defect formation due to electrons) and not the rule. With smaller bandgaps, there can still be metastable states produced, perhaps with a relatively long lifetime. In principle, excitation can lead to macroscopic processes, possibly involving phase changes. Such processes are often the concern of the user of a device, possibly through the switching of ferroelectric or ferromagnetic systems, or pinning regions of them into an unswitchable state. Diffusion-controlled processes lie between the atomistic and macroscopic scales. They may involve surface topography, grain structure, space charge, or the strain fields associated with dislocation arrays. Diffusion-controlled processes will be the focus of our attention. We shall see that excitation-enhanced diffusion processes are relatively common. Excitation can be achieved in various ways, including electron–hole recombination.

Do these processes offer useful control? What we shall see is that some classes of process give control (e.g. what we shall describe as 'local heating' mechanisms) only through the existence of an activation energy. In these cases, one must choose the duration and temperature of a treatment so only the right species moves. Other approaches involve the electronic control of a key process, and some control over the diffusion rate of a specific species is possible. Potentially, at least, this could be valuable if that step is rate-determining. If so, there could be several possible means of control:

(*a*) Supply of energy at some selected wavelength, e.g. when a defect can be excited into a state with different properties. This new state could include simple charge state changes, and the excitation might be by an electron beam as well as a photon. This form of excitation can lead to very fundamental questions about the nature of the transition state and the vibrational phase coherence of particular quantum states.

(*b*) Specific excitation of an infrared-active vibrational mode, causing enhanced motion of a very specific nature.

(*c*) Control of the spatial direction of a process using a bias field. There will often already be a built-in bias in many cases, such as that arising from a surface space charge layer, or a dislocation strain field. In other cases, a bias field will be built up as charge is trapped in deep traps. Simply modifying the diffusion rate in some regions will itself lead to an altered concentration pattern (Frederickson & Karakashian 1995).

On rare occasions, there are more complex cases where the control is subtle, as in the isotope dependence of resistivity induced by hydrogen implantation into GaAs.

In most cases, one recombination event can produce at most one diffusion jump. This means that rather a large number of recombinations is needed to cause significant matter transport. For ultraminiature semiconductor devices, this small number of jumps may be achieved with numbers of carriers which are within the standard operating ranges. Moreover, there can be substantial electric fields which bias the motion. In other cases, the numbers of carriers may be too small and the dimensions of specimens too large for significant effects. Thus, in glasses for the storage of radioactive species, the ionisation levels are usually far too small to bring significant numbers of particular species to the surface and so give selective leaching (Stoneham 1980, unpublished).

7.2 Enhanced diffusion

7.2.1 Characteristics of enhanced diffusion

Most enhanced diffusion operates through the enhancement of electron–hole recombination close to a defect which then makes a diffusion jump. This requires two basic conditions. First, a mechanism must exist to speed the diffusion step by the release of energy close to the mobile species. In some cases, momentum transfer may be a factor; a possible instance is the enhancement of Ge diffusion on an Si surface under bombardment with low energy (15–65 eV) noble gas ions. In that case (Ditchfield & Seebauer 1999) the strong dependence of activation energy and pre-exponential factor on temperature suggest a combination of mechanisms. Secondly, the controlled local energy release usually needs a means to enhance minority carrier densities beyond the values for thermal equilibrium. It is the changes in carrier density (whether by optical excitation, by ionising radiation, by injection, or some other indirect mechanism) which increase the numbers of electrons and holes which can recombine to provide the energy which enhances the diffusion.

In many ways, recombination-enhanced diffusion in semiconductors, with their relatively low bandgaps, has many parallels with defect formation for systems with higher bandgaps. If the gap is sufficiently large that defects like vacancies and interstitials can be formed during excitation, even if only transiently, then further forms of diffusion enhancement are possible, e.g. by increasing the number of vacancies available for a vacancy mechanism to transport some atomic species.

Table 7.1. *Examples of enhanced diffusive processes.*

Device	Phenomenon	Mechanism proposed	Reference
1. CW lasers; lasers for optical communication	Rapid degradation (hours at room temperature)	Seeded dislocation	Hayashi (1981)
	Gradual degradation (years)	Non-seeded dislocation loops; climb by interstitial condensation, accelerated by stress	
2. Injection lasers (buried heterostructure)	Pulsation	Macroscopic regions or surfaces which are non-radiative, rather than amplifying	Henry (1980a,b)
3. Double heterostructure lasers	Large dark spots	Recombination at scribed surface. Minority carriers move to groove by current flow in uncontacted samples containing p–n junctions	Henry & Logan (1977)
	Dark line defects (GaP, GaAs)	Growth of dislocation network. Climb in active region. Dislocation gives non-radiative transition	Kimerling (1978) Petroff & Hartmann (1974) Hutchinson & Dobson (1975) van Vechten (1975a,b)
	Catastrophic damage	Local melting from heat generated by recombination. Epitaxial recrystallisation leaves frozen-in defects	Henry *et al.* (1979)

4. Light-emitting diodes	Suppression of luminescence	In GaP (Zn, O) dissociation, 0.75 eV and 0.92 eV killer centres	Dean & Choyke (1977) Dapkus & Henry (1976) Wight (1977)
	Dark line defects. Abrupt degradation. Dislocation and interface recombination	Three-dimensional dislocation dipole networks. Epitaxially recrystallised material made non-radiative. Generation of dislocation loops etc by glide at high temperatures, limited by carrier diffusion.	Petroff & Kimerling (1976) O Ueda *et al.* (1980) Wight (1977)
5. Tunnel diodes	Fall in peak current during operation	Enhanced Be diffusion broadens width of space charge region	Uematsu & Wada (1991)
6. IMPATT devices (GaAs:Pt Schottky barrier)	Degradation	Interface reactions at high temperatures; recombination enhanced reactions at low temperatures	Kimerling (1978)

One characteristic of enhanced diffusion is a diffusion rate which is either athermal, or has a slower dependence on temperature than the thermal process. As a result, accelerated ageing tests at high temperatures can give unduly optimistic ideas about lifetimes. Thus, Bellamy & Kimerling (1978) found that an interface reaction with an activation energy of 1.6 eV controlled the degradation in accelerated ageing tests of the GaAs:Pt Schottky barrier structures for IMPATT devices. In typical operating conditions (lower temperatures and avalanche conditions), the rate-controlling step had an activation energy of only 0.3 eV. The consequence of the dependence on carrier densities and the altered temperature dependence is that enhanced processes hardly ever occur 'on the shelf'. They depend on the excitation rate, and so occur primarily in operation. Radiation-enhanced diffusion can occur during radiation processing. In some cases, this results from increased defect concentrations (vacancies and interstitials); these cases are not discussed here. In other cases, it is the carriers produced by ionising radiation which enhance defect motion. A classic example is the motion of the interstitial in Si, which is mobile at the very lowest temperatures studied. A further example is the demonstration (Lengel *et al.* 1996) that charge injection by a STM tip leads to vacancy migration on GaAs (110). In this case, the energy is supplied by the recombination of carriers injected from the STM tip with carriers from the bulk of the GaAs.

The consequences of enhanced diffusion are often harmful. Thus dislocation networks acting as non-radiative recombination centres can grow during operation in the active regions of GaAs heterostructure lasers and GaP light-emitting diodes. Just as defects like thermal donors are a problem because they emerge inconveniently during processing and device manufacture, so does enhanced diffusion often become a problem because of the products which develop during operation. However, we note later (Section 7.4.3) that enhanced diffusion can be beneficial in rapid thermal processing, e.g. in recrystallisation following radiation amorphisation.

7.2.2 Routes to enhanced diffusion

Carrier injection is one of several ways to bring about excitation enhancement of diffusion or of atomic motions. Large injections of energy are sure to have some effect. In some laser annealling experiments, melting will occur. If so, most species will have diffusion constants of the order expected for liquids (one jump per vibrational period is typical, 10^{-4} cm^2/s). There will also be multiple electronic excitations which might create an electron–hole plasma, which will modify all the interatomic forces through the altered screening. In special cases, direct excitation of specific defects is possible.

A good example is the (Zn, O) pair in GaP. GaP (Zn,O) was one of the earliest donor–acceptor pair systems to be used extensively for light-emitting diodes. The diodes degrade under forward bias operation (Bergh 1971; Dapkus & Henry 1976; Henry & Dapkus 1976). The donor–acceptor pair nature of this system means that the populations of (Zn, O) pairs at specific spacings can be identified. Feenstra & McGill (1982) observed the way in which nearest-neigbour pairs ($m=1$, where m labels the shell at which the partner in the pair is to be found) dissociated and reformed as a separated pair ($m>1$). They found that photo-induced dissociation shows Arrhenius behaviour over more than two orders of magnitude. The activation energy is 0.60 ± 0.07 eV, lower than the thermal activation energy of 2.6 ± 0.6 eV. The distribution over m develops into a distribution which falls off rapidly with m (their Fig. 7); the distribution is presumably the result of a series of reconfiguration processes, some involving reformation of $m=1$ pairs which dissociate again. It is not certain whether Zn or O is the species which moves; there may even be some more complex intermediate state for the first step.

Another example of light-induced migration is observed for laser-evaporated Ag in thin film of Xe (Kometer & Schwentner 1996). The jump probability appears to rise from about 10% to nearly unity at high temperatures. Precisely what mechanism is involved is not clear. One part of the behaviour must relate to the large vibrational energy available following the excitation process, but such descriptions need caution. There is a metastable state which has a generation probability of about 1.5%, which may provide a possible intermediate state in the migration process.

Other routes involve vibrational excitation. A tempting idea is the direct excitation of a specific vibrational mode by infrared excitation, as in laser isotope separation by the so-called molecular route. In a solid, this is harder to achieve, since the desirable mode may be delocalised in the presence of the host atoms surrounding the selected, critical, atoms. Effective excitation will be easier if it is a local mode which is to be excited. Light atoms are responsible for the commonest local modes, notably H. For such light atoms, the shape of the (anharmonic) potential energy surface should make vibrational excitation more effective. This is because the light ion can be excited to an energy higher than the barriers between sites. In such cases, the ion should have a relatively long mean free path. Ions with such energies may even contribute to thermal diffusion, and may explain the exceptionally high rates seen for H in liquid iron (of order 10^{-2} cm^2/s, about two orders of magnitude greater than for a jump of an interatomic spacing every lattice vibration period; see Oates *et al.* (1978)). The large difference in mass between H and most host atoms means that collisional energy exchange between them is inefficient, which is an added bonus.

7.2.3 Understanding enhanced diffusion

For classical diffusion in thermal equilibrium (Chapter 3), there are two main approaches to understanding the rate. In both, attention is concentrated on a 'saddle-point' region, the lowest energy part of the watershed in the adiabatic energy surface between the initial state and the final state. In the first group (which one might call 'thermodynamic' approaches), one asks about the probability in thermal equilibrium for the system to be in a region of phase space associated with this saddle point. In the second, alternative, group (which one might call 'first-passage' approaches), one asks about the probability of excursions in a reaction coordinate to beyond some critical value corresponding to this saddle point. As noted, the thermodynamic approaches seem to work best in solid-state diffusion, and there are good reasons for this, associated with the number of modes and the nature of the reaction coordinate. In much smaller systems, such as molecules, the first-passage approaches can be successful. We remark that, in first-passage approaches, the highest prefactors, expressed as a frequency, will be bounded above by the largest vibrational frequency, which is certainly inconsistent with experiment for many solid-state systems.

When we turn to excitation-enhanced diffusion, we are dealing with highly non-equilibrium situations. The thermodynamic approaches are much harder to apply. One can use molecular dynamics, and this will be the main tool we discuss. An alternative is to use an extension of the second 'first-passage' theories. This can be helpful qualitatively, but presents other problems.

The cases on which we concentrate will involve semiconductors in which electron–hole recombination (perhaps by some indirect process, including exciton processes) releases energy locally at a defect or impurity during recombination. Some of the energy available from recombination leads to enhanced diffusion, or switches a defect from one bistable form to another, or generates a defect such as an antisite defect. We shall discuss especially the diffusion case. The atomistic understanding of recombination-enhanced processes is far from complete. There is a widely-held phenomenological picture; useful as it is, it has important problems, and is certainly not correct in detail. The problems themselves have been resolved in outline, and await critical tests using some of the powerful techniques now available.

We remark that different mechanisms (and so different theories) lead to predictions of temperature dependences which tend to lie between two extremes. We recall from Chapter 2 that the predicted diffusion rate varies with temperature as $\exp(-E_T/\langle\epsilon\rangle)$, where $\langle\epsilon\rangle$ is kT for thermal equilibrium, and is proportional to the vibrational energy in the reaction coordinate. If this vibrational energy is increased by a fraction of the recombination energy, then we expect a dependence on the (bulk equilibrium) temperature of the form $\exp[-E_T/(kT + X)]$. This is a non-Arrhenius dependence, although it will mimic an Arrhenius dependence with reduced activation energy over a modest temperature range. No simple 'hot-spot' model will yield an Arrhenius dependence on the host solid temperature. Alternatively, if the excitation causes electronic structure changes which alter the energy surface, one expects a Arrhenius behaviour with a reduced barrier, $\exp[-(E_T - X)/kT]$.

7.2.4 *Types of enhanced diffusion*

For more detailed discussions, we shall group processes as follows:

Local heating and related models The conventional wisdom is that the dominant process in enhanced diffusion is what we shall call 'local heating', i.e. the result of enhanced local vibrational amplitudes. 'Hot-spot' and 'phonon-kick' mechanisms are examples. In the form usually presented (but not the form usually observed), non-Arrhenius behaviour is implied. The consequences of local energy release can be modelled by molecular dynamics; existing results (Masri & Stoneham 1984) show it is hard to understand why this is efficient. One problem is that the forces on the atoms associated with the local energy release need not be in the most effective direction to drive a jump. A further problem is rapid dephasing, with energy dissipated among other vibrational degrees of freedom (this phenomenon means that any modelling must treat enough atoms for dephasing to be effective; it is not usually a demanding condition).

Local Excitation and related models An alternative process we shall term 'local excitation', in which the key steps make use of an energy surface corresponding to an excited electronic state. The technical issues are discussed later. A particular case involves alternate charge state changes for a system whose atomic configuration changes with charge state, the Bourgoin–Corbett mechanism. In principle, such processes could be sorted out by self-consistent molecular dynamics. Defect formation from an exciton state in alkali halides is a clear example of a local excitation mechanism: defect formation is not induced by the vibrational energy released on exciton recombination.

7.3 Local heating models ('hot-spot' or 'phonon-kick' mechanisms)

In local heating, the recombination energy is converted into vibrational energy in the reaction coordinate for the process of interest. Some of the recombination energy and some thermal energy are used to achieve a passage over the potential barrier to motion. Much work in this field uses a simplistic, but descriptive, argument along these lines:

(*a*) Assume a defect with a level in the gap ϵ below the conduction band. Let us assume it is a donor which is neutral, so ϵ is the energy to remove an electron and place it in the conduction band. Its normal diffusion process has activation energy A.
(*b*) Assume a hole is injected. There is then a possible reaction in which the electron on the donor and the hole recombine, releasing energy of order $R = E_{\mathrm{gap}} - \epsilon$.
(*c*) Assume this energy R becomes vibrational energy in the reaction coordinate. Less thermal energy is needed for a jump to occur, so the observed activation energy will be reduced to $A - \alpha R$, where α is an efficiency factor, often taken to be unity.

The simplicity of the argument, as just presented, is its only real virtue. It is both misleading and naive, with problems addressed in the fuller discussions by Tully *et al.* (1975), Markvardt (1980), Markvardt & Landsberg (1981), Stoneham (1980, 1981b), Masri & Stoneham (1984), and Sumi (1985). For a more recent review, see Yassievich (1994).

7.3.1 The model of Weeks, Tully, and Kimerling

Weeks *et al.* (1975; see also Kimerling 1978) use chemical reaction theory, framing their discussion in terms of a 'defect molecule' embedded in a host lattice. They estimate a reaction rate which contains an efficiency η describing the fraction of recombination events which result in a successful reaction (diffusion jump). This efficiency η is written in the form:

$$\eta = (k_{\mathrm{D}}/k_{\mathrm{L}})\,[|(E_T - E_{\mathrm{R}})|/E_T]^{n-1}.$$

The factor involving the thermal activation energy E_T and the recombination energy E_{R} expresses the probability that energy distributed among n defect molecular modes can find its way to the reaction coordinate (Weeks *et al.* used the term 'critical reactive *mode*' which we avoid intentionally, as the reaction coordinate is most unlikely to be a dynamically-independent normal *mode*). The factor $k_{\mathrm{D}}/k_{\mathrm{L}}$ is the ratio of the rate of energy flow within the 'defect molecule' to the rate of loss to the surrounding lattice.

7.3.2 The model of Masri and Stoneham

An alternative estimate of an efficiency (albeit defined in a slightly different way) was given by Masri & Stoneham (1984), who used molecular dynamics so as to examine the consequences of many modes. Their argument was as follows. So far as the nuclear motion is concerned, the electronic recombination transition which provides energy locally can be represented by the sudden application of a force and changes in interatomic forces.

The applied force itself corresponds to the gradient of the difference in energy between the energy surfaces of the initial and final states, as in simple configuration coordinate diagrams. If the mode Huang–Rhys factors S_{n0} are written in terms of these forces $\mathbf{F}_n$ as $S_{n0}=[\mathbf{F}_n^* (1/\hbar\omega^3)\mathbf{F}_n]/2$, then there is a 'phonon kick' in a direction in configuration space $\mathbf{T}=\Sigma_n \mathbf{F}_n (1/\hbar\omega^2) \mathbf{e}_n$, with $\mathbf{e}_n$ the corresponding eigenvector of mode n. The reaction coordinate for the diffusion jump is $\mathbf{R}$, and one measure of the efficiency will be $g=\mathbf{R}\cdot\mathbf{T}$, which indicates how well the kick is directed along the reaction path; $g^2=(\mathbf{R}\cdot\mathbf{T})^2$ measures the energy transfer (Stoneham 1980).

These features show clearly in simple molecular dynamic models, where the forces can be chosen. The results suggest it will be the exception, not the rule, to be able to use the recombination energy efficiently. Molecular dynamic models, like those of Masri & Stoneham, also make it clear that coherent motion in the reaction coordinate is transient. This is obvious, of course, although models with a single frequency (or even with two or three modes) will mislead one into expecting longer-term coherence. As noted, there will be dephasing, which is related to the frequency dispersion $(\langle\omega^2\rangle-\langle\omega\rangle^2)/\langle\omega^2\rangle$ of the reaction coordinate. The dephasing therefore leads to a further loss of efficiency, since the enhanced vibration will not always survive long enough for thermal effects to provide advantage. The dephasing is represented in part in the (k_D/k_L) factor in the treatment of Weeks *et al.* Masri & Stoneham did attempt to check the temperature dependence by fitting their result using α and β in the general form $\exp[-(E_T-\alpha E_R)/(kT+\beta E_R)]$, but their statistics were not adequate to show whether there were serious deviations from Arrhenius behaviour.

7.3.3 *The model of Sumi*

Sumi (1984) developed an analytical description which introduced two important features. First, he recognised that one should consider the whole cycle of events in the enhanced motion: the minority carrier capture, the capture or recombination processes with a majority carrier, and any other intermediate steps for the specific example. Rate equations involving these processes determine the relative populations of the various charge states or, more generally, electronic states such as excited states, of the mobile species. The diffusion rate from enhanced diffusion will involve carrier cross-sections, carrier velocities,

Table 7.2. *Sumi's results for the quantum yield Y (his eqs. (4.24)–(4.26). Here we use* $g=|g|$, $kT''=(\hbar\omega/2)\coth(\hbar\omega/2kT)$, $\delta\omega^2=(\langle\omega^2\rangle-\langle\omega\rangle^2)$, $a=gE_A^{1/2}-E_P^{1/2}$, $p=gE_P^{1/2}-E_A^{1/2}$. *The conditions are slightly different from those summarised in his Fig. 1 (which relate only to the activation energy, and are taken from his earlier paper (Sumi 1982); these earlier conditions are given in brackets. The limiting case of Weeks et al. (1975) corresponds roughly to* $g=1$, *unit efficiency, and activation energy* E_A-E_P.

Condition	Prefactor	Activation energy
$a\gg kT''$ $(E_Ag^2>E_P)$	$(a/g)(\langle\omega^2\rangle/\delta\omega^2)^{1/2}(E_P/E_A)^{1/2}$	E_A-E_P
$a\gg kT''$, $p\gg kT''$ $(E_A/g^2>E_P>E_Ag^2)$	$[E_P(1-g^2)^3\,kT'']^{1/2}ap(4\pi)^{1/2}$	$[E_A^{1/2}-(g^2E_P)^{1/2}](1-g^2)$
$p\gg kT''$ $(E_P>E_A/g^2)$	1	0

and free carrier populations. These state populations are also found for other models for which the thermal diffusion rate depends on the defect state, whether excitation state or charge state.

Secondly, Sumi considered the motion of the system on a two-dimensional energy surface described by the energy-accepting coordinate Q_P (corresponding to unit vector **T** of the Masri & Stoneham model above) and the reaction coordinate Q_R (corresponding to Masri & Stoneham's unit vector **R**). He evaluated the rate for a system starting at a coordinate for carrier capture to reach the critical reaction coordinate for the reaction (diffusion step) to occur (Table 7.2). This is done using a steepest descent method. Sumi included in his discussion the damping of the motion because of the dephasing due to frequency dispersion in the reaction coordinate, essentially the effect seen in the Masri & Stoneham results from molecular dynamics. Table 7.3 shows that experimental results for quite a few systems can be fitted with parameters which are very reasonable.

7.3.4 Other general issues

How big can the reduction in apparent activation energy be? Sumi's emphasis on the full cycle of carrier capture and recombination steps recognises that it is possible to devise situations for which the whole bandgap, E_G, might be exploited, rather an optical recombination energy, since the cycle overall involves the recombination of an electron and a hole and the displacement of a mobile atom.

Table 7.3. *Sumi's analysis of recombination-enhanced defect annealing in n-GaAs$_{1-x}$P$_x$ at low injection levels. The analysis makes assumptions about the (small) thermal activation energies for multiphonon minority and majority carrier capture.*

		Activation energies in eV			
Composition (x)	Level	Thermal	Observed enhanced	For yield (analysis)	g
0	*E*2	1.75	0.98	0.88	0.38
		1.34	0.43	0.33	0.65
0.35	*E*3	1.44	0.51	0.41	0.52
0.45	*E*3	1.47	0.37	0.27	0.62
0.45	*E*1	1.70	Athermal		
0.45	*E*4	1.70	Athermal	0	≥0.94
1	*E*2	1.35	Athermal	0	≥0.79
1	*E*3	1.35	Athermal	0	≥0.81
1	*E*5	1.30	0.62	0.52	0.34
1	*E*6	2.10	0.83	0.73	0.56

A first general point is that we have assumed that the important effect of the recombination energy is to enhance motion in the reaction coordinate. There is another possibility: *the recombination energy may lead to an alteration of the barrier to diffusion,* rather than to motion in the reaction coordinate. This is well known as an idea. Indeed, the idea is part of the analysis of the quantum diffusion of H in fcc metals (Flynn & Stoneham 1970) in the understanding of the differences of behaviour between bcc and fcc hosts. In the quantum diffusion case, it is a matrix element, rather than a barrier, which is affected. A classical analysis was given by Bier & Astumian (1993), who considered a model in which a barrier fluctuates randomly with a characteristic rate γ between a higher energy and a lower energy. They used both a Langevin equation and a simpler and less general rate equation method. What emerged is that, for very slow flips (γ small) the mean time for first passage is slow, and given by twice the harmonic mean $2Kk/(k+K)$ of the faster rate (lower barrier) K and slower rate (higher barrier) k. For very fast flips (γ large), the mean time for first passage is given by the arithmetic mean $(K+k)/2$. Between these extremes is a regime of even faster diffusion, which occurs when the mean time for first passage is of order γ. One would expect this mechanism to give Arrhenius behaviour, but it is hard to see how to relate the reduction in apparent activation energy to the recombination energy.

A second general point is tempting but actually ineffective. This is *stochastic resonance,* a phenomenon in which the effects of a weak periodic signal becomes enhanced (rises above a threshold value; 'enhancement' needs careful definition) by added noise with the power spectral density $S(\omega)$. In certain circumstances which are not too restrictive, the enhancement is greatest for a value of $S(\omega_0)$ at the frequency ω_0 of the periodic signal and with an amplitude which is of the order of the rate predicted by Kramers' theory for first passage when the signal is zero.

One simple way of thinking of this is to recall the non-resonant absorption processes, like internal friction and dielectric relaxation. For these, a weak periodic signal has a maximum of absorption at a frequency of the order of the rate induced by an incoherent thermal motion. In non-resonant absorption, the signal is so weak as to have no effect, and it is the noise which is of interest; in stochastic resonance, the signal is larger, but it is the signal, rather than the thermal noise, which is usually of interest. Fuller consideration suggests that stochastic resonance is not a mechanism which can contribute significantly in enhanced diffusion. We mention it as a reminder that there are subtle phenomena in the dynamics of complex systems, and we should recognise that many of the best studies are still naive.

A third point may be much more significant. *Usually we cannot rule out the involvement of electronic excited states of the defect.* In semiconductors, as in ionic solids, changes of electronic state may be occurring. If so, we should identify and seek such electronic states. After all, most theories include only a few of the excitations (standard restrictions of unit cell or cluster size are especially limiting). In particular, are there any states of a sort which would be general enough to need consideration for a wide range of defects?

The answer is yes, there is a general class of states which are not usually seen in calculations because of commonly-made working approximations. Consider a neutral defect $[X]^0$. Then *there are always Rydberg states* of the form $[X^+e^-]$. There is a theorem (Mott & Gurney 1948 p. 82; Stoneham 1975 Chapter 9) which requires there to be an infinite number of such states. The theorem needs only the existence of a long-range Coulomb potential of X^+ which falls off inversely with distance at large distances. The centre does not need to be a shallow centre. Indeed, the GR2–GR8 states of the neutral vacancy in diamond involve Rydberg states of this sort (Mainwood & Stoneham 1997). The loosely-bound electron in some of these $[X^+e^-]$ states will contribute little to the energy surface, which will be rather similar to that for X^+ (the screening of long-range effects by the electron will be modest). So, in suitable cases, if the saddle point for $[X]^0$ lies above that for $[X^+e^-]$, then there could be diffusion with a lower activation energy. Clearly, one can imagine elements of this idea combined with the usual local heating mechanisms. Suppose an *exciton* (or an e–h pair) recombines at $[X]^0$. Then approximately ϵ of the energy could be used to change the electronic state, and $R = E_{gap} - \epsilon$ is available, as before, for vibrational or other enhancement of diffusion in this new state.

7.4 Local excitation models, including the Bourgoin–Corbett mechanism

We discuss next the cases where recombination causes electronic excitation to a state for which there is a reduced barrier to thermal diffusion or similar motion. Examples for defects in halides include F centre diffusion (Härtel & Lüty 1963; Lüty 1968, 1971; Rascon & Alvarez-Rivas 1983; Hughes & Jain 1979; Salminen *et al.* 1996; Reichling *et al.* 1996), and reorientation of the F_A centre (Lüty 1968) and the V_k centre (e.g. Murray & Bethers 1969). There are indications of charge-state dependence of diffusion in other systems, including anion vacancies and interstitials in MgO (Scholz & Ehrhart 1993). For silicon, examples include interstitial aluminium Al_i, muonium (Mu; in its neutral state, Mu comprises a muon plus an electron) and the Si self-interstitial. In some cases, the lower barrier occurs for an electronically-excited state, in others for a different charge state (a special case of a different excited state), and for further systems, the enhancement may involve a metastable state. Enhanced motion by local excitation enhancement and metastability will often be found together. Since what is described is simply thermal diffusion over a new energy surface, we expect Arrhenius behaviour when there is local excitation enhancement. There is no reason why the reduction in apparent activation energy has to be less than the recombination energy, although this may often be the case.

Let us assume for the present that the adiabatic energy surfaces are appropriate and sufficient to describe behaviour; generalisations to include tunnelling are possible, but complicate matters. We shall label the ground-state energy surface A. Its lowest energy is E_{Ag}, and the saddle point of the relevant energy barrier has energy E_{Ax}. The barrier for adiabatic energy surface A is thus $\Delta_A = E_{Ax} - E_{Ag}$. Suppose further that there is an adiabatic energy surface for an electronically-excited state B. Since this is an excited state, its minima E_{Bg} must lie above those of the ground state surface A; we shall write $E_{BA} = E_{Bg} - E_{Ag}$. The corresponding saddle point on surface B has energy E_{Bx}, and the barrier for motion on that energy surface is $\Delta_B = E_{Bx} - E_{Bg}$.

We can imagine several distinct cases (Fig. 7.1). We emphasise that there is no reason why the atomic positions for the minima in states A, B should correspond to the same configuration. Nor need the saddle points for the two surfaces correspond to the same atomic geometries.

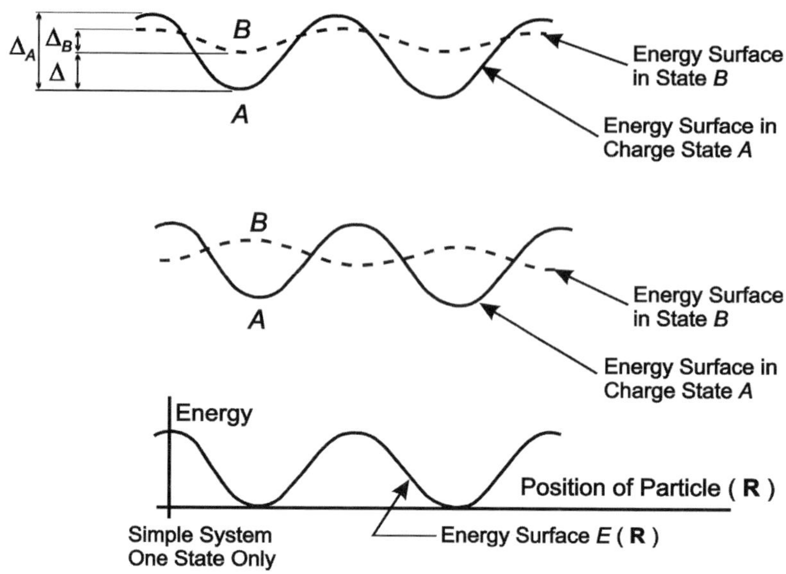

Fig. 7.1. Adiabatic energy surfaces, showing the several different relationships of excited and ground states.

7.4.1 *Case I: Energy extrema at the same site*

In alkali halides, the ground states and excited states of the F centre are associated with the same anion site; the same is true for the two halogen sites of the V_k centre. The lattice distortions differ significantly between ground and excited states, of course. The excited F centre (centred on an anion site) has both a substantial symmetrical relaxation and a dynamic Jahn–Teller distortion. The two halogen ions of the V_k centre (centred on an interstitial site) are pulled closer in the bonding ground state and pushed apart in the antibonding excited state. Both systems are from a wider group of defects and impurities for which the excited (B) and ground (A) electronic states have energy surfaces with minima at the same site.

If there is to be enhanced motion, the lifetime of the excited state needs to be long enough for jumps to be reasonably likely. We might guess that the relevant diffusion rate would be the diffusion constants on the upper and lower energy surfaces, weighted by the fraction of time for which each is occupied. This can be misleading, however, because most excitation will lead also to vibrationally-excited states, and 'hot diffusion', analogous to 'hot luminescence', may be important.

A further question is whether or not the diffusive jumps, starting from a minimum on energy surface B, actually take place via the saddle point on that surface. There may be cases where E_{Ax} lies between E_{Bg} and E_{Bx}, when it is possible that state-switching may occur, so as to reduce the effective activation energy in the excited (B) energy surface. Is the excited-state saddle point on surface B important in ordinary thermal diffusion (without excitation)? There are some systems for which one might expect two energy surfaces to be important even in thermal diffusion (Stoneham & Sangster 1985). If the saddle point for surface B has an energy E_{Bx} which lies below that at E_{Ax} for surface A, which saddle point will be the most important in thermal diffusion? Will the observed activation energy correspond to $E_{Ax} - E_{Ag}$ or to the smaller $E_{Bx} - E_{Ag}$?

These questions takes us back to the discussion of diffusion processes in Chapter 2. In two simple intuitive (and so inexact) pictures, the lower energy could be important. On the one hand, suppose that the reaction coordinate for diffusion is very close to being a normal mode, and that exchange of energy between modes is very slow. The natural description to use is the Rice–Slater dynamical theory, or an analogous first-passage theory. Suppose further that switches of state between A and B can take place very readily. As the system moves towards the saddle point along surface A, it will switch to B with high probability, probably near the cross-over, where the energies are the same on each surface. The thermal activation energy expected will be $E_{Bx} - E_{Ag}$, smaller than that on surface A alone. On the other hand, we may approach the diffusion process in a way more like reaction rate theory (Vineyard theory). In thermal equilibrium, there will always be a certain population on energy surface B. This needs only those changes of state to maintain equilibrium populations; the transitions need not be associated with any transport event. We need to look at the probability that the system is in a particular region of phase space near the alternate saddle points and ground states. It is primarily these probabilities which determine whether the most important saddle point is that on energy surface A or B.

7.4.2 Case II: Energy surfaces with extrema at different sites

The energy surfaces have distinct minima if, for example, the ground-state energy surface A has a minimum at a tetrahedral interstitial site in the diamond structure, and excited state B has a minimum at a bond-centre or hexagonal interstitial site. Long-range motion may be possible by moving alternately from energy surface A to energy surface B. This will be the favoured path in thermal equilibrium only if E_{Bg} lies below E_{Ax}, of course, but it will always be a possible alternative channel. Normally, there is a small number of ground-state sites to which the excited system can decay, and this number is determined by lattice geometry. Excitation from specific site A_1 on surface 1 can lead to relaxed excited states at one of specific sites $B_1, B_2, \ldots$; decay from one of these B_n will generate the ground state at one of the sites A_m on surface A. For example, a hexagonal site in the diamond structure has two adjacent tetrahedral sites, so one might expect that excitation would lead to a 1 in 2 chance that the system will make a diffusive jump.

The muon shows some striking examples of this mechanism. There are three important states of the positive muon (μ^+) in semiconductors: μ^+ (or Mu^+), $[\mu^+e^-]$ (or Mu^0, muonium) and $[\mu^+e^-e^-]$ (or Mu^-). These have stable or metastable states at certain of the interstitial sites in semiconductors, notably the tetrahedral site T, and the bond centre site BC. When dealing with a system which has a finite lifetime (for the muon, this is a few microseconds), there is no guarantee that the ground state is achieved.

There is a rich variety of behaviour. One could say that muon motion in semiconductors seems to be able to show every mechanism of enhanced diffusion except local heating. For Si, the high-temperature diffusion is controlled by switching between $Mu_T{}^0$ and $Mu_{BC}{}^+$; the neutral muonium is more stable at the BC site, where it is essentially immobile. In GaAs, the same process can contribute to diffusion. For Ge, the neutral state is stable or metastable at both T and BC sites (the energies are very similar at these two sites), so that the diffusion processes can involve $Mu_{BC}{}^0$, $Mu_T{}^0$ and $Mu_{BC}{}^+$. $Mu_{T(Ga)}{}^-$ is mobile with an activation energy of 0.68 eV, without evidence for enhanced diffusion. The electron/hole charge cycles seen are between $Mu_T{}^-$ and $Mu_T{}^0$ (which would enhance diffusion by exploiting a lower activation energy, if at all) at high donor concentrations, and between $Mu_T{}^0$ and $Mu_{BC}{}^+$ in weakly n-type and semi-insulating GaAs. Muons in Si show one other important phenomenon, in that they interact with the ionisation which they cause as they are implanted. This possibility was recognised early on (Stoneham 1979c), and has been analysed by elegant experiments exploiting applied electric fields (Storchak *et al.* 1997). It appears that $Mu_T{}^0$ is formed promptly, whereas $Mu_{BC}{}^0$ is formed by the delayed capture of radiolytic electrons.

Photo-induced reorientation in alkali halides is seen in so-called Type II F_A centres, even at very low temperatures, whereas the reorientation of other F_A centres is thermally activated. The F_A centre (Lüty 1968) consists of an impurity alkali adjacent to an F centre (a neutral halogen vacancy). When the alkali is the small Li ion, the centre has a low symmetry, and such centres are said to be of Type II.

Another example, although less certain, concerns the anomalous heat generated in AgCl and AgBr following photoexcitation (Kondo *et al.* 1997). Heat generation can continue for hours after excitation. Fu & Song (1999) suggested that there is a critical diffusion step for which the activation energy depends critically on the state of a diffusely-bound electron.

7.4.3 The Bourgoin–Corbett model

A special example of Case II occurs when the energy surface *B* corresponds to a different charge state to energy surface *A*. The muon in Si provides one such example. This, the Bourgoin–Corbett mechanism, was perhaps the first serious model for enhanced diffusion. The mechanism involves successive alternate captures of electrons and holes; these take the defect from one charge state to another, and so from one site to the next. The Bourgoin–Corbett (1978) model was first suggested as an explanation of the athermal motion of the self-interstitial in silicon at the very lowest temperatures, as in the generation of vacancies and interstitials by electron irradiation at helium temperatures. The mechanism was first validated from self-consistent calculations by Masri, Stoneham & Harker (1983), and later confirmed by many others by more powerful methods. More recent measurements (Hallen *et al.* 1999) imply an effective value of 0.065 eV for the migration of the Si self-interstitial below room temperature, which strongly suggests that enhanced diffusion is occurring.

Very rapid diffusion of the silicon interstitial is also observed at higher temperatures in the rapid thermal annealling of low-energy implants of B into Si (Cowern *et al.* 1997). These implants, aimed at shallow junction formation for the next generations of complementary metal-oxide–semiconductor (CMOS) technology, are followed by a rapid thermal anneal (10 s at 1000 °C). Most of the implanted B remains where it is expected to be from Monte-Carlo calculations. However, there is an extended tail to the distribution of B, and experiments involving trapping layers show that both B and Si interstitials diffuse rapidly. The diffusion constant deduced for the Si interstitial at room temperature is high: 10^{-7} cm^2/s, four orders of magnitude greater than previous results. The precise mechanism is not yet clear.

In any mechanism based on charge state changes, one expects some dependence on Fermi level. However, in low-conductivity systems, we note that thermal equilibrium over charge states may be set up only slowly. For MgO which has been X-irradiated, for instance, recovery times can be hours or days. It is therefore essential to monitor actual charge states in experimental studies, rather than to presume equilibrium with a well-defined Fermi energy. Even in silicon, centres like the positive vacancy V^+, which shows charge disproportionation (negative *U*), can survive for long enough to be studied.

Baraff *et al.* (1983) suggested the enhanced motion of Si:Al_i also involves a change of charge state, and is therefore also of this type. In their model, both the normal thermal process and the enhanced process involve charge switches. Different charge states are involved for thermal and enhanced diffusion. The ideas are a natural extension of those proposed by Bourgoin & Corbett. Baraff *et al.* (1983) argued that the normal and enhanced processes are these:

Normal thermal process:	usual site:	Al_T^{2+},
	saddle point:	Al_H^{2+} or $Al_H^{3+} + e_v$;
Enhanced process:	usual site:	$Al_T^0 + e_c$,
	saddle point:	Al_H^-.

The subscripts T and H represent, respectively, the tetrahedral and hexagonal interstitial sites in the Si lattice; e_c is a conduction band electron, and e_v a valence band electron. Carrier injection shifts the charge state so as to give diffusion on a different energy surface. In both the normal and the enhanced processes, the most likely state at the saddle point has a different charge state from that at the equilibrium site. The range of charge states (3+ to 1−) raises a number of technical problems. Since these technical challenges are general ones, we give the breakdown of the terms which are needed to assess the possible enhancement of diffusion. In essence, Baraff *et al.* calculated two key energies at the T site:

$$3e_c + Al_T^{2+} \rightarrow 2e_c + Al_T^{+} \qquad E_G - 0.1 \text{ eV},$$

$$2e_c + Al_T^{+} \rightarrow e_c + Al_T^{0} \qquad 0.0 \text{ eV},$$

and four key energies at the H site:

$$e_v + 3e_c + Al_H^{3+} \rightarrow 3e_c + Al_H^{2+} \qquad -0.1 \text{ eV},$$

$$3e_c + Al_H^{2+} \rightarrow 2e_c + Al_H^{+} \qquad E_G - 0.3 \text{ eV},$$

$$2e_c + Al_H^{+} \rightarrow e_c + Al_H^{0} \qquad E_G - 0.5 \text{ eV},$$

$$e_c + Al_H^{0} \rightarrow Al_H^{-} \qquad E_G - 0.7 \text{ eV}.$$

The numbers listed in each case are theoretical, with a certain accuracy (the precision might be 0.1 eV; the accuracy is harder to estimate). Overall, the enhanced process has a barrier less than that for the normal thermal process by $2E_G - 1.5$ eV, using the numbers given. But which gap should be used? The predicted gap, using the same local density approximation, is only 0.6 eV, and would yield a larger energy for the enhanced process. The experimental gap is 1.2 eV, and would yield an enhanced process with an activation energy appropriately smaller (by 0.9 eV) than the normal thermal process.

7.4.4 Analogous systems: Metastability and reorientation

Excitation can modify certain defects in ways which need not lead to diffusion. The defect usually remains associated with the same lattice site. For example, there are several cases for which a low-symmetry defect reorients on excitation.

The initial and final states are related by a symmetry operation, and so have the same energy as each other. In other cases, there will be different local structures possible. There is no guarantee that diffusion will be influenced in any way, although some metastable states may have a role as intermediates in a diffusion path. Often, however, it may be possible to align defects, and so introduce optical or other anisotropy.

For the V^- centre in ZnSe, the hole is localised on one Se next to a Zn vacancy. Excitation for that geometry is to a charge transfer state involving the other three Se ions neighbouring the vacancy; a localising relaxation occurs to leave the hole on just one of these distinct Se ions. Another example is the reorientation of the F_A centre. For Jahn–Teller systems (see, e.g., Fig. 8.2 or Fig. 8.8 of Stoneham (1975)) the initial energy surface might correspond to a tetragonal distortion along the x axis, and the excited state to a surface with minimum energy for a y or z tetragonal distortion. Similar situations are possible for small polarons localised near to defects, and indeed the V^- centre could be regarded as an example.

Metastable defects show great variety (e.g. Watkins 1990). They can occur in excited states, as noted in our discussions of the self-trapped exciton and close defect pairs. They can occur as alternate solid-state structures, whether amorphous or crystalline, including the many forms of SiO_2 (we remark that the thermal donors in Si, in which oxygen is a major component, also appear to show bistable forms). Organic systems are rich in alternate structures, such as the many forms of polyacetylene. There are also systems analogous to those which show enhanced diffusion, having at least two related but different structures. Watkins remarked that the thermally-activated barriers observed under optical and electrical minority carrier injection are *less* than those observed in charge state change experiments, which indicates parallels to enhanced diffusion. Examples of metastable systems include the following:

1 Charge transfer systems Charge transfer, usually of the form $X^{M+} + Y^{N+}$ going to $X^{(M-1)+} + Y^{(N+1)+}$, is the origin of colour in many minerals and glasses. Charge transfer underlies the photochromic centres in materials like CaF_2, where there is a reversible charge transfer between a rare earth and a photochromic defect (the two forms are $RE^{2+} + PC^+$ and $RE^{3+} + PC$). If there is a charge state dependence of the motion energy of the components between which charge is transferred, this may show through altered defect reaction rates. An example is the oxygen Frenkel pair in MgO (Scholz & Erhart 1993), where there is evidence of ionisation-assisted recombination.

2 Close pairs of defects with alternative relative positions Examples include several defects involving carbon interstitials in silicon. One is Si:(C_i, D), with

D a Group V donor, such as Sb. There are four distinct configurations of C_i–Sb pairs. There are some analogies with donor–acceptor pairs. The C_i can have 0, 1− or 2− states; there may be some control depending on how the sample has been cooled. Another case is Si:(C_i, C_{Si}). There are two configurations, related by a simple bond switch. The A configuration is an (Si, C) interstitialcy, not far from (100); the B configuration has both C atoms on site, with an Si interstial not far from their join.

3 On-site, off-site pairs Here some part of the defect switches between an interstitial configuration and an on-site position, such as an antisite defect. An example is the EL2 centre in GaAs, which is related to the antisite, As_{Ga}. In one model, As_{Ga} is metastable when the As moves off-site, giving a Ga vacancy and As_i as the alternative configuration. The model (occasionally described as extrinsic Frenkel pair formation) is still debated; in another model, the centre comprises As_{Ga} and As_i; the metastability corresponds to two alternative positions for the arsenic interstitial As_i.

4 Shallow–deep forms In these cases, the defect can occur with more than one relaxed state, these states having different ionisation energies. Thus CdF_2:In can be either shallow or deep (Langer 1990). The DX centre in $Al_xGa_{1-x}As$ has different forms depending on the Al concentration. For $x<0.22$, the shallow A configuration is stable, and the material is n-type. When $x>0.22$, the deep B configuration is stable, and the material has high resistivity. There are at least two models. In model 1, both A and B configurations are neutral, with a shallow–deep instability. In model 2, B is D^- and A is D^+ (+2e). Chadi & Chang (1988) argued that this may involve the ejection of a substitutional donor onto an interstitial site, which has also been described as extrinsic Frenkel pair formation.

7.5 How can the mechanisms be distinguished from each other?

We have remarked that many experiments have been analysed using simplistic local-heating (phonon-kick) models, rather than Sumi's model or some other more sophisticated approach. Ignoring mere simplifications, e.g. an inappropriate value of the recombination energy R because the one-electron picture implied is inadequate, the most important issues are these. First, observation indicates an unexectedly high efficiency factor α in the use of the recombination energy R (see Table 7.2). The major difficulties with $\alpha \approx 1$ (and addressed in the papers by Sumi and others) are the relationship of the reaction coordinate to the direction of the phonon kick, the fact that the reaction coordinate is not a normal mode, so there is rapid dephasing to damp the energy available to assist a jump process, and the Arrhenius behaviour usually seen. The simple and common picture implies a first-passage description of diffusion. The excess vibrational energy would be added to the existing thermal vibrational energy, leading to a temperature dependence of the enhanced diffusion of the form $\exp[-A/(kT+\beta R)]$, not the Arrhenius form $\exp[-(A-\alpha R)/kT]$. In all, the simplest picture will not do.

We have also remarked on cases in which there is a mixture of electronic excitation and vibrational enhancement of diffusion. When a new case of enhanced diffusion is identified, especially for a complicated system, the mechanism can only be identified unambiguously through careful arguments.

7.5.1 Consistency arguments

Any suggested mechanism must satisfy some basic points. The system must be stable when there is no enhancement. The energies should balance properly, so that the recombination energy is sufficient (allowing for a degree of inefficiency) and the bandgap must be wide enough to correspond to the reduction in the apparent activation energy. If the enhancement saturates, this should happen at the expected intensity, given credible cross-sections. Only in a few cases has consistency been considered seriously.

Perhaps the fullest discussion to date has been for interstitial Al in Si (Troxell *et al.* 1979; Baraff *et al.* 1983), which indicates some of the successes and some of the problems in this area. Troxell *et al.* deduce that one extra jump occurs for approximately every 10 electron–hole recombination events. This is a high efficiency, although it is both arguably reasonable and less than that claimed in some other systems. For laser-evaporated Ag in a thin film of Xe, Kometer & Schwentner (1996) observed light-induced migration. They stated that the jump probability rises from about 10% to about unity at high temperatures. Precisely what mechanism is involved is not clear; their own description involved the large kinetic energy in the electronic ground state following recovery to the ground state, but the ideas were not developed fully. There is a metastable state which has a generation probability of about 1.5% (falling with temperature) which they noted as a possible intermediate position in the migration process.

7.5.2 Reasonableness arguments

Arguments concerning the *reasonableness* of the process are more debatable. Often they will concern the efficiencies of processes: are the pre-exponential factors sensible? It is widely and reasonably held that pre-exponential factors for conventional allowed processes involving lattice vibrations (whether diffusion or non-radiative transitions) lie in the range 10^{10}–10^{14} s^{-1}. This follows from various plausibility arguments. It is not credible that processes transfer energy to lattice vibrations faster than a (net) rate of one event per vibrational period (see Section 3.5). Slower rates are possible, of course. One example is that rates can be very different if spin-forbidden, since spin flips can be very slow and rate-limiting. Another example is that excited states may be occupied only transiently. Suppose that excitation occurs every τ_x and the lifetime of the excited state is τ^*; suppose the process has rate R in the normal state and R^* in the excited state. Then we expect the overall rate to be of the order of the weighted mean of R and R^*, subject to certain qualifications. If so, the excited state process will have a pre-exponential factor reduced by something of the order $\tau^*/(\tau_x+\tau^*)$.

7.5.3 *Are charge state changes possible and significant?*

Suppose that a carrier capture process or other electron transfer event is rate-determining. Suppose there is a change in charge state of the 'dynamic' sort. Then we must be aware of orders of magnitude such as these for normal semiconductor systems. The time in transit of the diffusion jump is less than 10^{-12} s. The capture time for free carriers is $(n\sigma v)^{-1}$, and the components have these typical magnitudes: carrier density n, 10^{16} cm^{-3}; capture cross-section σ 10^{-16} cm^2; thermal carrier velocity v 10^7 cm/s. Thus the capture time is 10^{-7} s, which is far too long to occur during the diffusion jump. Clearly, instead, one should think in terms of an equilibrium ensemble in which there is a population of defects in the charge state favoured at the saddle point, not of a dynamic process in which the ion captures an electron on its trajectory.

In transition metal oxides, the situation is different. Transient changes of charge state during a diffusive jump are a possibility. Stoneham & Sangster (1985) identified NiO:Mn as one case where this seems likely. In normal situations, both for Mn next to a cation vacancy and in an otherwise defect-free lattice, Mn^{3+} is stable relative to Mn^{2+} and an adjacent hole (which was treated as Ni^{3+} replacing Ni^{2+}; clearly, now that it is known that the hole spreads over the oxygens in NiO, there could be some changes, but these do not alter the basic ideas). At the saddle point for motion by a vacancy mechanism, the stable situation is Mn^{2+} and Ni^{3+}. Assuming this conclusion to be correct, what can one say about the characteristic time for the charge state change. In this case, however, it is not free carrier capture, but motion of a valence band hole from one ion to a neighbour. This is determined by the bandwidth ΔE_v, of order

0.5 eV, and the characteristic time $h/\Delta E_v$ is of order 0.01 ps. Since the jump time is between one and two orders of magnitude longer, it seems that electron exchange will not be a limiting factor, and that a dynamic switch of energy surfaces is possible.

7.6 Issues in enhanced diffusion: Further discussion of mechanisms

In this section, we ask whether there are cases where we could have sensible control of a mechanism, as opposed to uncontrolled rate changes.

7.6.1 Competing processes: Isotope effects in electrical isolation

Enhanced processes are, by definition, relatively rapid, so it is perhaps surprising to find situations in which they have serious competition. One example is when there is another channel for rapid electron–hole recombination. An example is found in the use of H isotopes to create electrical isolation in GaAs (Steeples *et al.* 1980). For these studies of isolation, resistivity in GaAs was increased by H, D, and T implants with energies and doses chosen to make the conditions as similar as possible. There is evidence that H, D, and T all end up in the same defect structures. It is not too difficult to make arguments suggesting that H should be more effective in causing isolation than D, which should be better than T. Nor is it difficult to make arguments suggesting the converse. Experiment shows neither of these monotonic trends: D is substantially more effective than H or T. The highest resistivities for D are more than an order of magnitude larger than those for H or T, and are achieved at doses two orders of magnitude smaller.

What is the explanation? In all cases, it seems that the residual damage (vacancies, interstitials, and their aggregates) causes the resistance, so that more vacancies and interstitials survive for D implantation. This means that the defects are eliminated more effectively with H or T implantation. Presumably this extra elimination of damage involves recombination-enhanced diffusion, to a large degree. One possible way that D could interfere with enhanced diffusion and defect annealing would be for D to be a far better recombination centre for electrons and holes than H or T.

How could this be, for the only differences are isotopic, not chemical? The major effect of changing isotope is on local mode energies. Infrared spectra show the expected differences in local mode energies. Since the local mode energies are large, if electron–hole recombination occurs at these defects, most of the energy will be taken up in these local mode vibrations. The remaining energy Δ_{res} will go to other degrees of freedom, such as bulk phonons (the

GaAs LO phonon has energy 0.036 eV, far less than the H local mode energy). A critical question is the size of Δ_{res}: the larger this energy, the less effective will be the defect as a recombination centre. The data show that for H, recombination could involve five local mode phonons and leave a large residual of 0.18 eV; for T, nine local mode phonons would be needed, with a substantial residual of 0.11 eV. For D, however, eight phonons leave a negligible residue, of order 0.01 eV. The close match of bandgap and an integral number of local mode phonons appear to be the basis for the dramatic effectiveness of D. We use the word 'appear' intentionally, since the analysis does not address some possibly serious issues relating to anharmonicity, nor does it address some of the additional small energies involved.

7.6.2 Dislocation growth and motion

Many enhanced processes show up through macroscopic defects, like dislocations or inclusions. Examples listed in the review of Stoneham (1981b) relating to devices include processes limited by carrier transport (like the so-called large dark spot), macroscopic inhomogeneities like precipitates or inclusions which lead to pulsations in injection lasers, and the production of point defects and dislocation kinks.

Many of the data for dislocation motion (Hirsch 1985) are described phenomenologically by an expression for the dislocation velocity which is thermally-activated and proportional to some power m of the resolved shear stress. When effects are seen which enhance dislocation mobility (or dislocation velocity when the stress is independently determined), we must be aware of several distinct types of mechanism.

Direct excitation of electron–hole pairs

If there is direct excitation by light or an electron beam, one can anticipate motion involving kinks or jogs along the dislocation, enhanced in the same ways that point defect diffusion can be enhanced. Thus Hirsch (1985) observes that bandgap light or electron irradiation enhances dislocation velocities in Si and GaAs, and remarks that this is attributed to non-radiative recombination at dislocations, possibly at kinks. The data of Maeda & Takeuchi (1983) for n-GaAs show the effect clearly, with a reduced activation energy for glide in the presence of the electron beam, and a rate which increases with the electron current. For the α (B(g), As type for 30° partials) dislocations, the activation energy is reduced from 1.0 eV to 0.29 eV; for the β (A(g), Ga type for 30° partials) dislocations, the reduction is from 1.7 eV to 0.6 eV.

One can also imagine point defect motion or reactions stimulated by excitation and contributing to the growth or climb of dislocations. In KCl, electron beam excitation leads to the growth of perfect dislocation loops (Section 5.3). The loops form from K_i and Cl_i created by interstitial Cl_2^0 molecules punching out cation–anion divacancies. If we regard the molecule in the divacancy as two adjacent defects, then the Cl within the cation vacancy would be an antisite Cl_K. Interstitial anions and cations are produced as halogen molecules create anion–cation vacancy pairs. Similar phenomena are seen in the III–Vs like GaAs (Hayes & Stoneham 1985, p. 175). Petroff & Kimerling (1976; also Petroff 1979) proposed an ingenious mechanism by which Ga excess and As Frenkel defects can lead to climb, which involves defects on both Ga and As sublattices. Stoneham (1986, unpublished) suggested antisites would be more effective: interstitial loops could result from Ga_i present from the Ga excess and As_i from a reaction like $Ga_i \rightarrow Ga_{As} + As_i$. Dislocation networks can act as non-radiative recombination centres, and grow during operation in the active regions of GaAs heterostructure lasers and GaP light-emitting diodes.

Excitation-enhanced motion of grain boundaries is observed in oxides, like MgO. Indeed, Kizuka *et al.* (1998) show that irradiation by 200 keV electrons leads to the motion of the [001] $\Sigma 5$ grain boundary. Three types of structural unit appear to be involved, and the transformation or translation of these units determines the migration speed. However, the electronic processes which determine this motion are not yet clear.

Excitation affecting pinning defects

Dislocation motion is often limited by pinning by point defects. If excitation makes these defects more mobile, or if excitation changes the defect state (such as charge) then the point defect interaction with the dislocation will be altered, and the dislocation mobility will change. Ossipyan (1983a,b) suggests that the large increase in flow stress on illumination of ZnSe stems from changes in the dislocation interactions with point defects.

Charge state changes

Suppose a change of Fermi level E_F affects dislocation motion in the absence of other excitation. Suppose in addition that point defect populations are not affected. We can envisage several different reasons for changes in dislocation mobilities. Most theories follow Hirth & Lothe (1982) in relating the dislocation velocity in the high-stress regime for long segments to the generation and diffusion of kinks. The relationship has the form

$$\text{dislocation velocity} = (A/kT)\exp[-(F_k + W_m - TS)/kT],$$

in which A is a constant, F_k is the kink formation energy and W_m the kink motion energy. For the 90° glide partials in Si, Öberg *et al.* (1995) predicted F_k to be 0.1 eV and W_m 1.8 eV. They found corresponding values for GaAs to be

$F_k = 0.07$ eV (α [As] partial) and $W_m = 0.3$ eV (β [Ga] partial); in both cases, the motion energy was significantly larger than the formation energy. For reasonable agreement with experiment, the entropy is of order $5k$–$7k$. The prefactor is linear in the applied stress, and has the form expected for forced diffusive motion.

A central question is whether or not kinks can exist in more than one charge state on the dislocations affected. A number of authors (Haassen 1975; Frisch & Patel 1967) have suggested that the line charge is a critical quantity. If the kinks cannot trap either electrons or holes, one should also consider whether they may change their charge state dynamically during the diffusion jump, as discussed above. The existence of traps depends on the extent to which dislocations reconstruct. For GaAs, for example, Sitch *et al.* (1994) found the As core less strongly reconstructed than the Ga core (which has some hole traps near the valence band edge), and argued that this explains the higher mobility of the As cores. They remarked that acceptors (donors) lead to a local breakup of the Ga(As) core reconstruction but strengthen the As(Ga) core reconstruction. Presumably, the pronounced locking effects of donors and acceptors are partly the result of their effects on the reconstruction.

Dislocation mobilities in silicon depend on the position of the Fermi level. Motion is affected with n-doping; for p-doping, the effects are anomalous (Hirsch 1985; Patel *et al.* 1976, 1977). Hirsch's analysis of the data (Hirsch 1980, 1981, 1983, 1985) was, however, broadly in accord with trapping occurring at localised states. R Jones (1980) did not find gap states for the relaxed ground state of 90° partials in Si. He then looked at the consequences of there being states in the gap for the saddle-point configuration. It becomes necessary to follow the reaction rate theory through in detail, which means looking at the partition functions for the ground and saddle-point states. The expression, which is complicated, leads to a thermal activation energy involving the new energy levels for the saddle point and the electronic chemical potential. For Si, many of the results duplicate those of Hirsch, and the activation energy predicted for intrinsic Si is close to the observed value of 2.1 eV. For p-Ge, the observed dependence of velocity on acceptor concentration is predicted. However, the ratio of velocities in p-Ge and intrinsic Ge predicted is larger than that observed by a factor of 2, although there are possible explanations.

In MgO, charge injection during electrical degradation involves dislocation production. Typical dislocation densities in as-grown MgO are about 10^5 cm/cc; this rises to about 10^{10} cm/cm^3 just before breakdown (Y Chen 1997, private communication). The process is not clear, apart from the need for charge injection. It seems likely that there is dislocation climb, driven by charged features (like kinks) on the dislocations.

Effects of thermally-generated electron–hole pairs

Another oxide which shows evidence for some sort of electronic excitation enhancement is UO_{2+x}. What is observed is that creep changes its character at the high temperatures at which UO_{2+x} has a high intrinsic electron–hole population. At low temperatures, creep (Herring 1950; Ashby 1972; Burton 1977; Brethean *et al.* 1979) is consistent with a rate linear in stress and limited by uranium ion diffusion. The activation energy is around 4 eV, and decreases with O excess x. Slagle (1984) noted there is a second contribution with much the same activation energy, but varying as stress σ^n with n around 4.5. When the temperature rises above about 2000 °C, there is a further contribution with a large activation energy, about 12 eV, which again varies rapidly with stress (n decreases from 6.5 at 2100 °C to 3.5 at 2600 °C). The mechanism is not clear, but presumably it is related to the large rise in electrons and holes at these temperatures. UO_{2+x} is an intrinsic semiconductor at high temperatures, as is shown by its specific heat and thermal conductivity. The electrons and holes are both self-trapped (so, to a first approximation, the hole can be thought of as U^{5+}, and the electron as U^{3+}). The effects on creep may well involve enhanced motion of dislocations, but it is not known whether the enhancement is because of available recombination energy, charge state changes, or some screening effect. There are several possible indirect effects, such as new phases developing at grain boundaries. Calculations by J H Harding (1980, unpublished) suggested that electron–hole pair production affects mainly the elastic constant c_{44} and the static dielectric constant, which will have some effect on creep.

7.6.3 Enhanced oxidation

The oxidation of silicon can be enhanced by an electron beam (Collot *et al.* 1985). The enhancement occurs at even the lowest energies available. The likely reason (see also Mott *et al.* (1989)) is dissociation of the mobile interstitial oxygen molecules, leading to species (perhaps O^0 or O^-) which are even more reactive or mobile. The negative ion would also be attracted towards the Si by the image interaction, since Si is more polarisable than its oxide (Stoneham & Tasker 1987). Theory (Szymanski *et al.* 2000; Stoneham *et al.* 2000a,b) predict that O_i^0 should be incorporated into the oxide network readily, in agreement with the isotope exchange observed by Collot *et al.* Other workers have argued that excitation stimulates the oxidation of metals by assisting the achievement of the Mott–Cabrera potential (Cichy & Fromm 1991; Martin *et al.* 1994). This potential is the difference between the Fermi energy of the metal and the energy of the acceptor state on the surface (which might be O_2 capturing an electron to form O_2^-). This potential may then assist the injection or transport of a species across the oxide film.

An analogous mechanism has been suggested by Irene & Lewis (1987) to explain the observed laser enhancement of silicon oxidation. Their idea was that the enhancement is due to the thermionic emission of electrons from the silicon into the oxide. Oxygen molecules are then dissociated, either at the Si/oxide interface or within the oxide. Similar ideas have been discussed for the oxidation of Fe (Liebbrant *et al.* 1992). Mesarwi *et al.* (1988) suggested that the laser enhancement of the oxidation of polycrystalline Ni is associated with the excitation of adsorbed oxygen molecules.

Electron beams stimulate the oxidation of Ni(111), but apparently for quite different reasons. Stirniman *et al.* (1995) suggested that the beam creates metastable nucleation sites with a halflife of perhaps 10 min. When the beam is on, the rate falls as the temperature rises; without the beam, the rate rises with increasing temperature.

Summary

Semiconductor crystals (III–V, Group IV) with relatively wide bands and small gaps No self-trapping occurs, and there is no defect formation except under special conditions. Instead, there is enhanced diffusion or enhanced structure change (such as amorphous to crystalline). Enhanced diffusion has several variants: local heating/vibrational enhancement; local excitation: excited states; the Bourgoin–Corbett mechanism: charge switching. Defects like dislocations can be affected, as well as point defects.

The vibrational motion must be considered in some detail. One must distinguish between the *reaction coordinate*, directly associated with change or with the electron–lattice coupling which drives change, and the *accepting modes* which take up the energy. There is an important role for cooling transitions (the loss of excess vibrational energy) and the role of excitation in modes other than reaction coordinate. The distinction between mode and corrdinate can prove important (Chapter 3).

Chapter 11 discusses effects at very high energy densities. Chapters 9, 10 discuss surface and interface phenomena.

Chapter Eight

Local lattice modification of amorphous materials by electronic excitation

Amorphous materials lack long-range order in one of several possible ways. What happens in an amorphous solid following electronic excitation may differ from what occurs when a crystalline solid is similarly excited. In this chapter, we compare the behaviours following excitation of crystalline and amorphous materials. In amorphous materials, electronic excitation induces photoluminescence and generates defects, as in crystals. Excitation of amorphous materials can lead both to distinctive gradual structural changes and to changes of macroscopic properties. In this chapter, we compare the processes induced by the excitation of specific materials in their crystalline and amorphous forms. Photo-induced processes in amorphous materials have been reviewed by Shimakawa *et al.* (1995) and by Morigaki (1999).

There are two major complications in referring to amorphous materials (see Stoneham (1987b), pp 4–6). The first issue concerns whether a solid is amorphous or not. Normally, an experimenter will call a solid 'amorphous' when there are diffraction rings, rather than spots. A theorist will say it is amorphous when there is no long-range order. These two descriptions are not equivalent. For example, quasi-crystalline and para-crystalline materials may show diffraction spots without having long-range order. Likewise, diffraction rings may indicate a powder or micro-crystalline material. The second complication is that whereas one can be sure what is meant by a crystalline system in thermal equilibrium, an amorphous material is not so well defined. For example, one might have amorphous states of an oxide formed in different ways (normal glass formation, splat-cooling, heavy-ion irradiation, growth on a substrate, etc), and these amorphous forms may behave in different ways. The amorphous oxide grown on silicon can change its local structure under compression. Semiconducting amorphous silicons are different from the metallic liquid Si. It is common to distinguish between several types of disorder: *alloy disorder*, as in a mixed alkali halide crystal like NaF_xCl_{1-x}; *hard-sphere disorder*, as in a liquid rare-gas solid; *coordination disorder,* as in a-Si when there are some silicon atoms with four neighbours, some with three-fold coordination, and so on; and *ring-size disorder*, where the closed rings in the structure vary in size (Fig. 1.3) In polymers, even crystalline polymers, there will be differences in length from one chain to another; in non-crystalline polymers, the cross-linking and places at which the chain bends will also vary from one molecule to another. In general, some of the differences will be differences of topology (ring-size disorder, coordination disorder) and other differences will depend on the extent to which carrier energies fluctuate from one site to another (both relaxation energies and the diagonal disorder of theoretical models). These differences have profound implications. The energy fluctuations from one site to another may determine whether or not self-trapping is possible at a site.

The word 'phase', in the precise meaning of a thermodynamically-defined phase, is usually inappropriate for amorphous systems. Indeed, Solomon (1998) describes a-Si:H as 'a metastable system in thermodynamic equilibrium', arguing that there is a 'hydrogen glass' embedded in a silicon skeleton. Nevertheless, there are consistent and reproducible forms which are phases in an informal sense which we shall use occasionally here. The intermediate incommensurate form of silica via which the α to β transformation of quartz proceeds (Bachheimer 1986) is an example. If we look at the ensemble of information (Navrotsky 1987) on the forms of SiO_2 shown in Fig. 8.1, there are systematic features in the excess enthalpy per SiO_2 unit as a function of its volume (Stoneham 1991). For both crystalline and amorphous forms, the variations from one form to another of excess enthalpy with volume are similar in order of magnitude to the elastic energy for the same change of volume under pressure. However, the minimum energy for an amorphous system occurs at a volume greater by about 25% than that for quartz, the minimum energy crystalline form, and the minimum excess enthalpy for an amorphous form is greater than that for quartz by about 0.25 eV per SiO_2 unit. This energy is similar to the value of 0.5 eV for the maximum stored energy in quartz irradiated by heavy ions (Antonini *et al.* 1978). Roberts *et al.* (1976) gave 0.09 eV for the stored energy in neutron-irradiated vitreous silica, and reported 0.05–0.1 eV for neutron-irradiated silicate glasses, and 0.13 eV after α irradiation. All these energies are significantly larger than the maximum energies which can be stored in crystalline materials through deformation to the plastic limit, or as electrical field energy close to breakdown, which are both typically 0.5–5 meV/atom (Chapter 6).

Topological descriptions

We shall concentrate on non-metallic amorphous materials, such as amorphous carbon (a-C) and silicon (a-Si), chalcogens (Se), and chalcogenides

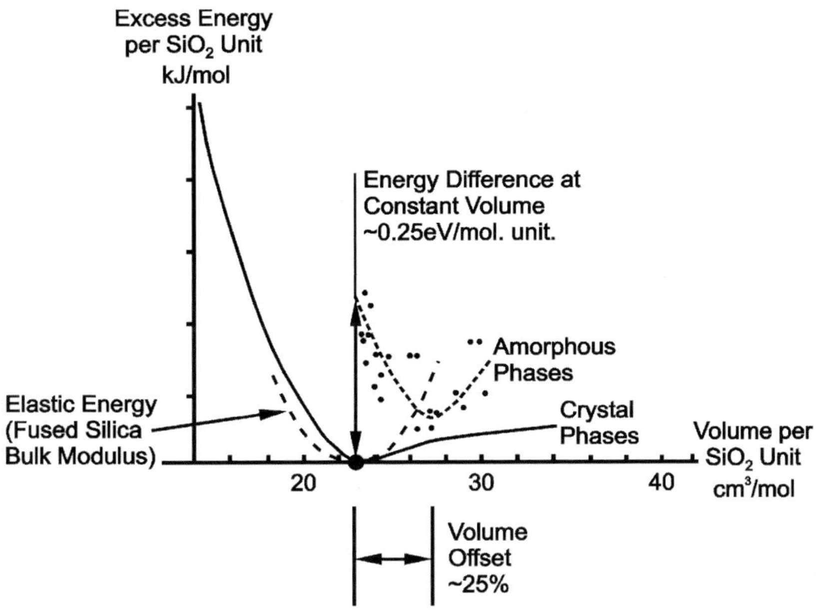

Fig. 8.1. Enthalpies and volumes per molecular unit for different forms of SiO_2 (based on Navrotsky (1987)). The amorphous forms are systematically less dense, and have a larger energy per molecule. Both crystalline and amorphous forms show trends in common with the elastic energy of vitreous silica.

(As_2S_3, and As_2Se_3) and fused silica (SiO_2). Sadly, there has been little work on ice. These are covalent, rather than ionic, solids and it is the saturatable directed bonds which make possible the arrangement in an open network. The covalently-bonded atoms usually have the same coordination numbers as those in the corresponding crystalline materials, although the symmetry is missing (Section 1.3.2, also Fig. 1.3). The idea of saturatable directed bonds leads naturally to the important *continuous random network* model, in which each atom has an ideal coordination. This model is simplistic, of course, but the continuous random network is a useful reference structure. The network description represents topological constraints, e.g. each Si should have four oxygen neighbours and each oxygen two silicon neighbours in an amorphous silica.

Clearly, real solids may not achieve this ideal topology. There is evidence for Si–Si bonds in silicas, for instance, in cases where there is no identifiable oxygen vacancy. In such cases, the description 'frustration' can be helpful. To give the simplest possible case, suppose there are atoms of types A and B, such that each A atom favours B neighbours, and vice versa. If there is an odd number of sites (say 3), then any arrangement will put some A next to A or B next to B. This is a useful reminder that, even if an amorphous random network has been constructed free from frustration or dangling bonds, there may still be bonds more vulnerable to reaction with some other species (e.g. in silica, (*a*) to water, (*b*) to carrier self-trapping, or (*c*) to forming defects which change refractive index, as in optical fibres).

When a crystalline solid is amorphised, a critical issue is whether or not the change is accompanied by the creation of dangling bonds. Coordination disorder will usually generate dangling bonds. Ring-size disorder alone will not normally create dangling bonds. The nucleation process for amorphisation is not clear, and there are few atomic models. One is the interstitial model of Torres *et al.* (1987), which also suggests how the two(tunnelling)-level defects emerge. The basis of the interstitial model is the so-called Jackson model of the interstitial in Si. This is best considered as the removal of four Si atoms (removing one Si and three of its nearest neighbours leaves ten dangling bonds) and their replacement by a ring of five Si atoms (whose ten dangling bonds saturate the others). This form is not the most stable interstitial, but could be metastable. It has two notable features. First, it creates non-standard ring structures. Secondly, there are two equivalent forms related by symmetry, and separated by a relatively low energy barrier. Tunnelling between these forms may be one source of two-level systems, and the way in which these develop during Si amorphisation is consistent with this. An analogous, but more complicated, situation can be devised for quartz amorphisation.

In a-C and a-Si, these dangling bonds can saturated by incorporating hydrogen atoms. The content of hydrogen in a-C:H and a-Si:H depends on the method of synthesis. There are strong topological constraints on the structure (Angus 1986; Phillips 1979; Thorpe 1983), giving a description which goes beyond that of a continuous random network. The number of constraints is related to m, the average coordination number. For each hydrogen atom ($m=1$) the number of constraints is $m \times m/2 = 0.5$ per H. For atoms with $m=2$ or higher coordination, the number of constraints is $(5m-6)/2$, i.e. 2 for oxygen ($m=2$), 4.5 for sp^2 carbon ($m=3$) and 7 for sp^3 carbon ($m=4$). If the average number of constraints per atom is 3, and if the atomic fractions of the species add up to unity, then there are two equations for the a-C:H compounds which give the proportions of sp^2 and sp^3 carbons uniquely for each atomic fraction X of hydrogen. For a completely-constrained random network, the average coordination (averaged over all species) is $2.4-X$, and the ratio (sp^3 fraction)/(sp^2 fraction) is $(8X-3)/(8-13X)$. The ratio is positive only for X between 3/8 and 8/13 (the lower limit can be reduced to 1/6 if the sp^2 carbons are paired by double bonds). The description is a useful framework (Angus & Janssen 1988), and is appropriate for a-C:H formed in many of the synthesis routes for diamond or diamond-like carbons. It provides a framework for understanding the dramatic variations in mechanical properties with the ratio of sp^2 to sp^3 carbons, from ultra-hard diamond (all sp^3) to soft, flaky graphite (all sp^2); similarly, varying the H/C ratio leads to solids ranging from tars to diamond-like carbons. However, the topological approach is far less helpful when one wishes to define defects, so that the role of excitation is hard to quantify. Those defects which are readily defined are those for which a topological description is not especially helpful, like dangling bonds, or interstitial species.

Underlying the topological approaches are several implicit ideas. One is that all bonds between two specific species are essentially the same: one Si in a-Si:H is the same as every other Si in that sample; each Si–O–Si group in a sample of SiO_2 is essentially the same as every other Si–O–Si group. We shall see that there are at least two ways in which this fails. Clearly, there is diagonal disorder, associated with fluctuations in the electrostatic potential at each site. Equally clearly, certain bonds may be more vulnerable to excitation. Less obviously, the variations in force constants from one site to another can be important, since these force constants determine the relaxation energies of self-trapping. A second implicit idea is that the system has been able to optimise the bonding patterns reasonably well: the system is in one of the nearly degenerate amorphous states whose energies are not much higher than the lowest energy state. This is unlikely for systems generated by radiation damage, where some of the processes are random, and where the topological constraints prevent reoptimisation of the structure. However, in other circumstances, such as the densification of glasses for radioactive waste immobilisation (Tinevella & Stoneham 1980; Stoneham 1994), radiation 'stirring' can speed the attainment of a relatively ideal state.

Topological models often assume implicitly that there are no dangling bonds. It is the case that when certain systems are amorphised by rapid cooling (chalcogens, chalcogenides and quartz, but not carbon or silicon) very few dangling bonds result. The concentration of dangling bonds with unpaired spins is less than 10^{14} cm^{-3} (Pfieffer *et al.* 1991). Even though the number of dangling bonds is small, such *fully-coordinated* amorphous materials do contain bonds vunerable to reactions with other species. In chalcogenide glasses, it has been suggested that the dangling bonds D form charged defects with paired spins by the reaction:

$$2D^0 \rightarrow D^+ + D^-.$$

The resulting paired spin is described as *coordination variation* (Street & Mott 1975; Kastner *et al.* 1976). For a negative-U system, the reaction is exothermic. The argument relates to the relaxation energies in the different states. In a rigid lattice, removing an electron from D^0 absorbs energy I; capturing an electron by D^0 releases energy A, so the reaction $2D^0 \rightarrow D^+ + D^-$ absorbs energy

$U_{electr} = I - A$. To estimate the contribution from lattice relaxation, we recall from Chapter 3 that there is a relaxation energy $F^2/2K$ from defect forces F and restoring force constant K. Suppose F is simply proportional to the number of electrons in the dangling bond (so F is 0 for the empty + state, f for the neutral unpaired-spin state, and $2f$ for the paired-spin negative state). The relaxation energy for $2D^0$ is $2(f^2/2K)$. The relaxation energy for $D^+ + D^-$ is $0 + (2f)^2/2K$. Overall, the relaxation energy change in the reaction favours charged defects and unpaired spins by $U_{elast} = 2(f^2/2K) - (2f)^2/2K = -f^2/K$. Unpaired spins will be rare if this relaxation term overwhelms the purely electronic term, and $U = U_{electr} - U_{elast}$ is negative.

Tunnelling levels and 'two-level systems'

Virtually all amorphous solids show low-energy vibrational motions, which lead to distinctive contributions to the specific heat and thermal conductivity at low temperatures. The origins of these vibrational motions are not clear. They may include quasilocalised vibrations, relaxation processes, and tunnelling, or other cooperative soft motions of a number of atoms (Schober *et al.* 1997). There is a successful phenomenological description in terms of systems which can tunnel between two states, hence the 'TLS' (tunnelling-level or two-level system) nomenclature. We are not aware of any evidence that the TLS actually influence materials modification. The TLS is a symptom of amorphisation under irradiation, but their direct influence on other defect processes seems slight.

8.1 Electrons, holes, and excitons in amorphous solids

8.1.1 The optical absorption edge

The optical absorption gap E_G is defined by the onset of optical absorption. Usually, the absorption cross-coefficient α is plotted as a function of photon energy $\hbar\omega$ in the form:

$$\hbar\omega\alpha(\omega) = A(\hbar\omega - E_G)^2$$

The band edges of amorphous materials are characterised by a band tail extending into the forbidden gap. The electronic nature of the band tail is crucial for the electronic excited states and is equally important in lattice modification induced by electronic excitation. In fully-coordinated amorphous materials, such as the chalcogenides and SiO_2, we might think of the amorphous solid as one in which every site has a perturbing potential which varies

Table 8.1. *Values of the Urbach coefficient* γ.

Material (crystalline unless stated)	Value of Urbach coefficient γ
InSb	5
CdS	2.17
Anthracene	1.7
PbI_2	1.5
TlCl	1.1
KCl, KBr, KI	0.8
SnO_2	0.7
α-SiO_2 (quartz)	0.7
c-GeO_2 (quartz structure)	0.7
a-GeO_2	0.6
a-$Ge_{16}As_{35}Te_{28}Si_{21}$	0.57
a-As_2Se_3	0.52
a-SiO_2 (10% smaller than for quartz)	0.54
a-As_2Te_3, a-$CdGeAs_2$, a-As_2S_3	0.49
a-Te	0.47
a-Se	0.44
a-GeTe	0.38

randomly from one site to another. If so, disorder will induce a tail of states deriving from the conduction and valence bands.

In these materials, the electronic nature of the band tail can be related in a fairly simple way to the band structure of the corresponding crystalline materials. On the other hand, if coordination variation is important, the valence band tail is mainly due to D^+ and the conduction band tail is primarily due to D^-. In a-C:H and a-Si:H, the states of the band tail will differ greatly from the band states of the corresponding crystalline materials. In a-C:H, the band tail appears to be due to the formation of sp^2 bonds (Robertson 1996); in a-Si:H, the band tail is similar to crystalline Si, but the top of valence band falls on hydrogenation.

The Urbach rule and band tails

In many solids, the optical absorption coefficient at the band edge varies with energy as $\exp\{-\gamma[(h\nu - E_0)/kT^*\}$, where T^* involves an effective phonon energy through a $\tanh(\hbar\omega/2kT)$ factor (Table 8.1). This is the so-called Urbach rule. Several quite distinct explanations have been offered, and indeed all may be correct for certain systems. Sumi & Toyozawa (1971) concluded that the exciton–phonon interaction was responsible. In their theory, the exciton is treated as being partly mobile and partly localised due to momentary trapping by lattice deformation; their theory embraces the possibility of exciton self-trapping. Dow & Redfield (1972) instead invoked the electric field ionisation of excitons by the fields associated with defects or with longitudinal optic phonons.

A later class of methods proposes a different explanation for the Urbach tail in amorphous, less-polar, solids. The model considers a random potential, based usually on a Gaussian distribution of the diagonal elements of a Hamiltonian, although sometimes a screened Coulomb potential is used. What is predicted in this approach is a density of states, not an optical absorption coefficient, although some important features may carry over to the absorption coefficient. The disorder is characterised by a length scale L and an energy scale E_1. The density of states is predicted to be proportional to $\exp[A(E-E_1)^n]$, where A is a constant, and where n tends to 1/2 for L tending to zero, and n tends to 2 for L tending to infinity.

If E_L is the energy needed to localise an electron in a well of width L, then n can be expressed in terms of E_1 and E_L (at least approximately). For the smallest physically-acceptable L (1–2 Å), n is of order 0.8. For 'reasonable' values (L of order a few ångstroms, E_1 of order an electron volt) n is about 1, the Urbach form, and not too sensitive to the details. For very large L, typically L bigger than 100 nm, then n can be of order 2.

However neat the explanation, the prediction is solely of the density of states. Since there *is* electron–phonon coupling (which is critical for the negative-U materials), and since lattice coupling certainly is important for the Sumi–Toyozawa and Dow–Redfield theories for crystalline materials, then it should affect the optical spectra of these amorphous systems as well. In particular, one expects a correlation between the spatial extent of a quasilocalised carrier and its energy. Presumably deeper states are more compact. Since the Huang–Rhys factor depends on the compactness of an initial state which is to be photoionised (Stoneham 1979a), this will be one of the several extra factors affecting the observed spectrum. Relaxation energies will correlate to some degree with diagonal disorder.

8.1.2 Motion of electrons and holes

Disorder affects the motion of electrons and holes in amorphous solids. One way to appreciate this is to start from a perfect crystalline solid. Suppose there is a perturbation associated with a single site. In simple tight-binding models, this perturbation will affect either the on-site energy or the bond energies associated with the site. If the perturbation is large enough, it will be possible to

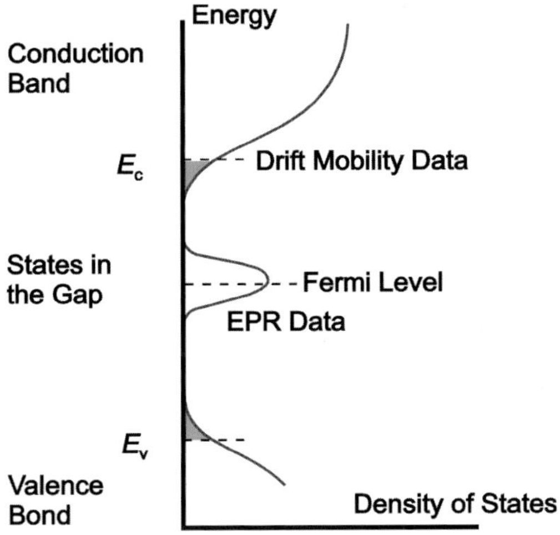

Fig. 8.2. Schematic density of states for an amorphous solid. Note the band tails (shaded) between the mobility edges (dotted) of states which contribute negligibly to charge transport.

trap a carrier, and there will be a level in the energy gap. In simple models, the threshold for an isolated trap can be written as a critical ratio of the perturbation at a site, ΔV, to the bandwidth $2B$. If there is a random fluctuation from site to site in mean square potential $\langle \Delta V^2 \rangle$ then, at a threshold value of $\langle \Delta V^2 \rangle^{1/2}/2B$, Anderson localisation can occur, in which there are no delocalised states. The threshold is much larger than the $\Delta V/2B$ needed for localisation by an isolated trap. We emphasise that the complete absence of delocalised states (Anderson localisation) is distinct from the common phenomenon of trapping, when a perturbation is merely strong enough to localise a single carrier (see Hayes & Stoneham (1985), p. 413).

Suppose we think of the amorphous solid as one where the perturbation varies from one site to another in a random way. This disorder will induce a tail of states deriving from the conduction and valence bands. Electrons and holes might be immobilised at those sites with the largest perturbations, corresponding to states in the tails of the density of states. Carriers in other parts of the bands may still be mobile. Qualitatively, we might expect the features to hold more generally, beyond the ideas of simple diagonal disorder. We introduce two definitions. First, the boundaries between the tails of trapped states and bands of mobile states will be called *mobility edges*. We label the mobility edge in the conduction band E_c and that in the valence band E_v; Fig. 8.2 depicts the density of states, including defect states. Secondly, if the perturbations are very large, above another threshold, there will be no delocalised (mobile) states at all, i.e., Anderson localisation.

The motion of electrons and holes in the band tail involves hopping. The individual jumps will be thermally activated. This can mean that the conductivity σ will show Arrhenius behaviour:

$$\sigma = \sigma_0 \exp(-W/kT).$$

The apparent activation energy W depends on the energy dependence of the density of the states in the tail. The origin of the hopping motion has been ascribed to thermal release of electrons and holes from the gap states (Davis & Mott 1970); Emin (1977, 1980) has argued for the importance of self-trapping. However, in amorphous solids, there is an additional feature related to the range R of a jump: there is a balance between tunnelling, favoured by short jumps, and the possibility of finding a final site within energy kT of the initial site. This balance leads to variable range hopping, with a characteristic $\exp(-A/T^{1/4})$ dependence. Variable-range hopping does not appear to be important in materials modification phenomena.

The self-trapping of holes is well established in amorphous SiO_2, in which the hole mobility is shown to be extremely small compared with the electron mobility (R C Hughes 1978). Self-trapped holes have been observed directly by spin resonance (Griscom 1989). Such self-trapped holes act as efficient traps for electrons; this leads to small electron mobilities, as observed. As we shall see later, excitons are self-trapped not only in several forms of SiO_2, but also in chalcogenides and Se. The fact that excitons are self-trapped does not require holes to be self-trapped. There is no experimental evidence that holes are self-trapped in quartz. It is certainly true that the holes are mobile: in α-quartz doped with 10^{-4} Ge per Si, ionising radiation leads to [Ge:h] centres even at low temperatures. However, both the existence of self-trapped holes in a-SiO_2 and theory suggest that the energy difference between the self-trapped and large-polaron states is small. Whichever state is lowest in energy, most of the charge transport would be in the mobile, large-polaron state.

8.2 Optical absorption and luminescence

Luminescence is a common features of non-metallic amorphous materials (Fisher 1985). Since the self-trapping of excitons is the cause of luminescence in (crystalline) α-quartz and in chalcogenides (Sections 4.4, 4.5), we must ask whether self-trapped excitons are also responsible for luminescence following band-to-band excitation of their amorphous counterparts. We shall also examine whether the luminescence from a-Si:H (which, like a-C:H, lacks a crystalline counterpart) is characteristic of self-trapped exciton luminescence.

Table 8.2. *Stokes shifts for various crystalline and amorphous materials. The table lists absorption energies (eV), emission energies (eV), and the Stokes shift as an approximate percentage of the absorption energy. Substances are listed in decreasing order of bandgap. Fig. 8.3 gives further information.*

Crystalline form		Amorphous form	
c-SiO_2	9.3 eV; 2.8 eV (70%)	a-SiO_2	8.7 eV; 2.3 eV (74%)
c-As_2S_3	2.6 eV; 1.3 eV (50%)	a-As_2S_3	2.5 eV; 1.2 eV (52%)
c-As_2Se_3	2.1 eV; 1.1 eV (48%)	a-As_2Se_3	1.9 eV; 0.9 eV (53%)
c-Se	1.84 eV (exciton); 0.73 eV (60%)	a-Se	2.2 eV; 0.8 eV (64%)
		a-Si	1.67–1.78 eV; 0.9 eV (48%)

Efficiency and Stokes shift

The luminescence can show a relatively large Stokes shift. Table 8.2 gives the Stokes shifts of several amorphous materials as a function of their optical bandgaps. Fig. 8.3 shows equivalent information, giving the luminescence energy as a function of gap. The Stokes' shifts of a-C:H and a-Si:H are significantly smaller fractions of the optical bandgap than those for a-Se, chalcogens, and a-SiO_2. Luminescence in amorphous materials has a high efficiency, possibly because there are no symmetry selection rules in amorphous solids.

Three different ideas have been suggested to explain the large Stokes' shifts. The first, the *small-polaron model* assumes that self-trapping occurs, whether of an electron, of a hole or of an exciton (Emin 1977). The second model, *the rigid-band one-electron model* (Street 1991) assumes that luminescence involves tunnelling recombination across the bandgap of an electron and a hole in the band tails. The third model is based on coordination variation, and proposes radiative recombination of an electron trapped to form a negative defect with a hole trapped to form a positively-charged defect (Street & Mott 1975). The high luminescence efficiency observed argues against this third model.

Luminescence of self-trapped excitons in amorphous materials

Of the major amorphous materials, exciton self-trapping is well established in SiO_2 and the chalcogenides. In comparisons of the luminescence in the crystalline and amorphous forms, the first major question is whether the luminescence is due to the recombination of self-trapped excitons in both cases. A second question concerns the differences in the luminescence in the crystalline and the amorphous material. Several factors make it hard to confirm STX luminescence in an amorphous material, and complicate comparisons with the

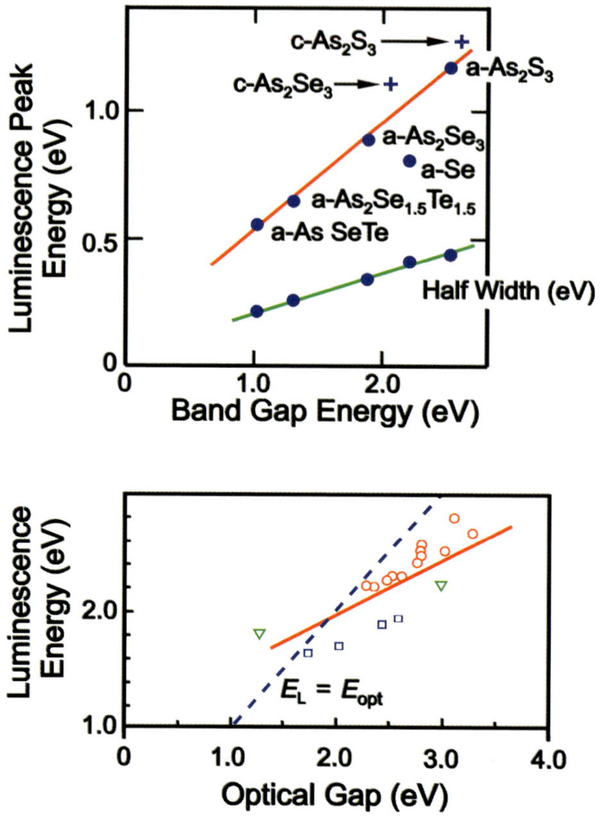

Fig. 8.3. Variation of the luminescence peak energy with the bandgap energy for various amorphous materials. The upper panel shows data for chalcogenides, for which the luminescence energy is typically half the gap; the Stokes shift is similar in magnitude (Table 8.2). The lower panel gives data for a-C:H.

crystal counterpart. The presence of the band tail makes the precise form of the excitation spectrum uncertain. Further, local structural fluctuations broaden the emission bands and lead to a multicomponent decay time. On the other hand, exciton self-trapping is favoured in amorphous structures because of the fluctuations, since some sites will always have features which encourage self-trapping. The fluctuations act rather like defects in extrinsic or defect-perturbed self-trapping. Just as Ge can cause extrinsic self-trapping in quartz, so fluctuations in a-SiO_2 can enable localisation. Overall, it is highly likely that intrinsic luminescence in a-SiO_2 and the chalcogenides involves recombination of self-trapped excitons.

The atomic and electronic structures of a-C:H and a-Si:H are completely different from those of tetrahedral C and Si crystals. Excitons or self-trapped excitons in the amorphous form will differ in nature from the non-self-trapped

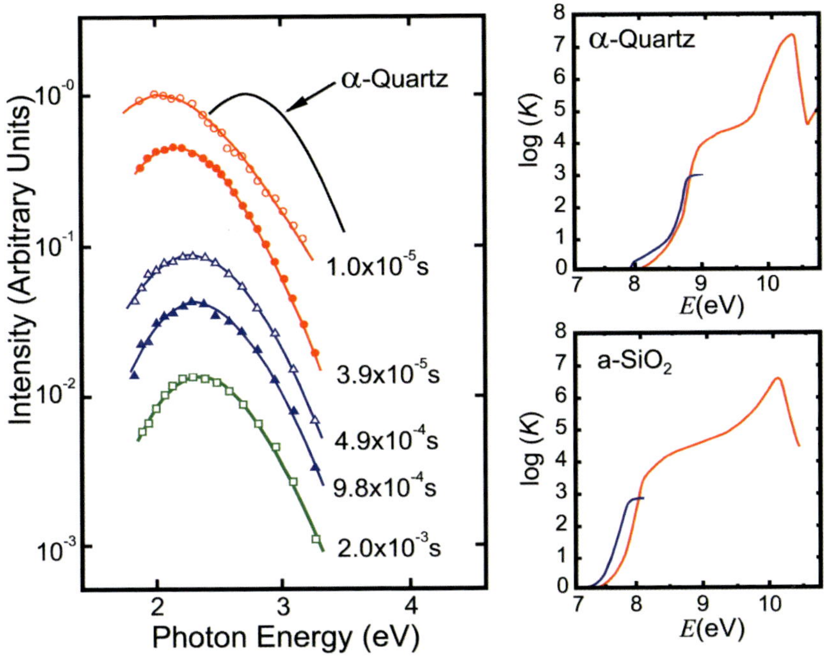

Fig. 8.4. The left hand panel shows the luminescence spectra of α-quartz and fused silica (C Itoh *et al.* 1988). The right hand panels show optical absorption spectra, the upper panel for α-quartz and the lower panel for fused silica (Bosio & Czaja 1993).

excitons in the closest crystalline analogues. The assignment of the origin of the luminescence is harder than for SiO_2 or the chalcogenides. In discussing the luminescence of a-C:H and a-Si:H, we shall exploit analogies with what we believe to be self-trapped excitons in a-SiO_2 and in the chalcogenides.

8.2.1 *Amorphous silicas: a-SiO_2*

Optical absorption and luminescence spectra in fused silica

The optical absorption spectrum, the emission spectrum for luminescence (Fig. 8.4), and the excitation spectrum for luminescence (Fig. 8.5) of fused silica are all very similar to those of crystalline α-quartz. Optically detected magnetic resonance (ODMR) gives direct evidence that luminescence in α-quartz arises from self-trapped excitons (see Chapter 4). No ODMR measurement has yet been done for fused silica, but there are three good reasons for the view that the luminescence of fused silica is also due to self-trapped excitons. First, the luminescence bands are in the same energy range for both amorphous and crystalline forms, and the excitation spectra are nearly the same. Secondly, those optical absorption bands which decay by emitting the characteristic luminescence are

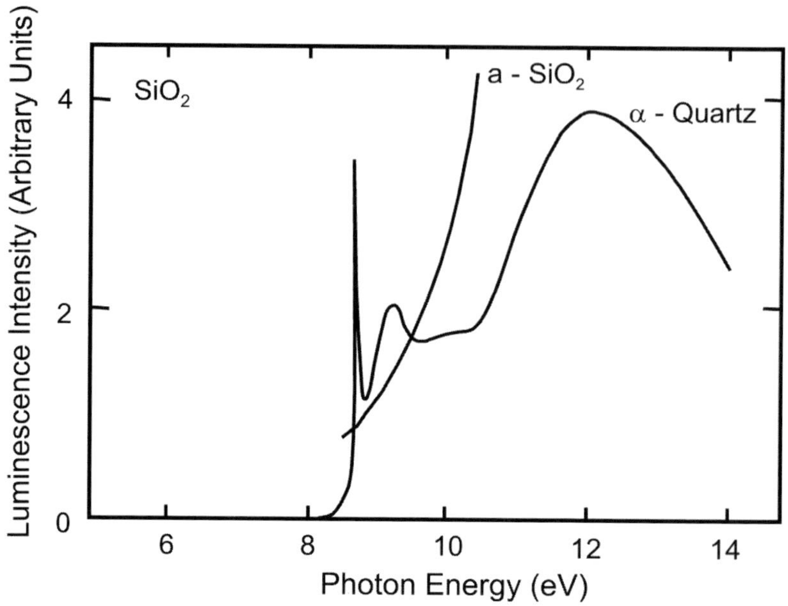

Fig. 8.5. Excitation spectra of the intrinsic luminescence in α-quartz (C Itoh *et al.* 1989) and in amorphous silica (Trukhin 1991).

the same for α-quartz and for fused silica (Fig. 8.6). Thirdly, the quantum efficiencies of formation of the luminescence centre, as detected from optical absorption measurements, are again almost the same in both crystalline and amorphous forms (T Tanaka *et al.* 1985), although the quantum efficiency of luminescence is much weaker in the amorphous phase, probably because of non-radiative transitions.

Comments on the relaxation energy for fused silica

The Stokes shift of the STX luminescence of fused silica is larger by 0.5–0.8 eV than that of α-quartz (see, e.g., Fig. 8.4). We suspect that there are two factors. First, the energy of an exciton, whether self-trapped or not, will vary from site to site in an amorphous structure. Both the core exciton and the normal exciton have relatively narrow absorption lines, so the energy variations from site to site must be small for the *unrelaxed* atomic positions. Thus the energy variations of interest must be associated with the relaxation energy, perhaps due to force-constant variations, which make it easier to move the oxygen for certain local geometries (see Fig. 8.7). These site-to-site variations affect both the stability (total energy) and the emission energy. Secondly, it seems likely that the exciton can diffuse for short distances to sites of lower energy. These sites of lower

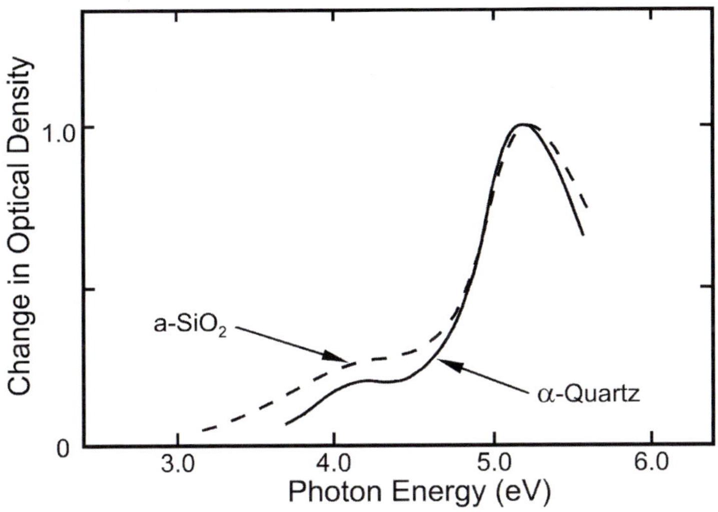

Fig. 8.6. Optical absorption bands for excitation from the lowest state of the self-trapped exciton to higher states for fused silica and α quartz (C Itoh *et al.* 1988).

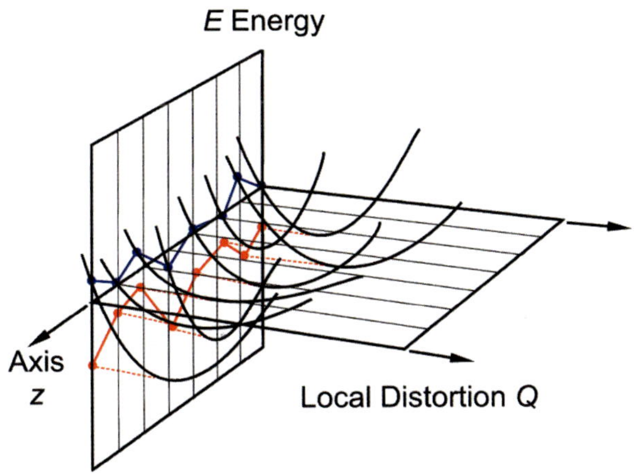

Fig. 8.7. Schematic diagram to show the difference between simple fluctuations of potential from site to site, and variations in relaxation energy associated with self-trapping. Imagine that the axis is a row of alternate Si and O atoms (–Si–O–Si–O–Si–) and that relaxation changes the Si–O–Si angle, as in self-trapping in quartz. The diagram shows variations in unrelaxed energies (the intersections of the curves with the vertical plane (blue curve)) and also in relaxation energies through variations in forces and force constants. The overall effect is shown as the projection of the minimum energies onto the vertical plane (red curve). The exploitation of these lower relaxed state energies shifts the blue luminescence from 2.7 eV in quartz to 2.1 eV in a-SiO_2.

energy are also the sites of larger relaxation energy, where the luminescence will have a lower energy. The actual magnitude of the relaxation energy difference from one site to another has not been calculated, but the difference in behaviours of holes in α-quartz and vitreous silica suggest the energy variations of a few tenths of an electron volt needed to explain the differences in exciton luminescence are not unreasonable.

Luminescence decay and configuration in fused silica

The optical absorption band of self-trapped excitons in fused silica (Fig. 8.8, T Tanaka *et al.* 1985) decays almost exactly inversely proportional to time t. The decay of the luminescence (Fig. 8.8) is similar, but with a higher rate at larger delays. The decay curves are typical of multiexponential decays, in which the concentration $n(t)$ of the self-trapped excitons at time t is given by:

$$n(t) = \Sigma n_i \exp(-t/\tau_i),$$

where n_i denotes the concentration of those components with a lifetime of τ_i. The luminescence intensity $I(t)$ can be expressed as

$$I(t) = \Sigma(\eta_i n_i/\tau_i) \exp(-t/\tau_i),$$

in which η_i is the luminescence quantum efficiency of the ith component. A comparison of dn_i/dt with $I(t)$ shows that there is a lower quantum efficiency for those components with a smaller lifetime.

Localised states in silicate and similar oxide glasses

In the wide-gap silica-containing oxide glasses and similar materials, there are several classes of localised states. The cations in the glasses themselves are conventionally grouped as network formers, like Si, Ge, Al, B, and network modifiers, like Na or K. The wealth of defect-related and impurity-related states is discussed in many places, and reviewed by Stoneham & Smith (1990). Other localised states have been analysed by Trukhin (1994, 1995), who has also reviewed self-trapped excitons in oxides such as GeO_2, Li_2GeO_3, $AlPO_4$, and $GaPO_4$ (Trukhin 1997). For the glasses, he concluded that, in the alkali silicates, the modifier states are below the network former states, resulting in localisation on the modifer. In the alkali germanates, localisation is on network former states. There appear to be six-fold (rutile-like) and four-fold coordinated regions, as shown from the Urbach tail. It appears that the self-trapped exciton occurs only in the four-fold regions. In the alkali phosphates and the alkali lead glasses, localisation is again on the glass former states, especially Pb^{2+} for the lead glasses. Trukhin observed polarised luminescence, so the absorption and luminescence come from anisotropic localised states. There is

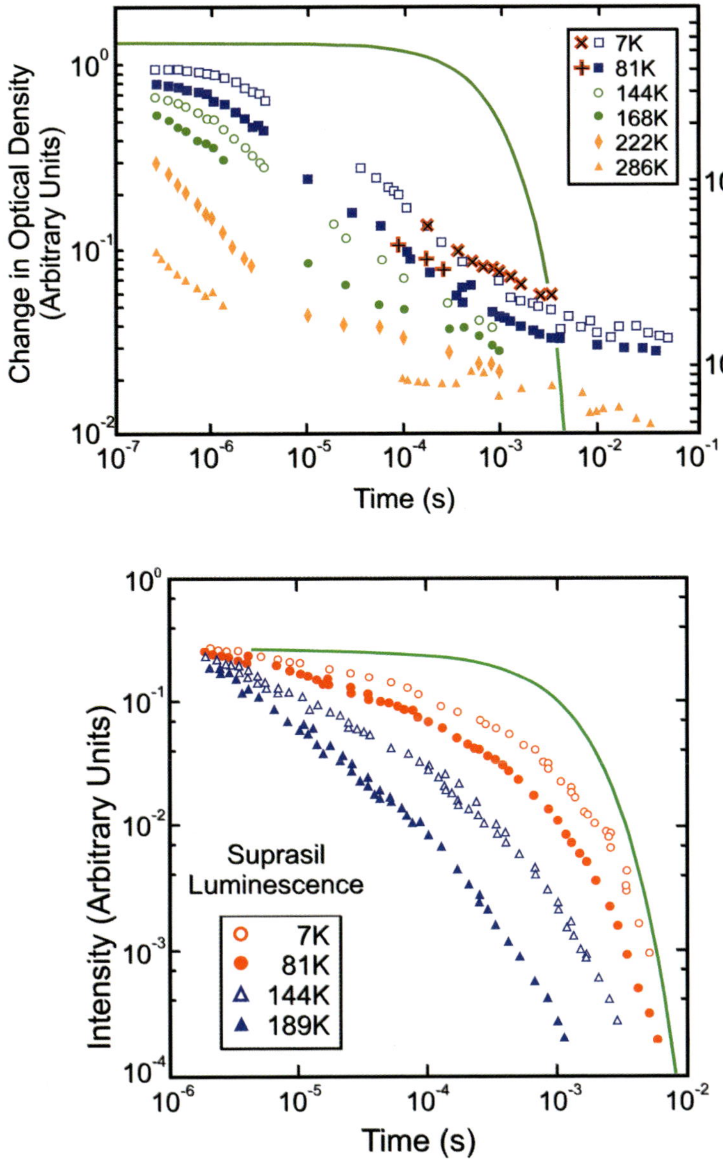

Fig. 8.8. Decay characteristics of the luminescence of fused silica at several temperatures (T Tanaka *et al.* 1985). The upper figure shows the decay of the absorption of the self-trapped exciton to higher states, and the lower figure shows the decay of the luminescence. The full lines are the corresponding values for α-quartz.

Table 8.3. *Comparison of the properties of tetrahedrally-bonded amorphous systems and the chalcogenide glasses.*

	Tetrahedrally-bonded amorphous systems	Chalcogenide glasses
Examples	Si, Ge, GeAs, CdGeAs	Se, Te, As_2S_3, As_2Se_3
Structure	Four-fold-bonded network	Two-fold bonds (chains)
Conduction	Band-like at higher temperatures, variable-range hopping at low temperatures	Generally band-like
Doping?	Si can be doped n-type or p-type	Apparently cannot be doped
Spin resonance?	EPR seen (less than 10^{20} spins/cc)	Weak EPR (less than 10^{15} spins/cc)
Defect structure	Valency and coordination defects, such as five- or seven-membered rings	Valence alternation pairs with negative-U behaviour

a triplet luminescence, with a long lifetime; there is also some evidence for a soft mode in thermal quench. One defect discussed is the L centre, for which the model is $O_3{\equiv}Si{-}O^-{-}Na$; the centre appears to exist in two other charge states, L^+ and L^-. We shall discuss Ge-doped silica glasses in Section 8.3.1.

8.2.2 Chalcogenides

There are many differences between the amorphous chalcogenides and the systems which we have been discussing. These are summarised in Table 8.3. The standard notation for the main intrinsic defects has the form C_N^{M+}, where N is the chalcogen coordination and $M+$ is its charge. The three main coordination defects are $D^0=C_2^{\,0}$, $D^+=C_3^{\,+}$ and $D^-=C_1^{\,-}$.

Is there a self-trapped exciton in amorphous chalcogenides?

Band-to-band optical absorption in a-As_2S_3 and a-As_2Se_3 leads to midgap luminescence (Fisher 1985 and references therein). The optical absorption edges and luminescence bands of a-As_2S_3 and a-As_2Se_3 are shifted to lower energies than for their crystalline forms (Fig. 8.9). The shifts of the luminescence and excitation spectra are interpreted as associated with fluctuations of the band edges in the amorphous structures. ODMR spectra tagged at the luminescence band for a-As_2Se_3 show that the luminescence arises from a triplet state (Depinna & Cavenett 1982; Robins & Kastner 1987; Ristein *et al.* 1990). Ristein *et al.* (1990) have shown that the broad background spectrum

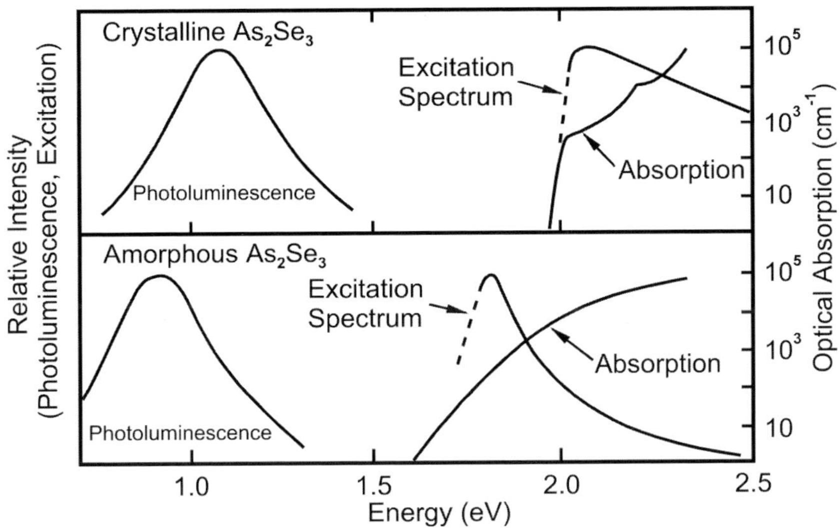

Fig. 8.9. Absorption, emission and excitation spectra of crystalline and amorphous As_2Se_3. Results for a-As_2S_3 are similar (see Fig. 8.11).

can be reproduced by a powder spectrum based on parameters for c-As_2Se_3. This indicates that self-trapped excitons are formed in a-As_2Se_3 with nearly the same structure as in c-As_2Se_3 (Section 4.5) and that the luminescence is due to the recombination of self-trapped excitons.

Luminescence in the amorphous chalcogenides

Several other features of the luminescence of a-As_2S_3 and a-As_2Se_3 show analogies to SiO_2. Thus, as well as large Stokes' shifts, the decay of the luminescence intensity follows a multiexponential decay (Fig. 8.10; Robins & Kastner 1987). The luminescence peak shifts to lower energies as the delay time increases (Fig. 8.11; Higashi & Kastner 1981). This result indicates multiple local structures of the amorphous material, even though the shift is opposite in sign from that for SiO_2.

Maruyama (1983) finds that the luminescence peak energy is roughly constant for excitation above the first exciton peak. However, the luminescence peak energy is reduced as the excitation energy is reduced below the first exciton peak. There appear to be exceptionally low energies for relaxed excited states at highly deformed sites. Murayama & Bosch (1982) find 40% polarisation memory remains after excitation of a-As_2S_3 at 2 K with linear polarised light. This memory shows that the luminescence is due to the local recombination of an electron and a hole (i.e. a localised exciton) formed in the fluctuating potential, as in several oxide glasses (Trukhin 1997). The result supports the

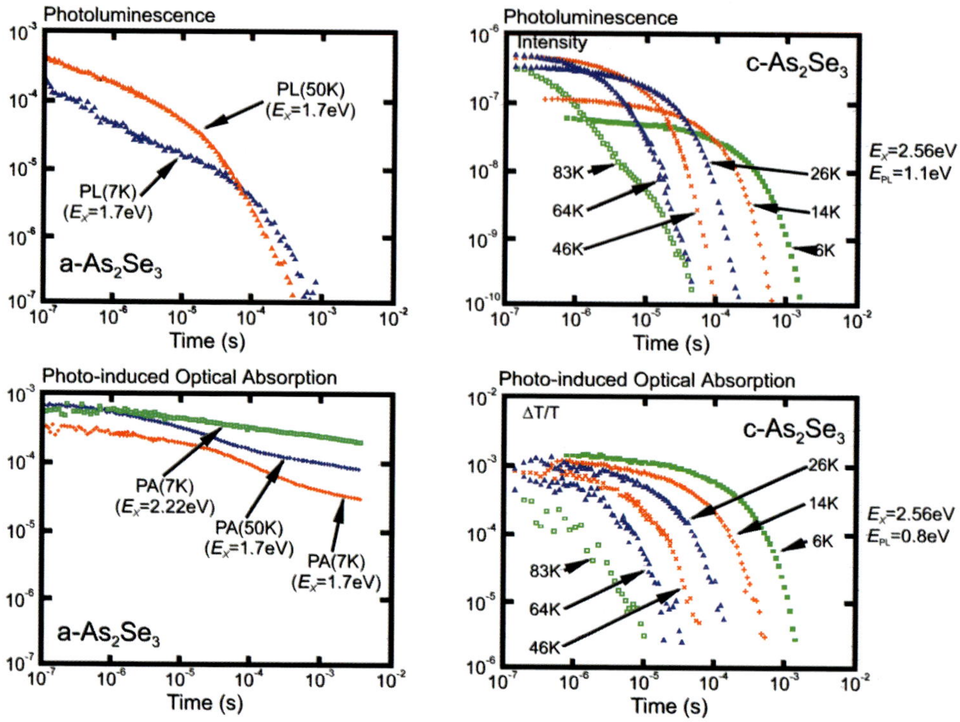

Fig. 8.10. The decay of luminescence intensity and of photo-induced absorption in the crystalline and amorphous forms of As_2Se_3 (Robins & Kastner 1988).

self-trapped exciton model of the luminescence. It is interesting that the degree of polarisation is reduced at higher temperatures, particularly for lower-energy photons.

8.2.3 Diamond-like carbon (a-C:H; DLC) and amorphous silicon (a-Si:H)

Crystalline diamond and crystalline silicon have a common structure and have much else in common, despite their many differences. The hydrogenated amorphous solids, the diamond-like carbons (a-C:H; DLC) and a-Si:H, have rather little in common with each other or with their crystalline counterparts. Perhaps the most important common feature of these amorphous systems is that their structures seem largely determined by topological rules. Further, in both materials, the bandgap depends on the method of preparation, more so for a-C:H than a-Si:H. Indeed, the characteristic structure for a particular composition of a-C:H will also depend strongly on how the sample was prepared. For example, a DLC formed by a process like chemical vapour deposition will be

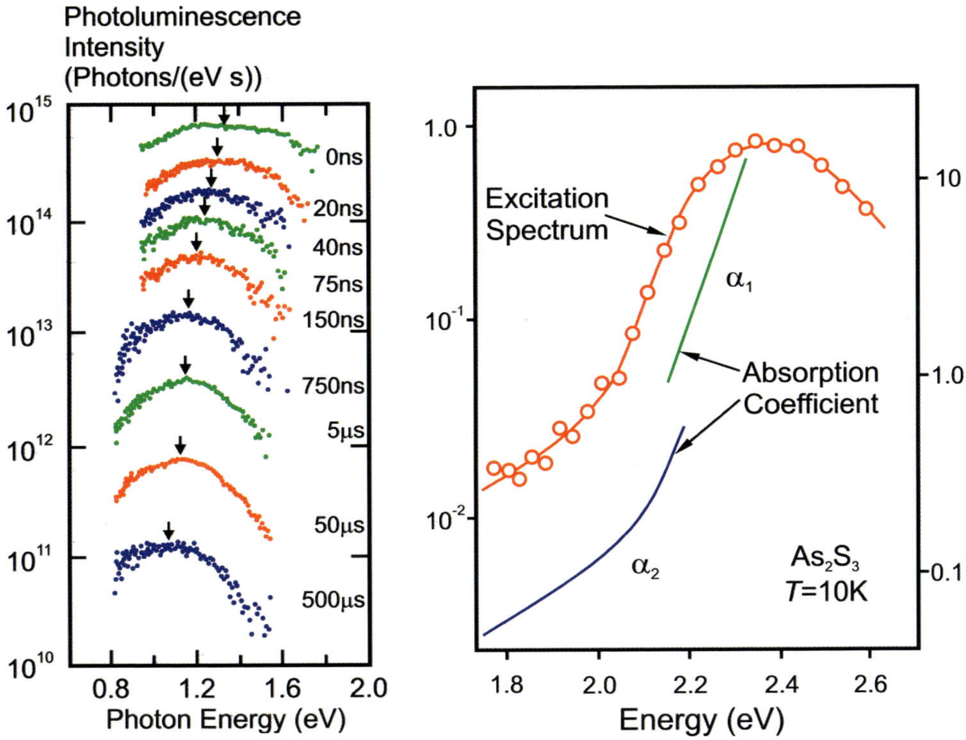

Fig. 8.11. The left hand panel shows the luminescence spectrum of a-As_2S_3 at various times subsequent to pulsed excitation (Higashi & Kastner 1981). The right hand panel shows the excitation spectrum and absorption coefficient.

very different from one prepared by removing H from a hydrocarbon by low-energy N bombardment.

a-C:H and analogous hydrocarbon systems

The topological models are of only limited value for the forms of a-C:H produced from hydrocarbons by the removal of H by low-energy N ions (Godwin *et al.* 1996; Stoneham *et al.* 1998b). In particular, tenacious coverage of components by DLC can be created on engineering components. The elastic and plastic properties of the layers depend on the choice of starting hydrocarbon. The H removal occurs through nuclear collisions, and leads to cross-linking, just as in conventional radiation damage of polymers. Of course, with most radiations, polymers and hydrocarbons may undergo scission or cross-linking, depending on the polymer. Roughly speaking, cross-linking is favoured when the chain carbons are linked to H, whereas degradation (depropagation, which regenerates monomers, or scission, with random breaks) occurs when they are linked to other carbons. Cross-linking dominates in polyethylene, polystyrene, poly-methylacrylate and polyvinyl chloride; it increases cohesion and resistance to chemical attack, and may lead to shrinkage. Degradation dominates in poly-isobutylene, polymethyl methacrylate and polyvinyledene chloride. Polymers can be chosen to be positive resists (shorter chains, more easily dissolved) or negative resists (cross-linking to give lower solubility) in lithography. We remark that oxygen is an important interstitial species because it can react with radicals created by excitation (e.g. Momose *et al.* (1992), Gillen & Clough (1992)).

These ideas provide a link between topological descriptions and excitation-induced processes. Some of the processes associated with electronic excitation can be described simply in terms of cross-linking, scission, and reactions with species like oxygen. The darkening of commercial polythene sheeting exposed to sunlight and to the atmosphere is an example. The treatment of teflon surfaces with an argon plasma in the presence of oxygen leads to peroxy radicals bonded to carbons in the cross-linked structure. There are structures like $-CF_2-CF(OO\cdot)-CF_2-$ (Chou *et al.* 1992). Ultraviolet radiation and exposure to the atmosphere can convert the cross-linked peroxy radical into the chain-scission peroxy radical. As the fluorine content falls and the oxygen content rises, the surface becomes more wettable.

In electrical breakdown, the products of damage are primarily a carbonaceous conducting region. The products of radiation damage (such as hydrogen) and the corresponding products from breakdown prove very similar across a range of polymers (Stoneham 1997a; Milinchuk & Tupikov 1989). This suggests that similar processes occur under applied electrical stress and under particle bombardment. Luminescence is seen from organic insulators (Jonnson *et al.* 1995, 1996a,b; Mary *et al.* 1996) under ultraviolet (photoluminescence) as post-discharge emission under cold plasma surface excitation (plasma-induced luminescence) and by an applied electric field in the absence of any gaseous discharge (electroluminescence). The luminescence in the 2.0–2.5 eV range is seen only when charge is injected (and so not in photo-luminescence), possibly involving conjugated groups as recombination centres. Such luminescence is used as a symptom of approaching electrical breakdown.

Differences between a-C:H and a-Si:H

One major difference between a-Si:H and a-C:H concerns sp^2 sites, the three-fold coordinated sites, as in graphite. As noted above, a-C:H includes sp^2 sites as well as sp^3 sites, and both must be included in topological descriptions. The optical gap of a-C:H is governed by the π bonds, probably in clusters of sp^2

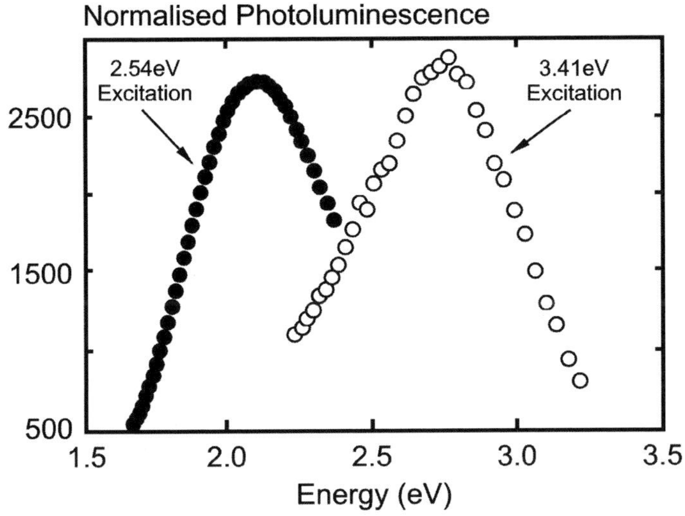

Fig. 8.12. Emission spectra of a-C:H induced by photoexcitation at 2.54 eV and 3.41 eV (after Silva *et al.* (1996)).

sites. On the other hand, Si is far less likely than C to exist as sp^2. The natural working assumption is that a-Si:H has sp^3 sites exclusively, so that the structure is the standard continuous random network of Si, disrupted only by H bonded to some of the Si atoms. The absorption edge is governed by the Si-H bonds, whose main effect is to shift the valence band edge down in energy, so increasing the bandgap (Robertson 1983).

Luminescence of a-C:H

The optical bandgap energy of a-C:H varies even more widely with method of preparation than that of a-Si:H. It is known that hydrogen-free a-C has a bandgap of 0.4–0.7 eV, and is probably largely sp^2 bonded. Hydrogenation introduces many sp^3 sites and increases the bandgap energy to 1.6–2.7 eV. Robertson & O'Reilly (1987) showed theoretically that, for a number of model structures with different configurations of sp^2 and sp^3 sites, the bandgap energy varies inversely with sp^2 cluster size.

The luminescence of a-C:H shows different features from that of a-Si:H, largely because of the importance of the sp^2 carbons. The photoluminescence energy depends on the sp^2 fraction (and so on the bandgap energy) and also on the photon excitation energy (Fig. 8.12; Silva *et al.* 1996). A correlation of the polarisations of exciting and emitted light has been noted (Gehan *et al.* 1996). Robertson (1996) used the rigid band-tail model to explain the Stokes' shift,

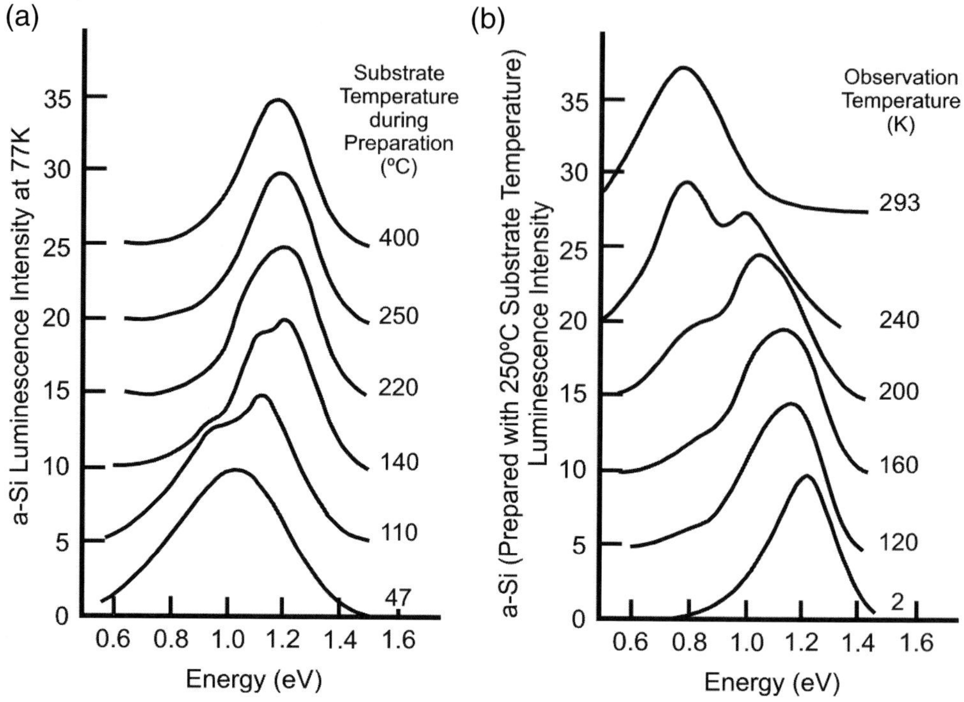

Fig. 8.13. Photoluminescence intensity of a-Si:H. (a) dependence on substrate temperature during preparation, with observation at 77 K; (b) dependence on observation temperature for samples prepared at a substrate temperature of 250 °C (data from Engemann & Fisher (1977) and Wilson *et al.* (1983)).

indicating that luminescence arises from the recombination of electron–hole pairs strongly localised in sp^2 clusters. Alternatively, the luminescence might be ascribed to small-polaron formation within the sp^2 clusters.

Just as for As_2Se_3, there is memory of polarisation of the luminescence. Again, the degree of polarisation falls as the emitted photon energy is lowered, i.e. as the relaxation energy increases. For a-C:H, the degree of polarisation is higher for material with larger optical bandgaps, and the slope of the loss of polarisation with respect to relaxation energy is also smaller. These results are ascribed to a smaller localisation radius for excitons for material with a higher optical gap or sp^3 content.

Luminescence of a-Si:H

Luminescence bands about 0.3 eV wide result from band-to-band excitation of a-Si:H. Hydrogen-free a-Si prepared by vacuum evaporation shows only weak luminescence, whereas a-Si:H prepared by the decomposition of SiH_4 shows broad luminescence spectra (Fig. 8.13). Three peaks may be seen, the lower-

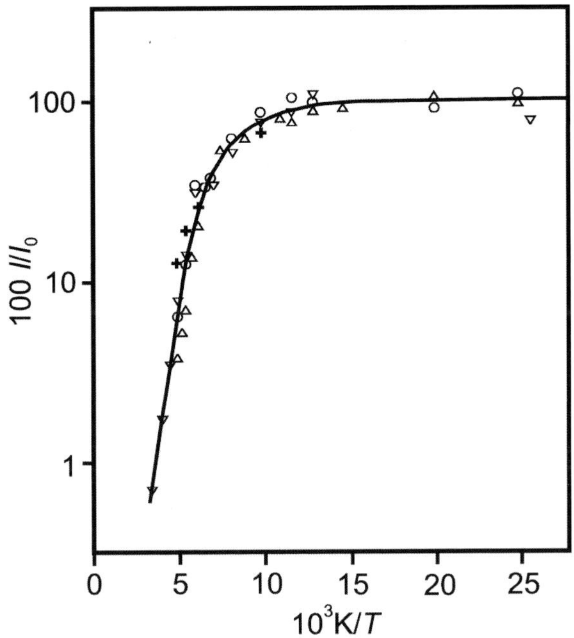

Fig. 8.14. Temperature dependence of the total luminescence intensity of a-Si:H (Street 1976).

energy band being most prominent at high temperatures. The highest-energy peak shifts with the time delay after excitation, as in the chalcogenides and in a-SiO_2, although the shift is in the opposite direction to a-SiO_2. The time decay of the luminescence is multiexponential, again similar to SiO_2 and the chalcogenides. The luminescence intensity is almost independent of temperature below 80 K, but decreases at higher temperatures (Fig. 8.14). The intensity is reduced by prolonged photon irradiation, which creates dangling bonds (Hirabayashi *et al.* 1981) and by ion irradiation (Engemann & Fisher 1977). We shall discuss later the reduction in luminescence on forming dangling bonds. The luminescence is also quenched by an electric field of about 10^5 V/cm (Street 1981).

The polarisation of the luminescence in a-Si:H has a memory with features similar to those for a-As_2S_3 and some oxide glasses. The degree of polarisation reduces as the emitted photon energy decreases, i.e. for larger excited state relaxation energies. For a-C:H, where the optical bandgap is bigger, the degree of polarisation is higher; moreover, there is a smaller loss of polarisation for a given relaxation energy. These results are ascribed to a smaller exciton localisation radius, and to a higher optical gap or sp^3 content. In a-C:H, the luminescence is less temperature-dependent, and is still strong at room temperature.

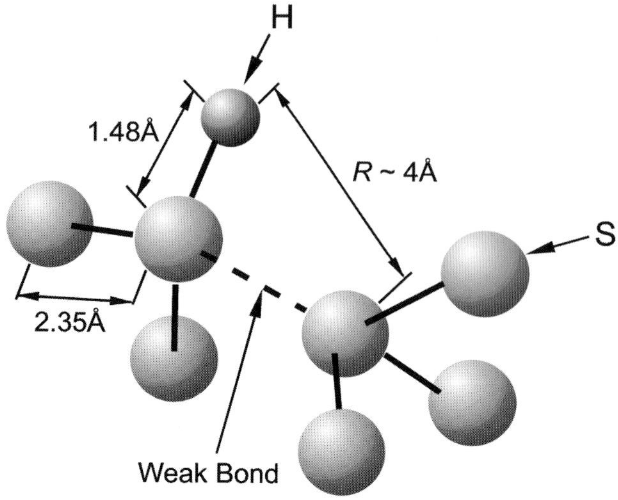

Fig. 8.15. A model of the recombination centre in a-Si:H (Morigaki *et al.* 1993). It can be regarded as a hole self-trapped at a weak Si–Si bond. The H is detected by ODENDOR.

The early ODMR work on a-Si:H is reviewed by Cavenett (1981). Depinna & Cavenett (1982) found negative EPR signals at around $g = 2.0$ by low-temperature detection at the 1.25 eV and 0.9 eV bands. More recently, Kondo & Morigaki (1993) used ODENDOR to investigate the hole centres which participate in distant radiative electron–hole recombination. They found a signal associated with a hole close to a hydrogen. It is suggested that the hole is trapped at a weak Si–Si bond close to the H; the trapping of an electron by the self-trapped hole forms a self-trapped exciton (Morigaki *et al.* 1993). Fig. 8.15 shows the structure of the recombination centre suggested by the superhyperfine interaction with a hydrogen atom.

Are there self-trapped excitons in a-Si:H?

All observations of a-Si:H photoluminescence point to useful analogies with a-SiO_2 and the chalcogenides. These suggest strongly that self-trapped excitons are formed in a-Si:H, as suggested by Morigaki *et al.* (1993). ODENDOR shows that the luminescence centre is associated with a hydrogen atom, implying that self-trapped excitons are formed following excitation at a Si–H bond.

8.3 Defect formation

In all the amorphous materials treated in this chapter, photons with energies greater than the bandgap energy create paramagnetic defects. These defects are

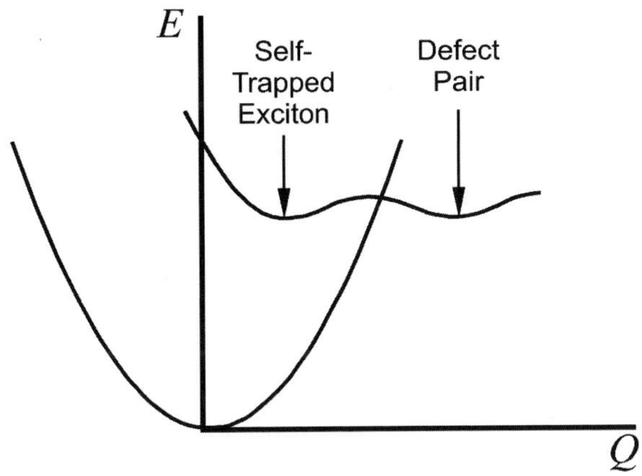

Fig. 8.16. Energy surfaces allowing the relaxation of an exciton to form a self-trapped exciton or a defect pair in amorphous materials.

annealed by heating. The formation of paramagnetic defects is the origin of the reduction of electrical conductivity by irradiation, first observed by Staebler & Wronski (1977) for a-Si:H, and known as the Staebler–Wronski effect. We shall argue that there are common features in this reversible defect formation in various amorphous materials.

Mechanisms of defect formation

The fact that the density of paramagnetic defects generated by ionising radiation reaches 10^{-20} cm^{-3} shows that bond breaking is involved in the formation of paramagnetic defects, not just the trapping of charge carriers by native centres. A natural interpretation of the bond breaking is to invoke exciton relaxation into a metastable state, as for alkali halides. Street (1977) has suggested that an exciton creates by bond scission a state with two dangling bonds, which he called the self-trapped exciton. We prefer to call it a defect pair or a metastable state, keeping the term self-trapped exciton for the luminescent state. The correlation between the self-trapped exciton and the metastable state is depicted in Fig. 8.16.

8.3.1 Amorphous silicas: a-SiO$_2$

Primary defects

The paramagnetic defects induced by electronic excitation of fused silica include E′ centres (Si dangling bond defects), non-bridging oxygens (NBO) and the peroxy radical (PR) (Weeks 1994). These defects have been well characterized by EPR, and their atomic structures are shown in Fig. 8.17. Common

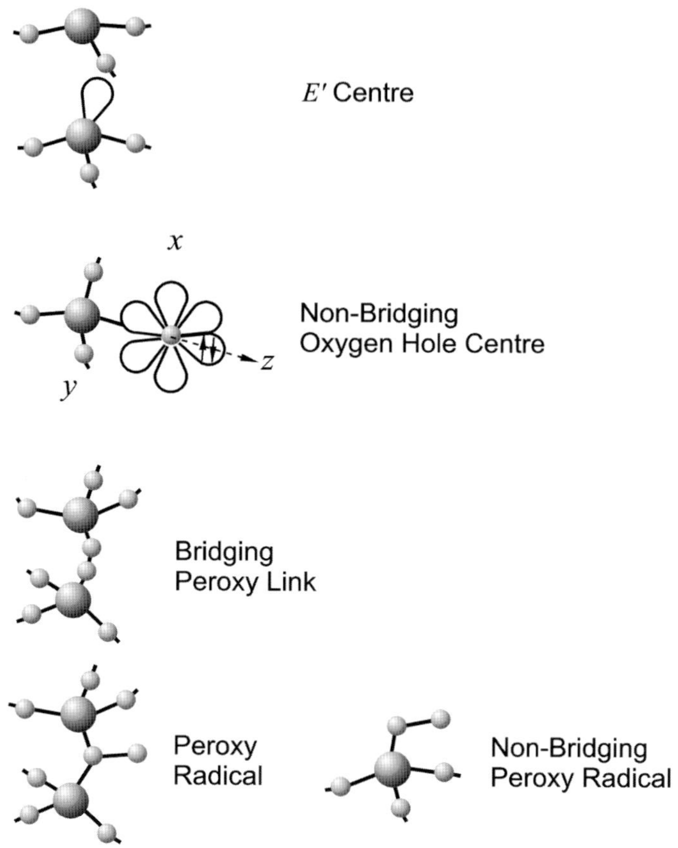

Fig. 8.17. Major paramagnetic defects produced by irradiating fused silica with ionising radiation, including E′ centres (Si dangling bond defects), non-bridging oxygens, and peroxy radicals.

impurities, like H, can be important, since the =SiOH species has an absorption at 7.4 eV, close to the intrinsic absorption edge (Morimoto *et al.* 1999a,b).

Defect yield

The yield of stable E′ centres in fused silica as a function of dose can be divided into *extrinsic* and *intrinsic* ranges (Imai & Hirashima 1994). The extrinsic behaviour, which occurs at low doses, is associated with impurities, while the intrinsic behaviour is believed to indicate bond breaking in the a-SiO_2 network. The yield of E′ centres in the intrinsic range is low (Devine 1990): for every silicon dangling-bond centre produced, there are 1000 photons emitted at the self-trapped exciton energy. This low E′-centre yield implies a high radiative recombination probability for self-trapped excitons. We do not know if E′ production needs special sites, or whether it is simply another relaxation channel,

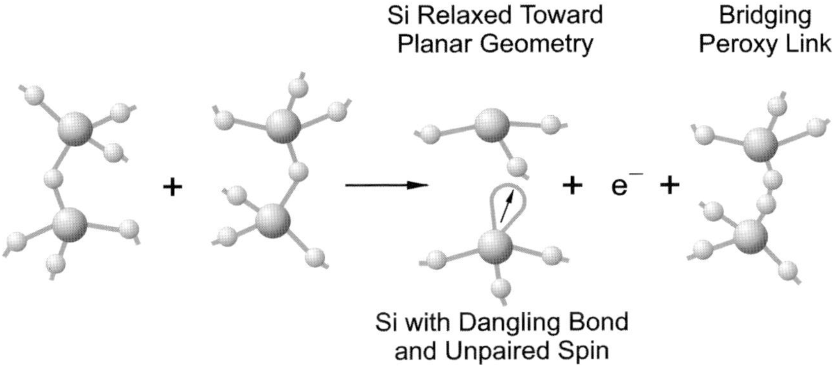

Fig. 8.18. A schematic model of E′-centre formation in SiO_2 with the creation of an E′–PR pair. Other mechanisms are discussed in the text.

as for alkali halides, although the branching ratio is very low for silica. No defect formation is observed in quartz, so it may be that E′ centres are only formed at special sites in the vitreous silica. We are not aware of information on the many other amorphous structures of different formation enthalpy and density (see Fig. 8.1).

Bond scission or vacancy-intersitital formation?

The standard picture of E′-centre creation in an a-SiO_2 network involves the production of an oxygen interstitial O_i (Griscom 1975; Hayes & Stoneham 1985). This interstitial may be incorporated into the a-SiO_2 network as a PR, forming an E′–PR (vacancy-interstitial) pair (Fig. 8.18). It has also been suggested that an exciton induces a Si–O bond scission to form an E′–NBO pair (Devine 1990; see Fig. 8.17). If E′–NBO pairs are indeed primary products, and if their production simply represents another relaxation channel with a low branching ratio, then one needs a mechanism that separates the NBO from the E' centre (Tinivella & Stoneham 1979, unpublished; Silin *et al.* 1980; Snyder & Fowler 1993). Another possibility is that the creation of a self-trapped exciton in a highly-distorted part of the amorphous structure might result in a pair comprising an E′ centre and an NBO stable against recombination; in this case, such regions would amount to an atomic fraction of about 1/1000 of the sample.

Imai & Hirashima (1994 and references therein) analysed existing results for the formation of stable paramagnetic centres by ionising radiation and by laser irradiation of fused silicas grown by several methods. They found that the yield, of E′ centres and of NBOs in the intrinsic range were proportional in the concentration range up to 10^{18} cm^{-3} (Fig. 8.19). This suggests that

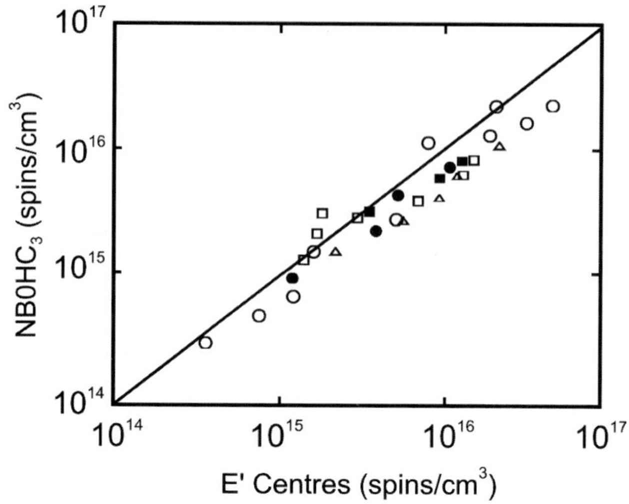

Fig. 8.19. Dose dependence of E′–NBO pairs (Imai & Hirashima 1994), showing approximate equality of numbers over a wide range of conditions. The different symbols denote results for amorphous silicas of different origin.

E′–NBO pairs are primary products. It is tempting to conclude that E′–NBO pairs are created by exciton relaxation only at special sites of the a-SiO_2 network. The reasons are these. First, densification enhances E′–NBO pair growth (Devine 1987; C Itoh *et al.* 1990) but reduces oxygen motion (Devine 1987). Densification is known to result in an increase in intermediate order but reduces ring sizes, which suggests that a 'defect' site is the origin of the E′–NBO pairs and of oxygen stabilisation. Secondly, the enhancement of the growth rate by densification at the very beginning of irradiation shows that densification increases the number of active sites for defect formation. Thirdly, an NBO is decomposed by emitting an oxygen atom around 150 °C (Griscom 1991), so it is unlikely that the E′ and NBO defects separate to become stable defects. The number of E′–NBO pairs is proportional to the square root of the dose. This suggests that rate of creation of E′ and NBO is proportional to the concentration of the pairs. If excitons are responsible for the reaction, only those excitons having energies above the mobility edge will cause the reaction. The square-root dose dependence can be explained if excitons are annihilated efficiently by the pairs. The yield of stable defect pairs is much smaller than that of the self-trapped excitons in both alkali halides and SiO_2, though for different reasons. In halides, most of the defects are transient, whereas in the amorphous silica the defect (E′–NBO) pairs can form only at a small fraction of sites.

Tsai & Griscom (1991) found that the E′–PR pairs are produced by intense laser irradiation. They showed further that the yield of PRs is higher when laser pulses are repeated with frequencies higher than 100 Hz. PRs are also formed when specimens including E′–NBO pairs are warmed to above the temperature at which NBOs become unstable; under these conditions, PR formation is accompanied by E′-centre destruction (Griscom 1991). It is likely that the E′–NBO pair is produced directly, and that PRs are formed as secondary products. Hosono *et al.* (1998) have also observed E′–PR pairs formed by proton irradiation; the dense local energy deposition induces dense excitation. Thus it appears that bond scission in a-SiO_2 due to electronic excitation leads to E′–NBO pairs, and that E′–PR pairs are formed only by dense electronic excitation.

Germanium-doped silica optical fibres

There are two circumstances in which excitation-induced defect production in optical fibres is important. The first (and undesirable) situation is that there could be darkening: a loss of transmission due to defects. The darkening may be the result of the signal transmitted along the fibre, or due to external radiation sources, such as gamma irradiation. The second (desirable) situation is that one may wish to write some pattern with an altered refractive index in the fibre, for instance a Bragg grating.

As regards loss of transmission, it is clear that impurities play a part. Pointer (1997) observes that purity and quality have a big effect on photo-induced defects, noting that many commercial doped silica core fibres show colour centre formation at power levels of 500 mW at 488–514 nm (2.41–2.54 eV). This colour centre formation is not observed in the purest-grade all silica fibres. Impurities known to have a role include H, F, Cl, and Ge. Irradiation by a fluorine dimer excimer laser (Hosono *et al.* 1999) shows that NBO hole centres (NBOHCs) are created in wet silica (120 ppm OH), whereas E′ centres form in dry (<1 ppm OH) silicas. F doping reduces the concentrations of NBOHCs or E′ by an order of magnitude, which suggests that F-doped silica glasses may be suitable photomask materials for lithography by F_2-excimer lasers. Many effects of H have been noted. Morimoto *et al.* (1999a,b) find that the photon energy dependence of stress induced by vacuum ultraviolet light suggests that absorption by hydroxyl is one key factor.

As regards controlled modifications of refractive index, there are three main types of photosensitivity (Albert 1998). In Type I, the refractive index increases slowly and monotonically at ultraviolet intensities well below the damage threshold. Type II modifications occur at the damage threshold for a single pulse from a ultraviolet laser. Type IIA is seen in glasses with relatively high Ge contents: at first, Type I behaviour occurs; the positive change is then erased, and a negative index change follows. Type II changes are less controllable than Type I, and so are less useful, despite the advantages of single-pulse writing.

Theories of the refractive index change are of two main types. One explanation invokes bleaching of the GODC (germanium oxygen deficient centre) absorption band near 242 nm (5.1 eV). The band is itself composite, comprising a 5.06 eV band from the neutral oxygen monovacancy (NOMV) and a 5.14 eV band from the germanium lone pair centre (GLPC). This *colour centre model* (Hand & Russell 1960) is sufficient to explain refractive index changes of about 10^{-4}, although explanations are still needed for the precise dependence on laser wavelength and intensity. The second explanation is based on *densification* of the oxide glass (Sceats *et al.* 1993).

A small number of defects has been postulated as responsible for Type I changes in silica fibres with modest Ge doping. Some of the defects suggested (Esaid *et al.* 1998; Fujimaki *et al.* 1998; Nishii *et al.* 1995; Sceats *et al.* 1993) are these:

(a) **The Ge E′ centre.** In silica, the E′ centre (which can take several forms) comprises a singly-occupied Si dangling bond. Formally, at least, one can create an E’ centre by removing O^-. The Ge E′ centre absorption in Ge-doped silicas is at 6.3 eV, with a 1.2 eV full width at half maximum. Its intensity anticorrelates with the GODC (see below) with laser excitation at 4.0, 5.0, and 5.6 eV (Fujimaki *et al.* 1998). The Ge E′ centre contributes to the formation of the photo-induced Bragg grating. Some authors have suggested it is formed from a Ge–Si defect in the original glass.

(b) **Ge electron-trap centres (GEC) (GEC = Ge(1), GODC = Ge(2))** There are at least two types, GEC at 4.6 eV (1.3 eV width) and GODC at 5.8 eV (0.79 eV width). The GODC centre is a positively-charged donor, and appears to be either the same as the GLPC at 5.14 eV (0.48 eV width) or closely related. The GODC centre anticorrelates with the Ge E′ centre and correlates with the absorptions at 4.5 eV (increase), 5.1 eV (decrease), 5.8 eV (increase) (Fujimaki *et al.* 1998) under KrF excimer laser irradiation at 5.0 eV; lasers at 4.0 eV and 5.6 eV have no effect. The generation of Ge E′ centres diminishs the concentration of $GLPC^+$ centres (Fujimaki *et al.* 1999).

(c) **The NOMV** This has a band at 5.06 eV (0.38 eV width). It has been suggested that a NOMV comprises a Ge–Ge bond. Shigemura *et al.* (1999) found the intensity of the 5 eV band to be an order of magnitude larger in sintered glasses prepared by a sol–gel route, as compared with a fibre preform of similar composition prepared by vapour-phase deposition.

The densification model recognises that ultraviolet excitation makes possible bond reorganisation, so that the oxide can lead to more stable network structures. Thus, in so far as there are defects, these correspond to concentration, composition, and topology fluctuations. Moreover, impurities are likely to be involved, since silicas of high chemical purity rarely show refractive index changes of more than abour 10^{-6}. The presence of hydrogen is beneficial, presumably because hydrogen saturates and stabilises dangling bonds created by excitation. This parallels the behaviour in irradiated polymers, where molecular interstitial oxygen plays a similar role. One problem with the simplest densification models, is that the gratings anneal out (e.g., Albert *et al.* 1995). If ultraviolet light really forces the system to a more stable structure, heating should reinforce the process; in fact, heating reverses it.

There is some evidence for both concentration fluctuations on the 8–10 nm scale (Hosono *et al.* 1996, for 10–15% Ge) and changes in ring statistics (topology changes identified by the 50 cm^{-1} Raman band) as six-member rings transform to smaller rings (Liu *et al.* 1997). There are also effects due to applied electric fields. In glasses subject to a strong electrostatic field, there is spatial modulation by intense light at the optical frequency and at twice that frequency, apparently due to relative phase effects in photoconductivity (Kazansky & Pruneri 1997). There are three main forms of 'poling'. The first (optical poling) uses ArF laser excitation to induce a large second-harmonic generation coefficient (Takahashi *et al.* 1997); poling consisted of applying a field of up to 30 kV/mm whilst irradiating with up to 10^4 pulses of 100 mJ/cm^2 of an ArF laser. The second (ultraviolet poling) involves irradiation with intense ultraviolet light at 193 nm. The third (thermal poling) involves heating to 300 °C in the presence of an electric field. This third process is very similar to anodic bonding (see Chapter 10) in that mobile defects (probably alkali or protons) leave behind defects which cause the poling phenomenon (Margulis *et al.* 1998).

Poling is observed in other silica-based glasses, including lead-silicate glasses (X C Long & Brueck 1999), which show large photo-induced refractive index changes at 248 nm, and borosilicate glasses. Prolonged exposure to infrared laser irradiation can erase the second-order non-linearity of thermally-poled borosilicate glasses, apparently through diffusion of photo-induced carriers (Montant *et al.* 1999).

Calculations for a number of model defects have been made by Stefanov & Raghavachari (1997) and by Zhang & Raghavachari (1995, 1997). Their results are broadly in line with what is observed. The Appendix to this chapter lists the bands which have at least provisional identifications in Si-based glasses. Accepted identifications of the several centres and optical transitions for the Ge-related centres are still lacking.

The writing of gratings in optical fibres involves a sequence of defect reactions which links some of these centres. The overall effect of the several processes is that some precursor involving at least one substitutional Ge is converted to a Ge E′ centre and an NBO centre. The reaction has at least two steps, with the Ge electron trap as an intermediate stage. Electron trapping (charge localisation) is obviously one process; it is likely that electron–hole recombination and the associated energy localisation is another critical ingredient. A possible sequence of reactions is: (i) the NOMV centre converts into a Ge E′ centre and an electron; (ii) the electron is captured by one or more types of centre containing substitutional Ge to create a GEC or GODC; (iii) the GEC or GODC transforms into Ge E′ and an NBO. The third step will need energy localisation.

Gamma and X-irradiation of optical fibres

The effects of hard radiation on fibres have been studied extensively, in relation to both civil and military nuclear applications. The effects of neutron and gamma damage are not expected to be additive, since displacement damage by the neutrons will be modified by subsequent solid-state reactions. Radiation hardening is one aim of these studies. Radiation may affect the core fibre and its clad to different extents. Griscom (1995) assessed the hardening of two types of optical fibres in the 415–1100 nm range. The first had an SiO_2 core and an SiO_2:F clad, with various levels of OH; the second fibre had an SiO_2:Ge core. These studies used high doses (above 1 MGy) of ^{60}Co gamma pre-irradiation. Various bleaching effects were noted. For pure silica, all the damage in the visible range in excess of a few tens of dB/km arose from the activation of precursor states, including Cl impurity and process-induced defects. Photohardening appeared to be due to photostimulated complexing of extrinsic centres with radiolytically-produced interstitial oxygens.

Mashkov *et al.* (1996) looked at the generation of E′ centres under X-irradiation. They argued that there is an irreversible creation of defects from network sites (which leads to power-law kinetics) coupled with the reversible activation of precursor sites (leading to stretched exponential statistics). A J Miller *et al.* (1996) examined X-ray induced defects through the dose dependence of luminescence and its relation to the spin resonance signal of the E_γ' centre. They concluded that the centres responsible for the 2.6 and 2.75 eV emissions were due to the same defect, and that this defect is related to the E_γ' centre.

Other (non-silica) fibres

Gamma irradiation also degrades fluoride fibres. Bilodeau *et al.* (1995) found that gamma irradiation (^{60}Co, 560 rads/h) causes a Raman signal to fall by a factor of 1000 in 20 minutes. We are not aware of experiments showing damage in alumina fibres, which are often crystalline. However, experiment does show

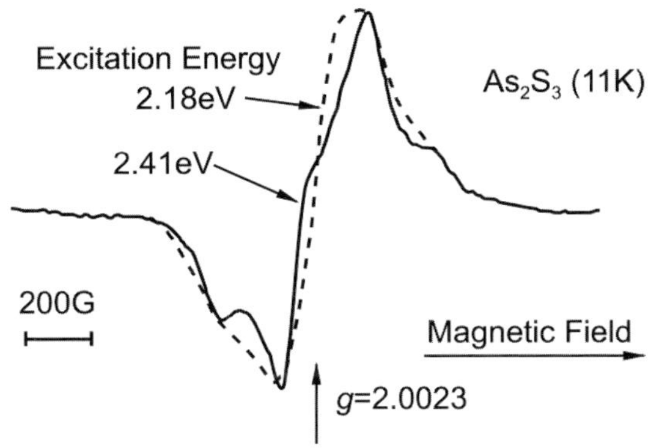

Fig. 8.20. EPR spectra of paramagenetic defects in a-As_2S_3 induced by weak excitation at 2.18 eV (below the bandgap, dashed line) and by prolonged excitation at 2.41 eV (above the bandgap, solid line)(after Biegelsen & Street (1980)).

the effects of phonon confinement within alumina fibres. Fokker *et al.* (1997) observed resonant phonon avalanches from E line excitation, and concluded there is a shorter time to avalanche development because of phonon confinement.

8.3.2 Chalcogenides

EPR and optical studies

Paramagnetic defects are produced in a-As_2Se_3 and a-As_2S_3 by electronic excitation, although defects are not produced in either c-As_2Se_3 or c-As_2S_3. The defects reduce photoconductivity and photoluminescence, and enhance optical absorption within the forbidden gap, just as for a-SiO_2. The EPR spectra of photo-induced defects in a-As_2Se_3, a-As_2S_3, a-Se, and a-As are structureless, with g factors near to 2. The signals are believed to arise from defect pairs, since heating reduces the signals reversibly. A series of experiments (Bishop *et al.* 1977) using photons with subbandgap energies and relatively weak intensities found saturation of the spin density at about 10^{17} cm^{-3}. Excitation with intense beams of photons with energies above the optical bandgap produces a spin density higher than 10^{20} cm^{-3} (Biegelsen & Street 1980). The shapes of the EPR spectra with band-to-band transitions at short times are not precisely the same as those from prolonged irradiation, as shown in Fig. 8.20. The EPR spectrum produced by band-to-band excitation anneals at about 130 K, whereas that produced by sub-bandgap photons is more stable. Further detailed studies by Hautala *et al.* (1988) show clearly that there are two types of EPR spectra. Band-to-band excitation produces both Type I and Type II, but subbandgap photons produce only Type II. Under band-to-band excitation, the yield of Type II defects is much higher than that of Type I defects. In experiments with lower photon energies, defect creation is limited to those sites showing a low-energy tail in the bandgap. Optical studies (as reviewed by Shimakawa *et al.* (1995)) support the presence of two types of defects.

The nature of defects and their production mechanisms

Type I defects are produced at concentrations above 10^{20} cm^{-3}. They do not arise from pre-existing native defects, but are due to bond scission by excitation. Type II defects are related to pre-existing defects excited by subbandgap photons. Their early saturation under irradiation with subbandgap photons is because defect formation involves the pre-existing defects which absorb photons of the band tail.

Hautala *et al.* (1988) studied the EPR spectra produced by band-to-band excitation of As_xS_{1-x} and concluded that both Type I and II EPR spectra arise from a pair of centres, comprising a hole localised on a non-bonding 3p S orbital and an electron localised on an As orbital. These results are consistent with the self-trapped exciton model by Street (1977), and correspond to the formation of an E′–NBO pair in SiO_2. Specific models for the centres formed by the bond scission have been suggested by Street (1977) and by S R Elliott & Shimakawa (1990).

The mechanism of defect formation by photoexcitation is not yet fully understood for the amorphous chalcogenides. There are two main interpretations. In the first interpretation, the formation of the Type II defects involves pre-existing defects, and Type I defects are broken bonds in highly-distorted parts of the amorphous structures, as for the E′–NBO pairs. In the second interpretation, Type II defects are broken bonds in highly-distorted parts of the amorphous structure, and Type I defects are vacancy–interstitial pairs, as for E′–PR pairs.

It has been claimed that forming Type I defects requires intense radiation. No intensity dependence has yet been measured, so it is not yet clear whether defect creation is a linear or non-linear process. In any case, we believe that the self-trapped exciton gives rise to the luminescence, and that the defect pair is a metastable state resulting from an exciton. Both processes involve bond scission. The bond scission occurs through the coupling of the exciton to the lattice. The self-trapped exciton and the defect pair are different configurations derived from bond breaking. Defect pairs can probably adopt a variety of structures at higher temperatures, where defects can have higher mobilities. The photoconductivity change, which has different annealing temperatures from Type I and II defects, is therefore considered to arise from these varied forms

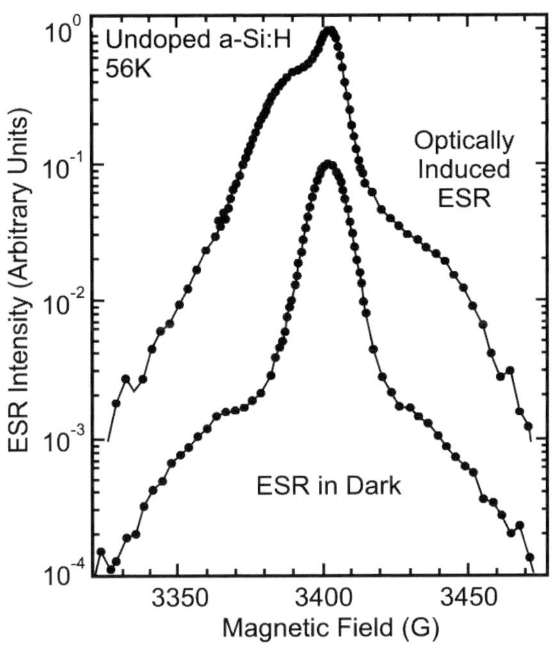

Fig. 8.21. Intensities of spin resonances in unirradiated and optically-irradiated a-Si:H (Yamasaki *et al.* 1990).

of defect pairs derived from excitons. The difference in annealing temperature may be due to differences in cross-sections for photo-induced charge carriers (see Shimakawa *et al.* (1992)).

8.3.3 Amorphous silicon: a-Si:H

EPR studies

Three types of defects have been identified by EPR for a-Si:H in the dark, under photon irradiation of previously unirradiated a-Si:H (Fig. 8.21). The spectrum in the dark with $g=2.0055$ shows hyperfine structure due to a single ^{29}Si, which has 4.7% abundance. This indicates that the spectrum is due to an electron or a hole on the dangling bond of a single Si atom (Stutzmann & Bieglesen 1989). On the other hand, the optically-induced EPR spectrum appears to be a composite, comprising two components with $g=2.01$ and $g=2.004$ (Knights & Biegelsen 1977; Street & Biegelsen 1980; Yamasaki *et al.* 1990, 1991). Yamasaki *et al.* (1990) analysed the hyperfine structures of the photo-induced EPR spectra and showed that both spectra are due to spins localised on a single Si atom. From the relation between the intensity of each signal and the position of the Fermi energy, they showed that the $g=2.004$

signal is due to an electron trapped at a positively charged dangling bond, and the $g = 2.01$ signal is due to a hole trapped at a negatively-charged dangling bond. Both positively- and negatively-charged dangling bonds become ESR sensitive under photon irradiation because of charge alternation. Prolonged photoexcitation increases these signals (Dersch *et al.* 1980; Hirabayashi *et al.* 1980; Stutzman *et al.* 1985) but only slightly for signals with $g = 2.01$ and $g = 2.004$ (Shimizu *et al.* 1989). Again, this is similar to a-SiO_2 and to the chalcogenides, and is consistent with defect formation by bond breaking through an excitonic mechanism, as suggested by S R Elliot (1979). The EPR signal at $g = 2.0055$ observed prior to photoexcitation is also enhanced by band-to-band excitation. The increased EPR signal is stable up to 470 K.

Other properties of photo-induced defects

The photo-induced defects in a-Si:H have properties similar to those in chalcogenides. There is a broad optical absorption in the gap (Amer *et al.* 1983) and a reduction of photoluminescence (Morigaki *et al.* 1980). However, the correlation between these property changes and the formation of the defects seen in EPR is not straightforward. Probably the EPR centres formed at low temperatures convert to non-paramagnetic defects, but continue to affect the luminescence decay and optical absorption. Photoconductivity changes are also seen following band-to-band excitation, as reviewed by Shimakawa *et al.* (1995). There is evidence for substantial numbers of charged, rather than neutral, defects (Schmidt *et al.* 1997, 1999). Other evidence suggests that small volume fractions of crystalline Si can suppress defect creation under moderate illumination, apparently because excess carriers recombine within the crystallites instead of creating damage in the amorphous matrix (Kamei *et al.* 1999).

Mechanism of defect formation

According to Stutzmann *et al.* (1985), the creation of 10^{12} dangling bonds needs 15 J (about 10^{20} eV) of photons of 1.9 eV: the yield is extremely small. It is likely that the defect formation mechanism is similar to that in chalcogenides. R Jones & Lister (1990), using density functional theory, predicted that a metastable state is formed by electronic excitation at the sites where a Si–Si bond is stretched, whereas the electronic excitation of the unstretched part is ineffective. This supports the idea that defect pairs are formed by excitation of highly-deformed sites. The predicted metastable defects had two gap levels: a filled one at $E_v + 2.1$ eV, and an empty one at $E_v + 3$ eV. Suggestions of possible atomic structures of the defect pairs have been made (S R Elliott 1978; Shimakawa & Elliott 1995; Morigaki 1988), but no detailed theory is available yet.

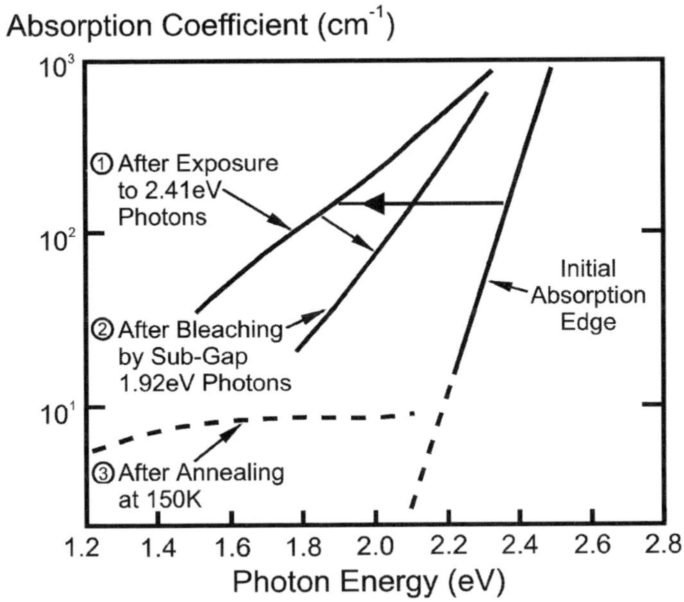

Fig. 8.22. Change of optical absorption (1) by prolonged irradiation of a-As_2S_3 with 2.41 eV photons (solid line) and (2) by short-time irradiation (dashed line) (after Biegelsen & Street (1980)). The optical absorption induced by prolonged irradiation is recovered partly by irradiation with subbandgap photons. The remaining part is called photodarkening as explained later.

8.4 Photo-induced structural changes: Photodarkening

What is photodarkening?

The phenomenon known as *photodarkening* is observed in fully-coordinated amorphous materials such as chalcogenides, but not in a-Si:H. It is a modification in the amorphous network structure, rather than formation of point defects. Photodarkening is not observed in *crystalline* chalcogenides. The most significant phenomenon is the change in absorption edge (Fig. 8.22), which can be annealed slightly below the glass transition temperatures. Photodarkening occurs both in bulk chalcogenide glasses and in thin films. However, that for bulk glasses anneals at the temperature at which photo-induced defects anneal (Hamanaka *et al.* 1977), which is much lower than the annealing temperature for thin films (Hautala *et al.* 1988). Furthermore, the volume change accompanying photodarkening in bulk glasses (Hamanaka *et al.* 1977) is smaller than that of thin films (K Tanaka 1990). Thus, substantial photodarkening is favoured in thin films. Probably this is related to the stress in the specimens, as pointed out by Kolobov & Adriaenssens (1994).

Photodarkening is also accompanied by a structural change observed in X-ray (K Tanaka 1975) and neutron diffraction (S R Elliot *et al.* 1982). The reversible changes of Raman signals (Frumar 1997) and X-ray extended absorption fine structures (EXAFS) (C Y Yang *et al.* 1987; Gladden *et al.* 1988) in parallel to photodarkening showed that the photodarkening of As_2S_3 is related to an increase in the number of As–As 'wrong' bonds by a few per cent. Since formation of As–As bonds and complementary S–S bonds should raise the optical absorption edge (Street *et al.* 1978), the formation of wrong bonds alone cannot be the direct cause of the shift of the absorption edge. Kolobov & Adriaenssens (1994) suggested that the formation of As–As bonds between layers distorts the layer structure, and it is that which reduces the bandgap energy. Certainly illumination increases the thickness of thin films and reduces their bandgaps (Kuzukawa *et al.* 1999). Kuzukawa *et al.* suggest that there is a change in the mesostructure as well, involving slip motion associated with the accumulation of electrons in conduction band tails. The photo-induced volume expansion shows both transient and a metastable components (Ganjoo *et al.* 1999), with a strong correlation between photocurrent and changes of thickness.

Phenomena related to photodarkening

There is a group of photo-induced phenomena related to photodarkening, which are accompanied with bond rearrangement. These include photo-induced crystallisation, amorphisation, and melting. Extensive studies of these phenomena have been carried out for Se. The crystallisation of a-Se is known to be induced by irradiation with photons of subgap energies although not at low temperature (Kim & Turnbull 1973; Stephens 1980; Hamou *et al.* 1985). The simultaneous irradiation with photons of energy above and below the bandgap energy reduces the crystallisation. This shows that irradiation with photons above the bandgap energy can cause amorphisation of the crystalline Se embedded in a-Se (Roy *et al.* 1998). Amorphisation and photo-darkening, characterised by EXAFS studies and by the formation of the a-Se Raman peak (Kolobov *et al.* 1997), are induced by irradiation with photons above the band-gap energy. Irradiation with photons of subgap energies at liquid nitrogen temperature can cause melting, and breaking of the bonds between Se chains (Poborchii *et al.* 1999). It is likely that irradiation with photons below the bandgap energy breaks bonds at highly distorted parts of the amorphous structure, leading to crystallisation, whereas irradiation with photons above the bandgap energy breaks the Se–Se bonds and leads to amorphisation. Structural change is also accompanied by bond scission. Local stress is probably one important factor in amorphisation, as we discuss later. One reason for this is very general, applying in cases as different from photodarkening as electrical breakdown in silicon oxide films (Jeffery *et al.* 1998; their interpretation is different but not inconsistent with the present point). This is that defect formation energies change with stress. The change with pressure, for instance, is defined by a defect volume, which is often similar to the unit cell volume. This dependence means that, for a general stress distribution, it will be significantly easier to form defects in certain regions.

Photodarkening with polarised light

Irradiation with polarised light, with energy either above or below the bandgap, causes dichroism in chalcogenide glasses and in oxide glasses (Zhdanov *et al.* 1979, Lyubin & Tikhomirov 1989). The phenomenon is thought to be similar to the dichroic bleaching of colour centres: continuing irradiation with polarised light destroys those colour centres whose transition dipole is parallel to the electric field vector of the polarised light. For amorphous materials, irradiation with polarised light excites regions where the electric vector of the incident light coincides with the direction of the transition dipole moment. If there is a photostructural change following recombination, the system will evolve such that the optical absorption intensity is no longer the same when measured with light polarised parallel and perpendicular to the incident polarisation.

The dichroism relevant to photodarkening is that induced by photons above the bandgap energy. The dichroism induced by low-energy photons may involve defect polarisation. In spite of controversy about whether the anisotropy and photodarkening are correlated, it seems likely that dichroism does correlate with photodarkening, as suggested by Kolobov *et al.* (1997) and by K Tanaka *et al.* (1996). One interesting observation (Kolobov *et al.* 1997) is that irradiation at a certain photon energy near the bandgap energy can induce dichroism at even higher energies. It seems likely that this is a result of bond rearrangement, with a higher-energy structure forming perpendicular to the direction of that broken by the excitation.

Mechanisms of photodarkening

A model of photodarkening, first proposed by Elliott (1986), suggests that photodarkening involves intramolecular and intermolecular bond breaking and perhaps bond weakening, and that these changes alter the interlayer structures. A slightly different idea was suggested by K Tanaka (1983), namely that the local structure is modified by the rotation of flexible bonds induced by electronic excitation. Given that the formation of wrong bonds accompanies photodarkening, the model that electronic excitation causes bond breaking and weakening seems more likely. The question remains as to what are the differences between the formation of luminescent centres, the formation of defect pairs, and photodarkening, since all involving bond scission due to electronic excitation.

We recall that photodarkening is dominant in thin glasses on substrates: the shift of the absorption edge in thin glasses is larger than bulk glasses, and anneals only near the glass transition temperature. Kolobov & Adriaenssens (1994) suggested that photodarkening occurs only when there is macroscopic strain in the material: bond weakening and bond relaxation allow relaxation of the residual stresses of films on substrates. Thus, photodarkening occurs as a result of bond scission by electronic excitation in amorphous materials under strain. Strain must be defined with care for amorphous systems, since they can (in principle, at least) respond to stress by changing bonding pattern. The strain discussed by Kolobov & Andriaenssens relates to a bonding pattern which is unchanged when the stress is applied.

Photo-induced processes in amorphous materials

We can think of all three classes of photo-induced processes in amorphous materials (luminescence, defect formation, and photodarkening) as the result of some form of bond weakening due to the formation of excitons. Self-trapped excitons, formed by a quantum efficiency close to unity, undergo either radiative or non-radiative recombination. The radiative transitions give the luminescence. Some of the excitons, probably those formed at rare, highly-distorted, sites, evolve into paramagnetic defect pairs or into metastable states. The metastable states lie on the adiabatic potential energy surface of an excited state, and anneal at relatively low temperatures. Photodarkening of thin films, however, involves many more sites than those yielding defects. The total area under the optical absorption curve due to photodarkening is far larger than that due to defects. Moreover, the fact that photodarkening in thin films only anneals near the glass transition temperature shows it is not due to metastable excited states. It is more plausible to assume that the states responsible for photodarkening lie on the adiabatic potential surface of the ground state, as suggested by K Tanaka *et al.* (1980) and Averyanov *et al.* (1980); see also Shimikawa *et al.* (1995). Presumably, they correspond to some topologically-modified form. It seems probable that photodarkening represents another recombination channel of the self-trapped exciton. Photo-induced crystallisation and amorphisation change the topology and morphology of the ground state, probably at the amorphous/crystalline interface (Legros *et al.* 1995).

Presumably the topology changes are determined by the nature of the excited state prior to recombination, e.g. whether it has interchain or intrachain character.

In amorphous structures, the ground-state adiabatic potential energy surface needs careful definition. Clearly, for some chosen bonding pattern, there will be a local minimum, and small displacements will follow Hooke's law. However, there will be many bonding patterns which have very similar energies, and each will have its own local minimum of energy. Let us refer to each specific bonding pattern and its associated topological structure as a *configuration*. Reconfiguring the bonding pattern is possible via excited states, and these are the processes of interest here. Suppose an exciton is formed at an initial configuration P. The adiabatic potential energy surface in the corresponding excited state is usually flatter, which means a smaller restoring force following displacements, which one can regard as bond weakening. If so, a transformation from the adiabatic potential energy surface for configuration P to that for another configuration P' can happen more easily in the excited state. The direction of the transformation will be to lower the stress. When recombination of the self-trapped excitons occurs, whether radiatively or non-radiatively, the amorphous network will have adjusted to configuration P'. The transformation from P to P' may involve displacements of several atoms, and will occur only near the glass transition temperature in the ground state. However, because of the bond reorganisation, the transformation in the excited state requires a small amount of energy, and may even be exothermic.

As pointed out by Kolobov & Adriaenssens (1994), there is strong mechanical stress after annealing of films on substrates, the usual starting materials for photodarkening. The stress plus rebonding model explains why photodarkening does not occur in crystalline materials. Kolobov & Bedel'baeva (1991) suggested that the reason for photodarkening in amorphous materials is because the site-to-site variations in these non-crystalline solids allow localisation of electronic excitation energy at special sites, even when this is not possible in the corresponding crystalline material. This particular site dependence is unlikely to be the explanation for the chalcogenides, since we know that exciton self-trapping (which localises electronic excitation energy) happens in both the crystalline and amorphous forms. An alternative explanation for the lack of structural change in crystalline chalcogenides is that recombination restores the system to its original atomic configuration, whereas some, at least, of the recombinations in the amorphous material lead to new structures. Consistent with this,

Kolobov & Elliott (1995) have shown that photoexcitation of crystalline As_2Se_3 does cause amorphisation when illuminated under a macroscopic stress. Presumably the stress is sufficient to ensure that some of the recombination events generate new structures.

Ice is interestingly different (Chakarov & Kasemo 1998), in that thin films of amorphous ice can be photocrystallised. The indications are that there is photo-induced tunnelling from the graphite substrate to defects in the ice, and the excitation stimulates ordering.

Photodarkening is not observed in a-Si:H. This may be because the exciton formation does not significantly weaken or break Si–Si bonds, or simply because there is too little energy involved: the photoexcitation energy is only 1.5 eV, significantly less than the Si–Si bond energy of 2.5 eV.

Charge transfer under excitation, especially γ excitation

One contribution to darkening of glasses comes from charge transfer. This is especially important for the windows for post-irradiation damage caves, which are subject to a strong gamma flux from one side. The 'Compton wind' redistributes electrons across the glass, so causing changes of charge state. One way to avoid darkening is to incorporate an impurity which has two charge states, neither of which absorbs in the visible. The standard choice is Ce, which is found substitutionally in the 3+ or 4+ charge states. If the 3+ ion is stable at a particular site, the 4+ will be metastable at that site. We remark that Ce also suppresses colour centre formation in borosilicate glass fibre up to 10^5 rad (Danilov *et al.* 1990). Experiment does not observe all the four absorptions expected (there are two charge states, and the initial state can be the metastable form or the stable form at its site). However, we can predict where the remaining absorption should be using simple theory. The bands observed for charge states at sites where they are stable are at 3.95 eV (width 0.65 eV) for the 3+ state and 5.17 eV (width 0.65 eV) for the 4+ state. The 4+ ion at a site stable for 3+ absorbs at 4.96 eV (width 0.72 eV). Where should the 3+ ion absorb at a site stable for 4+? Imagine transferring an electron from the valence band to the conduction band by the process

$$e_v + X^{N+} \rightarrow X^{(N-1)+} \rightarrow X^{N+} + e_c.$$

If we assume the polarisation energy is the same for the two types of site listed, then we expect (3.95 eV + 4.96 eV) = 8.91 eV = ((unseen? energy) + 5.17 eV) which would give a 'missing' component at 3.74 eV, possibly hidden by the 3.95 eV band, i.e. the two 3+ absorptions overlap. A further check comes from the width data. If p is a typical phonon energy then we can use the halfwidths to deduce a relaxation energy of $(\text{halfwidth})^2/(5.545p)$, or about 1.6 eV for $p=$

0.05 eV. This would give the energy gap for direct excitation as $8.91-(2\times1.6)$ $=5.7$ eV, which is not unreasonable.

8.5 Ion-beam-induced structural changes

For an amorphous material, the phenomena which can be called 'structural change' unambiguously are crystallisation and the amorphisation process itself. Both can be driven by excitation. The crystal/amorphous interface of Si under ion irradiation moves in either direction (crystalline to amorphous, or amorphous to crystalline), depending on the ion flux and on the irradiation temperature. But there can be clear changes in structure even in amorphous solids, and these can be induced by ion beams. The best-studied examples are the ion-induced modifications of a-SiO_2 and silicate glasses, which first densify and then expand; other glasses, like those for the vitrification of radioactive wastes, can expand or contract. The role of volume changes can be important, as shown for the electron beam amorphisation of coesite (Gong *et al.* 1996). Both electronic excitation and elastic encounters induce structural change in SiO_2. There is clear evidence for bond reorganisation; this may imply bond breaking.

In ion-beam induced amorphisation, several mechanisms are possible, depending both on the system and on the beam and dose. Discussions are normally concerned with whether defect creation is the only cause of amorphisation, or whether there is an important role for hot zones (hot spots in some descriptions) within collision cascades. These issues arise again in Chapter 11.

8.5.1 Ion-induced crystallisation and amorphisation of silicate glasses

Fused quartz

It has been known for many years that amorphous SiO_2 first densifies, then expands, under ion irradiation. The relation between the density and fluence for various ions can be scaled either by the elastic stopping power for low-energy heavy ions (Eernisse & Norris 1974) or by the electronic stopping power for electrons and high-energy light ions (Dellin *et al.* 1997; Presby & Brown 1974). Primak (1984) recognised differences in behaviour during the non-uniform irradiation of v-silica: with ion irradiation, post-irradiation stress relaxation occurs, whereas with electron irradiation, there is no such relaxation.

Two mechanisms have been suggested for the processes induced by elastic encounters. The first idea (Eernisse & Norris 1974) was defect creation. A later suggestion was of hot spots, as for ion-induced amorphisation. Subsequent experimental observations favour the hot-spot model. First, vacancy formation saturates at a dose ten times smaller than the saturation of the density change. Secondly, the densification goes along with the change in the etching rate, change in reflectivity index, and shift of infrared absorption spectra (Devine 1994). This indicates that defect creation cannot be the direct cause of densification, but leaves the possibility that densification by elastic encounters occurs by collision energy localisation in collision cascades. The rate of densification per absorbed energy by electronic excitation is 1/1000 of that by elastic encounters.

Studies of infrared absorption and of Raman spectra suggest a large reduction of bonding angle; this, with the enhancement of chemical etching rate, is not consistent with plastic densification. Densification is therefore ascribed to the formation of some new local structure, analogous to the crystalline porosil structures. It has been argued that, when the energy imparted is smaller than the displacement energy, transient bond scission in collision cascades permits local rebonding, developing into structural change. This mechanism appears to operate for electronic excitation as well, where bond scission is associated with exciton formation. The reasons for the small efficiency for electronic excitation are not clear. The same ratio, 1/1000, for the relative efficiencies of electronic and elastic encounters applies both for amorphisation and for defect formation. It would seem that the same bond scission in SiO_2 that produces defects is also responsible for amorphisation.

The yield of amorphisation by energetic heavy ions with electronic stopping powers higher than the stopping power for amorphisation of SiO_2 is much higher than for light ions. The mechanism is presumably the same as track registration (Section 11.6).

Radwaste glasses

One possible containment method for high-level radioactive waste is vitrification. The waste, plus certain glass-forming materials, is melted, and placed in a container. Once it has cooled as far as it will, consistent with the heat generation from the radioactivity, there is a question as to whether it will expand or contract with time. Both situations are found, depending on the glass composition. Tinnevella & Stoneham ((1980); see also Stoneham (1994)) offered a simple argument to predict whether expansion or contraction would occur. It exploited the fact that, for normal glasses, processed in standard ways, the density can be predicted as a function of composition. We may call this the 'ideal' density ρ_0; the full formula includes much empirical information but, roughly, it corresponds to close-packed oxygens with interstitial cations.

Suppose that the actual density of the glass is ρ just after it has cooled. Tinnevella & Stoneham suggested that the major effect of the radioactivity is as a 'stirrer', providing a means for the structure to move towards that with the ideal density. Thus glasses for which $\rho > \rho_0$ should expand with time, and those with $\rho < \rho_0$ should contract. This simple rule is consistent with all the glasses for which data were available. They also found that the rate at which the density changed correlated to some extent with glass viscosity, the changes being most rapid for the less viscous systems.

Amorphisation of sapphire

Most of the examples of amorphisation relate to high-power regimes, discussed in Chapter 11, or to surface or interface phenomena, discussed in Chapters 9 and 10. We note here that sapphire is amorphised by a 308 nm XeCl excimer laser above the sapphire threshold at about 0.7 J/cm^2 (Pedreza *et al.* 1995). Laser treatments enhance adhesion of metals, both below and above this threshold. The enhancement is best in an oxidising atmosphere. For Au, the untreated surface has very weak adhesion which is improved only by excitation above the melting threshold. For Cu and Ni there are effects below the threshold as well. Cao *et al.* (1994) demonstrated that a pulsed laser can generate γ-alumina on sapphire. At higher intensities, Canut *et al.* (1995) noted that the irradiation of sapphire by megaelectron volt U ions leads to disorder related to electronic stopping power.

Appendix: Basic defects in glasses

Details of defects in silcate and related glasses are dispersed in the literature. Here we summarise information about many of them. Silica is a basic ingredient of most glasses. The nature and variety of its defects will have parallels in more complex glasses. 'Silica' is, however, not well defined. In particular (*a*) there are several crystalline forms of ideally pure silica; (*b*) there are many sensibly distinct amorphous forms of ideally pure silica, e.g. vitreous, thermal oxide, radiation- amorphised. There are also many commercial forms, like Suprasil (various types) which are strictly impure amorphous silicas; (*c*) there are numerous forms of impure silica which are 'nominally' pure. The most persistent impurities are Al, Ge, possibly C, and certainly H. Indeed, there are at least six forms of hydrogen in the most hydrogen-free silicas; as an illustration, the effects of moisture show in the oxidation of silicon down to less than 1 ppm of water vapour in dry oxygen. Alkali species like Na are also common impurities. Assertions that given silica is 'pure' should be treated cautiously. In the

Defects common to several glasses (such as borate, silicate and phosphate glasses).

Holes trapped on oxygens (often with impurity)	2.0–2.6 eV 2.5–3.0 eV	(This band depends on alkali)
Alkali electron centre	1.5–2.4 eV	Depends on glass former, with the energies borate > > silicate > phosphate)
Electron trap	4.1 eV	Silicate glasses only, but apparently all of them
Electron trapped in a bonding orbital of glass former with neighbouring alkali	5.1 eV	

Defects giving absorption bands in SiO_2.

E eV	*W* eV	*f*	Name	Suggested description
8.2				Observed after gamma irradiation
8.0				Observed after gamma irradiation
7.6	0.5	0.65	E band	Not a single defect. One component appears to be associated with the peroxy (Si—O—O) defect, characterised by spin resonance. Another view is that an O vacancy is responsible. Seen in both pure and alkali doped silica. A hole trap
7.15	0.8		D band	Appears after heavy ion irradiation. Not the same as the D_0 band
6.1			C band	Seen in impure silica; sometimes confused with the E_1' absorption, similar in energy. The C band appears to be associated with aluminium impurity
5.85	0.6	0.14	E_1' band	Observed in pure or lightly-doped silica. Hole trapped at a neutral oxygen vacancy
5.5	0.4	0.28	E_2' band	E_2' band for 'pure' silica
5.3	0.4		E_2' band	E_2' band for alkali-doped silica. Apparently the E_1' centre (the neutral oxygen vacancy plus a hole) with a proton (pure silica) or an alkali cation (alkali doped silica)
5.1	0.4		B_2 band	Seen after heavy ion or neutron irradiation; possibly a neutral oxygen vacancy
4.7			D_0 band	Seen in pure silica
4.0	1.2		B band	An electron trap found in impure silica
2.9	1.5		A_3 band	This gives smoky quartz its colour. Associated by spin resonance with a hole localised on an

Defects giving absorption bands in SiO_2 (cont.)

E eV	*W* eV	*f*	Name	Suggested description
				oxygen at a bridging site between substitutional Al and a neighbouring silicon
2.5	0.8		A_2 band	Both A_2 and A_3 are associated with Al in impure silica. There is a third band at 1.9 eV. The other A centres may involve alkali.
2.0	0.33 for high OH; 0.26 for low OH		Partly NBOHC	NBOHC. Some non-paramagnetic electron trap, perhaps a three-fold coordinated Si which has trapped a pair of electrons

Emission bands in silica: Apparently intrinsic bands.

6.70 eV	Seen during steady-state electron (2.5 MeV) irradiation of high purity silicas
4.28 eV	Apparently associated with electron–hole recombination at oxygen vacancies (in effect, an $e^- + E'$ reaction); possibly due to two-fold coordinated Si. The band centred at 4.4 eV is complex, and may be due to a structural transition from an oxygen vacancy to the two-fold Si form (Boscaino *et al.* 1999)
2.82 eV	Seen in crystalline and amorphous silicas. This blue luminescence shows complex behaviour and dependence on excitation and impurities. Some possible correlations with the E_1' centre.
2–3 eV	Self-trapped excitons. The spread in energy is partly due to a large spread in relaxation energies
2.65, 2.3, 2.0, 1.7 eV	Observed in Si-implanted thermal oxide. The 1.7 eV band appears to be associated with centres at the nanocrystal Si/oxide interface
1.91 eV (0.65μm)	Seen in vitreous silica after neutron or ionising radiation. Decays with two characteristic times, namely 18 and 2.2 μs. Excitation band centred on 4.7 eV (in neutron-irradiated silica glass, there is also luminescence at this energy with an excitation band at 2 eV). Possibly associated with the NBOHC, at least when the OH content is high, otherwise with some non-paramagnetic electron trap like a three-fold coordinated silicon (Skuja *et al.* 1996).

following lists, observed absorption bands are in decreasing order of energy. The models are often speculative. *W* is the full width at half maximum; *f* is the oscillator strength.

Summary

Amorphous halides and halide glasses Probably the halogen molecular ion X_2^- continues to play a critical role, as for the crystalline halides. As noted in Chapter 5, intrinsic defects are not formed in fluoride glasses of Al, Y, and alkaline earths; impurities like O seem necessary.

Oxide glasses with closed-shell cations, like borosilicate or aluminosilicate glasses The self-trapping of electrons and holes to give STE and STH is more favourable than in the corresponding crystalline system, e.g. c-silica versus a-silica and oxide glasses. Exciton self-trapping is the origin of luminescence and defect formation. Topological changes and sequences of defect reactions lead to density changes in lithography optics and to the creation of Bragg gratings in optical fibres.

Oxide glasses which include transition metals STE, STH are often regarded as charge state changes when there are transition metals in the glass. The main effect is charge transfer; this process can permit control of colour.

Amorphous semiconductors There are common features for crystalline and amorphous systems, as well as differences: luminescence; defect formation and photodarkening are all initiated by STX. The materials can be very different, ranging from DLCs and a-Si:H to chalcogenide glasses. Phenomena of interest include photodarkening (which is not seen in the corresponding crystalline forms) and the Staebler–Wronski effect (a reversible defect formation phenomenon). Organics and polymer systems undergo photo-induced cross-linking and scission.

Chapter Nine

Atomic emission and surface modification

9.1 Energy absorption near surfaces

Electronic excitation at solid surfaces can be induced in two main ways. The first is electronic excitation in the bulk, followed by direct energy transfer or by the diffusion of excitons or of electron–hole pairs to the surface. The second route involves the excitation of surface states or surface defect states. Such states are especially important when the excitation energy is less than the bulk bandgap. However, the division into two classes is not clear-cut, for there will usually be a space-charge region of a few nanometres in thickness; energy absorption in this region may make an important contribution to near-surface excitation. De-excitation processes at surfaces may follow the same pattern as in the bulk. The modification of local surface structures will often be simpler and easier than in the bulk. What will be different are reactions involving specific surface species, including adsorbed species, and processes which involve the emission of atoms, ions, or molecules. Our emphasis on the modification of the underlying material continues. We shall therefore note, but say little about, processes such as the laser removal of silicon chloride species during plasma etching (C C Cheng *et al.* 1994).

Several terms are used for the emission of atoms or ions, depending on the rate and the application: desorption; sputtering; ablation. In this book, *desorption* usually refers to reemission of adsorbates on surfaces, whereas *sputtering* is the emission of atoms or ions constituting the host material. The distinction in literature can be obscure. The *sputtering yield* is often defined in terms of the thickness of the material removed per unit number of excitations. The *quantum yield of emission* is often defined by the number of emitted atoms per unit number of absorbed photons near the bandgap energy. The sputtering yield of a compound can be obtained only under conditions in which all constituent atoms can be emitted (whether thermally or by excitation); the sputtering yield is governed by the emission yield of those constituent atoms having the lowest quantum yield. Selective sputtering describes a stage in which certain species are preferentially removed. The removal of a massive number of atoms using laser irradiation is called *laser ablation*.

There have been many studies of the excitation of adsorbed species (e.g. Palmer & Rous 1992; Sanche 1997; Franchy 1998), especially for metal or rare-gas substrates. These systems lie outside our main theme, but illustrate some important points. For example, we note the several ways by which electron excitation can lead to desorption, apart from direct excitation by inelastic scattering. First, of course, there can be capture of the electron into an excited state. The excited state may evolve through some surface reaction. Walton *et al.* (1998) found that, after Al has been exposed to oxygen to about one monolayer coverage, light stimulates O^- emission, apparently through the photodissociation of $(AlO)^-$. Other processes lead to secondary electron emission. Franchy listed a number of photodissociation reactions on ionic surfaces, such as the removal of I from CH_3I on alumina or of S from OCS on LiF(100), and also photostimulated reactions like $2HBr \rightarrow H_2 + Br_2$ on LiF(110) or $2NO \rightarrow N_2O + O$ on Si(111). Nagesha & Sanche (1998) noted that, if the substrate has a negative electron affinity (as does v-nhexane) then the image interaction greatly increases the low-energy electron capture cross-section of adsorbed species. Values can be 2–4 orders of magnitude larger than for the same adsorbed species on a positive electron affinity substrate (like c-Kr).

Secondly, the captured electron may be reemitted, leaving the surface (including any adatoms) in an excited vibrational state. Sanche, in particular, noted the efficient vibrational excitation of molecular oxygen by low-energy (2–16 eV) electrons. Again, there are possible surface reactions which can be stimulated by vibrational excitation (Chapter 7).

Thirdly, there can be long-lived resonances, rather than bound states. Such a temporary electron attachment, if followed by dissociation into a neutral and an ionic species, is known as *dissociative electron attachment*. Dissociative attachment appears to be the cause of low-energy stimulated desorption of molecular hydrogen from amorphous ice (Kimmel & Orlando 1994). There is a 6.3 eV threshold, well below the 7.3 eV first excited state of condensed water. It is attributed to the $H^- + H_2O \rightarrow H_2 + OH^-$ (and a similar reaction for the deuterated species), primarily at the surface. The yield below 11 eV is attributed to the dissociation of Frenkel excitons; above 11 eV, the recombination of electrons with H_2O^+ or H_3O^+ species seems to be most important.

A fourth situation is found for the desorption of molecules adsorbed on semiconductor surfaces, reviewed by Ho (1994). One interesting example is the desorption of NO from the Si(111)7$\times$7 surface. The excitation spectrum shows a peak at 320 nm, suggesting that hot electrons are contributing to the desorption process.

9.1.1 Perfect surfaces

A perfect surface is not the ideal termination of the bulk drawn in some texts. The atomic and electronic structures of surfaces can differ substantially from that in the bulk. There is always some reconstruction or polarisation at the surface, and there will often be a dipole layer at any interface. The displacement of atoms near a surface perpendicular to that surface is referred to as *relaxation*; when different surface species move outwards by different amounts (as do halogen and alkali on [100] surfaces of alkali halides), this is known as *rumpling*. Displacements parallel to the surface, and especially those which alter the periodic symmetry of the surface, are referred to as *reconstruction*. Relaxation and reconstruction reduce the surface energy. In experiments, it is important to know if reconstruction happens on a timescale fast or slow relative to the measurement.

Reconstruction in ionic materials primarily minimises the Coulomb and polarisation energies. Sometimes faceting occurs, e.g. a (110) surface may transform into (100) and (010) facets. In covalent materials, reconstruction often eliminates dangling bonds; this is one of the reasons why simplistic surface-state calculations are not helpful. One consequence of reconstruction is that surface atoms are no longer all equivalent. Within a periodic surface structure, there may be some special atom which can trap a carrier more easily, or be removed more readily. As regards energy localisation or charge localisation, that at the surface must be localised both in depth and laterally. The electric dipole associated with most surfaces (whether intrinsic or associated with space charge) gives rise to the well-known band bending of semiconductor studies, and usually ensures that one type of carrier can be confined near the surface. The magnitudes of fields at oxide surfaces (with rumpling) can be both predicted and observed. For the cation site on MgO(100), 2.9×10^8 V/cm is observed, compared with 1.9×10^8 V/cm predicted. Predicted values for NiO(100) are 6.6×10^8 V/cm at the cation sites and 2.0×10^8 V/cm at the anion sites. Other contributions to confinement close to the surface can result from the difference between bulk and surface Madelung energies, or from polarisation (image) terms. The Madelung constants do not differ much, but the Madelung energy itself is large. For MgO(100), relaxation also affects the Madelung energy significantly from the unrelaxed surface value; the shifts are 1.2 eV for the cation and 1.5 eV for the anion.

Many oxides show few, if any, signs of surface excitons. In particular, there are no signs of exciton states significantly shifted into the bandgap for TiO_2, SnO_2, Ti_2O_3, or V_2O_3. For MgO, CaO, and SrO, there is a clear shift: for MgO, there is a surface component at 5.7 eV, well below the 7.8 eV bandgap. One possible explanation relates to materials with negative electron affinity, since an electron has lower energy in the vacuum than in the conduction band. A surface exciton in such a system will naturally form with the electron lowering its energy by concentrating outside the surface, and the hole pulled to the surface by this electron (P A Cox & Williams 1986). Lateral confinement is more complex, in that it may be helped by the surface reconstructions, which may result in a relatively narrow band involving only a fraction of the surface atoms. Surface features like steps will encourage lateral localisation, and we discuss such features later. Whether or not MgO is indeed such a negative electron affinity system is not clear. Direct measurements (not always on clean samples) give a positive affinity; evidence from the observed stabilities of transition metal ions suggests a negative affinity.

The conditions for self-trapping are modified near a surface, but there is very little information to give a clear picture of conduction. Experimentally, sapphire is reported to have a high surface conductivity (Chase 1994b). If a surface has a negative electron affinity (which may be so for MgO, and could be so for sapphire) then electrons just outside the surface could be bound in high-mobility states. However, these would be very sensitive to surface defects and microstructure.

9.1.2 Near-surface defects

We have been discussing the surface as if it involved only the outermost layer of atoms. We remark that near-surface defects and surface-intersecting defects can also be important, and especially those in a space-charge layer near to the surface. Our remarks refer to the 'surface', implying the external surface of the solid. Most of the phenomena we describe occur with only minor modifications near to internal interfaces, or near extended defects like dislocations.

There are two main sources of differences between near-surface regions and the bulk. The first difference is intrinsic, independent of impurities. In the bulk, any defect process preserves electrical neutrality. With Schottky disorder, negative vacancies and positive vacancies balance charges. At the surface, however, different numbers of negative and positive vacancies could occur, depending on their individual formation energies. Obviously, there will be neutrality over the whole crystal, but local charge imbalance is possible. A characteristic length for the space-charge region is given by the Debye–Hückel screening length L, with the value $[8\pi Z^2e^2c_0(T)/\epsilon kT]^{1/2}$ for defects of net charge Z and bulk concentration $c_0(T)$ in a host of dielectric constant ϵ. For $\epsilon = 10$, $T = 300$ K and $c_0 = 10^{16}/\text{cm}^3$, L is about 2.5 nm. If there are positively and negatively charged defects with formation energies g^+, g^- then, at distance x from a surface, the potential is given by (Flynn 1972)

$$|e|\phi(x) = (g^- - g^+)\,[1 - \exp(-x/L)]/2.$$

This leads to a potential difference of $(g^- - g^+)/2$ between surface and bulk. There is a dipole layer which leads to a surface excess of the defect with the larger formation energy. In AgBr, the (111) surface is negatively charged, the charge density being about 2×10^{11} charges/cm^2; the charge density on the (100) surface is a factor of 20 lower.

The usual discussions of space charge ignore image interactions. Yet these can affect the distribution of charge near an interface, especially when there is a large change of dielectric constant (Duffy *et al.* 1995a,b). These papers emphasise that space charge can be created by moving electrons (and so changing charge states of defects or impurities) rather than moving ions (of course, both electrons and ions may move, cf. Mott & Gurney (1948), Chapter IV, Section 7). A consequence is that interfacial energies can be modified by optically-stimulated charge transfers. A further consequence (Stoneham & Tasker 1985a,b) appears to be the Elovich kinetics seen during the elimination of F^+ centres when MgO microcrystals adsorb oxygen (Nelson *et al.* 1968). In principle, a polarisable solid will have an attractive image interaction for charges outside the surface. This could lead to image–charge states, although they are not usually important in the present context. What is perhaps more important is that the image interaction favours self-trapping, since charge localisation enhances the polarisation energy terms.

In dynamic (non-equilibrium) situations, such as fracture, but including thermal ion emission in some cases, it is possible for regions of high net charge to appear (see, e.g., Buttman *et al.* (1994), Dickinson *et al.* (1993)). These can affect strongly the emission energies and charges of emitted species. The electrostatic forces due to charged defects can be measured with atomic force microscopy; this techniques also allows local measurements of ionic conductivity (Bennewitz *et al.* 1997).

The second difference is extrinsic, depending on impurity segregation, which alters the impurity concentration near the surface. If the impurity needs charge compensation, we expect the charge-compensating defects to dominate over intrinsic defects at low temperatures (the so-called extrinsic region) and thermally-induced defects to dominate in the higher-temperature, intrinsic region. In particular, the net charge may change between these regimes. In principle, an isoelectric temperature may be found, at which the surface has no net charge. In AgCl and AgBr, divalent cations lead to a surface that is negative relative to the bulk by about 0.2 eV at low temperatures. At higher temperatures, Ag vacancy generation becomes more important. No isoelectric temperature is seen for the free surfaces of AgCl or AgBr up to 200 °C, apparently because the free energy for cation interstitials is less than that for cation vacancies. An isoelectric temperature is observed for dislocations in the Ag halides.

In systems which readily go non-stoichiometric, notably those with cations which can exist in several charge states (like FeO or UO_2), the near-surface region will usually differ in composition from the bulk. In some cases, a new phase may form. Even in traditionally stoichiometric oxides, like MgO, thin regions of different stoichiometry can occur. Thus Hebard *et al.* (1986) reported an oxygen-rich MgO with low electrical loss, which is surprising in view of MgO's strong tendency to remain stoichiometric.

9.1.3 Surface defects

Experiments yield *signals*, characteristic energies or spectra associated with some defect species. From these signals and other information, models are deduced: specific assignments with defined atomic structures and charge and spin states. These models must be treated with caution: there are few surface defects whose identities have been established as well as have those of standard bulk defects. There is a further problem, in that many theoretical approaches have attempted only neutral, paired-electron defects. Common defects on oxide surfaces are reviewed by Stoneham & Tasker (1985b).

Anion vacancies

In the F_s^+ centre, the electron is trapped in the surface vacancy itself. In transition metal oxides, including titanates, the electron is trapped on a neighbouring cation, giving e.g. a Ti^{3+} ion.

The F_s^+ centre on MgO(100) is perhaps the best-established defect centre, since it has been the subject of comprehensive spin resonance studies (Nelson & Tench 1964, 1967; Tench & Nelson 1966; Nelson 1966; Nelson *et al.* 1967; Tench 1971) and self-consistent theory (Sharma & Stoneham 1976; Kassim *et al.* 1978; Underhill & Gallon 1982). The optical absorption is less certain: the predicted energy would be hidden by the absorption of the bulk F^+ centre. There is a 2 eV band seen in reflectance, and also a 2.3 eV electron energy loss signal; both, like the spin resonance signal, disappear on exposure to oxygen gas. The F_s^+ centre is stable thermally to about 400 K (Yanagisawa & Huzimura 1981; Zecchina *et al.* 1975). The F_s^0 centre has been discussed theoretically (Kantorovich & Gillan 1997), as has the F_s^{2+} vacancy resulting from the removal of an O^{2-} ion.

The Mg^+ and Mg^{2+} ions observed in laser ablation by Dickinson *et al.* (1993) appear to result from Mg ions above an F_s^+ or F_s^0 centre. The discussion (Yanagisawa, 1986) of photodesorption of chemisorbed oxygen molecules suggests reactions of molecular oxygen with surface F centres, for instance $O_2 + F_s^+ \rightarrow O_2^- + F_s^{2+}$ or possibly $O_2 + F_s^o \rightarrow O_2^- + F_s^+$.

Cation vacancies

Spin resonance of the V_s^- centre has been reported (Eley & Zammitt 1971; Tench 1972; Tench & Kibblewhite 1973).

Low-coordinated oxygens

There is no doubt that corners, edges, ledges and kinks on ledges, and places where grain boundaries or dislocations intersect the surface provide sites of low symmetry and high local electric fields. What has proved hard has been the identification of specific signals associated with these sites. The 2 eV and 2.3 eV bands mentioned above might be related to these low-coordinated oxygen sites.

Adsorbed oxygen

Adsorbed oxygen species are especially important. On the one hand, oxidation processes usually involve an adsorbed oxygen species, which may then move into the oxide and diffuse to react at an internal interface (as in Si oxidation) or react with a cation which has diffused out (as in Ni oxidation). On the other hand, there is good reason to believe that solid-state gas sensor action is controlled primarily by species such as O_{2s}^- (Moseley, Stoneham & Williams 1991). An illustration of how this might happen can be seen by noting that it is exothermic for a MgO:Fe^{2+} to transfer an electron to a surface oxygen molecule, yielding O_{2s}^- and MgO:Fe^{3+}, which can then affect the electrical properties in a measurable way. O_{2s}^- is seen on cubic oxides, including MgO, CaO, SrO, and on CoO and CdO as films on another oxide as substrate, but is not reported on NiO; it is also seen on ZnO and on various sesquioxides (α-Al_2O_3, γ-Al_2O_3, La_2O_3, Sc_2O_3 and Y_2O_3) but apparently not on Fe_2O_3, and on various dioxides (those of Sn, Ti, Ce, Th, Zr, and Si). On MgO, O_{2s}^- is stable to 600 K (Yanagisawa & Huzimura 1981; Cordischi *et al.* 1978). The optically-induced (254 nm, 1 μW/cm^2) desorption of oxygen from MgO (Yanagisawa 1986) is attributed to electron–hole production, with the holes reacting with the adsorbed molecular ions to give the very weakly-bound O_2. Excitation of the bulk V^- band does not lead to desorption.

Another suggestion is that the peroxy defect O_2^{2-} will be found associated with defects at surfaces. Thus D E Williams & Pratt (1998) have proposed that peroxy can be associated with a surface vacancy and an impurity Sb on the surface of SnO_2, where it plays a role in gas sensor operation. It is found that one O_2 molecule leads to a $4e^-$ response. If one of the oxygens O^0 recombines with a vacancy next to an Sn^{2+} then $2e^-$ are accounted for, and the local surface has simply returned to the perfect structure. Some other process, not yet identified, is needed to pull $2e^-$ from sites which contribute to the electrical characteristics of the sensor. There is evidence of a surface trap 1 eV below the conduction band (and so 4–5 eV below the vacuum level).

Hydrogen-related species

Hydrogen is just one of the common surface impurities. Other ubiquitous elements are C and O. Sometimes a useful description is that H saturates dangling bonds. This is the case for SiO_2, where O–H and Si–H bonds are both possible. The removal of H leads to an electron trap, which is one of the forms of degradation of oxide in microelectronic applications. On MgO, spin resonance shows modified F_s^+ species. These H-associated anion vacancy centres appear to involve surface hydroxyl. The initial stages of oxide dissolution frequently involve the creation of OH^- species. Surface features change on a mesoscopic scale during hydration.

Impurity segregation

Impurity concentrations can be high at surfaces. This can be associated with the slightly different Madelung constant, or with the easier surface relaxation, amongst other reasons. Even isovalent impurities can segregate strongly, as in MgO:Ca,Sr,Ba. In most cases, equilibrium segregation is involved. Equilibrium is achieved most easily after slow cooling to allow equilibration. At high temperatures, especially those for which kT exceeds the segregation energy, the degree of segregation can be substantially reduced. Occasionally one can encounter non-equilibrium segregation, which is favoured by rapid quenching from high temperatures. In this case, a vacancy–impurity complex diffuses much faster to the surface than the impurity itself. At the surface, the vacancy is eliminated, leaving the impurity (see, e.g., T M Williams *et al.* (1976)).

All surface properties are sensitive to contamination, and suitable precautions must be taken. For example, sputtering experiments should be performed sufficiently fast that the fraction of surface contaminants remains small during measurements. Adsorbed species should not be regarded solely as a nuisance.

They can be exploited in a variety of ways, from dyes used to absorb energy selectively to the species like O_2^- which are central to sensor action.

9.1.4 Real surfaces: Recognising imperfection

Real surfaces, STM and AFM images remind us, are rough, with characteristic textures. There are scale lengths associated with this roughness. One is the root mean square height $\langle h(x)h(x)\rangle^{1/2}$, $h(x)$ being the height at position x relative to some average reference surface. Another measures the length scale on which the height changes. For many surfaces, $\langle h(x+X)h(x)\rangle$ falls off as $\exp(-X/L)$, where L is the correlation length. Clearly, it can matter whether L is large or small relative to some beam width, whether the beam is a laser spot or a current of focused electrons. Dislocations and grain boundaries which intersect the surface define another length scale. L can be controlled within certain limits. Indeed, as we remark later, laser smoothing can be effective. Thermal diffusion also tends to smooth the surface. In diffusion-controlled processes, the nature of the diffusion equation shows that the features which vary rapidly in space are the ones which are eliminated first. In the simplest case, the rate of diffusion damping at which a hypothetical sinusoidal height variation with wavevector k has time constant $(Dk^2)^{-1}$. However, surface processes can either roughen or smooth a surface (see, e.g., Sofield & Stoneham (1995) for data for Si/SiO_2).

9.1.5 Surface topography

Charge and electric fields are certainly associated with steps and defects. It is not always clear where the charges are localised in experiments which have longer length scales. Cleavage can leave charge patches, as can contact between interfaces, including rubbing. In such cases, it is often hard to determine where the charge is and how it can be manipulated. Crystals grown from solution show emerging screw dislocations associated with the growth process. Typical dislocation densities are 10^5 cm/cc, so typical spacings are a few tens of microns. Electron microscope studies of CaF_2 (Rasagni *et al.* 1983) show scale lengths of a few tens of nanometres. Scanning probe microscopy data show defect features on surfaces. Only a few of these features can be identified as specific structures, such as divacancies on Si(100).

When we are concerned with the localisation of energy or charge to some particular surface sites, we need to consider kink sites, steps, and special sites on these steps. Such features on ionic surfaces will often result in large electric fields (see, e.g., Tasker & Duffy (1984); see also Stoneham (1987a) and Section

9.1.1). Kinks on steps on halide surfaces may well be important, since isolated kinks will always be able to bind a carrier. This follows from two early results. One is the result of Seitz (Section 5.2.3), that a kink on a step will have a non-integral charge. A kink on a (100) step on a (100) NaCl surface has a charge $Q = +e/2$ or $-e/2$. Trapping of carriers will not eliminate the Coulomb field. The second result, due to Mott & Gurney (1948, Section 5.1) is that the halfintegral charge is fully sufficient to bind an electron or hole. Indeed, in the absence of other defects, there will be an infinite number of bound states. Topography can change the surface bound states significantly, as isolated kinks can bind carriers.

On a random rough surface, there will usually be a kink opposite in sign not far away, which will inhibit binding (see, e.g., Stoneham (1975), Chapters 9, 25). Nevertheless, it is clear that surface states of ideally-terminated crystals are likely to be meaningless at best as a description of real surfaces. We remark that the charges of kinks are integral for MgO and other simple oxides, unlike halides. The Seitz argument suggests non-integral charges are possible when there are ions of odd valence. This would suggest different behaviour for oxides containing Al, for example, although H or other monovalent ions may be able to eliminate the effective non-integral charge.

Oxide surfaces are often prepared from the melt, followed by cleavage. There are many alternatives: laser ablation and other methods are used to prepare oxide films; thermal decomposition of carbonates or oxides leads to fine particles. These various methods lead to a rich variety of structures. Thus MgO formed from the carbonate forms flaky (cornflake) structures, whereas MgO from the hydroxide forms nodular (cauliflower) structures (Kingery 1983). Cleaved surfaces have been found to show patches about 20 nm across and 1 layer high (Abriou *et al.* 1996). This implies a certain step edge length per unit area, but the number of kinks is harder to estimate. The surface topography, as seen by scanning probe microscopy, is sensistive to moisture (King *et al.* 1996).

It is a general theorem (Tasker 1983; see Hayes & Stoneham 1985, p. 342) that certain surfaces should not occur, in that the extended surface has infinite surface energy. The (111) surface of MgO and NaCl is an example. Yet experiment continues to claim that these surfaces exist. Sometimes this is naive interpretation of ambiguous experiments, e.g. the failure to recognise faceting of an unstable surface into regions which render it stable. For the (111) MgO surface, alternate patches of Mg and O would be stable, whereas an outer layer solely consisting of Mg or of O would be unstable. Other ways in which nominally-unstable surfaces can exist are discussed by Stoneham (1981a). A general point worth making is that surfaces under excitation can be maintained in non-equilibrium states. These states can include high electric fields (about 60 V over 5 nm) as has been found by Ribeiro *et al.* (1997a,b, 1998a,b) in studies of laser ablation.

9.1.6 Excitation of surface states

Local surface modifications can be induced by electronic excitation. The excitation can involve intrinsic surface states or surface defect states. It can result from the direct absorption of photons by surface states or by absorption of photons by bulk band-to-band excitation, followed by the transport of charge carriers to the surfaces. The direct absorption of photons by intrinsic surface states can create electron–hole pairs or excitons, whose motion is confined to two dimensions on the surface. Self-trapping of two-dimensional excitons will localise electronic excitation energy laterally on the surface, and can be the cause of local surface modifications. Similarly, as in the bulk, defect excited states can also be the cause of local lattice modification. The defects can be excited directly, or by interaction of electron–hole pairs or excitons created in the bulk or elsewhere at the surface.

Whether surface modification or sputtering originates from surface or from bulk excitation depends on the incident particles and the diffusion length of charged carriers. For photons of energy less than the bandgap, the only means of energy transfer to the solid are optical absorption by defects or multiphoton band-to-band transitions. Since the multiphoton processes have small cross-sections, photo-induced surface processes will be initiated primarily by absorption at defects on surfaces. For ionising radiation and for photons with above the bandgap energy, the penetration depths of the particles and the diffusion length of charge carriers compete. If the diffusion length is shorter, surface processes are governed by photoabsorption in the surface layers. Photons with energies greater than the bandgap have typical penetration depths of order 100 nm, so that diffusion of charge carriers plays a role in many cases.

9.1.7 Surface excitation following bulk excitation

Suppose that excitons are generated within a mean free path of the surface. Their recombination is governed by the recombination rate at that surface. We may define their *surface recombination rate* s as the net rate at which electron–hole pairs and excitons are absorbed by the surface per unit concentration of photo-generated carriers. If the capture process is rate-determining, the surface recombination rate is

$$s = \sigma_S v_{th} N_S,$$

where σ_S is the trapping cross-section, v_{th} the thermal velocity, and N_S the concentration of surface recombination centres per unit area. Suppose further that the electron–hole pairs undergo ambipolar diffusion, keeping the local electron and hole concentrations the same, and that the charge carriers are generated uniformly in the bulk at a rate R_B. The overall concentration $n(x)$ at a depth x of electron–hole pairs and excitons can be obtained by solving this diffusion equation with a boundary condition:

$$L^2 (\mathrm{d}n/\mathrm{d}x)_{x=0} = s\tau n(0),$$

with τ the lifetime in the bulk and L the bulk diffusion length. The annihilation rate R_S of electron–hole pairs per unit area of the surfaces is given by

$$R_S = sn(0) = R_B\, s\tau\, L/(s\tau + \mathrm{L}) = R_B\, s\tau/(1 + s\tau/L).$$

If we write $X = (s\tau/L)$, then the fraction of pairs annihilating at the surface R_S/R_B is $LX/(1+X)$. Evidently, if L and $s\tau$ are similar in magnitude, a substantial fraction of electron–hole pairs and excitons generated within depth L will be annihilated on the surface. R_S approaches $R_B s\tau$ when the bulk diffusion length L is much bigger than $s\tau$ and is $R_B L$ for smaller values of L. The product of the surface recombination rate and the lifetime, $s\tau$, can be expressed as:

$$s\tau = \sigma_S N_S/(\sigma_B N_B),$$

where N_B is the concentration of the bulk recombination centres and σ_S and σ_B are the recombination cross-sections of surface and bulk recombination centres respectively.

The probability of annihilation of an electron–hole pair generated at a depth x can be calculated by solving a diffusion equation. The solution of such a diffusion equation with a sharp Gaussian source function gives the fraction of the energy dissipated at the surface. Figure 9.1 shows for various values of $X = s\tau/L$ the fraction of the energy released at the surface, having been deposited initially at depth x. A substantial fraction of the energy is dissipated at the surface, a part of which will cause sputtering. Evidently, the energy deposition at the surface gives a larger contribution as $X = s\tau/L$ increases. In laser ablation (Chapter 11), where the role of grain boundaries and dislocations becomes important, similar principles apply.

The recombination of electron–hole pairs at a surface can take place in several ways. Some correspond to the mechanisms which occur in the bulk,

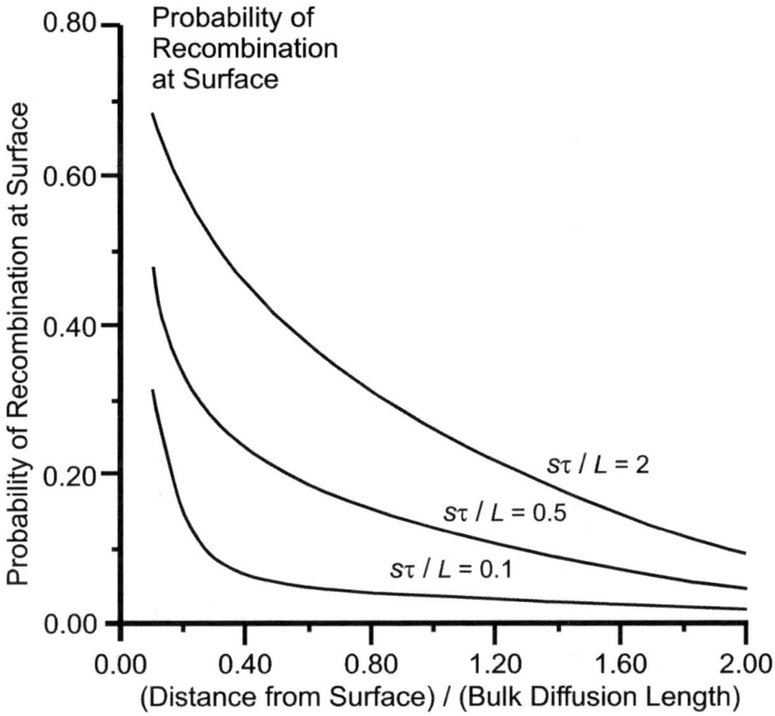

Fig. 9.1. The fraction of electron–hole pairs generated at a depth x (expressed in units of the diffusion length L) which recombine at the surface, for several values of $s\tau/L$; s is the surface recombination rate and τ is the lifetime of an electron–hole pair.

including radiative recombination. The process could involve self-trapping; it will often involve a defect site. When self-trapping is the rate-determining step for recombination, all sites can act as recombination sites, so $N_S/N_B = d$, where d is the lattice constant. Thus $s\tau = \sigma_S d/\sigma_B$. If $L \gg s\tau$, the recombination rate at the surface is $\sigma_S R_B d/\sigma_B$.

9.1.8 Exoelectron emission

Secondary electrons are commonly observed following ionisation of core electrons by X-rays or electron beams. Those electrons observed externally are often Auger electrons, which may have been scattered elastically or inelastically several times before leaving the solid. The actual distribution depends on how deep the excitation penetrates as well as on how strong are the scattering mechanisms (see Section 1.2.2). In NaCl and KBr, the spectrum for electrons in the 1–5 eV energy range suggests that acoustic phonon scatter dominates for electrons in the conduction band with kinetic energies of 2–3 eV (Ausnees *et al.*

1990). In electron microscopy, secondary electrons are those which leave the specimen with energies less than 50 eV, and typically 1–10 eV. It is no surprise that high-intensity excitation leads to exoelectron emission, and we shall discuss this in Chapter 11.

Secondary electron spectra have been evaluated for a wide range of systems by Henke *et al.* (1979), including LiF, NaCl, KCl, LiBr, RbBr, NaI, KI, RbI, CsI, AgCl, CuI, PbI_2, CdS, PbS, Se, and Ge. Analysis of the energy loss following soft X-ray excitation gives useful information on plasmon energies and electron affinities. Electron affinities for oxides and halides are known only poorly, and may vary from point to point on the surface because of impurities or surface topography.

Exoelectrons are those emitted during some solid-state process, especially (but not only) those situations for which there is no external energy beam. Emitted electron spectra can give information about defects. Thus Galii & Tsai (1985) discussed CsBr and CsBr:Cd exoelectron spectra and yields, and the effects of ultraviolet and electron irradiation. They suggested as mechanisms Auger processes and recombination of colour centres. Hopman *et al.* (1997) remarked on a dose-dependent negative potential on rutile for electrons with energies above 2 keV. Using secondary electron information, they concluded that primary 1500 eV electrons incident on rutile give an electron-stimulated O desorption cross-section between 4×10^{-3} and 4×10^{-2} Å^2.

As illustrations of the range of processes which lead to exoelectron emission, we give some examples associated with silica-based systems. These include: (i) exoelectrons associated with mechanical damage during optical breakdown (Smirnov 1986); (ii) exoelectron emission during oxidation after mechanical treatment of the Si surface (Devochkin & Ivanov 1987); (iii) exoelectron emission during the cooling of porous Si, apparently from the oxidation of bare Si exposed by stress during cooling (Krylova & Petrukhin 1996); and (iv) suprathermal electrons observed during pulsed laser irradiation of silica (Bastiani *et al.* 1997). Bastiani *et al.* found that, when a photon pulse (120 fs, 50 mJ, 800 nm pulses from an Al_2O_3:Ti laser, giving 4×10^{16} W/cm^2 pulses) was incident on a silica target at 45°, suprathermal electrons with energies of 20 keV were produced at the specular angle. The electron energy could be increased to 180 keV with a prepulse and chosen delay. No explanation was offered, but apparently the electrons come from the plasma produced.

Ferroelectric ceramic cathodes can yield exoelectron current densities as high as 100 A/cm^2. The process appears to be plasma-assisted, with initiation of the surface flashover plasma either from field electron emission at triple junctions or electron emission caused by polarisation reversal (Shur & Rosenman 1999; Shur *et al.* 1996).

9.1.9 Luminescence from surfaces

The luminescence of surface excitons is of several different types. There is resonant luminescence, with essentially the same energy as exciton optical absorption Such resonant luminescence has been detected from semiconductor surfaces (Lester *et al.* 1987; Halas & Bokor 1989). There is luminescence shifted modestly from the absorption energy (MgO, CaO, SrO), perhaps because the oxide has a negative electron affinity (Cox & Williams 1986; Stoneham & Tasker 1985b). Luminescence shifted to a greatly reduced energy has been noted, but it is not clear whether this is due to self-trapping, as in the bulk, or associated with surface defects to which the exciton has transferred its energy.

Non-radiative transitions may dominate on surfaces unless the surfaces are carefully prepared. No evidence has been reported yet of STX luminescence on surfaces. It is unlikely that STX luminescence will be observed on alkali halides, since these self-trapped excitons collapse with the emission of a halogen atom, leaving a halogen vacancy behind. If an exciton on the surfaces of wide-gap materials couples strongly to the lattice, non-radiative transition seems to dominate. This is surprising, in that multiphonon non-radiative transitions become less and less probable at large energies (Chapter 3), whereas radiative transition probabilities increase with emission energy. Presumably the non-radiative channels which are so effective are either Auger channels or involve atomic displacements. Large phonon energies can be associated with wide bandgaps (e.g. MgO, diamond) and this can increase the probability of multiphonon transitions, since fewer phonons are needed.

9.1.10 Local lattice modification on surfaces

Surface modification involves the displacement of an atom away from its site. There are two main cases, as shown in Fig. 9.2. In the first, the atom is still bound on the surface. This metastable case (Fig. 9.2(b)) needs thermal activation before emission occurs. Whether emission happens as a consequence of exciton formation depends on whether the bandgap energy is higher than the energy to emit an atom. There is sufficient energy for emission in wide-bandgap materials, but not for semiconductors. In the second case (Fig. 9.2(a)), the atom is emitted without any potential barrier. The antibonding adiabatic potential surface shown is known for desorption of adsorbates as Menzel–Gomer–Redhead mechanism (Menzel & Gomer 1964; Redhead 1964).

As in the bulk, there is a further distinction: local lattice modification can occur by recombination either at perfect sites on surfaces or at defect sites. Modification by recombination at perfect sites requires strong electron–

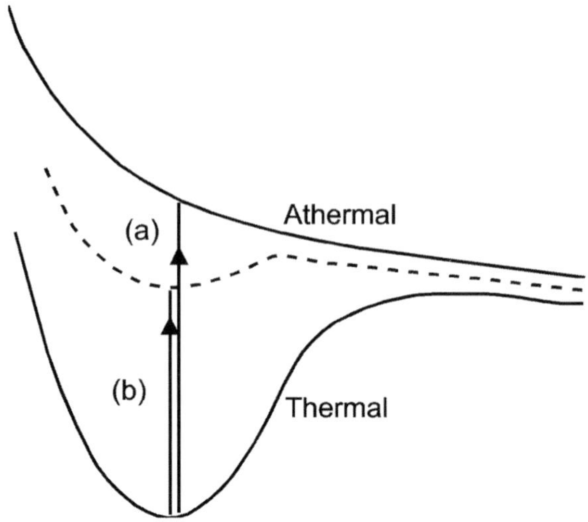

Fig. 9.2. Schematic configuration coordinate diagrams for atomic emission from surfaces by electronic excitation: (a) athermal and (b) thermal. If the potential barrier is sufficiently high, the configuration remains in a metastable state on the surface.

phonon coupling, and will usually need exciton self-trapping. The excitonic mechanism of atomic emission from a surface leads to emission from a specific sublattice, e.g. the halogen sublattice in halides. Modification of surfaces on the mono-layer scale leads to a non-stoichiometric surface, for which the effects of electronic excitation will be very different from that of the stoichiometric surface. Unless the stoichiometry is restored during irradiation, perhaps by thermal evaporation, the phenomena subsequent to the stoichiometry change will be those of the non-stoichiometric surface. Atomic emissions initiated at defect sites dominate in materials in which excitons are not self-trapped. Even in the materials in which excitons are self-trapped, defects can be the source of emissions for photons below the bandgap energies, since photons are absorbed only by defects. An atomic emission originating from a defect site will produce a new vacancy nearby and may enhance the emissions. It has been suggested that vacancy-initiated emission produces vacancy clusters on a surfaces, and therefore can be a cause of massive emission of atoms from a surface (Okano *et al.* 1993).

There are two types of measurements of the surface modification: those of the modified surface, and those of the emitted atoms, ions, or molecules. Measurements of the surface include the sputtering rate, measured by the thickness of material removed by a given amount of radiation, and the various surface characterisation techniques. These techniques include Auger electron spectroscopy (AES), photoelectron spectroscopy (UPS or XPS), electron energy loss spectroscopy (EELS), STM, and so on. Such spectroscopic techniques are especially useful for the analysis of non-stoichiometric surfaces from which there has been only a partial emission of the component species. Emitted charged ions and neutral atoms present different problems. Species emitted from surfaces are mostly neutral, except for those produced by core excitation or intense laser irradiation. Emitted neutrals require ionisation before measurement, either by an electric discharge or a laser, and measurements are less sensitive than for emitted ions. It is often the case that only the ions are measured. These are only a small fraction of the emitted particles, so the information they provide on sputtering is incomplete.

9.1.11 Core excitation on surfaces

Photons with high enough energies can ionise core electrons. The subsequent non-radiative Auger transitions which refill the core holes generate holes in the valence bands. Feibelman & Knotek (1978) suggested that the production of core holes on the surfaces of ionic crystals results in normally-anionic species with net positive charges, hence the emission of positive ions by Coulomb repulsion. Since such an Auger transition occurs within a mere 10^{-15} s, an ion can only be emitted if the holes in the valence band stay longer than the time for atomic motion (the period of characteristic lattice vibrations, typically 10^{-13} s). Cini (1985) has argued that the hole lifetime can be long enough if the Coulomb repulsion energy exceeds the valence bandwidth. The Feibelman–Knotek idea has parallels with the Varley mechanism, but is more specific as regards mechanism, and leads to important tests relating to threshold energies. Further, we shall describe several examples of ion emissions by the Feibelman–Knotek mechanism, whereas we know of no cases of bulk defect formation due to the Varley mechanism.

The quantum yield for emission by the Feibelman–Knotek mechanism is small, typically 1/1000. One should note that electrons capable of inducing core excitation can also excite valence electrons. Thus, only emissions which show a clear core threshold should be assigned to the Feibelman–Knotek mechanism.

9.1.12 Laser excitation of surfaces

There are two distinct cases of interaction between laser beams and surfaces. In the first, typical of wide-gap insulators, the laser energy is smaller than the bandgap. In the second case, typical of semiconductors, the laser energy exceeds the gap. As we discuss in Chapter 11, laser damage and ablation can happen in both cases. Just as in conventional photon irradiation, neutral atoms are emitted at low intensities. Emissions of ions are observed as the laser intensity increases. At extremely high intensities, massive amounts of material are removed from the surface, accompanied by signs of melting. We discuss in this section the electronic processes induced by medium-intensity laser beams interacting with surfaces.

Wide-bandgap materials

When the bandgap is larger than the photon energy, the material will appear transparent at the start of a laser pulse for single-shot irradiation, and for the first few laser pulses for multi-shot irradiation. However, a small number of photons are absorbed at defects in the bulk and on surfaces, including point defects, dislocations, and steps and kink sites on surfaces. Defect excitation can include excitation of electrons trapped by a defect, excitation of a valence electron into the defect level, and the generation of a localised exciton. At high intensities, resonant multiphoton excitation should also be taken into account (Matthias *et al.* 1987). Within a laser pulse of fluence f, each species in the solid is excited sf times, where s is its average cross-section. Defect excitation often leads to a metastable state through interaction with the lattice, and its lifetime can be as long as the laser pulse. For a laser pulse of 0.1 J/cm^2, approximately 10^{18} photons/cm^2, sf can be more than unity. Cascade excitation at defect sites can occur, as well as resonant multi-photon ionisation.

Both resonant multiphoton excitation and cascade excitation produce changes in the material. First, they produce free electrons, which are then heated directly by the laser (Epifanov 1974). Cascade excitation of defects, particularly those which form two holes on a defect sites, can create further defects, and these new defects can result in additional optical absorption. Furthermore, the interaction of free electrons with the lattice and electron–hole recombination cause a temperature rise. This temperature rise during irradiation with laser pulses has been confirmed by infrared techniques (Cielo 1984) and by photothermal spectroscopy (Olmstead *et al.* 1983). The phenomena of laser ablation certainly involves a temperature rise, as discussed in Section 11.3. Here we concentrate on the so-called electronic processes in surface–laser interactions of wide-bandgap materials. Usually they involve defect excitation, although the effects of laser pulses have been observed for excitation above the bandgap energy.

There seem to be two consecutive processes in laser ablation of wide-bandgap materials: heating and photolytic processes. In materials laser-irradiated at energies below the bandgap, the photolytic processes and heating must be initiated at defects or impurities, or at band tails in amorphous materials, since these can absorb photons. For this purpose, the surface itself may be an absorber. Defect-initiated photolytic processes on surfaces result in anion emission. This leads to the creation of an anion-deficient surface. Heating can occur by defect ionisation, followed by free-electron heating and Auger-type and phonon-emitting non-radiative recombinations. Since the products of photolytic processes are volatile, the surfaces after vaporisation of photolytic products are presumed sharp. On the other hand, the surfaces ablated by heating alone should be rough, with debris, because only a part of the material which is heated above the vapourisation temperature is actually emitted.

Only a small fraction of the incident photons is absorbed in the surface layers at the beginning of a laser pulse. Direct heating of the materials cannot be expected at that stage unless the number of pre-existing surface defects is extremely large. Only as defects accumulate from photolytic reactions will heating increase. In the early stages of irradiation, because there is so little optical absorption, photolytic processes dominate. Photolytic processes increase the number of defects and volatile products, and increase the optical absorption coefficient. Since photolytic processes need a relaxation time of 0.1–1 ps, heating is more effective for shorter laser pulses if the fluence is the same, or for higher fluence if the width is the same. This tendency is enhanced if the photolytic processes rely on two neighbouring excitons, as may be the case for halides and SiO_2, in which lattice relaxation and free-electron heating compete.

Semiconductors

Photons with energies larger than the bandgap are absorbed strongly, generating electron–hole pairs. A typical penetration depth is 100 nm. For intense laser beams, further photons are absorbed by free electrons, which gain kinetic energy. Thus, early in a laser pulse, the electron temperature exceeds the lattice temperature. The lattice temperature is raised through electron–phonon coupling within about 100 fs. It seems that most of the energy imparted by the laser beam is converted to lattice vibrational energy, apart from energy possessed by electron–hole pairs. It is generally believed that this temperature rise causes laser ablation and massive emissions of atoms from semiconductor surfaces, just as for metal surfaces. However, we shall see that electron–hole pairs can also modify semiconductor surfaces. In this chapter, we discuss the modification of semiconductor surfaces by electronic excitation; the links to laser ablation will be discussed in Chapter 11.

9.2 Sputtering and surface modification of halides

9.2.1 Excitons and holes on surfaces

Carrier (electron or hole) self-trapping at a surface differs from that in the bulk in several ways. First, the surface Madelung potential is different. The difference is only small when expressed as a fraction, but is usually a few tenths of an electron volt less. This aids localisation to the surface layer for holes on anion sites or electrons at cation sites, and so usually assists self-trapping. Secondly, the free space outside the crystal is not polarisable. This means that charge localisation is discouraged, in that the polarisation energy advantage of localisation is reduced (an effect equivalently described as a repulsive image interaction). However, the constraints on relaxation are less severe, and there may be opportunities at the surface for more significant rebonding. Thirdly, there is the question of band line-up. Deep in the bulk, the conduction band of the crystal lies below the vacuum level by the *electron affinity A*, which is typically 1 eV (but which may be positive (conduction band below vacuum) or negative (conduction band above vacuum)). Inevitably, the important line-up will depend on any band bending or surface rumpling. If A is negative, an electron in the conduction band can lower its energy by moving outside the crystal. The analogous situation for a hole has a clear difference, for a hole cannot move into the vacuum. For an exciton at a surface with negative electron affinity, the electron can move outside, remaining bound to the surface hole. In such a case, the exciton energy is likely to be smaller than in the bulk, and localisation in easier (Cox & Williams 1986). Otherwise, exciton self-trapping at the surface will be affected by the same issues as carrier self-trapping, although the consequences are less obvious. The image terms will be less important, since the exciton is neutral overall, for instance. As regards the kinetics of localisation, the parameter $s\tau/L$ for excitons in halides is expected to be large, since σ_S/σ_B may well be much larger than unity.

MacLean & Duley (1984) report what they believe to be an amorphous oxide MgO on a LiF substrate. The oxide film shows no detectable exciton until after an 800 °C anneal. One interpretation would be that the exciton has self-trapped at special sites in the amorphous oxide, as in amorphous silicas, and so it is shifted and broadened; alternatively, it might have other decay channels.

Self-trapped excitons at halide surfaces should exhibit the same instability as that which converts a bulk [V_k:e] self-trapped exciton to an off-centre self-trapped exciton. At the surface, this instability may give rise to the emission of a halogen atom, leaving behind an F centre (Fig. 9.3). The exciton instability at the surface results in the atomic emission only from the anion sublattice, analogous to defect formation in the bulk. As noted later, the excitation of alkali halide surfaces leads to sputtering with a very high quantum yield. The primary process of this sputtering is halogen atom emission. The preferential sputtering of halogens is a crucial factor which governs the overall sputtering yield.

The localisation of charge can lead to the emission of either halogen ions or

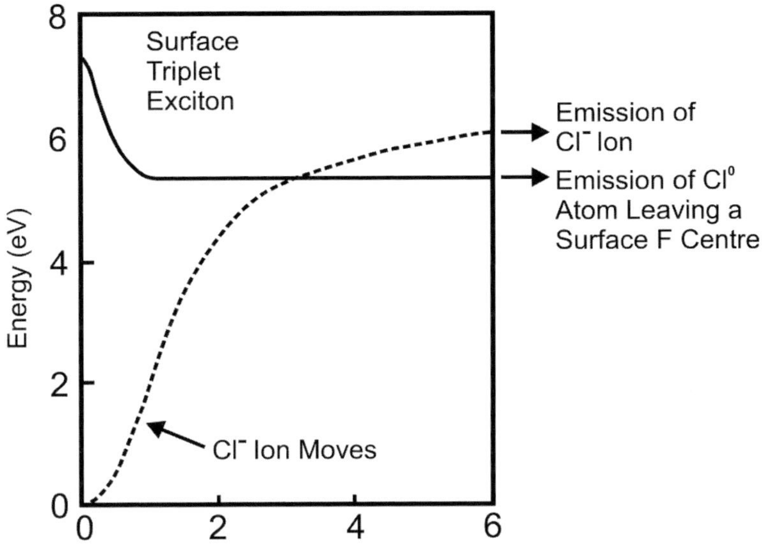

Fig. 9.3. Schematic energy surfaces, after Li *et al.* (1992) and Puchin *et al.* (1993a). The full line shows the emission of a halogen atom following the creation of an exciton; a surface F centre is left behind. The dotted line shows the energy surface for removal of a halogen ion in its ground state. The abscissa gives the distance from the surface in the unit of the nearest neighbour distance.

alkali ions. Figure 9.4 illustrates two possible mechanisms following the localisation of two holes on a halogen, giving it a net positive charge. Localisation at the outer surface can lead to halogen emission; localisation in the second layer may cause alkali emission.

9.2.2 Mechanisms for the sputtering of alkali halides

Agullo-Lopez & Townsend (1980) have suggested two possible mechanisms of halogen emission. The first is a *replacement collision sequence*, initiated by an exciton instability. The second is *H centre decomposition* on the surface. Overeijnder *et al.* (1979) suggested that, in addition to an exciton instability, there is a *hole instability* leading to an emission of a halogen atom. The idea of a hole instability was supported by Green *et al.* (1989b), who suggested that a hot hole reaching the surface converts its kinetic energy into bond-breaking energy, and leads to an F centre and halogen atom emission. No microscopic mechanism for the energy conversion has been suggested. It is unlikely that a self-trapped hole at the surface has sufficient energy to decompose into a halogen vacancy and an emitted halogen atom. Overall, in view of the results

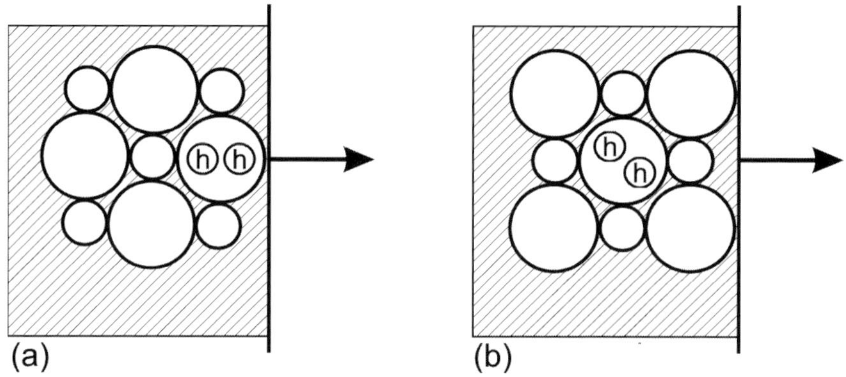

Fig. 9.4. A schematic diagram of emission from an alkali halide surface caused by Coulomb forces following the localisation of two holes on anion X^-, creating an X^+. In (a), the anion is expelled from the outer surface; in (b), the anion in the second layer pushes out a cation in the surface layer.

of newer calculations, we consider that the energetic atoms are emitted from surface excitons and thermal emissions arise from the F–H pairs generated near the surface.

Energy conversion from the exciton to emission of halogen atoms

X Li *et al.* (1992) and Puchin *et al.* (1993a) simulated the relaxation of an exciton near the surface of NaCl, using an embedded-cluster Hartree–Fock approach. They found a self-trapped exciton instability at a surface: a one-centre exciton on the surface top layer decays into a surface F centre as a halogen atom is emitted from the surface. If an exciton can be localised initially, then it will generate an F centre and an emitted halogen atom. Figure 9.3 shows a projection of the adiabatic potential energy surface for a one-centre self-trapped exciton on the outermost surface layer.

Puchin *et al.* (1993a) also calculated the energy of a self-trapped exciton and of an H centre produced near the surface. Figure 9.5 gives the adiabatic potential surface of an extra halogen atom on the surface; each minimum below the second monolayer corresponds to an H centre configuration. The H centre has an energy *higher* than a halogen atom emitted from the surface. An F–H pair formed from an exciton near the surface can evolve into a near-surface F centre and an emitted halogen atom. Deeper into the crystal, other defect reactions can occur, such as F–H recombination, and halogen emission becomes less probable. Similar calculations have been made for NaBr, KBr, and RbBr (Song & Chen 1994; L F Chen *et al.* 1994).

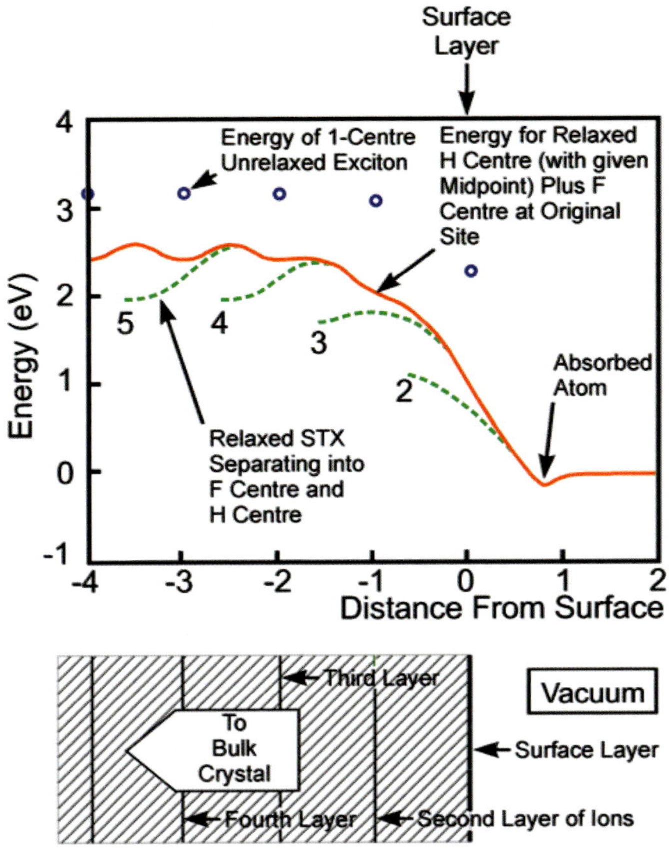

Fig. 9.5. The energy of a Cl atom near a NaCl surface. The solid curve is the adiabatic potential energy surface for an isolated H centre; the coordinate represents its geometric centre. Curves 2, 3, 4, and 5 are for a relaxed self-trapped exciton formed in layers at various depths from the surface, as it transforms into an F–H pair. The energy of the F centre has been subtracted, so that these curves merge with the solid curve when the F–H interaction is no longer significant. The open circles give the unrelaxed one-centre self-trapped exciton energy for the corresponding layer.

Clearly, there are two excitonic mechanisms for halogen atom emissions. There is a *direct channel*, namely the emission of a halogen atom from a one-centre exciton in the top layer. There is also *H centre decomposition* at the surface. The surface will be a sink for the H centres generated within a diffusion length of the surface. Even though surface F centres will recombine with some of these H centres, we should expect that both processes are effective in the sputtering of alkali halides. Likewise, both mechanisms should operate in other halides, such as alkaline earth fluorides.

Excitation spectra for sputtering

The excitation spectrum for the sputtering of alkali halides shows clearly that both excitons and electron–hole pairs can cause emission, supporting the exciton mechanisms suggested. Typical excitation spectra near the band edge are shown in Fig. 9.6. Most experimental results indicate an energy threshold close to the exciton energy, although sputtering starts a few tenths of an electron volt below the exciton peak, which may be associated with a surface exciton peak or to defects on or near the surface.

Emission of halogen atoms from alkali halides

The electronic excitation of alkali halide surfaces induces emission of atoms, neutral halogens, and alkalis (Palmberg & Rhodin 1968; Szymonski *et al.* 1985; Kanzaki & Mori 1984). As expected, halogen atom emission is primarily by the excitonic mechanism already described. A number of experiments show that halogens are emitted in preference to alkalis. First, as irradiation continues, the surface is enriched in alkali. This has been observed by Auger spectroscopy (Palmberg & Rhodin 1968; Szymonski *et al.* 1985). Secondly, the sputtering rate increases with increasing temperature (Fig. 9.7), saturating at high temperatures where evaporation of alkali is fast enough to keep the surface stoichiometric (Al Jammel & Townsend 1973; Szymonski 1990). In the range where the rate inceases with temperature, there is an Arrhenius energy of 0.2 eV, which is interpreted as the energy of vaporisation for surface alkali atoms. Thirdly, the sputtering yield falls as the irradiation is continued. For electron irradiation of LiF, Wurz & Becker (1989) estimated that the dose at which the yield largely vanishes is that needed to form a surface Li monolayer. Obviously, halogen atom emission is reduced as the surface becomes almost entirely alkali, with fewer and fewer surface halogens available for emission. Further, groups of metal atoms will usually cause additional Auger processes for deexcitation, which will overwhelm the excitonic processes causing emission. At higher temperatures, halogens can replenish the surface layer rapidly, restoring its halogen content to close to stoichiometry. Szymonski *et al.* (1985) showed that the NaCl surface is almost stoichiometric above 300 °C under conventional electron beams. The yield at high temperatures may be considered intrinsic, characteristic of the stoichiometric surface.

Detailed modelling of desorption following core excitation of NaF (Green *et al.* 1989c) emphasises the importance of Auger processes, and has suggested a role for hot holes in both neutral atom and positive ion desorption.

Hyperthermal and thermal halogen emissions

It has been known since the 1970s (Overeijnder *et al.* 1979) that one component of the halogen atoms emitted during electronic excitation of alkali halides

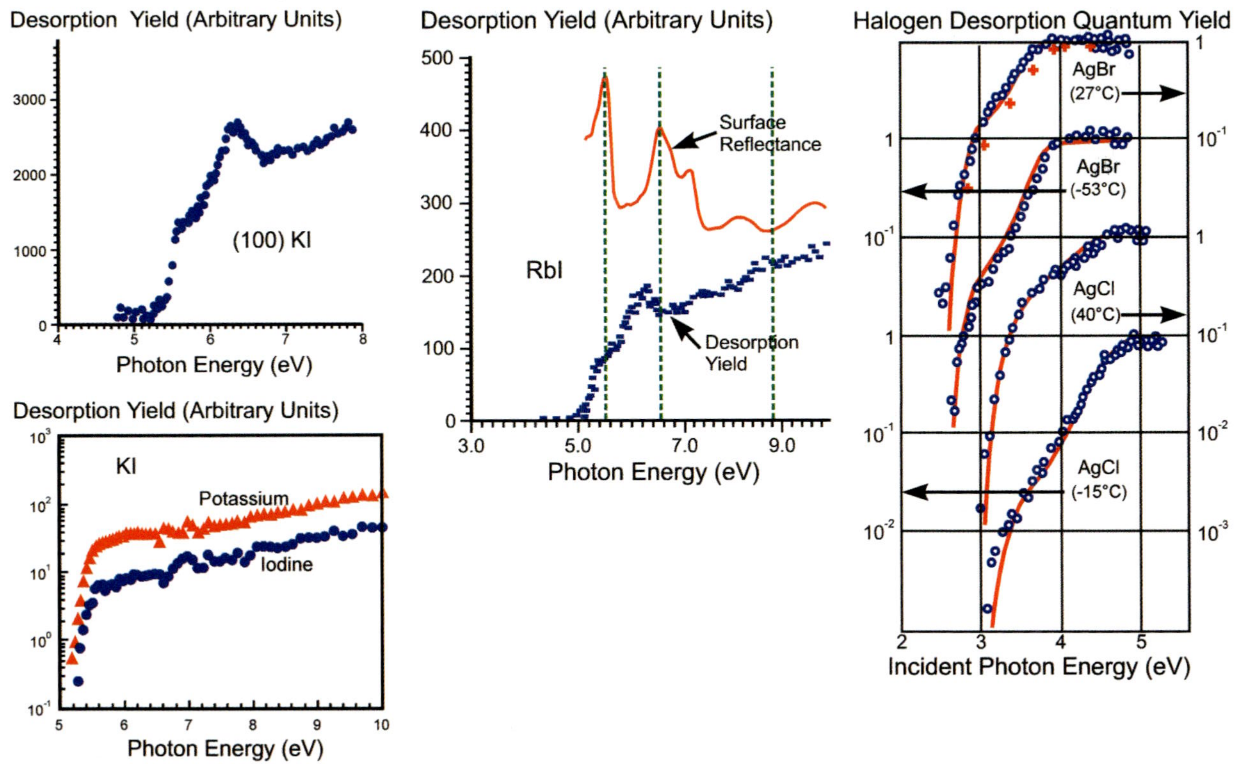

Fig. 9.6. Excitation spectra for sputtering of alkali halides KI (Szymonski 1994) and RbI (Kolodziel *et al.* 1996) and silver halides AgCl and AgBr (Kanzaki & Mori 1984). The solid lines for the silver halide results indicate the optical absorption curves.

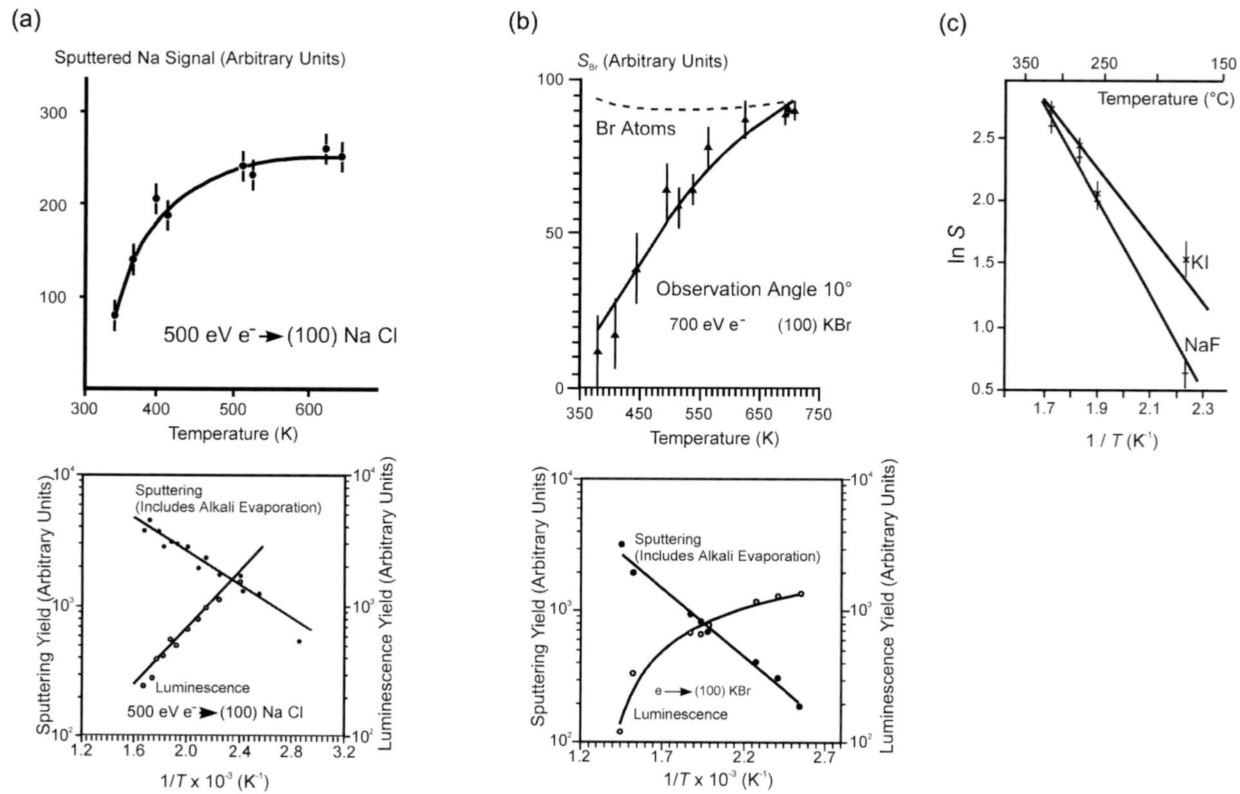

Fig. 9.7. Temperature dependence of the sputtering yield of alkali halides. Evaporation of metal is usually present as well (Szymonski *et al.* 1985, 1990; Al Jammal & Townsend 1973) (a) 500 keV electrons on NaCl[100] (b) 700 keV electrons on KBr[100]; (c) NaF and KI.

has an energy of about 0.1 eV. Time-of-flight studies of the energy distribution of halogen atoms emitted by electron irradiation of well-characterised alkali halide surfaces provide useful guides to the mechanisms. The main points from the experiments are these:

(1) For several alkali halides, the time-of-flight spectra consist of two peaks (Fig. 9.8). The *thermal* peak can be fitted to a Maxwell distribution for the surface temperature. The other, *hyperthermal,* peak has a maximum kinetic energy of about 0.2 eV (Szymonski 1991b).
(2) The particles emitted in the thermal peak have an angular distribution described by a cosine function; the hyperthermal peak is dominated emission in the ⟨100⟩ direction (Postawa & Szymonski 1989), irrespective of whether a (100) or (110) surface is irradiated (Szymonski 1991a,b).
(3) As the temperature is raised, the total yield increases. Likewise, the number of atoms emitted in the thermal peak rises. However, increasing temperature reduces the number emitted in the hyperthermal peak (Fig. 9.9; Kolodziej *et al.* 1992a).
(4) The dependence of the yield on electron energy is more complicated. For the hyperthermal component, the yield increases with increasing electron energy, reaching a maximum around 900 eV. For the thermal part, the situation is less clear.
(5) There is a systematic variation of the hyperthermal part of the yield from one alkali halide to another (Kolodziej *et al.* (1996); Fig. 9.10). Basically, a hyperthermal component is seen for K and Rb halides (KCl, KBr, more weakly for KI; RbCl, RbBr, RbI) and is weaker or not seen for Na halides (weak for NaF; absent for NaCl, NaBr, NaI).

Mechanisms for thermal and hyperthermal emissions

A natural interpretation of the hyperthermal emissions is that they start from one-centre excitons at the surface, whereas the thermal component results from an instability of surface H centres, as suggested by Szymonski (1980, 1993). Self-trapped excitons produced below the second layer can generate F–H pairs; thermal processes then convert the interstitial halogen atom into an emitted halogen atom. Clearly, the halogen emitted does not need to be the same halogen as that initially displaced.

The first important test of the model is the observed dependence of the yield on electron energy (point (4) above). Our discussion of exciton annihilation at surfaces showed a maximum yield for the highest concentration of excitons generated within the bulk diffusion distance L of the surface, typically 10 nm. Postawa *et al.* (1993) showed that 900 eV incident electrons (which give the maximum yield) deposit their energy within 10 nm of the surface. Although Postawa *et al.* assume that the hot holes are responsible for the hyperthermal halogen emissions, their results appear to be equally consistent with diffusing excitons.

The role of exciton diffusion in sputtering was suggested to explain the dependence on incident electron energy of the sputtering yield of alkali halides

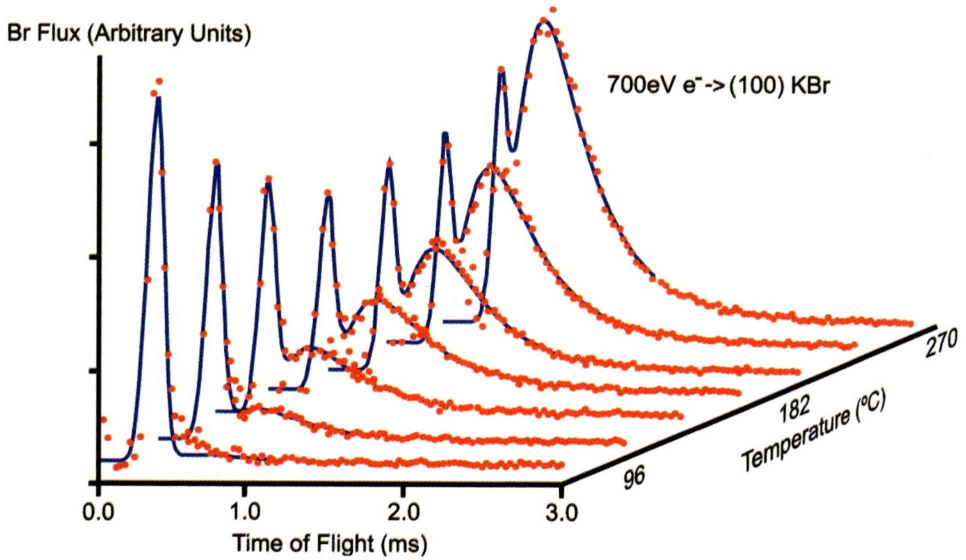

Fig. 9.8. Time-of-flight spectra of halogen atoms emitted from KBr(100) under 700 eV electron irradiation (Szymonski 1991a,b).

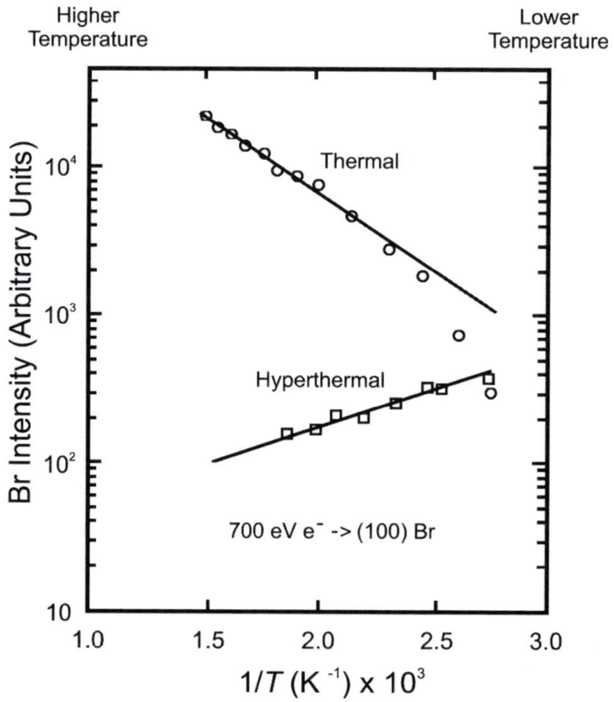

Fig. 9.9. Temperature dependence of the yield for the hyperthermal and thermal components of halogen emission during irradiation of KBr with 500 eV electrons (Kolodziej *et al.* 1992a).

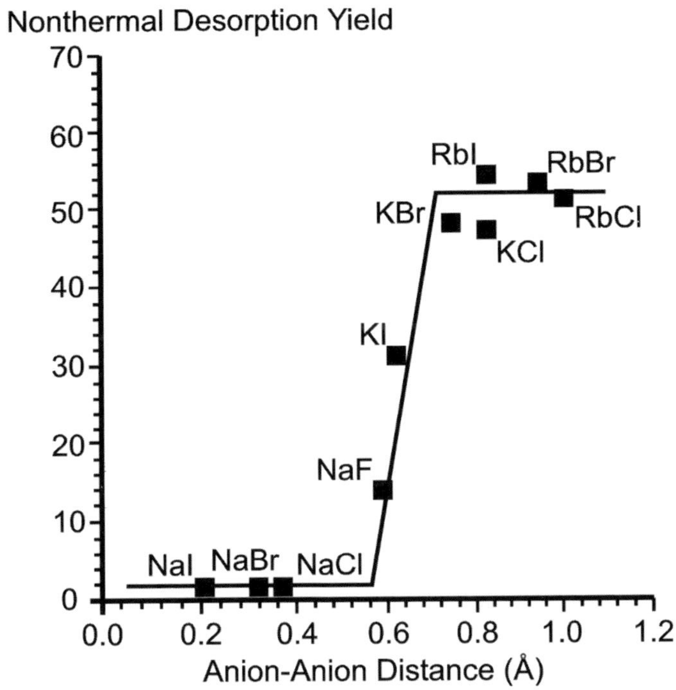

Fig. 9.10. The relation between the yield of the hyperthermal part of halogen emissions and the halogen–halogen distance in the lattice (after Kolodzeij *et al.* 1996).

(Al Jammal & Townsend 1973). The energy dependence for sputtering is similar to that of the hyperthermal component for halogen atom emission, but has a maximum around 500 eV (Fig. 9.11) rather than 900 eV. The peak energy shift comes from the contribution of the thermal component, for which other secondary factors are involved.

The second important test of the one-centre exciton model comes from the dependence of the yield on halogen–halogen separation (point (5) above). A one-centre exciton on the outermost surface can lead to either halogen atom emission or formation of a self-trapped exciton by combining with a neighouring halogen ion. One major factor is the halogen–halogen distance, since this influences the force driving the formation of an X_2^- molecular ion. Thermal fluctuations will favour the formation of an X_2^- molecular ion from a one-centre self-trapped exciton, in agreement with the observed fall in yield of the hyperthermal component with rising temperature (point (3) above).

The thermal component, emission from the decay of surface or near-surface H centres, will be influenced strongly by non-stoichimetry or by the presence of F or F-aggregate centres. The yield should fall as the surface is enriched with alkali, and so higher yields are expected for higher temperatures and increasing beam intensity.

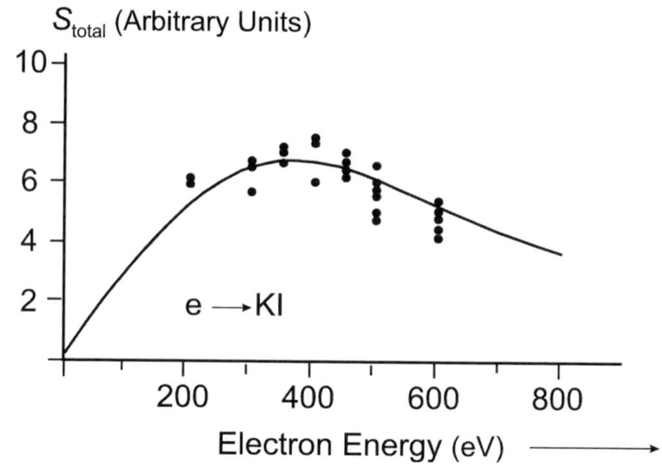

Fig. 9.11. Dependence of total yield on electron beam energy for KI (Al Jammal & Townsend 1973; Szymanski 1980).

The absolute sputtering yield

The absolute sputtering yield has been obtained by measuring the weight loss induced under electron or ion beam irradiation (Table 9.1). The energy loss occurs at a significant depth: a 500 eV electron creates about 200 electron–hole pairs over a depth of about 20 nm, with the greatest energy loss at 4 nm. It does not seem possible to explain this high yield observed without assuming exciton diffusion (Al Jammal & Townsend 1973). The high sputtering yield implies that the surface acts as the sink for the H centres or interstitial halogen atoms. The near-surface F centre yield should be higher than that in the bulk, for the same reason.

The overall yield for sputtering is a monotonic function of the Klick–Rabin factor S/D, rising as the Klick–Rabin factor falls. This is quite different from the yield for bulk defect formation, which has a maximum and then decreases as S/D falls. This difference relates to the difference in decay mode for a self-trapped exciton in the bulk and on the surface. The space to incorporate an interstitial halogen is not critical on the surface, and so the yield is higher as S/D decreases (Elliott & Townsend 1971).

Emission of alkali atoms

The sensitivity of conventional sputtering experiments for alkali halides is low, so that it is necessary to remove several layers of atoms. The surface becomes alkali-rich and forms alkali colloids because of the preferential emission of halogen atoms. At low temperatures, as irradiation proceeds, this change of composition leads to a fall in halogen yield, and hence to a reduction in alkali

Table 9.1. *Sputtering yield of alkali halides per 500 electrons.* The results are listed in order of increasing Klick–Rabin ratio, *S*/*D*.

Material	Klick–Rabin parameter	Yield (molecules per particle)	Reference
NaBr	0.28	2.5	D J Elliott & Townsend (1971)
KI	0.33	2.5	D J Elliott & Townsend (1971)
NaCl	0.36	14±4	Szymonski *et al.* (1985)
NaCl	0.36	1.25	D J Elliott & Townsend (1971)
LiF	0.39	2.1	Neidhart *et al.* (1995)
KBr	0.47	1.3	D J Elliott & Townsend (1971)
KCl	0.60	1.3	D J Elliott & Townsend (1971)
NaF	0.70	0.5	D J Elliott & Townsend (1971)

yield. At higher temperatures, the alkali colloids can be evaporated thermally, and the surface can remain essentially stoichiometric during sputtering. It is not known whether the alkali emission is a result of electronic excitation or mere thermal evaporation; it is possible that surface diffusion is involved, since this has an activation energy around 0.1–0.2 eV.

Kubo *et al.* (1993) measured Na atom emission from NaCl induced by a 400 eV, 1 μs, electron pulses. They had submonolayer sensitivity, and could show that the Na atom yield below 200 °C reduces as surface damage is accumulated. This suggests that the accumulation of Na atoms reduces the overall sputtering yield. However, Na emission from fresh specimens has an essentially constant yield before it starts to decrease by damage accumulation. This implies that Na atoms do not come solely from regions where the halogen is depleted. Below 100 °C, the emission of Na atoms from fresh surfaces is delayed by a few microseconds after an electron pulse. The source of the Na emission induced by band-to-band excitation is not yet clear: it may involve excitation at alkali-rich defect sites, such as dislocation jogs (Puchin *et al.* 1993b).

Feibelman–Knotek or Varley mechanisms on perfect surfaces of alkali halides? It is controversial as to whether the Feibelman–Knotek mechanism is effective for stoichiometric alkali halide surfaces. The mechanism has a low probability, and is only effective at energies which also cause valence excitation. N Itoh *et al.* (1989b) have shown that the Feibelman–Knotek process is energetically favourable for alkali halide surfaces, although the Varley mechanism in the bulk alkali halides is not favourable (Chapter 5). Walkup *et al.* ((1986); see also the more extensive calculations of Green *et al.* (1989b)) carried out a classical molecular dynamics simulation, and argued that it was the lattice relaxation followed by formation of two-hole localisation which made the repulsion

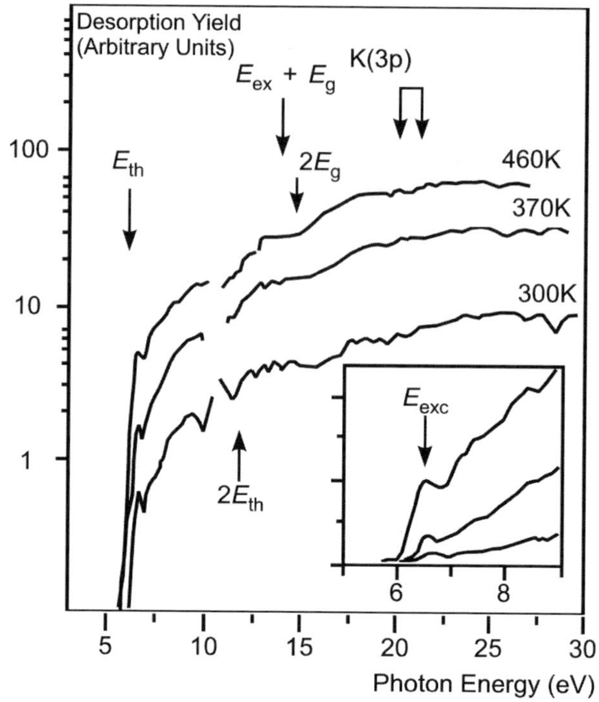

Fig. 9.12. The dependence of the desorption yield of KBr on photon energy, after Zema *et al.* (1997). The optical absorption spectra are shown at the bottom for comparison.

energy ineffective. Spectroscopic studies for a wide photon energy range by Zema *et al.* (1997) show no evidence of the enhancement of emission at the alkali core excitation energy above 370 K (Fig. 9.12). Probably the alkali-ion emissions seen after core excitation are characteristic of an alkali-rich layer. Whether the emission from an alkali-rich layer has a significant contribution from the Feibelman–Knotek mechanism is unclear.

Sputtering induced by core excitation

The irradiation of alkali halides by photons with energies big enough for core-electron excitation creates electron–hole pairs as well as core holes, and so can create alkali-rich surfaces. The so-called core effects do not necessarily represent the characteristics of stoichiometric surfaces. Tolk *et al.* (1982) measured photons emitted from excited alkali atoms ejected following core excitation of LiF, NaCl, and KCl. As shown in Fig. 9.13, the yield is essentially negligible below the core threshold energy, but starts to increase once the photon energy crosses the core excitation energy. Evidently, the emission of excited atoms originates from core excitation. Similarly, ions are emitted under electron irradiation at energies higher than core-electron energies (Pian *et al.* 1983;

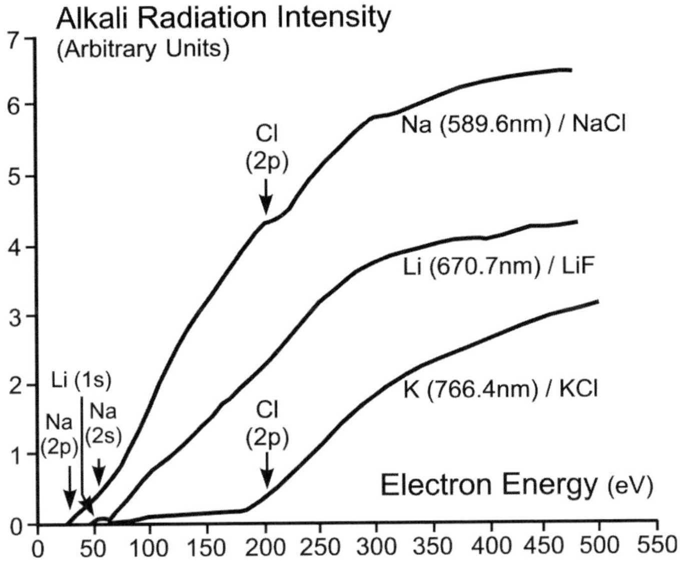

Fig. 9.13. Electron-energy dependence of the intensity of radiation from excited state alkali atom for LiF, NaCl, and KCl (after Tolk *et al.* (1982)). The sample temperatures varied between 250 °C and 300 °C.

Parks *et al.* 1983). However, Taglauer *et al.* (1986) found that the yield of excited- and ground-state alkalis anticorrelate with respect to sample temperature for NaCl: the ground-state signal rises and the excited-state signal falls with increasing temperature above 300 K (Fig. 9.14). As described earlier, the increase in ground-state yield is due to evaporation of surface alkali. It is likely that the emission of excited atoms by core excitation occurs only from alkali-rich surfaces. Walkup *et al.* (1988) concluded that the small observed amounts of excited Na and Na^+ obtained from NaCl at sufficiently high temperatures and low electron beam current densities arose solely from collisions of electrons with emitted ground-state Na.

The measurements of ion and excited atom emissions are made with a large beam current, sufficient to remove several layers of ions. Sputtering can continue over several layers only if atoms can be removed from the alkali-rich layers which develops as halogen is removed. Thus existing experiments do not give direct evidence for Feibelman–Knotek or Varley mechanisms on perfect surfaces of alkali halides.

N Itoh *et al.* (1989b) have shown that the Feibelman–Knotek process is energetically favourable for alkali halide surfaces, although the Varley mechanism is not favourable. We have already noted that the Varley mechanism does not work in the bulk of alkali halides. Walkup & Avouris (1986) carried out a classical molecular dynamics simulation, and argued that it was the lattice

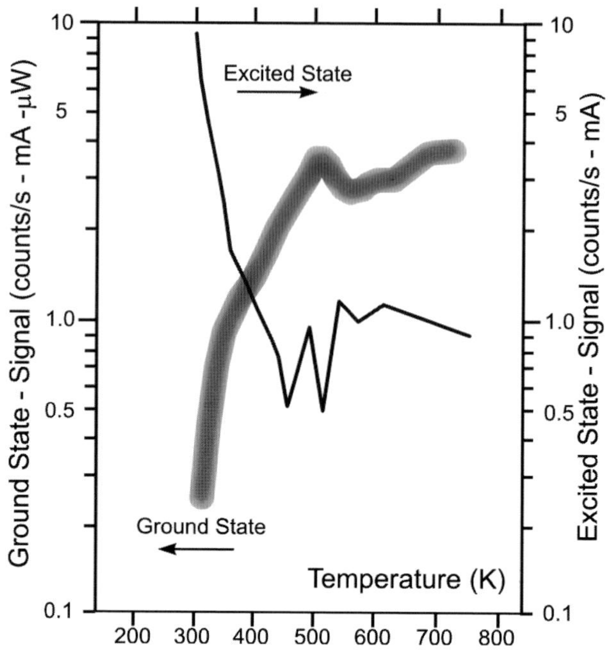

Fig. 9.14. Temperature dependence of the yield of excited and ground-state alkali atoms emitted from NaCl upon irradiation with photons of 14–73 eV.

relaxation followed by formation of two-hole localisation which made the repulsion energy ineffective.

Laser-induced processes on halide surfaces

Laser-induced sputtering by photons of sub-bandgap energies has been measured for halides. Initially, there must be single-photon absorptions at defect sites or by multiphoton band-to-band excitations. Multiphoton excitations create electron–hole pairs, which cause preferential emission of halogen atoms, hence producing F centres and metal colloids, which have defect levels within the gap. Such levels could be the source of the single-photon absorption and resonant multiphoton absorption, as suggested by Matthias *et al.* (1987), and would enhance the emissions. The surface is thus continuously modified during a laser pulse. The number of atoms removed is an integral of the number of atoms emitted from the surface through its stages of progressive surface damage. The temperature rise during a laser pulse due to non-radiative transitions and free-electron heating will also cause the emission of metal atoms, which compensate the halogen atoms emissions from the excitonic process. All of these complex cumulative processes must be taken into account to understand laser-induced processes.

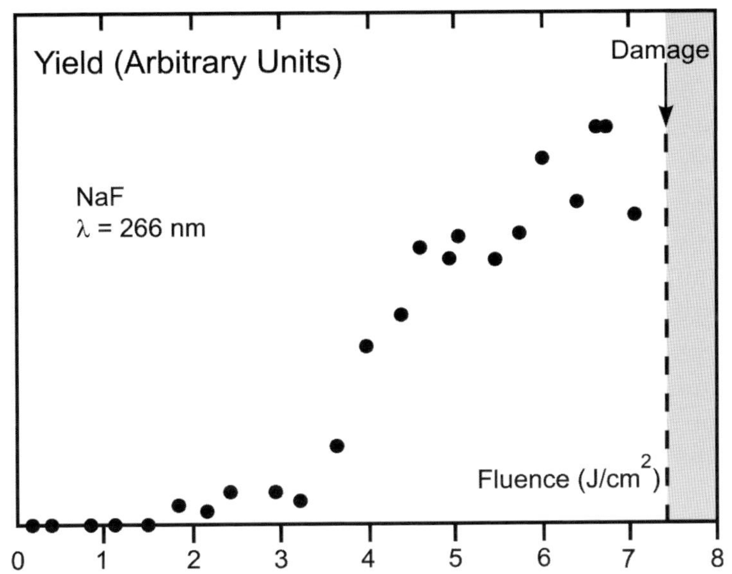

Fig. 9.15. Laser-induced emission yield of Na atoms from NaF below laser-ablation threshold (after Chase *et al.* (1994a)).

An example showing the importance of defect absorption is given by an earlier experiment, where laser intensity was not sufficiently strong. Schmid *et al.* (1975) measured the intensity dependence of the emission yield using a Nd:YAG laser (0.532 μm) and found that the emission is induced by multiphoton band-to-band excitation at lower intensities. At higher laser intensities, however, the power index for the yield–intensity relation is doubled. This result has been interpreted in terms of an enhancement of multiphoton band-to-band excitation by resonance to the surface F centre level created by the multiphoton absorption (Haglund & Itoh 1994). This is an example of the case where an nth-power yield–intensity relation does not necessarily imply n-photon excitation: cumulative effects can make apparent power dependence higher. A similar alteration of the yield with higher laser intensity is shown in Fig. 9.15. Although the final cause of surface destruction by laser irradiation might be due to heating, possibly by free-electron heating (S C Jones *et al.* 1988), it cannot true that all input energy is absorbed as heat from the very beginning of a laser pulse.

Ion-induced sputtering of alkali halides

The sputtering of alkali halides by ion beams in the kiloelectron volt energy range involves both electronic excitation and elastic encounters. The sputtering induced by elastic encounters is largely independent of temperature, while that induced by electronic encounters depends on temperature. Probably elastic encounters dominate at lower temperatures, with the contribution of electronic encounters becoming increasingly significant at high temperatures. The contributions of electronic and elastic encounters can be distinguished by measuring the dependence of the yield on the incident energy and the most probable energy of the emitted atoms. The energy deposited to electronic encounters should increase monotonically with incident energy, while that to elastic encounters should have a maximum near 10 keV (Chapter 2), depending on the mass of the incident ions. The most probable energy of the emitted atoms is proportional to the square of the inverse of the incident energy for the elastic encounters.

The energy dependence obtained at room temperature shows the characteristics of elastic encounters (Navinsek 1965; see also N Itoh 1976) and the velocity distribution of Na from a NaCl surface by He ions also follows that expected for elastic encounters (Miyagawa 1973). Although alkali atoms sputtered by heavier ions at room temperature include a temperature-dependent component characteristics of electron-induced sputtering, this disappears at low temperatures (Szymonski & Vries 1981). Husinsky *et al.* (1988) measured the Doppler broadening of the luminescence emitted from excited Na atoms sputtered by electron and ion beams, and showed that the velocity distributions were the same. However, it appears that the sample was heated under the high beam currents used in their experiments.

9.2.3 Sputtering of other halides

Emission of halogen atoms from alkaline-earth fluorides

Ca-rich CaF_2(111) surfaces are formed by irradiation with electron or photon beams. They have been examined by photoelectron spectroscopy (Karlsson *et al.* 1986), by EELS (Saiki *et al.* 1987; Miyake *et al.* 1996; Miura 1995), by He atom scattering (Miyake *et al.* 1996) and by RHEED and EELS (Miura *et al.* 1991). The absence of new He diffraction peaks and RHEED patterns after irradiation indicates that the atomic structure of the top surface has not been altered. However, an increase in the intensity of He scattering indicates that the underlying layers are modified. Photoelectron spectroscopic studies show a reduction in the peak due to F^- ions and the formation of new levels in the bandgap; the formation of an ordered array of F centres has been suggested (Karlsson *et al.* 1986). The intensity of the peak in the gap saturates after irradiation, indicating that the yield for removal of F^- ions falls as radiation proceeds. The quantum yield for removal of fluorine measured with 36 eV photon irradiation is almost unity, which implies that the process is not due to core excitation, but due to formation of electron–hole pairs. It is likely that excitonic fluorine emission creates a Ca-rich layer, which suppresses the excitonic process.

Cation emission by core excitation

Irradiation with photons capable of core-electron excitation of CaF_2 induces the emission of halogen ions. Halogen ion emission will continue until a monolayer of halogen atoms has been removed, as noted already. The positive ion emission under core excitation, if it is observed, must be from a Ca-rich surface layer, not from a stoichiometric surface. For further studies of the reaction on surfaces, detection of emitted atoms or ions with submonolayer sensitivity is essential.

Laser sputtering of calcium fluoride

The laser-induced sputtering of CaF_2 has been reviewed by Reif (1989) and Matthias & Green (1990), and there has been much work (Reichling 1998; Reichling *et al.* 1998a,b; Reichling 1995) including photoemission studies (Huisinga *et al.* 1998) and modelling of features like steps and oxidation processes (like reaction with oxygen to form CaO) affecting the Ca metal which is produced by F loss (Puchina *et al.* 1998).

The overall ion emission yield for a 518 nm laser pulse shows a high power dependence on laser intensity. The intensity dependence of the electron yield measured on the same specimen shows that the yield has a quadratic dependence at low laser intensity. Evidently, the electrons arise from the two-photon excitation of defects at low laser fluences. At higher laser fluences, the electron yield has a higher power dependence. This result does not necessarily mean that a larger order multiphoton excitation is responsible for the emission, but can be related to electron emission from defects generated at an earlier stage of a laser pulse. Ion emissions are seen at the laser fluences at which the higher power electron yield is observed. This $n = 10$ dependence has been interpreted as a consequence of multiphoton excitation in resonance with defect levels (i.e. with photon energy such that an integral number of photons provides the right energy for the transition). It may be that the emission of ions is from secondary products on the surface, such as metal colloids. A variety of processes must be taken into account. F centres and their aggregates can form by multiphoton band-to-band excitation, possibly in resonance with surface defect levels. Pairs of electrons and halogen-ion vacancies may be generated by the ionisation of the F centres and their aggregates. Free electrons generated by band-to-band excitation and by F centre excitation will be heated by optical excitation. Ca emission should follow the excitation of F centres and their aggregates, analogous to Na emission from NaCl (Ca emission has not been proved for CaF_2). There will be the formation of Ca colloids and their evaporation by heating. All of these processes enhance the emission yield of both atomic and ionic species, bringing in the stage of plume formation.

In the laser sputtering of BaF_2(111), there is an anticorrelation of Ba and F which suggests the laser sputtering occurs layer by layer (Fallgren *et al.* 1986).

Following electron beam or photon metallisation (Huisinga *et al.* 1998) there is the surprising result that, even though the surface appears metallic to the eye, there are no photoemission features characteristic of the Ca Fermi surface. This appears to follow from the band structure selection rules for emission normal to a (111) surface on which Ca has grown epitaxially. The formation of a metallic band at the Fermi edge is clearly visible in metastable ion excitation spectroscopy (Reichling 1998; Reichling *et al.* 1998a,b).

Magnesium and aluminium fluorides

Processes similar to those in CaF_2 may occur for other fluorides. Certainly there are suggestions that the deep nanoholes created by convergent nanosized electron beams in MgF_2 (Broers *et al.* 1981) and AlF_3 (Isaacson & Murray 1981; Kratschner & Isaacson 1986) films occur by direct sublimation (Broers 1988; see also Chapter 11). We remark that the insensitivity of CaF_2 and MgF_2 to sub-bandgap ultraviolet light has made these materials favoured choices for future lithography optics and for ultraviolet light source windows, respectively.

Photo-induced sputtering of Ag halides

Photons with above bandgap energies give sputtering yields of almost unity for halogen atoms from the surfaces of AgCl and AgBr (Kanzaki & Mori 1984). The dependence of the yield on photon energy parallels the optical absorption coefficient, showing that there is an efficient photolytic process to emit halogen atoms through the formation of electron–hole pairs at the surface. As noted in Chapter 4, bulk self-trapped excitons exist only in AgCl, not in AgBr. Further, the self-trapped exciton in AgCl is described as [Ag^{2+}:e], without a significant off-centre displacement. Even if an [Ag^{2+}:e] exciton forms on the surface, there is no obvious route to the emission of a halogen atom. One way to resolve this difficulty is to assume that self-trapped excitons of [X^{2-}:e]-type form on surfaces of silver halides, as in bulk alkali halides, and that the relaxation processes are similar to those in alkali halides (Section 9.2.2).

The formation of the [V_k:e]-type self-trapped excitons on silver halide surfaces might be favoured for several reasons. First, the low coordination number of Ag^+ on surfaces could lower the d-states from the top of the valence band, making their admixture less effective in self-trapped holes or excitons, and hence favouring the formation of the [V_k:e] type self-trapped excitons. Secondly, after all hole sensitisers are filled with holes, the products of the surface photolytic process may be hole traps and interstitial silver ions Ag_i^+. This hypothesis is able to explain the continuing colloid growth in a silver halide grain until a large fraction of the silver has been converted to silver metal.

Possible mechanisms of photolytic decomposition of AgBr

Photolytic decomposition at grain surfaces can be described by the loss of a Br atom to form a surface F centre, F_s. The reaction cannot continue without some means to stabilise the latent image, such as the Mott–Gurney mechanism. This involves electron traps to nucleate silver colloids, and holes, which are trapped by surface F_s centres. Overall, the reaction is one in which an electron and two holes generate a cation interstitial, a halogen atom, and a divacancy at the surface:

$$\text{perfect surface} \rightarrow F_s + Br^0,$$

$$F_s + h^+ \rightarrow Ag_i^+ + Ag_{sv}^-\ Br_{sv}^+.$$

The halogen atom, Br^0, is emitted from the surface. Combining this with the equations for the Mott–Gurney mechanism (Section 5.2.3), creation of an electron–hole pair leads to the formation of Ag^0 at the electron trap and the loss of a halogen atom into the surrounding gelatine. The reaction suggested above is tentative, and awaits theoretical studies of the creation of an Ag_i^+ from an F centre by hole trapping. An alternative reaction to sustain colloid formation or photolytic decomposition by prolonged irradiation involves the positive ion vacancy Ag_{sv}^-, the complementary pair to Ag_i^+:

$$\text{perfect crystal} \rightarrow Ag_i^+ + Ag_{sv}^-,$$

$$Ag_{sv}^- + h^+ \rightarrow Br^0 + Ag_{sv}^-\ Br_{sv}^+.$$

Although this reaction itself is plausible, it needs the generation of Ag Frenkel pairs. For a perfect bulk crystal, the thermal activation energy is 1.2 eV in AgBr, and is too high for normal bulk thermal defects to supply the Ag_i^+ ions. Platikanova & Malinowski (1978) suggested that Ag_i^+ is generated by hole trapping, without specifying the trapping sites. According to the model suggested here, single-crystal silver halides irradiated in a vacuum will have silver-rich surface sites. The surfaces will be the source of Ag_i^+, which replenishes the loss of Ag_i^+ nucleated at dislocations or at grain boundaries by the Mott–Gurney mechanism and ensures continuing photographic sensitivity under prolonged irradiation. Thus, colloid growth in silver halides is completely different from that in alkali halides, in which it is the coagulation of the F centres (Section 5.2.3).

9.3 Sputtering and surface modification of oxides

Those phenomena most likely to demonstrate excitonic sputtering have been observed in SiO_2. Yet most oxide surfaces are hard to sputter by photons and ionising radiation. Nevertheless, the sputtering of oxides through electronic excitation can be caused by irradiation with intense electron or laser beams, or by irradiation with energetic heavy ions which generate a high density of electron–hole pairs. In most cases, laser beams of sub-bandgap energies, which cannot create electron–hole pairs directly, can still heat the surface layers and cause congruent evaporation. The question remains as to how an initially transparent surface becomes one which can absorb photons and become a source of heating. Core excitation can remove anions, but sputtering (the removal of many layers of material) needs a means to remove cations as well. The possible evaporation of volatile species should always be born in mind. For oxide films on silicon, for example, the SiO species is volatile, and the oxide is easily destroyed when oxygens are removed by excitation or when the oxygen pressure is low. Low-field breakdown of capacitors also correlates with the temperature and pressure regime for which SiO evaporation is significant. This can be used to advantage. For example, an electron beam (16 nA, 30 kV) can remove some oxygen from a 5 Å film of oxide on Si at room temperature. Subsequent thermal processing at 720 °C leads to selective decomposition, and permits patterning (Watanabe *et al.* 1997).

9.3.1 Emissions by core excitation

The emission of O^+ ions by core excitation is seen for MgO by excitation at 80 eV (2p hole of Mg^{2+}; Gotoh *et al.* 1989) and 510 eV (1s hole of O; Kurtz *et al.* 1987). Gotoh *et al.* also found generation of 400 nm photoluminescence centres by core excitation, and ascribed it to the generation of F centres by core excitation near surfaces.

The O^+ emission from TiO_2 by the excitation of Ti 3p electrons was reported for the first time as evidence for the Feibelman–Knotek mechanism (1978). Kurtz *et al.* (1988) showed that the emission yield and its angular dependence were influenced by annealing, although the detailed correlation between these properties and the surface structure has yet to be fully analysed.

It appears that O^+ emission by core excitation is effective from perfect surfaces of oxides. Walkup & Kurtz (1988) carried out a molecular dynamics simulation of O^+ emission by core excitation, and found a high emission probability for O^+ ions produced by Auger transitions (unlike the situation for alkali halides) and a strong site dependence of the trajectories. Souda (1999) has measured O^+ desorption from oxide surfaces during He^+ bombardment. Desorption is weak for MgO(100), Al_2O_3(0001) and SiO_2 surfaces, but much stronger for the TiO_2(110) surface. The process leading to O^+ ions appears to involve several factors: an O 2s core hole, strong hybridisation of Ti(3d) and O(2p), an intraionic Auger process, and charge exchange between the He 1s and O 2s states. The Feibelman–Knotek mechanism appears to occur in covalent oxides, where there is emission of anions, rather than cations (Ramaker 1983). O^+, H^+, and OH^+ ions are emitted from SiO_2 by excitation of the O(2s) level around 32 eV (Knotek & Houston 1982), and further enhanced emission occurs by excitation of the O(2p) level around 103 eV (Ramaker 1983). The Cini–Sawatzky condition for the core holes to remain localised has been shown to be satisfied theoretically for SiO_2 (Ramaker *et al.* 1982).

9.3.2 *Surface modification of quartz*

Excitonic sputtering of quartz

Akazawa *et al.* (1992a,b) have observed the rather efficient sputtering of fused quartz by synchrotron radiation in the energy range 10–10^3 eV, peaked at 100 eV. The sputtering yield was temperature-dependent, and followed an Arrehenius relation with an activation energy of 0.87 eV. The sputtering rate at the highest temperature observed can be as high as 0.2 A/s at a synchrotron beam current of 100 mA, which corresponds to roughly one atom removed for every 10^{+4} incident photons. If we take the penetration depth of the incident photons as 1000 layers, which is an underestimate, then one atom is emitted for every 10 photons in the outermost surface. Akazawa *et al.* suggested that sputtering occurred by photodecomposition into O and SiO followed by the evaporation of SiO. Given the experimental evidence (Section 8.3.1) that oxygen interstitials are produced from two neighbouring excitons in the bulk, it is likely that the decomposition is induced by production of two neighbouring excitons. Although the yield is independent of beam current, most secondary electrons, including photo and Auger electrons, have energies of less than 100 eV and therefore have small mean free paths, and produce high-density electron–hole pairs. The sputtering yield of Si by the same synchrotron beam is 1/50 for Si_3N_4, it is 1/10 of that for SiO_2. The difference may be related to the presence or absence of self-trapping of excitons.

Many oxides are poor conductors, so that STM is difficult. Experiments on oxide films on Si are possible, however. N Li *et al.* (1998) found that a low-energy electron beam from an STM (100–150 eV, far less than the 30–300 keV typical of electron microscopes) etches the oxide. Windows down to 70 nm can be achieved at 700 °C with currents from 5 nA to 1 μA.

Excimer laser irradiation (248 nm) of thermally-grown SiO_2 on Si also produces defects detected by EPR and Auger spectroscopies. It is after annealing at 760 K that emissions of oxygen can be detected through the formation of E′ centres and SiO_x structures (Fiori & Devine 1984). Since similar reactions do not take place in bulk a-SiO_2, the emission of oxygen from the surface is the cause of the production of SiO_x. This phenomenon appears similar to that observed on irradiation with synchrotron radiation, although the photon energy in this case is below the bandgap energy. The photons of this energy range will be absorbed by the band tail or by defects in the amorphous network, and can cause reactions similar to those induced by electron–hole pairs. Thus, as for alkali halides, the defect yield and hence the emission yield can be high when the surfaces act as a sink for a product of a photolytic reaction.

Laser excitation of SiO_2

Photolytic processes appear to play a role in laser ablation of SiO_2. Sugioka *et al.* (1995) showed that smooth ablated surfaces with less debris were created by simultaneous irradiation of fused quartz with laser pulses of 8 ns and 266 nm and with those of 10 ns, including photons above bandgap energy. They observed a 15% enhancement of optical absorption of 266 nm photons by simultaneous irradiation, but the ablation threshold was 1 J/cm^2 in both cases. Although the results can be interpreted in other ways, it does seem that irradiation with a laser pulse capable of band-to-band excitation creates self-trapped excitons, and that these can be excited by 266 nm photons to induce a photolytic reaction. The effects of irradiation with two types of laser pulse disappear when the 266 nm laser pulse is delayed from the bandgap excitation pulse by more than 4 ns (J Zhang *et al.* 1997). Since the lifetime of the self-trapped excitons in fused quartz is distributed, this experimental result alone cannot determine the mechanism, without a similar experiment for α-quartz. Preferential emission of O from surfaces of amorphous and crystalline SiO_2 occurs during ablation by a 126 nm laser capable of band-to-band excitation (Takigawa *et al.* 1990; see also Ohmukai *et al.* 1999). Enrichment by Si is found using X-ray photoelectron spectroscopy.

Lithography by laser beams

Repeated laser-pulse irradiation on a single spot on a surface can create a hole, a technique used for micromachining. The material removed by each pulse increases as the irradiation proceeds. For instance, ablation of SiO_2 for the first shot of a series of 24 ns, 248 nm laser pulses occurs only above 10 J/cm^2. After 20–40 shots, ablation requires just 5 J/cm^2, and the ablation rate is 2 μm per pulse (Ihlemann & Wolff-Rottke 1996). The incubation stage probably leads to defect generation, and these defects absorb photons in the later stages. If all photons in a single shot were absorbed within the depth ablated in that shot, about 10 eV would be absorbed per atom after the defects have been created. The major cause of ablation is probably evaporation. Ashkenasi *et al.* (1998) found that the average ablation rate increased with the number of pulses for the first 10–50 pulses. The increase was greater for subpicosecond pulses than for picosecond pulses. The increase for subpicosecond pulse irradiation was ascribed to heating following the excitation of bulk defects generated by earlier pulses.

9.3.3 Laser excitation of other oxides

Laser excitation of MgO

No photolytic processes have been observed for MgO surfaces under ionising irradiation or synchrotron radiation (Kuntz *et al.* 1987). Thus MgO surfaces are stable with respect to electronic excitation, unless intense laser beams are used. Dickinson *et al.* (1994) have observed forward-oriented emissions of Mg^+ ions in the energy range of 10–15 eV by irradiating MgO with nanosecond 248 nm laser pulses of about 1 J/cm^2. They suggested that the emission arose from Mg^{2+} ions attached on surface F centres. Since the Mg^+ emission yield is a superlinear function of fluence, it appears that the defect structure responsible for the emission of energetic Mg^+ builds up during laser irradiation, by an electronic process, not by heating. According to Shluger *et al.* (1991c), multiple excitation at surfaces induces the emission of O, leaving an F centre behind. This suggests that the surface will be Mg-enriched via defect-localised multiple excitation, probably forming metallic colloids as well as F and F^+ centres. The energetic Mg^+ will be emitted from the Mg-rich surface. It could be the emission of an Mg ion on an F centre, as suggested by Dickinson *et al.*, or alternatively due to a Coulomb explosion of a multiply ionised Mg cluster by incident photons.

Laser excitation of Al_2O_3

Dreyfus *et al.* (1986) observed the emission of AlO molecules on exciting Al_2O_3 by excimer laser pulses. They observed vibrational structure in the luminescence emitted from the molecules. The temperature of the molecules, derived from the intensity ratio of the vibrational structure, was substantially smaller than the melting point of alumina. The emission is not consistent with a thermal mechanism, and has been ascribed to an electronic process (Dreyfus, *et al.* 1986; Kelly & Dreyfus 1988; Gutfield *et al.* 1986). It has been shown that the emission yield is related to the surface defect concentration. The emission is observed only with intense laser pulse irradiation, so the emission is probably due to multiple excitation of defect sites. The Al_2O_3 surfaces ablated by nanosecond excimer laser pulses show no debris, suggesting that the electronic process dominates in ablation, whereas those surfaces ablated by picosecond laser pulses include more debris.

Lithium niobate

Desorption of $LiNbO_3$ molecules occurs at a threshold intensity well below that for atom or ion emission (Affatigato *et al.* 1994). This molecular yield is affected by intensity, pulse rate, and history; it can be quenched, apparently either by hot electron emission causing molecular dissociation or by the altered stoichiometry of the surface. In the ablation regime, O is sputtered preferentially. At the higher intensities, the main effects are increased absorption due to changes in stoichiometry and higher temperatures, and defect accumulation in the near-surface region.

Oxide superconductors

Rosenberg & Wen (1988) noted that synchrotron radiation leads to loss of molecular O_2 from YBCO. The mechanism is not certain. Despite the fact that oxide superconductors have metallic conductivity (albeit bad metallic conductivity in the normal state), there is evidence for behaviour under optical excitation which is strikingly similar to that for insulating oxides (Stoneham & Smith 1991). Variants of excitonic mechanisms may occur in near-surface insulating regions. Oxygen desorption can be used as a means of changing the doping level of oxide superconductors in a controlled manner (Schwaller *et al.* 1999). There are several advantages: the crystallinity is not affected, and the doping level can be monitored by photoelectron spectroscopy.

9.4 Semiconductors

Photo-induced surface processes for elemental, III–V and II–VI semiconductors are significantly different from those of halides and oxides. First, exciton self-trapping does not occur in these semiconductors. Secondly, the bandgap energy is insufficient to break bonds or create defects. Because the bandgap is so low, no atomic emission will result from a single electron–hole pair, even if it localises at a defect site. However, there is unambiguous experimental evidence from STM that atoms are indeed emitted non-thermally from

Si surfaces by irradiation with nanosecond laser pulses. The emission yield is a superlinear function of laser fluence, so the emission is not caused by single excitation, but is due to some localised multiple excitation. Two models have been suggested. One involves multiple excitation of an antibonding excited state; the other requires fusion of excitons through some atomic displacement. In both cases, the energy localised is more than the bandgap, and is large enough to cause atomic emission. However, exciton fusion by itself may not suffice to cause emission without some suitable defect structure, surface reconstruction, or impurity. Certainly, neither excitonic molecules nor multiple bound excitons (nor even exciton gases) are reported to cause significant atomic processes in bulk semiconductors.

9.4.1 Atomic emissions from semiconductors by laser irradiation

Emissions of neutral constituent atoms induced by nanosecond laser pulses from Si(111)7×7, Si(100)2×1, GaAs(100)1×1, GaAs(111), GaP(111), and InP(100) surfaces have been measured using resonance ionisation spectroscopy. This technique is highly sensitive, so that emissions of a mere 10^{-5} layers can be detected. For these surfaces, it is unlikely that reconstruction leads to potential fluctuations large enough to induce localisation, except for the Si(111)7×7 surface. Atomic emission is observed from all of these surfaces, with a yield which falls on repeated irradiation. This self-decaying emission occurs at laser fluences of about 10% of the ablation laser fluence, and is ascribed to emission from adatom-type defects, which are removed by the emission process. Increasing the laser fluence to about 50% of ablation threshold leads to an emission which is almost unchanged on repeated irradiation. This is ascribed to emissions associated with steps, since (cf. Section 9.1) these cannot be eliminated rapidly by taking atoms away. In this fluence range, the emission yield is substantially suppressed for photon energies just below the bandgap energy (Okano *et al.* 1993). For direct gap semiconductors, the suppression occurs resonantly near the bandgap energy, whereas the suppression extends over about 1 eV above the band edge in indirect gap semiconductors. The results have been explained in terms of energy transfer from a surface excited state to a bulk excited state. These results, like the STM results, demonstrate that purely electronic emissions take place at surface defects.

The yield of atomic emissions detected by submonolayer sensitivity increases rapidly above a certain threshold fluence; above the threshold, the yield continues to increase and ultimately reaches ablation. Although it is obvious that the emissions below this threshold are due to electronic processes, it is not clear whether the sudden increase in yield on reaching ablation is due to heating or involves electronic excitation. Okano *et al.* (1993) suggested that atomic emissions from sites neighbouring surface vacancies increase the active surfaces sites for atomic emissions and surfaces. They ascribed the sudden increase to surface-vacancy associated emissions. We discuss this problem in Chapter 11.

We should emphasise two important aspects of laser-induced processes in semiconductors. First, since there is no self-trapping, localisation occurs only at defect sites, such as point defects, dislocations, or kinks on surfaces. Secondly, the generation of a single electron–hole pair does not lead to lattice modification. Laser-induced electronic processes in semiconductors involve multiple excitation, initiated by thermalised electron–hole pair generation or by defect excitation.

High-energy photons can lead to desorption in further ways. The studies of Hoffman *et al.* (1999) pointed to two routes to desorb H ions from the (100) surface of diamond using photons in the 280–340 eV range. One process involved secondary electrons created in the bulk; this process was the main route for stimulated desorption of H^-. The other process led to H^+ desorption following C(1s) core ionisation.

Processes involving more than one excitation step

In semiconductors, there are two types of mechanisms possible for laser-induced processes. The first type is multiple excitation of a defect during a laser pulse, leading to defect modification. The second type is the multiple excitation of electrons generated by band-to-band excitation or by defect excitation. We shall discuss the first type in detail later. The second type has been discussed particularly for metals, where Gadzuk (1991) has suggested that an anti-bonding state embedded in the continuum may be reached by multiple excitation of free electrons. The same idea has been used to explain laser-induced emission of adsorbates on metal surfaces by picosecond laser pulses (Misewich *et al.* 1992).

The electronic excitation energy necessary to create a surface vacancy is more than the bandgap energy and so is more than the bandgap, and so more than is available from a single exciton. Localisation of a carrier or exciton is not itself a problem, since it might happen at sites where there is some favourable surface reconstruction or surface defect. However, the special problem is how to localise even more energy, corresponding either to two or more electron–hole pairs or to a special and more highly-excited state. The term 'bound exciton' is sometimes mentioned; useful though it can be, one must recognise that a bound exciton is simply an excited state of a defect; the term is unhelpful unless the bound exciton is indeed similar to the free exciton, but localised by a perturbation.

A first class of routes is simply for laser irradiation to create two associated, localised, electron–hole pairs. This might occur (I.1) by the creation of a second electron–hole pair adjacent to a site D already perturbed by a localised electron–hole pair ([Deh] + $h\nu \rightarrow$ [Deheh]), or (I.2) by the capture of a mobile exciton at the site already excited ([Deh] + [eh] $\rightarrow$ [Deheh]), or (I.3) even by formation and capture of a biexciton ([D] + [eheh] $\rightarrow$ [Deheh]). The second electron–hole pair could be bound by the one-electron local potential of the defect or reconstruction; alternatively, it could be bound to the existing electron–hole pair, just as the excitonic molecule [eheh] is stable. Panchratov & Scheffler (1995) have discussed the consequences of two-exciton localisation on the GaAs(100) surface, suggesting that an As atom might be displaced from the surface. Hattori *et al.* (1992) suggested that some sort of cascade excitation might lead to atomic emission. Clearly, the formation of two associated electron–hole pairs is not a complete answer: it ensures energy is available, but does not guarantee forces to drive emission. Indeed, exciton condensates in the bulk show no signs of driving displacements. The holes must be in specific bonding orbitals, or the electrons in specific antibonding orbitals, which implies some spectral selectivity.

A second route is best described as multiple excitation of a defect, just as an electron–hole pair may be generated near a self-trapped exciton. For a relaxed excited state [Deh] of a defect D with a level in the bandgap, there are three types of excitation: (II.1) the excitation of an electron (giving [De*h]), (II.2) the excitation of a hole (giving [Deh*]), and (II.3) the creation of an exciton perturbed by the relaxed excited state (Deehh). The role of the second excitation is two-fold. First, there is more energy available to drive a reaction. Secondly, there will be destabilising forces if the electron is excited to an antibonding orbital, or a hole is created in a bonding orbital, or if the two holes are localised and reduce the ionic charge. One major mode of lattice relaxation will be to push an atom away from the surface.

A third route requires at least one electron to be excited into an antibonding orbital, or creates a hole in a bonding orbital. These excitations ensure that there are forces of the right type to drive emission. How is enough energy made available? One possibility (III.1) is that the energy needed for these excitations is large (much more than the bandgap) and, for some reason such as correlation, the excitation survives in a way which permits emission. A second possibility is that more carriers could be localised. There might be two localised holes in bonding orbitals, or two electrons in antibonding orbitals (Khoo *et al.* 1993; N Itoh & Nakayama 1982). The forces come from the natures of these states; the energy comes from recombination of these carriers. But with which carriers do they recombine? Suppose there are two holes in a bonding orbital. We should consider three cases: (III.2) involving [Dhhee] (this case is closely related to routes I and II), (III.3) involving [Dhhe], and (III.4) involving [Dhh]. We show only bound carriers, and we assume for simplicity that the electrons do not contribute to the forces driving emission.

In case (III.4) there is a question as to how, despite the Coulomb repulsion, two holes are trapped at the same special site D. Is the electron lattice coupling providing a negative-U interaction, or is it another mechanism?. In routes (III.2), (III.3) the electrons can screen the hole–hole interaction; indeed, for this purpose, free-carrier screening would suffice. Such screening is part of Sumi's (1991) suggestion that an increase in the density of excitation in the degenerate valence band could overcome the potential barrier for trapping the second hole. Several of these mechanisms can be combined. For example, two-hole localisation (III.4) may involve a hole in the valence band and a hole of a relaxed electron–hole pair [Deh], to form [Dhhe]. The results of Kanasaki *et al.* (1998) provided a possible example. The excitation spectrum for the non-linear emission of Si from the Si(111)7$\times$7 surface has a peak at the excitation energy of a back-bond electron. This suggests the possible formation of a core exciton, creating a relaxed [eh], which is combined with an energetic hole in a densely populated degenerate valence band. Sumi's theory predicts the shape of the yield–fluence relations from the Si(111)7$\times$7 surface. The mechanisms for laser-induced emissions will vary, depending on the material, but we may expect that imperfections can make localisation feasible, and that several excitations will often be needed to supply sufficient energy.

Even though there are major uncertainties as to the mechanism of photo-induced emission of atoms from semiconductor surfaces, there are good reasons for accepting that multiple excitation is possible, and that there may need to be two or more holes localised in bonding orbitals (or two or more electrons in antibonding orbitals). Likewise, it seems important to have pre-existing sites D which are different in some sense, whether from reconstruction or from pre-existing defects.

The way in which localisation occurs is far from clear. Lattice deformation is one feature, but how delocalisation is averted is less obvious. Cascade excitation assumes that the antibonding state is reached after a series of excitation–relaxation processes. Is such an excitation path realistic? It needs large cross-sections, and the emission from the surface is likely to be related to a localised defect state. If two holes are localised, what has happened to the electrons excited to produce them? Are bound biexcitons important? The emission of surface atoms following electronic excitation offers an attractive means of surface smoothing and modification in semiconductors (N Itoh *et al.* 1995); further studies of this problem will be of interest.

X H Chen *et al.* (1997) studied the modification of the Si(111)7$\times$7 surface using 193 nm laser pulses. It appears that a single photon can excite an antibonding state in the continuum, causing atomic emission with very low quantum efficiency. Similar modifications of semiconductor surfaces, including Si(111), can be induced by relatively low-energy electrons (90–2000 eV). Again, the effects are linear in intensity (K Nakayama & Weaver 1999; Han *et al.* 1999). In this case, it is not clear whether the antibonding states are reached by a single excitation, or through the dense excitations produced by the low-energy electrons. These studies raise concerns about whether observation affects the system being studied more widely than previously thought (K Nakayama & Weaver 1999). On Si(111)7$\times$7, the adatom layer vacancies increase with beam energy; on Si(100)2$\times$1, dimer vacancies are formed and ad-dimers move to steps.

9.4.2 *STM observation of photo-induced atomic emission from Si surfaces*

Inspection of the surface of Si by STM after photo-induced emission can provide evidence for removal of an atom from the surface without melting. Two STM experiments have been performed, one for the Si(111)7$\times$7 surface, and the other for the Si(100)2$\times$1 surface. The (111)7$\times$7 surface has substantial atomic relaxations, and photo-induced emissions are possible from the reconstructed surface without defects. For the (100)2$\times$1 surface, defects are involved.

Ishikawa *et al.* (1992) have shown that the surfaces of Si(111)7$\times$7 surfaces after 500 nm laser irradiation below the ablation threshold include 7$\times$7 unit cells with a larger number of vacancies than before irradiation. The vacancies generated by laser irradiation include single vacancies, divacancies, and trivacancies, but there are no signs of melting: the atoms surrounding these vacancies are unaffected. Vacancies are more likely to be produced at the central sites of the 7$\times$7 structure. Kanasaki *et al.* (1998) found a correlation between the creation of vacancies and the emission of atoms: the number of vacancies created is approximately equal to the number of neutral atoms emitted, and shows the same fluence dependence. Photoexcitation causes the emission of neutral Si. The emission yield is a highly non-linear function of laser fluence. They showed that the excitation spectrum for the emission has peaks at 2.0 and 2.6 eV, which suggests that the excitation of an electron from the adatom back-bond to the unoccupied adatom dangling bond drives the emission.

X H Chen *et al.* (1997), in their study of the effects of 193 nm laser pulses on Si(111)7$\times$7, found that the probability of emitting an Si atom neighbouring a vacancy was higher than for the original relaxed surface. K Nakayama &

Weaver (1999) found an increasing yield with energy for electron excitation in the range 90–2000 eV, suggesting a possible role for the diffusion of electrons and holes to the surface.

Convincing evidence for photo-induced elimination of atoms from the Si(100)2$\times$1 surface has been obtained by Xu *et al.* (1995, 1996). They showed that laser irradiation of the Si(100)2$\times$1 surface removes a monolayer of atoms, and reveals the atomic layer below the original surface. The new surface is not reconstructed, and is much less affected by laser light. The original top layer is not removed completely, however, but leaves debris on the surface. It is not known yet whether the lack of reconstruction of the newly exposed layer is due to the debris or due to the way the surface was revealed. Clearly the elimination of what were the outermost atoms is an electronic process, since there are no signs of heating on the new surface. The presence of defects on the original surface makes the removal process more efficient, whereas the newly-revealed surface is more stable because it has fewer defects.

Han *et al.* (1999), using STM, found that irradiation with 90–2000 eV electrons created single-layer vacancies on Si(100) and GaAs(110) surfaces. The cross-section was about 10^{-20} cm^2, similar to that for Si(111). Vacancy generation appears random on terraces, and is not associated with pre-existing defects. However, the rate of vacancy generation increases as vacancies are created, showing that pre-existing defects can be effective. The mechanism of energy localisation for the defect-free surface is not clear.

9.4.3 Photo-induced surface modification of compound semiconductors

The surfaces of compound semiconductors are modified by laser irradiation at relatively low intensities. Thus, the reconstructed structures of GaAs(100), InP(100) (Moisson & Bensussan 1983), GaP(111) (Kumazaki *et al.* 1987) and CdS(100) (Brewer *et al.* 1990) are converted to the 1$\times$1 structure by laser irradiation above a certain threshold. Laser irradiation at higher fluencies converts the 1$\times$1 structure to an amorphous form. Auger electron spectroscopy of the laser-irradiated, unreconstructed, surface of GaAs reveals a change in surface stoichiometry (J P Long *et al.* 1992): scanning electron microscopy shows that Ga islands are formed. Vivet *et al.* (1995) showed that, below the threshold fluence for surface stoichiometry conversion, there are defect-initiated atomic emissions. It appears that the threshold for altering the surface stoichiometry is close to the ablation threshold detected at submonolayer sensitivity. Vivet *et al.* interpreted the conversion as due to temperature rise. Alternatively, it seems likely that the alternation of stoichiometry is initiated from the vacancy-related emissions, as discussed in Section 9.4.1.

Pre-existing defects are important for the (110) surfaces of GaAs and InP, and presumably for other III–Vs. STM studies of GaAs (Han *et al.* 1999) show that 2.3 eV photons produce Ga and As vacancies only in the vicinity of existing vacancies. Electron irradiation produces Ga and As vacancy pairs randomly on terraces. In similar STM studies on InP (Kanasaki *et al.* 1999), in which In and P vacancies are produced on terraces, generation is enhanced by existing vacancies. Clearly, the localisation of energy at surface defects plays a role in atomic emissions from surfaces. It is still uncertain as to whether electronic excitation can lead to emission from perfect (defect-free) semiconductor surfaces.

9.4.4 Scanning probe microscopy-induced removal of atoms

Scanning probe microscopy (SPM) tips can remove atoms from surfaces, deposit atoms on surfaces, and enhance the migration of defects. SPM is potentially a powerful tool for manipulating individual atoms on surfaces. Some of the effects are mechanical, in that they rely on short-range repulsive forces between tip and surface atom. Other effects are associated with van der Waals energies; we remark that van der Waals interactions lead to tip instabilities which can interfere with manipulations (Sutton & Pethica 1988). Still further effects result from the electric field of the tip (Ramos *et al.* 1990). Atoms of Xe on a Ni surface have been successfully removed and deposited (Eigler & Schweizer 1990). There are observations of the removal of atoms from Si(111) (Lyo & Avouris 1991; Uchida *et al.* 1993) and from Ge(111) surfaces (Becker *et al.* 1987). Lengel *et al.* (1996) noted the migration of anion and cation vacancies on the GaAs(100) surface under the application of SPM tips.

One especially interesting case (Lyding *et al.* 1996a,b) is the UHV STM patterning of Si(100)2 $\times$ 1:H. It appears that the tunnelling electrons stimulate H desorption. This is a depassivation mechanism, in that the dangling bond exposed can trap carriers; repassivation is possible with ammonia. There is a strong isotope effect (Lyding *et al.* 1996a,b) seen in what may be analogous experiments in a field-effect transistor (FET). One H can be removed for every 4×10^5 electrons, whereas one D is removed for every 10^8 electrons. This would allow a decrease in the FET channel length, which would give greater advantages than increasing FET lifetime. However, it is not completely clear that the STM and FET cases are equivalent (van de Walle & Jackson 1996; see also Chapter 10).

The key question is what is the mechanism of atom removal? It is certainly accompanied by bond breaking; what is less clear is whether it is the same mechanism as desorption induced by electronic excitation. The tunnelling electrons injected are in a energy range less than 10 eV and the current is 10^{12}

electrons per second through an area of atomic dimensions. Several mechanisms have been suggested, mostly based on excitation processes induced by injected electrons. Stockbro *et al.* (1998) suggested that desorption under negative sample bias is due to vibrational excitation due to inelastic scattering of tunnelling electrons with the Si–H 5σ resonance. This explains the power-law dependence on current. It also explains the maximum desorption rate found at 7 V as corresponding to the maximum fraction of inelastically scattered electrons at the onset of the field emission regime.

Other ideas are based on field evaporation (Lyo & Avouris 1991) and on thermal atom transfer (N D Lang 1992), both assisted by lowering the activation energy through the electric field or chemical bond formation. The simplest case is the desorption of hydrogen from the Si(001)3×1 surfaces, for which there is an antibonding state for Si–H bond breaking (Shen & Avouris 1997). For electrons with energies higher than 6 eV, the probability of the hydrogen desorption does not depend on voltage and current, and is 10^{-4} per electron. For electrons with energies below 6 eV, the yield is a superlinear function of current, indicating that multiple excitation is needed to remove an atom. Because of the relatively long lifetime of the vibrational excited states of the Si–H system, it has been suggested that vibrational cascade excitation causes the desorption.

The description for desorption from the Si(001):H surface supports the view that the electronic or vibrational excitation by the tunnelling electrons causes atomic ejection. The mechanism of the removal of Si atoms from the Si(111) surface and of Ge atoms from the Ge(111) surface is still controversial. N D Lang (1992) applied the model of thermal atomic transfer, while Lyo & Avouris (1991) suggested field evaporation, the ionisation and desorption of individual atoms from the surface of a material by the application of strong electric field. The key factor here is that yield from (111) surfaces does not depend on current, as it does for H desorption from Si(001) surface. Desorption by STM tips has some features in common with that due to laser irradiation, because of the high-density excitation. In fact, the centre adatoms of the reconstructed Si(111)7×7 surfaces are removed at a rate 1.6 times higher than the corner adatoms by STM tips (Uchida *et al.* 1993), while the same ratio for laser irradiation is 2.2–3.6 (Kanasaki *et al.* 1997). The threshold voltage for desorption by STM tips 3 eV is also close to the threshold photon energy of 2.7 eV (Kanasaki *et al.* 1998). The enhancement of the motion of anion and cation vacancies when an STM tip is applied has been interpreted in terms of radiation-enhanced diffusion, caused by tunnelling of minority carriers to defect levels followed by recombination (Lengel *et al.* 1996). It would be of interest to compare sputtering by STM tips and lasers in a wider range of materials. The advantage of laser beams is that they can cause desorption near specific types of defect sites over a wide area, whereas the STM technique can eliminate specific sites in a defined region.

Summary

Surface defects, including adsorbates and near-surface defects, are important. The precise surface reconstruction matters, since specific sites may be more readily modified than others. The surface topography is very important, not least because of charges associated with steps. Exoelectrons will lead to surface charging, which modifies the evolution of the defect populations.

Halide crystals with closed-shell cations (NaCl, CaF_2) Atom or ion removal can take several forms, depending on factors such as core versus valence excitation. Removal by valence excitation usually involves the self-trapping of excitons followed by the energy transfer for atomic emission, while core excitation can involve the Feibelman–Knotek mechanism. Atom removal by valence excitation is analogous to defect formation in the bulk: in effect, an interstitial halogen atom is ejected from the surface. A role for the X_2^- species seems likely in some cases. Other significant cases involve defect-related excitation, and the excitation of more than one exciton.

Amorphous halides and halide glasses Al/Y alkaline-earth fluoride glasses are discussesd in Chapter 5. No intrinsic defects seem to result from 193 nm excitation, although there is extrinsic defect formation associated with O and perhaps P. Atom removal by valence excitation for AgCl is similar to that for alkali halides, although the Ag cation can change charge state. There appear to be no data for $PbCl_2$.

Oxide crystals with closed shell cations (MgO, ZnO, Al_2O_3, usually SiO_2) The Feibelman–Knotek mechanism occurs, but the excitonic mechanism does not appear to be effective. Sputtering may involve defect-related excitation and excitation of more than one exciton (see Chapter 11).

Oxide crystals with cations which can have several charge states ($Fe_{1-x}O$ and other transition metal oxides) STH, STE can both exist. No clear examples of STX are known, but may be involved in radiative recombination afterr a charge-transfer excitation. Non-stoichiometry effects can be important.

Semiconductor crystals (III–V, Group IV) with relatively wide bands and small gaps The excitonic mechanism is not effective. Defect-related excitation and excitation of more than one exciton play roles in laser-induced atomic emissions and surface reactions. It appears that excitation of specific back bonds causes atomic emissions although the yield is very small.

Chapter Ten

Interface reactions induced by electronic excitation

Interfaces control a variety of technological applications of materials, and their characterisation, synthesis, and modification determine how well these interfaces perform. Interfaces have several roles of direct interest to us. First, interfaces provide sites where electronic excitation energy can be localised. This makes it possible to modify interfaces without affecting bulk properties. Secondly, at an interface atoms from the materials on both sides of the interface meet. Bond rearrangements at the interface can affect microscopic structures. Thirdly, interfaces include a variety of defects which do not exist in the bulk. The structures of these defects are also susceptible to electronic excitation. Some of these defects are associated with mismatch in lattice parameter between the media on either side of the interface. Fourthly, a wealth of new interfaces is being created by methods such as ion-beam sputtering. These include tens or hundreds of alternating layers, for example of Si/SiO_2, Al_2O_3/SiO_2, Ta_2O_5/SiO_2, TiO_2/SiO_2, HfO_2/SiO_2 or MgF_2/SiO_2 (see, e.g., Howe (1999), or Reichling *et al.* (1998a)). Reichling *et al.* suggest that most induced laser damage occurs at the substrate/film boundary. These multilayers can be crystalline or amorphous, and can be constructed so that layer thicknesses match a particular wavelength (typically, $\lambda/4$ layers with a $\lambda/2$ cap might be used). In some cases, self-organisation has been seen. Since these systems are used in laser applications, their performance under excitation is important. Finally, we note that interfaces can evolve in interesting ways when the interfacial free energy is lowered by mixing.

The possible modifications of an interface by excitation are diverse. Some are local modifications, such as changes in defect populations and changes in bonding patterns for specific atoms. Local changes may stimulate specific rebondings or diffusive jumps at the interface. Another consequence of interfacial changes can be defect passivation, a reduction in the recombination velocity or electrical activity of the interface, which is welcome in device technology.

Other modifications extend over large regions of the interface, perhaps causing motion of the interface and the growth of one phase over another. Charge redistribution near the interface can enhance adhesion (Section 10.4). Typical examples of excitation-induced interface motion include movement of the crystalline/amorphous interface, and the oxidation process.

In this chapter, we describe various cases in which interfaces are modified by excitation. We shall concentrate on two types of interfaces. One is the synthesised interface, such as one where one material has been deposited onto the surface of another material, or where the surface layers have been modified by irradiation, ion implantation, or chemical reactions. The other type of interface is harder to characterise, namely that developed by some reaction, such as oxidation, or a catalytic reaction. The existing information on modification of interfaces by electronic excitation is less complete than that for our other chapters, but rapid progress is expected. We shall say little here about the nucleation of new phases within an existing phase, even though an interface and its energy are involved. Allied processes are bond scission and cross-linking in polymers, or the creation of diamond-like carbons by N-bombardment of vacuum oils.

10.1 Atomic and electronic structures of the interfaces

Interfaces are formed by a variety of methods: epitaxial growth, chemical reactions, ion implantation at high densities, the deposition of atoms on surfaces, and so on. Both sides of the interface may be aligned crystallographically, as for epitaxial growth on a crystalline substrate; other interfaces may be crystalline but misaligned, or between crystalline and amorphous solids. Some interfaces are sharp and abrupt on an atomic scale. Stacking faults are an example in which the sequence of atomic planes is modified. In other cases, the boundary cannot be defined accurately because flexibility in the structure or composition changes through interdiffusion lead to an intermediate region which extends over many layers. Localised states at interfaces have long been known to affect electronic properties, for example by the pinning of the Fermi level (J Bardeen 1947). Interface electronic structures are best defined for epitaxially systems, especially in semiconductor superlattice structures. Space charge built up at an interface can profoundly affect the electronic properties (Section 9.1).

10.1.1 Interface atomic structures

We begin with a very simple description of some of the main classes of interface. Real interfaces show many of the features we discussed for surfaces in

Chapter 9. Real interfaces can be rough, rather than flat, and the steps and kinks may be important in defect processes. The interfaces may reconstruct. They may have electrical dipole moments associated with them, so there is a term in the energy when an electron is transferred from one side to the other. Image interactions may favour defects in particular charge states near to an interface between media of differing dielectric constant. There will often be impurity segregation to linear or planar defects. Associated with each planar defect, like a grain boundary, there may be a cloud of defects or impurities or simply ions in unusual charge states, and the Debye–Hückel screening length is often a good guide to the scale of this atmosphere. Around any mobile interface, such as a dislocation array, there may be debris generated during dislocation motion. These atmospheres and defect debris are able to absorb subbandgap light, with consequences which we discuss in Chapter 11. Interfaces, unlike surfaces, can separate an amorphous region from a crystalline one, a situation which is made complicated by the varied ways a solid can be amorphous.

Grain boundaries and interfaces within a single material

Suppose one takes a perfect crystal, like NaCl or MgO, and cuts it on a plane (*hkl*). A twist grain boundary can be made (in principle) by rotating one half about the normal to the plane by angle θ and rejoining the two halves. To make a tilt boundary, material would be shaved away (or added) to change one face to angle $+\theta/2$ to the original and the other to $-\theta/2$ before rejoining. The tilt and twist grain boundaries are special cases of more general planar interfaces, both within single crystals and between one crystal and another.

Low-angle tilt grain boundaries can be considered as an array of edge dislocations. As θ increases, several new features become more obvious. First, the Madelung potential will vary substantially from site to site within the repeat units. This has interesting consequences electronically. As Duffy & Stoneham (1983) have shown, it may be possible to have a negative-U situation, with charge disproportionation. Certainly, it is easy to understand that grain boundaries can be relatively effective in electrical conduction. Secondly, the larger-angle grain boundaries have sizeable channels through which rapid diffusion might be expected.

Twist boundaries introduce a new feature. If we choose the (100) plane of NaCl, and rotate one half about the normal, when we try to rejoin the two halves, we will find some cations in one half are directly opposite cations in the other half. Twist boundaries prepared as described will always have close encounters of like charges and high energies. How is this resolved? The simple solution (Duffy & Tasker 1986) is that there will be an intermediate plane, different in structure, between the two rotated components (note we are describing a thought experiment; the way grain boundaries are formed is another matter). In this intermediate plane, there are ions missing from sites which would be occupied in a perfect crystal plane. These should not be called vacancies, since they are a natural part of the lowest-energy structure (if they were vacancies, they would have negative formation energies).

The grain boundary structures will inevitably have electronic consequences. One simple example is that, in electron microscopy, the grain boundary appears to have an altered inner potential relative to the bulk (Rühle & Sass 1984). A significant part comes from the change in Madelung potential across the grain boundary region (D M Duffy, 1985, private communication). Grain boundary motion can be enhanced by excitation, as noted in Chapter 7 (Kizuka *et al.* 1998; Kizuka 1999).

Interfaces between two different crystalline materials

The surfaces of two different crystals will usually not match exactly. If the mismatch is small, and if one of the crystals is thin, there could be epitaxy, with the limited mismatch simply taken up by distortion without defects. If the mismatch or crystal thickness is greater, the elastic strain energy will be more than sufficient to create defects, and misfit dislocations are a common consequence. Many of the structures can be understood in terms of pairwise interactions across the interface: structures should favour strong attractive bonds and eliminate strong repulsions. An example of a small mismatch is the CaF_2/Si(111) interface, with the Si(111) plane facing the Ca-terminated CaF_2(111) plane.

The structure also depends on how the system has been prepared. There will be a difference between an oxide deposited from a vapour phase and an oxide of the same composition formed by an oxygenous species diffusing through the outer oxide to react at the substrate/oxide interface. When a film is deposited, there are several possible mechanisms of growth. There may be layer-by-layer growth (*Frank–van der Merwe,* as for many semiconductors on semiconductors, or rare gases on graphite), growth as islands (*Volmer–Weber,* as for metals on ionic substrates), or growth of a layer followed by island growth (*Stranski–Krastanov,* as for metal on semiconductor or gas on metal systems). The form observed depends not only on the species, but also on the relative rates of processes such as diffusion and deposition (Venables *et al.* 1984).

The sharpness of the interface is a further issue. ‘Sharpness’ is not always well defined. Just because the two media are different, there may be effects of the Madelung potential, or from the differences in dielectric constants through polarisation (image charge) effects, and these can cause shifts in core levels even when there is no modification of structure. Rumpling or relaxation at the interface may give it a dipole moment, which can cause band-bending, or influence space charge. But, normally, a sharp interface is one at which the composition changes abruptly, without any intervening atomic layer which might be regarded as some sort of 'intermediate layer' or 'transition layer'. Sometimes, there can be a region (possibly simply space charge) in which there is an excess of mobile defects. This appears to be the reason why alumina powder enhances the ionic conductivity of LiI (Wade *et al.* 1979). For the annealed, mutually soluble system KCl/KBr, an extended shift of exciton peaks has been observed (Ejiri *et al.* 1990), whereas this shift is not seen for the annealed, mutually insoluble system RbCl/KBr. Sometimes an intermediate layer may result from chemical reactions between the two media. Interfacial complexes may form (Phillips 1978). Intermediate layers are also a means of joining two media with large mismatch. Suppose there are two media with an intermediate layer of thickness L. Suppose further that the mismatch between the two media is measured by a difference in length per repeat unit of X, such that the relevant component of strain is X/L. Then per unit area of interface, we have these terms in the energy: (*a*) interfacial energies for the interfaces which bound the intermediate layer; these are essentially independent of L; (*b*) a term proportional to L because the intermediate layer is a different structure or different in composition; (*c*) a strain energy term of the form $Lc(X/L)^2/2$, with c an elastic constant. The dependence on L has the approximate form $A + BL + C/L$, which can have an optimum thickness (in many cases, $L=0$ is lower in energy). Elastic energies are especially important in superlattice systems.

For our purposes, the key questions concern energy deposition near the interface, and processes which occur more readily near the interface. Almost certainly, the misfit dislocations will allow optical absorption from subbandgap light. Any segregated defects and space charge will probably allow different defect processes, especially those associated with charge trapping. If the dielectric constants of the two crystalline media differ, self-trapping will be enhanced near to the interface in the medium of lower dielectric constant.

10.1.2 Interfaces between crystalline and non-crystalline solids

Interfaces for a single chemical composition The interface between a crystalline phase and its melt is an example of a crystalline/amorphous interface (see Fig. 10.1). However, it may not be the only example, even for an element. Crystalline Si can be amorphised by radiation, for instance, and is a semiconductor, whereas liquid Si is a metal. The way that a-Si is formed during irradiation is not completely clear, but probably involves the generation of point defects

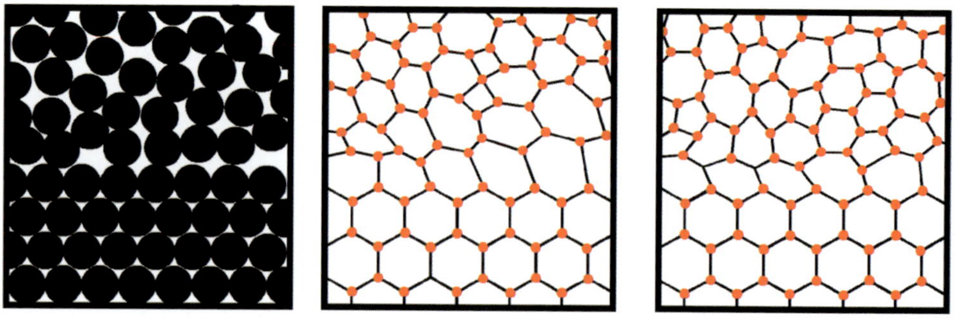

Fig. 10.1. Idealised models of crystalline/amorphous interfaces, describing hard-sphere disorder (left panel), coordination disorder (centre panel) and topological disorder (right panel) (see Fig. 1.3).

which change the topology (producing five- and seven-membered rings, perhaps) which act as nuclei for amorphisation. It is even less clear what happens during the amorphisation of oxides, since there may be compositional fluctuations as well as topological changes.

The issue of mismatch is more complicated in these cases. For Si, for instance, the interatomic spacings in a-Si and c-Si may be essentially the same, but the density can differ. If irradiation amorphises a region inside a crystal of Si, then there will generally be a density mismatch, and there will be stresses across the interface. These stresses can relax in several ways, e.g. by creating Si interstitials which move away (changing the local atom number density) and by reconstructions to minimise the strain energy. The reconstructions and relaxations will normally alter the topology of any continuous random network. The rate at which such reconstructions occurs has been studied very widely for silica-based glasses. The converse process, the crystallisation of glassy ceramics, also may involve nucleation of crystalline regions and their growth. Again, there will be interfacial stresses, which will influence the growth of the crystalline regions and which will be partly relaxed by topological change in the amorphous regions. There are indications that topological changes can be speeded by excitation. An example (Tinivella & Stoneham 1980; Stoneham 1994) is the 'radiation stirring' of the radwaste glasses for the vitrification of radioactive waste. Complex oxide glasses, such as those based on silicates, have 'ideal' densities which depend in a systematic way on their composition. There is a fairly well-defined quasisteady state, even though the glass is arguably not in equilibrium. In fact, this 'ideal' density corresponds roughly to oxygen close-packing (Stevels 1944), implying some medium-range order. Radwaste glasses may contain nearly 100 elements. How does their density change once the glass leaves the chemical plant in which it was prepared? How does this change with composition? After the initial solidification and short-term cooling, it appears the glasses simply expand or contract towards their 'ideal' density. The rate is determined partly by their viscosity and partly by radiation-associated mechanisms.

Structures for the Si/oxide interface

The oxide on silicon performs several critical roles in microelectronics. It passivates the surface, eliminating many of the trap states. It is an insulator, separating conducting or semiconducting regions. It is a dielectric, a component of a capacitor. For these reasons, the Si/SiO_2 interface has been studied intensively. Some of the early studies have lost their relevance, since there have been changes in demands for cleanliness and purity, and because the drive to miniaturisation has led to different growth conditions. Oxides of a few nanometres thickness are becoming standard, rather than the tens of nanometres a decade ago. It is no surprise that the structure depends on the crystallographic surface, such as Si(100) or Si(111). Nor is it surprising that the structure depends on the manner of preparation and subsequent thermal treatments. Thus the photoemission extended fine structure technique (Sieger *et al.* 1996) suggests that Si^{2+} and Si^{3+} are distributed only at the interface, implying only a monolayer of intermediate oxide, whereas Grunthaner *et al.* (1986) conclude that Si^{3+} is distributed throughout the thin SiO_2 layer. Such differences are attributed to differences in growth procedures and to the different extents to which H is involved. Yet certain basic questions remain unresolved, even for chemically-pure oxide (see Mott *et al.* (1989), Sofield & Stoneham (1995) and Stoneham & Sofield (1998) for relevant reviews).

First, is the oxide the dioxide SiO_2 everywhere? There is evidence from oxidation using ^{18}O, from the pulsed laser atom probe, and from other experiments (Stoneham *et al.* 1987; Mott *et al.* 1989; Sofield & Stoneham 1995) that there is often an oxygen-deficient region close to the Si/oxide interface. This leads to the 'reactive layer model', in which interstitial molecular oxygen is able to react with the oxide close to its interface with the Si. There is evidence for Si=Si bonds associated with P_b centres (X Zhang *et al.* 1994), and there may be other defects contributing to non-stoichiometry.

Secondly, has the oxide the 'same' structure throughout its thickness? Many workers presume that a continuous random network is an adequate model, and that each O has two Si neighbours, each Si four O neighbours. This is probably largely true. But, even when non-stoichiomatry and defects are ignored (and it is likely that there are defects which can take several charge states, as shown by the observed charge trapping showing $1/f$ noise (Kirton & Uren 1989), there is evidence for topology changes (e.g. ring size statistics) under stress. Evidence for several structures (or structural elements) emerges from infrared studies.

The variation in structure through the oxide is implied by the asymmetry in defect production between the oxide adjacent to the Si and that near the outer surface during long-term degradation. In the long-term degradation, defects are produced by some process involving the energetic electrons injected into the oxide. The differences between the oxide at the Si side and that at the gas side are not surprising, since oxidation occurs at or close to the Si/oxide interface. This means that the outermost oxide was that formed first of all, with no constraints from oxide above it. The oxide close to the Si/oxide interface is the most recently formed, and it is created in what must be essentially a quasisteady non-equilibrium state. This has an obvious implication: there is no reason to believe that the interface is the *equilibrium* interface of lowest free energy. It may well be a structure with a local minimum of energy. The oxide close to the Si/oxide interface is the latest oxide to form in thermal oxidation, and it corresponds to a quasisteady non-equilibrium state.

Thirdly, is the oxide homogeneous from one place to another across the interface? The answer is no, of course; there are steps, there are P_b centres (Si dangling bonds pointing into the oxide) and surely other intrinsic defects. There may even be crystalline regions as well as amorphous regions. If oxidation occurs only close to the Si/oxide interface, any defects or crystalline regions will stay at about the same distance from the oxide/gas interface as oxidation proceeds. Both the oxide/gas and Si/oxide interfaces are rough in ways which can be influenced by the growth process (see Sofield & Stoneham (1995)). This roughness contains information on the growth process.

Almost all experimental data are consistent with an interface between crystalline Si and amorphous oxide. The volume per Si in the oxide is about twice that per Si in the silicon, so there is a major mismatch. Models by early workers (like Pantelides & Long (1978)) showed that it does not need large strain energies to force a match of an amorphous oxide to Si (of course, this does not show that such a structure occurs). Those experiments which hint at an intermediate layer of oxide with some sort of different properties (noted above) are mainly consistent with an interface between c-Si and some amorphous oxide. Only in special cases is there evidence for crystalline oxide over some (not necessarily all) of the interface.

For the Si/oxide interface, we shall be interested in two types of processes influenced by excitation. One is the oxidation process itself (Section 10.5). The other is electrical degradation due to charge injection from the Si or from an electrode put on the oxide/gas interface (Section 6.3).

10.1.3 Defects at interfaces

It is no surprise that defects at interfaces have similarities with bulk and surface defects. The standard intrinsic bulk defects are to be expected: vacancies, interstitials, perhaps antisite defects. The standard intrinsic surface defects, like kinks on steps, will have their analogues. Impurities will have states somewhat altered, e.g., effective-mass like donors and acceptors will occur, but with altered properties because the electron or hole will have a different effective mass and experience a different dielectric constant in the two media.

The number of genuinely new defects is relatively small. The most important case is the P_b centre at the Si/SiO_2 interface. At the $Si(111)/SiO_2$ interface, the P_b centre is best thought of as an Si dangling bond pointing into the oxide. Similar centres occur at the Si(100) interface, with more complex and less well-understood geometries. Analysis (X Zhang *et al.* 1994) of the spin resonance data of Stesmans & Vanheusen (1991) and Stathis *et al.* (1991, 1992) shows that there must be a $Si=Si$ bond in the oxide directly above the dangling bond. The P_b centre is readily saturated by H, and this has important technological consequences. The density of P_b centres can be quite high, if those with H as well as those without are included. This number correlates with interfacial stress, suggesting that these centres are one of the ways of dealing with mismatch (Stesmans 1993).

10.2 Energy and charge deposition near interfaces: The modification of interfaces

10.2.1 Energy deposition by elastic encounters

Energy can be deposited at interfaces either as the kinetic energy of atoms or as electronic excitation energy. When ion beam methods are used, the range of the beam can be optimised to maximise collisional or electronic energy losses at the interface. In certain cases, such as channelled beams, or focused collision sequences, the interface will disrupt the highly directional atomic motions, and lead to energy deposition at the interface.

Collision cascades which develop near interfaces cause so-called ballistic mixing (Sigmund 1972). Component atoms of both sides of the interface are mixed by the sequence of replacement collisions. The cascades also lead to defect formation, excited electrons, and heating near to the interface. All of these will tend to enhance diffusion processes. Another effect is seen from the observation (Wang *et al.* 1982) that the mixing rate is much higher for the

Cu/Au interface than for the Cu/W interface, even though the atomic masses are nearly the same. The difference is ascribed to the thermodynamic driving force for the Cu/Au interface, such that mixing lowers the free energy. This is one of several cases in which the thermodynamic driving force enhances mixing at interfaces; examples include insulators and semiconductors.

10.2.2 Energy deposition by electronic excitation

Electronic excitation can occur at interfaces in several ways. There may be direct excitation at the interface, perhaps at a defect site, by interaction with incident photons or electrons, or with secondary excitations (such as excitons, or electrons and holes) which have migrated to the interface. Photon excitation is of particular interest for interface modification, since it can be possible to arrange that photons are absorbed primarily at the interfaces. One way is to use synchrotron radiation which matches the core ionisation energy of the atoms at the interface. A second way is to use photons with energies less than the bandgap energy of the material through which the light passes on its way to the interface. This is important in laser ablation, which we discuss in Chapter 11.

The photoexcitation of an interface transfers charge between the two materials. Mihaychuk *et al.* (1999) showed that 1.55 eV and 1.75 eV photons passing from Si to SiO_2 transfer electrons across Si/SiO_2 interfaces, and induce a transient electrical field. They found a residual electric field from the trapping of some of the electrons at oxide defects.

Ruby (Cr-doped alumina) shows unusual behaviour after laser pumping in a modest electric field. The splitting of a sharp optical R line shows that there are domains of strong electric field (350 kV/cm) along the c axis, different regions having different signs of field (Basun *et al.* 1984a,b,c, 1985, 1986, 1987a,b,c, 1989; Kaplyanskii 1991 and 1998 private communication). The behaviour is related in some way to the axial polar symmetry of defects in alumina; phenomenologically, it is described in terms of a photoelectric instability. The process needs ions of at least two charge states (Cr^{3+} dominates, but some 2+ or 4+ ions are needed) since the charge probably moves by hopping between an excited state of the 3+ ion and an ion in a different charge state. The precise shapes and sizes of the domains are not clear, but one obvious idea is that charge has separated to grain boundaries. It is possible (Stoneham 1994) that the field amplification process is related to the radiation induced electrical degradation of alumina, which is significantly enhanced by a small applied field (Chapter 6).

Given that large electric fields can build up at surfaces during laser ablation (up to about 10^8 V/cm, see Chapter 11), it is probable that there is substantial charge transfer during collisional mixing of insulators.

10.3 Photo-induced processes at interfaces

Quantum-confined luminescence is a major research activity, with considerable technological potential. Materials modification at interfaces influences its effectiveness in several ways. First, modification can be used to tailor and tune the optical gap. Secondly, the efficiency of luminescence in nanoscale materials is influenced by the interface. There is a continuing debate about whether the luminescence of porous Si is due to recombination in the nanoscale Si or at an interface. As for amorphous materials, photoexcitation at interfaces is probably localised, perhaps because of the presence of misfit defects. It is natural to expect photo-induced processes at the interfaces to be characteristics of the transition layer. Broad-band luminescence is likely, both because of possible large electron–phonon coupling, and because of structural inhomogeneity within transition layers adjacent to an interface. Near the interface, therefore, photoexcitation may induce bond rearrangements, causing defect formation, changes of interface structures, and alteration of the thickness of transition layers. In this section we deal with photo-induced processes at existing interfaces. In subsequent sections, we shall discuss phenomena involving the formation and evolution of interfaces, like mixing and adhesion.

10.3.1 Luminescence at interfaces

Quantum confinement

Quantum-confined luminescence in the superstructures of GaAs-related compounds is well established. In a carefully-prepared Si/SiO_2 superlattice, the luminescence bands with a peak energy in the range 1.7–2.3 eV show the predicted dependence on thickness (Lockwood *et al.* 1996).

Attempts have been made to modify the interfaces of semiconductor superlattice structures. Early efforts to improve the interface properties with laser beams failed (e.g. Ralston *et al.* 1987). Irradiation with excimer laser beams from the front surface essentially heated the whole sample. For $Al_xGa_{1-x}As$/GaAs, the result was the formation of an intermixed alloy. Irradiation with heavy ion beams, followed by rapid thermal annealing, proved more sucessful. An increase in the exciton energy in the confined layer was obtained by irradiation of GaAs/AlGaAs quantum wells with 35 keV As beams (Elman *et al.* 1989). There appears to be a limit to the thickness change due to the introduction of radiation damage. Irradiation with the proton beams was more successful. An energy shift of 200 meV was attained by irradiation of GaAs/AlGaAs quantum wells with 40 keV protons to a dose of 4×10^{16} cm^{-2}. The contribution of electronic excitation in this reaction is not known yet.

Luminescence of porous Si

The origin of the luminescence emitted at energies in the range 1.4–3 eV is still controversial. The three main interpretations involve quantum confinement, novel Si environments, and oxide layers, each of which explains some of the data. The quantum-confinement model has problems because of the lack of correlation between particle size and the luminescence energy (Prokes & Glembocki 1994). Koch *et al.* (1993) and Qin & Jia (1993) suggested that electron–hole pairs, generated in quantum-confined Si nanoparticles, move to SiO_x layers to cause luminescence. Several experimental results favour recombination in the SiO_x layer as a model of for the 1.7 eV luminescence band. First, oxidation enhances the luminescence band (Qin *et al.* 1996); secondly, the luminescence is reduced by the loss of hydrogen from the oxide layer. The hydrogen probably passivates the Si dangling bonds (Cooke *et al.* 1996). Caldas *et al.* (1997) have noted that small H-terminated Si clusters have an instability in the excited state, leading to energy localisation.

Interfaces of Si precipitated in a-SiO_2

Another way to fabricate Si nanostructures involves the implantation of Si into a-SiO_2, followed by annealing above 1100 K. This produces much the same luminescence band as that observed in porous Si (Shimizu-Iwayama *et al.* 1993; H Z Song & Bao 1997). The luminescence is seen only after the elimination of defects in SiO_2 by annealing. Growth of the Si cluster does not influence the peak energy of the band. The luminescence appears to arise from defects in the highly disordered Si/SiO_2 interface. No luminescence is seen associated with the well-prepared interfaces of the Si/SiO_2 superlattice: probably ideal Si/SiO_2 interfaces will not luminesce.

10.3.2 Modification of interface structures

Changes at an interface can affect the material on both sides. The best-known examples are for MOS (metal-oxide–semiconductor) devices, for which the effects of ionising radiation are reviewed by Dresendorfer (1989). Photo-excitation causes degradation of these devices. Excitation with photons above 9 eV generates electron–hole pairs in the oxide, SiO_2. Electrons migrate to the Si, leaving less mobile holes in the oxide, so the oxide has a positive charge. The electrical properties are altered in part by band-bending associated with this charge.

The excitation of MOS devices with photons between 4.9 and 9 eV has a quite different effect. Electrons excited from the Si populate interface traps, making the oxide negatively charged with respect to Si. There is no degradation associated with photons of less than 4.9 eV (Ling 1994), i.e. 4.9 eV is the threshold energy to excite an electron from Si to an interface trap. The formation of negatively-charged traps has been verified by several methods, including current–voltage characteristics (Kasabov *et al.* 1992) and gate-to-drain capacitance measurements (Ling 1994). The traps influence the bulk properties of Si, such as carrier lifetimes. The lifetime depends on the irradiation with 4.9 eV photons in a very complicated way (e.g. Cheng & Ling 1998 and references therein) and is considered to be controlled by band-bending associated with the change in Fermi level.

One important defect found at the (111) Si/SiO_2 interface is the P_b centre (Section 10.1.3), an sp^3-dangling bond on a three-fold coordinated Si, the bond being directed into the oxide (Poindexter & Caplan 1983). P_b centres can be passivated by treating an oxidised Si surface in a hydrogen atmosphere at about 200 °C (Brower 1988). Apparently H_2 diffuses through the SiO_2 layer and reacts with the P_b centres as

$$P_b + H_2 \rightarrow P_bH + H.$$

P_b centres are known to be generated at room temperature during irradiation (Lenahan & Dressendorfer 1984) and are also generated by hot electrons (see Chapter 2) at the interface (Brower & Meyers 1990; Griscom 1992b). It has been suggested that the atomic hydrogen generated by electronic excitation reacts with passivated P_b centres:

$$P_bH + H \rightarrow P_b + H_2.$$

However, the opposite reaction

$$P_b + H \rightarrow P_bH,$$

induced by hot electrons, is five times as efficient as the depassivation reaction, so only 20% of the P_bH precursors can be depassivated.

As noted in Chapter 9, there is a strong isotope effect (Lyding *et al.* 1996a,b). It is far harder to remove D than H from the Si/oxide interface. There are two models for this behaviour. Lyding *et al.* suggested an analogy with the isotope effect seen in the STM-stimulated removal of H from the Si(111) surface, basing their model on vibrational excitation in the Si–H stretch mode. Implicit here is the assumption of a long vibrational lifetime (slow cooling, in the sense of Chapter 3). On the other hand, van de Walle & Jackson (1996) observed that the Si–H bending modes would have the right properties, in that the Si–D bend will be rapidly de-excited because its frequency is very close to the bulk transverse optic (TO) frequency at the X point of the Brillouin zone. The Si–H has no such close match to a bulk frequency, and the bend modes could be excited so as to form an intermediate metastable state with the H in a bond centre. In both pictures, there is an implicit assumption that there is a process which excites a particular (stretch or bend) mode. Whether this is true or not depends on the precise electron transitions involved. For example, if the H is bound to a P_b centre, and if the electron is transferred from below the Si−H bond to the −Si=Si− bond which lies above it (see X Zhang *et al.* (1994)), based on the spin resonance work of Stesmans & Vanheusen (1991) and Stathis *et al.* (1991, 1992)), then one might expect substantial excitation of the stretch mode. However, the two mechanisms can coexist.

The Si/CaF$_2$ interface

Photochemical reactions cause difficulties in the photoemission spectroscopy of Si/CaF_2. The photoemission spectra show alternating changes of character during the irradiation, and the analysis of the spectra must allow for these changes. No further changes in the spectra are seen after the evolution of the metallic Ca spectrum on the CaF_2 surface, so the photo-induced reactions at the interface, if any, should have a higher efficiency than that the emission of atoms from the CaF_2 surface. Three interfaces states have been found by Bouzidi *et al.* (1994), who ascribed them to Ca atoms at three different sites, following calculations by Salehpour *et al.* (1991). A major cause of the alternation is the emission of the F atoms from the surface (Section 9.2.3).

10.4 Mixing and movement of ions

10.4.1 Mixing by elastic encounters

Ballistic mixing is observed for many metal/metal and metal/semiconductor interfaces. The recoils from binary collisions cause relocation of atoms on both sides of the interface. The theory of ballistic mixing (Sigmund 1972) explains the mixing rate for a wide variety of systems.

Nevertheless, there are systems which show mixing rates much higher than predicted from ballistic mixing theory. These cases prove to be those interfaces for which the enthalpy is lowered by mixing. Johnson *et al.* (1985) have measured the correlation between the enthalpy change on mixing and the mixing rate under ion irradiation; they obtained the effective kinetic energy of constituent atoms as 1–2 eV. Molecular dynamics simulations for NiAl (Zhu & Averback 1995) suggest that mixing occurs in hot zones of the collision cascade, consistent with experiment. This work points out that, whereas only perhaps 25 atoms are knocked on by a 10 keV recoil atom, the number of atoms participating in the mixing is as large as 700. Given these results, it is generally believed that the cause of mixing is the thermal spike (see Chapter 11), rather than a knock-on process. The process of mixing ends 10^{-11} s after the start of a collision cascade (cf. the timescales noted in Chapter 2). Bond rearrangement processes should be largely complete within this time, as should diffusion processes involving transient heating or defects.

In our terminology, ion-induced mixing is caused by bond rearrangement in the hot zone of a collision cascade, where the average atomic kinetic energies are 1–2 eV. This corresponds to a temperature of order of 10^4 K, much higher than the melting temperature, so material in the hot zone may be considered to have melted homogeneously without needing local nucleation of a liquid phase. This description may be too simple for cases like Si, where the solid is a semiconductor and the liquid a metal. We should also be concerned that the molten region may have lost electrons transiently, and so have a net positive charge.

The thermodynamically-driven mixing of insulators with metals can be induced by low-energy ion irradiation. Thus SiO_2 can be mixed with Ti, Zr, Nb, Al, and Te. An enthalpy rule is often used to interpret such mixing. If the sum of the differences in the enthalpies for each reaction is negative, the enthalpy rule predicts mixing; otherwise no mixing should occur (Farlow *et al.* 1985). The rule implies possible bond rearrangement, but not whether long-range mixing is to be expected. Mixing is predicted for all reactions noted above, but there are many exceptions to the enthalpy rule, and in both directions. For example, Al/SiO_2 does not mix, although the enthalpy rule suggests mixing. Probably the issue is the difficulty in diffusing oxygen through the Al_2O_3 layer formed by mixing. On the other hand, mixing does happen for Si/ZrO_2, even though it is unfavourable thermodynamically. This is not understood.

10.4.2 Mixing by electronic excitation

Mixing caused by dense electronic excitation has been observed for the Si/Fe interface. Bauer *et al.* (1997) used Mössbauer spectroscopy for ^{57}Fe to make microscopic observations of mixing in Si/Fe multilayers. In the fluence range where tracks did not overlap, mixing creates either a magnetic amorphous SiFe or a paramagnetic amorphous Si with dilute Fe. Which result is obtained depends on the stopping power. Amorphous SiFe occurs when incident particles have stopping powers in Fe above 20 keV/nm, which corresponds to a stopping power of 3 keV/nm in Si. Amorphous Si with Fe occurs for incident particles with stopping powers between 8 and 37 keV/nm, which would correspond to stopping powers in Si of 8 keV/nm. The authors suggest that the mixing is due to heating of the multilayer. An alternative view, since heavy ions of the stopping powers described above do not melt Si or Fe, is that the mixing is due to the creation of high-density electron–hole pairs at the interface. Mixing of interfaces involving insulators associated with electronic excitation occurs only for energetic heavy ions above a certain threshold stopping power (Chapter 11).

10.5 Radiation-enhanced adhesion

10.5.1 Phenomenology

It is found that metal–insulator adhesion can be enhanced by radiation (optical, especially ultraviolet or gamma irradiation; electron beams; glow discharge; ion beams). There are several possible mechanisms. The first is ion-beam mixing, through elastic collisions near the interface. Experiment rules out this mechanism in many cases. One can select the ion (H,He,C,O,Si,Ni, but especially F, Cl) and its energy so as to control the relative proportions of nuclear and ionisation energy deposition, which can give some extra information. Secondly, ionisation of certain insulators, like polymers, can be changed by radiation in ways which give a chemical contribution to adhesion. One obvious idea is that free radicals might be created, and then interact more strongly with the metal. In many polymer–metal systems, the work of adhesion is dominated by the many weaker interactions between polymer strands which determine the work necessary to pull a strand from the entwined molecules. These weaker interactions will be enhanced in cases where cross-linking occurs. However, in many systems, none of these explanations is viable. Nor is any explanation based simply on van der Waals interactions. A simpler and more general explanation appears to hold. The extra adhesion is Coulombic, and arises because of the charged species in the insulators and the way in which they polarise the metal. This is the so-called image interaction picture (Stoneham & Tasker 1985a). This picture is a natural generalisation of Stoneham & Tasker's explanation of the systematics of metal–oxide adhesion, and is in line with much of the current experimental data.

The quality of the data is not always good. Experiments tend to be simple and qualitative (notably the so-called Scotch tape and cotton bud tests, but also peel tests, pull tests, topple tests, and scratch tests). The effects are sometimes small, and there are some justified concerns about the cleanliness of the surface oxide (which may lead to the apparent enhancement of metal–metal adhesion). There may be separate radiation issues, like changes in the brittleness of the insulator. Nevertheless, calculations (Stoneham & Tasker 1985a; Duffy *et al.* 1995b) predict effects which are of the right order of magnitude. Since the charges in an insulator can be manipulated, perhaps optically, or possibly by ionising radiation, the mechanism offers a route to controlled adhesion. For example, since the vacuum is unpolarisable, defects near a free insulator surface will have the lowest energy when in low nominal charge states or neutral. In the presence of a metal, however, high nominal charge states are favoured because of the high metallic polarisability. Switching charge states will change the interfacial energy. Doping with ions which can occur in a variety of charge states can increase the size of the effect. It is no surprise that heat treatments influence the enhancement.

We shall not discuss cases in which metal and oxide may react. Some effects have been reported, but adhesion is usually reasonably strong even without enhancement. In such cases, it is important in any description to know which surface is limiting. A metal M in contact with an oxide AO may form some new phase (A/M)O, and failure (if it occurs at an interface) is governed by the weakest of the interfaces between the new phase and either M or AO. An analogous phenomenon was reported by Clark *et al.* (1984), who found ion beams were beneficial as a low-temperature processing method for Nb, V, Ti, Hf, and Zr on silica, apparently because of interfacial silicide formation, e.g. $M_x + SiO_2 \rightarrow M_ySi + M_{x-y}O_2$. For Nb ($y=3$) the peel test gave 2 g/cm^2 instead of 0.5 g/cm^2.

10.5.2 Summary of existing data

We shall only describe a fraction of the work, namely that for relatively unreactive metals on oxides. Data for crystalline oxides (including appropriate native oxides or microcrystalline oxides) are separated from those for amorphous oxides or glasses.

Relatively non-reactive metals (Au, Ag, Cu, Pt, Pd) on crystalline oxides

Cu on polished alumina He and Ne ion beams have been used to compare ionisation and nuclear components of radiation-induced changes (Baglin *et al.* 1984). There is no evidence of ion-beam mixing, although nuclear energy deposition (Baglin *et al.*'s Fig. 5) and electronic energy deposition are involved (Baglin *et al.*'s Fig. 7).

Cu/Cr, with about 2 nm of intermediate native oxide This showed some enhancement (Böttiger *et al.* 1984). There may have been some ion-beam mixing; nuclear deposition had a big effect, whereas electrons alone had only a modest effect.

Pt on alumina; Pt on yttrium-stabilised zirconia (Mitchell *et al.* 1984; J S Williams 1986; Sood *et al.* 1985; Budhani *et al.* 1986) He ions and electron beams gave modest enhancement, with some dependence on the metal film thickness. This shows that secondary effects, like those induced by strain energy, can be important. Budhani *et al.*'s work with plasma-deposited Pt showed that annealing was harmful, in line with a role for strain.

Au on the native oxide on Ta (Stoksad *et al.* 1986) The use of C, O, Si, Ni ion beams with 2.85 MeV per nucleon showed that a large electronic stopping power (S_e) led to a low threshold for radiation-enhanced adhesion. The effect saturates. There was evidence for ion migration: a higher threshold is needed if one waits after bombardment before testing.

Pd and Ag on alumina; Au on tantala; Au on ferrite (Tombrello 1984) Results using Cl and Ar ion beams suggest a specific amount of electronic energy liberated at the interface. A link with the Richardson (thermionic emission) equation was sought. Griffith *et al.* (1982) also discuss Au on c-cut and a-cut sapphire.

Cu on MgO (Vuoristo *et al.* 1986) Here the oxide was sputtered onto a number of substrate metals, including 'oxygen-free' Cu and Cu with 1% Cr and 0.1% Zr. Reactions involving oxides of Cu were evident. Most cases probably involved a thermal oxide layer of some sort, which introduced the issue of thermal expansion.

Relatively non-reactive metals (Au, Ag, Cu, Pt) on glasses or amorphous silica

Cu on fused quartz or glass ceramic (Corning MACOR, Si/Al/Mg/K/B oxides/F) (Baglin *et al.* 1984) For fused quartz, Ne ions are more effective at lower doses and He ions are at higher doses. This may imply initial nuclear effects and subsequent electronic ones. There were pronounced effects of heat treatment (Baglin *et al.'s* Fig. 8). Nuclear deposition was more significant for the glass ceramic.

Au, Ag on Si thermal oxide and silica (Werner *et al.* 1982; Tombrello 1983a,b) Tombrello suggested a thermionic emission argument, with new bonds (in the image picture, new charged species) proportional to the thermionic current, and the current determined by the Richardson equation $J = (T^2)\exp(-W/T)$, and hence a dependence on work function difference W and on T, the temperature. Jacobson & Jänsson (1983) found that 12 MeV S^{3+} ions significantly improved the adhesion of Au which had been RF sputtered onto the glass. Griffith *et al.* (1982) noted that the bonding of Au on silica and on glass improves with a delay between irradiation and test.

10.5.3 How to interpret radiation-enhanced adhesion

This mass of data, together with data for other classes of system, is broadly in line with the ideas noted at the start of this section. Usually there is no ion-beam mixing. Excitation influences adhesion by physical and chemical processes which do not occur merely on placing the materials in contact. Chemical reactions are critical only in limited situations, such as those involving organics, when the decomposition of an oxide occurs, or when the radiation simply permits a reaction which would otherwise only happen at a higher temperature or on a longer timescale.

Uncertainties about the system often inhibit understanding of what happens. Native oxide can be important, and forms on Si and on many metals. The oxide may be too thin to be worrisome, yet sufficient to affect adhesion. Even a monolayer of O or H could have its effect. If the oxide is etched off, there can be residual ions (like F^- from HF) which themselves affect adhesion. The role of contamination, e.g. by C, is hard to establish from current data. There can be secondary reactions, especially if there are mobile species like H or Na. Motion of such species is known from the removal of H from traps at the Si/SiO_2 interface and the motion of Na to the cathode in electrostatic adhesion, where the Na may react with moisture or air.

There is evidence for mechanical effects, probably associated with strain energy in the film and the substrate. These include effects on the thickness of the film, but the data do not suffice to tell if they are associated with volume changes in the film. Stress relief can be reduced in some cases by ion-beam assisted deposition, a somewhat different effect which can lead to dependence of the film on bias voltage. Ion-assisted deposition changes the grain morphology of MgF_2 on silica (S D Jacobs *et al.* 1986). Since processes can be carried out at different temperatures when using radiation, there may be some scope to avoid problems with thermal expansion.

Overall, there is evidence for a defect contribution to adhesion, as in the image interaction picture. There are several recurring points. First, nuclear energy deposition is often needed, yet there is a dependence on the electronic energy deposition shown by the dependence on the electronic stopping power. This suggests that defect production is essential, and that the electronic redistribution of charge contributes to the effectiveness of the process. The so-called thermionic effects (Werner *et al.* 1982) are consistent with a picture of defects produced near the surface and charge adjusting to lower the total energy. Secondly, there is evidence for migration processes (sometimes defect migration, sometimes charge migration, sometimes uncertain) so that beneficial heat treatments can be envisaged. Defect migration may be why ageing can influence adhesion.

The natural question is whether or not there are systematic ways to select and control adhesion. Overall, the image charge description seems a significant part of the behaviour, although the issues of stress and of contamination are important. The tools available are primarily those of defect engineering, with excitation as one of the possible options. The basic process in enhancement and control can involve any of the three main steps: pretreatment, irradiation, and post-treatment.

The simplest pretreatment is cleaning, whether mechanically, or by a laser, an ion beam, or a plasma method (with or without a bias voltage). One possibility is to try to move charge before irradiation, either by removing H or O from a semiconductor surface, or by moving sodium ions in a glass electrically. Thus Bozack *et al.* (1986) showed a prebombardment of Si with ions creates active sites which enhance the surface's adsorption of propylene; this enhancement can be switched off by exposure of the surface to atomic hydrogen, an effect the authors describe as 'capping the sites.' Another example (Pedreza *et al.* 1995) may involve several mechanisms. They found that sapphire is amorphised by a 308 nm XeCl excimer laser above a threshold (about 0.7 J/cm^2). This laser treatment enhances adhesion, both below and above this threshold. The enhancement is best in an oxidising atmosphere. For Au, the untreated surface has very weak adhesion (0.1 MPa) and is improved only by excitation above the melting threshold. For Cu (untreated, 13 MPa) and Ni (untreated, 32 MPa) there are effects below the threshold as well. What could be the mechanisms? In the case of Ni and perhaps Cu, there might be a chemical contribution of some sort, although this would not ensure stronger adhesion. One possibility might be enhanced image interaction contributions through charge redistribution. Another possibility is that there is mechanical pinning where grain boundaries have been eroded, combined with a more intimate surface contact. There could be alternative alumina phases formed, which contribute directly by better adhesion, or indirectly because the removal process drives endothermic phase changes. Cao *et al.* (1994) noted that pulsed laser action generates γ-alumina on sapphire.

The options for irradiation are very varied: ion beams, electrons, photons (especially ultraviolet or X-rays). Relatively gentle methods might be practical in those systems for which there are excitonic methods of defect production. We note cases in which the effect is polymer curing; a monomer is painted on and can penetrate surface roughness before polymerisation, so that adhesion is mechanical rather chemical.

Post-treatment will normally exploit heat or light. One aim will be the annealing of unwanted damage. Another aim is the optimisation of defect charges. This includes enhancing adhesion in selected regions only of the interface. It may be helpful at times to have adhesion which can be switched off, if only temporary bonds are needed.

10.5.4 Related phenomena: Anodic bonding

The approach in anodic (Mallory) bonding is strikingly simple: place the surfaces in contact; apply a static electric field; wait; switch the field off (Pomerantz 1968; Wallis & Pomerantz 1969; Wallis 1970; Boroin 1973; Landis & Younger 1979; Younger 1980; Taylor 1983; Anthony 1983; Albaugh 1986). In most cases the direction of the field does matter, though the reasons are partly indirect (e.g. if Na is brought to the interface of a glass, it will probably form a weak layer of NaOH by reaction with atmospheric water, and it is this weak layer which will break) so the metal to be bonded is usually the anode, and the method is known as 'anodic' bonding.

There are two types of contributions. One view asserts that the bond formed is just like other bonds, i.e. there are chemical changes (for those who like chemical models) or changes in charged defect populations which lead to image charge contributions (Stoneham & Tasker 1985a; Duffy *et al.* 1995a,b). Thus, for a steel in contact with a glass, the natural diffusion processes would be a rapid diffusion of alkali cations or protons towards the cathode, and some slower diffusion processes near the anode. These might move intrinsic point defects in the glass towards the anode, or lead to diffusion of some Fe species from the steel. It is these defects which might be further affected by optical excitation, whether to speed the adhesion, or to redistribute charge so as to maximise the image energy terms. We remark that this class of argument, with the emphasis on electrostatics, appears to be widely applicable, even if it is not always the complete explanation. The phenomena it appears to cover include the poling of optical fibres (Chapter 8), the adhesion of ice to mercury (Petrenko 1998; Petrenko has also suggested that electric fields can be used to enhance the adhesion of car tyres on ice) and the control of misting on dental mirrors.

A further contribution arises because the surfaces are only partly in contact at first with regions between asperities where no contact is made (Anthony 1983). When the field is switched on, the energy of the electric field can be optimised by eliminating these air-filled regions. This needs plastic deformation (which takes time); elastic deformation would be of no avail after the field was turned off. So the field acts as a stress. Wallis & Pomerantz (1969) estimated the stress as about 1000 psi for a field of 4×10^6 V/cm, which is probably as close to breakdown as can be used sensibly. The stress pushes the surfaces into more intimate contact. As described, the sign of the field does not matter, but there can be secondary factors.

As for radiation-enhanced adhesion, most of the effects are on the non-metal side, though some of the deformation could be on the metal side. The mechanism has features which depend on temperature. Higher temperatures will speed up the process, but may give problems from differences in expansivity.

10.6 Oxidation and chemical reactions at interfaces

Suppose a material M is exposed to an oxidising gas, and that thermal oxidation does occur. Once an oxide layer has formed, there are four basic processes. For convenience, we shall assume that the actual reaction between M and O occurs at the interface between M and its oxide. In Step 1, an oxidising species must stick to the oxide surface. In Step 2, it must be incorporated into the oxide in some mobile form. In Step 3, this mobile species must diffuse across the oxide. Diffusion is often the rate-limiting step for any but the thinnest oxides. For Step 4, there will be a reaction with M at the M/oxide interface. This last step is exothermic.

Excitation is important only if it influences the rate-determining step. If it is diffusion which is influenced, we might look to some of the explanations of Chapter 7. However, for silicon oxidation, it appears that a different effect is important. In all the cases of which we are aware, the effects of excitation on oxidation are associated with a change in the oxidising species, such as from O_2 to O or to some oxygenous ion.

The standard reference model (the Deal–Grove model) for silicon oxidation assumes that there are two rate-determining processes in series. One process is diffusion-driven by a concentration gradient of the mobile species (O or Si) and with a diffusion constant which does not depend on oxide film thickness. The other process is an interface reaction, whether at the Si/oxide interface or the oxide/gas interface, and with a rate independent of oxide thickness. This model describes most of the data for films thicker than a few tens of nanometres.

However, there is clear evidence that the model fails for low oxygen pressures, for low temperatures, for oxidation by certain species other than molecular oxygen, and when there is excitation. The failure is most clearly seen by using the dimensionless parameter $g = -\mathrm{d}(\log v)/\mathrm{d}(\log x)$, where x is the oxide thickness and $v = \mathrm{d}x/\mathrm{d}t$ the growth velocity (Stoneham & Tasker 1987). In the Deal–Grove picture, g lies between $g = 0$ (interface control, whether by Step 1, Step 2 or Step 4) and $g = 1$ (diffusion control). But, as Fig. 10.2 shows, g exceeds 1 for low oxygen pressures and low temperatures. These are precisely the conditions for the production of thin films. When there is excitation by very low-energy electrons (Collot *et al.* 1985) values of g as high as 10 are found (Sofield & Stoneham 1995; see Fig. 10.2).

The likely explanation follows from two distinct arguments. A natural idea is that oxidation is enhanced when the diffusing oxygen species is changed from O_2 to some form such as O^0 or O^-. In particular, enhancement usually occurs when it is possible to create atomic oxygen or oxygen ions. This can be done in at least four ways. The first is to create oxygen atoms or ions in the gas phase (perhaps by creating a plasma), or to provide a catalyst (like an alkali metal, Starnberg *et al.* (1988)) which naturally causes dissociative adsorption of molecular oxygen. The second is to create an oxygenous species, like ozone, which easily decomposes to give atomic oxygen. The third method involves an electron, exploiting a process such as $O_2 + e^- \rightarrow O + O^-$ (or even O_2^-). The electron might be injected thermionically from the Si substrate (Irene & Ghez (1987); Irene & Lewis (1987); similar ideas for Fe oxidation have been suggested by Liebrant *et al.* (1992)); it might come from a scanning tunnelling microscope probe (Welland & Murrell 1993); or it might come from a very-low-energy electron beam (Collot *et al.* 1985). A fourth method uses ultraviolet light to break up the oxygen molecules (Young & Tiller 1987; Glachant *et al.* 1991; Craciun & Boyd 1993).

How does the altered oxygen species affect the oxidation rate? One way is surely diffusional. However, the deviation from Deal–Grove kinetics rules out simple diffusion or interface reaction control, since the value of g should lie between 0 and 1 for any single oxidising species. Most of the alternative explanations require a charged oxidising species. For Si (unlike Ni, for example) the model of Cabrera & Mott (1948) does not seem adequate. The Cabrera–Mott assumptions are that: (*a*) electron tunnelling from the Si to adsorbed oxygen occurs, driven by a chemical potential difference; (*b*) tunnelling ceases when the back field due to the (now charged) species balances the chemical potential difference; and (*c*) the field enhances the step in which the oxygenous species is injected into the oxide. The predictions of this model lead to several problems. The ideas of Stoneham & Tasker (1987), as extended by

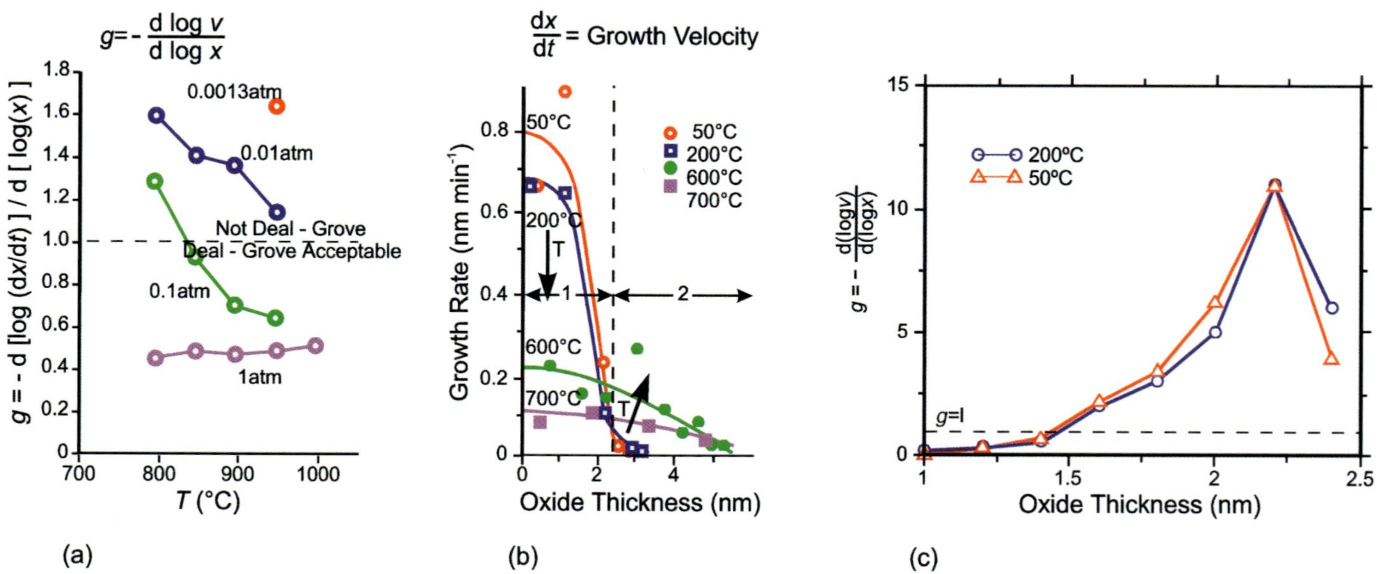

Fig. 10.2. (a) The logarithmic derivative of oxidation velocity (g) versus thickness for oxidation of silicon. (b) Effects of electron excitation on oxidation velocity. (c) Analysis of the lower-temperature curves indicates g values sharply peaked near 2 nm thickness.

Torres *et al.* (1995), seem more likely and provide a coherent view of a number of results.

The Torres *et al.* analysis addresses other problems in the oxidation behaviour for thin oxide films. The problems are that: (*a*) oxidation appears to occur in a layer-by-layer manner, yet the oxidation occurs on terraces, not at steps; (*b*) Deal–Grove growth kinetics $x(t)$ are not seen; and (*c*) the roughness oscillates with oxide thickness. A description consistent with all these features has been given by Torres *et al.* (1995; see also Sofield & Stoneham 1995, Stoneham & Sofield 1998). Their mechanism requires an interfacial step which does depend on the local oxide thickness. Specifically, it relies on the much greater polarisability of Si than of SiO_2. This, in turn, biases the sticking (adsorption) of the oxidising species or the incorporation from that adsorbed site into an interstitial site of this species. The bias is such that the oxidation takes place preferentially where the oxide is locally thin, which encourages what can be described loosely (since the oxide is amorphous) as layer-by-layer growth. For a charged oxidising species, the image interaction does indeed bias the sticking and incorporation steps by about the right amount.

There is a further issue, since experiments with different isotopes (Collot *et al.* 1985) show that electron excitation leads to isotope exchange between mobile oxidising species and the network oxygens. In normal dry oxidation, the majority of the mobile oxygen species are assumed to be interstitial oxygen molecules which do not exchange. Detailed calculations (Szymanski *et al.* 2000; Stoneham *et al.* 2000a,b) show that certain of the charged species should be incorporated into the network, in line with both the experiments and the underlying ideas of Torres *et al.*

The Cabrera–Mott mechanism contains important ideas for the earliest stages of the oxidation of metals. As noted, the model leads to a characteristic kinetics for the early stages of oxidation, and has been confirmed for thermal oxidation of Ni (see Atkinson (1985)). Clearly, excitation could enhance this process. It might do so by ensuring tunnelling continues to thicker oxides; alternatively, excitation might maintain a population of charged adsorbed ions. Both these mechanisms would alter the kinetics, as well as enhance oxidation. Effects of this sort have been discussed by Ebinger & Yates (1998) for Al oxidation under bombardment by a 100 eV electron beam in the presence of water vapour. Stirniman *et al.* (1995), however, suggested that the stimulated oxidation of Ni(111) is associated with the creation of metastable (about 600 s lifetime) nucleation sites by the electron beam; secondary electrons may be involved.

Nucleation appears to be important in at least one other case, namely the effect of an STM on the oxidation of Al(111). Brune *et al.* (1993) found that oxygen islands are formed via a mobile precursor ('hot adatoms', which move over 8 nm), followed by the nucleation of grains until there are grains of about 2 nm size over the surface.

10.7 Amorphisation and recrystallisation

Crystalline Si can be amorphised by ion irradiation at low temperatures, forming a crystal/amorphous interface at the end of ion range. The interface can be moved either inwards or outwards in the temperature range between 200–600 K under further irradiation of the specimen by ion beams which have a longer ranges. Conversion to crystalline Si (interface moving to the crystalline side) occurs at higher temperatures and lower fluxes, whereas conversion to the amorphous form occurs otherwise. The motion is an interface reaction due primarily to elastic encounters (Priolo & Rimini 1990). Motion of the interface is governed by the precipitation of defects generated on both sides of the interface. Goldberg *et al.* (1999) have shown that amorphisation of crystalline Si is nucleated preferentially at a thin band of dislocations, and that amorphisation does not occur under the same conditions without such a nucleation. This suggests that amorphisation is nucleation-limited: it proceeds at nuclei at interfaces between crystalline and amorphous until the whole system is amorphised.

Molecular dynamics studies (Jose Caturla *et al.* 1995) of the effects of elastic energy deposition on the two sides of a c-Si/a-Si interface indicate that crystallisation is induced by energy deposition on the a-Si side transferring less than the threshold energy. Amorphisation can occur only by elastic encounters on the c-Si side transferring more than the threshold energy. To make a connection with the defect model, the result is interpreted in terms of a difference in threshold energies for defect generation in the crystalline and amorphous sides of the interface.

A somewhat different approach has been suggested by Wang *et al.* (1995), who argued that amorphisation occurs by knock-ons in c-Si, whereas crystallisation of a-Si occurs in the hot zones generated by subthreshold elastic encounters. The ideas are very similar to the hot-zone mechanism of thermodynamically-driven mixing: the enthalpy is lowered by crystallisation, but is increased by amorphisation. We shall discuss hot-zone mechanisms further in Chapter 11. Donovan *et al.* (1989) reported that, for the recrystallisation of ion-implanted a-Si on 100 Si, there is a homogeneous heat release of 5.1 kJ/mole and an interfacial energy of 13.4 kJ/mole. Given an enthalpy difference, Wang *et al.* suggested that a-Si can crystallise without elastic encounters above the threshold, while amorphisation needs defect formation. Wang *et al.* (1995) explained the temperature and flux dependences in terms of the divacancies generated under ion irradiation: the divacancies inhibit crystallisation and enhance amorphisation. In fact amorphisation and crystallisation rates can be scaled by a parameter, $\phi^{1/2}\exp(E/2kT)$, where ϕ is ion flux and E is the activation energy for divacancy migration. The factor 2 in the exponent appears because of the bimolecular recombination of divacancies. The molecular dynamics studies of Weber *et al.* (1999) indicated that collisions of 5.5 eV, much smaller than the displacement threshold, can enable recrystallisation by eliminating specific types of defects formed at the interface.

Interface motion is generally believed to be due to the generation of defects which diffuse to the interface (Jackson 1988; Wang *et al.* 1994). These defects can be characterised. For ion-induced crystallisation and amorphisation, there seems to be a mobile defect which can decay by binary recombination. The defect has a 1.2 eV activation energy in amorphisation and a 1.5 eV energy in thermal crystallisation. These numbers may be compared with 2.7 eV for the rate of thermal recrystallisation of a-Si. Sinke *et al.* (1989) reported that explosive crystallisation of a-Si is induced by a pulsed laser: there is triggering and propogation. The defect motion has further effects; for example, Polman *et al.* (1997) noted segregation of Er at the a-Si/c-Si interface.

Summary

Collisional energy input and electronic excitation stimulate modifications of phase changes (like amorphous to crystalline), with little suggestion so far of exciton involvement. There are opportunities for exploiting excitonic processes which involve decomposition processes.

Halide crystals with closed-shell cations (NaCl, CaF_2) Excitonic processes lead to halogen atom loss, leaving alkali atoms behind at interfaces; reactions of halogens at interfaces are expected.

Amorphous halides and halide glasses Excitonic processes probably cause interface reactions.

Halide crystals in which the cation can change charge state (AgCl, $PbCl_2$) The photographic process in AgBr appears to require dislocations or interfaces.

Amorphous halide or halide glass in which the cation can change its charge state

Oxide crystals with closed shell cations (MgO, ZnO, Al_2O_3, usually SiO_2) Enhanced adhesion has been observed under high-density excitation.There is a photo-induced instability in ruby.

Oxide glasses with closed-shell cations, like borosilicate or aluminosilicate glasses Enhanced adhesion is observed (anodic or Mallory bonding)

Oxide crystals with cations which can have several charge states ($Fe_{1-x}O$ and other transition metal oxides)

Oxide glasses which include transition metals

Semiconductor crystals (III–V, Group IV) with relatively wide bands and small gaps Enhanced oxidation, e.g. of Si; this can lead to control of the interface with the oxide. For mixing and doping, see Chapter 11.

Amorphous/crystalline semiconductor interfaces The growth of oxide on silicon can be enhanced by excitation. Recrystallisation, and the a-Si to c-Si transformation take place at the amorphous–crystalline interface.

Chapter Eleven

High excitation intensities

11.1 Introduction

So far, we have concentrated on situations in which the number of excitons is low, so that behaviour can be analysed in terms of individual electrons or holes, or excitons, or perhaps biexcitons or small exciton complexes. This type of description holds to modest levels of excitation. Thus Tanimura & Itoh (1988b, 1990) discussed exciton fusion, in which a free exciton and a self-trapped exciton in RbI transform to either an F–H pair or a luminescent singlet self-trapped exciton. Yet there are important cases for which this class of description is dubious. The numbers of excitons, or electrons and holes, can even be of the order of the number of atoms. Such a massive degree of excitation needs a different description. For example, the host interatomic forces will be changed severely. If, as a result, the host is unable to resist shear, a description in terms of a liquid state or a plasma is tempting. The energy per atom can easily be large enough to cause evaporation or exoelectron emission, with associated electric fields. In the cooling process, the system may evolve into a different solid structure, possibly a non-equilibrium form. In this chapter we discuss some of these situations. It is hardly surprising that we cannot give a detailed dynamical and microscopic picture of the behaviour. However, for practical applications, we can make do with less information.

The situations we discuss are ones where local energy densities can be tens of electron volts per atom, equivalent to about 10^5 J/cm^3. These are the conditions which lead to laser ablation, fission track formation, and other major reorganisations of the solid (for amorphisation, see Chapter 6). Such energies are far higher than the excitonic and electron–hole liquids seen at low temperatures in semiconductors like Si and Ge, where the energy per host atom may be only 0.01–0.1 eV and does not lead to any significant materials modification. In oxides like ZnO, electron–hole plasmas are seen for excitation at 355 nm (3.5

eV; the exciton energy is 3.27 eV) at above 800 kW/cm^2 (Bagnell *et al.* 1998) at temperatures as high as 550 K. No damage is reported.

The means by which the energy is delivered has an enormous effect on the outcome. This depends partly on the timescale of energy delivery, relative to other timescales of the system, and partly on whether the excitation is primarily to the electronic system or primarily to nuclear motion. The timescale may affect the place at which energy is deposited. In laser ablation, for instance, the plasma-like plume created above the surface will absorb or reflect much of the incident energy. Femtosecond pulses excite the solid far more efficiently than nanosecond pulses. The critical timescale for an oxide like alumina is about 1 ps.

Some of the differences from case to case can be subtle. Thus, polished surfaces of diamond of natural isotopic composition are blackened by graphitisation under ArF excimer light at about 2 J/cm^2. However, with diamond which is isotopically pure, consisting solely of ^{12}C, no graphitisation occurs, even under focused beams at 30 MJ/cm^2 (Deshmukh *et al.* 1994). The difference appears to stem from the fact that the isotopically-pure diamond has a thermal diffusivity 50% larger than that with the natural isotopic composition. The diamond to graphite transition is an example of the phase transitions which can occur on strong excitation. Often such phase transitions will be to some transient state which need not be stable in equilibrium (e.g. in laser annealing, Section 11.2; see also Sokolowski-Tinten *et al.* (1998)). In the case of diamond, there can be further complications related to the presence or absence of hydrogen surface termination and also the the presence or absence of an atmosphere. In the synchrotron radiation etching of diamond (Ohashi *et al.* 1996) the supply of reactive oxygen radicals is important. Rothschild *et al.* (1986) observed etching even without gas present, but note that gases improve the morphology of the etched features.

Clearly, thermal phenomena are very important. By thermal processes, we mean those processes which can be described without invoking excited states or non-standard charge states of atoms or ions. Thermal descriptions may describe one aspect of the behaviour completely, yet may leave gaps. Thus Meldrum *et al.* (1998) gave a rather complete description of the amorphisation of silicates $MSiO_4$ (M = Zr, Hf, Th) by assuming a transient liquid state in displacement cascades, the liquid state allowing new phases (like ZrO_2) to form. They found no need to invoke excited state phenomena, in that the liquid state appears to have the properties of the normal melt. However, one reason that ambiguity can arise in discussing just one mechanism is, for example, that it might be that defects associated with electronic excitation (as known for silica) occur outside the regions which have been in the liquid state.

One important comparison is between laser and ion-beam excitation. Energy deposition by heavy ions occurs in a time of order 1 fs. A zone along the path of the ions is excited, either electronically or vibrationally, depending on the ion velocity. A limited region along an ion path is excited to a catastrophic condition in a short time. Relaxation of the electronically or vibrationally excited states follows. This involves a temperature rise (increases in nuclear kinetic energies), diffusion, and the modification of local atomic structures. Energy deposition during irradiation by a laser pulse occurs within the time width of the pulse, typically 100 fs to 10 ns. During the pulse, the state of the target changes. The target is undamaged initially and electronically in its ground state. At the end, the sample can be drastically damaged. The primary process of energy deposition is electronic excitation, which occurs within 10 fs. This is followed by lattice modification through the electron–lattice coupling. The electronic excitation changes the optical absorption coefficient, and hence the subsequent energy absorption. The lattice is heated, and the optical absorption coefficient is altered as optically-induced processes continue. Electrons, ionised from defects or impurities, can gain further energy from the light pulse, and may be emitted from the free surface. The result is what can be a substantial electric field near the surface. The target surface changes topography. This is not usually a major issue, unless there is a systematic stoichiometry problem, or unless large particles (like grains) are flung off and are able to land on the target. Laser-induced processes can be heirarchical. The thermal and electronic changes produced by the early stages of the pulse modify the response of the later stages, an effect which becomes increasing important for longer laser pulses.

An extreme example of these issues is observed for high-intensity (2×10^{16} W/cm^2) ultrashort (150 fs) laser pulses (Hutchinson *et al.* 1998). Ions with kilo-electron volt energies can be created. Such pulses on clusters of 100 or more Xe atoms both ionise the atoms and heat the free electrons rapidly. Charge separation leads to rapid expansion of what is essentially a miniplasma. Up to 90% of the energy is transferred to the ions, which can be in very-high-charge states (typically 20+, but up to 40+ has been noted). Studies of ultrashort pulses in the 10^{13}–10^{18} W/cm^2 range lead to nuclear kinetic energies in the 1–1000 eV range, and show that, at high laser intensities, all target materials appear to reach a 'universal plasma mirror' state (Price *et al.* 1995).

11.1.1 Thermal models of modification

Thermal models offer a simple macroscopic description. This view often gives a useful rationalisation of experimental results, such as the dependence on stopping power of the radii of tracks produced by energetic heavy ions, or the laser ablation threshold. The thermal model assumes that all energy deposited becomes thermal energy very quickly, within 0.1 ps, and that a temperature is established rapidly. The model deals with energy flow using standard thermal constants, but often with additional parameters. Yet these standard parameters can fail (Dragnea & Bourguinon 1999), possibly in ways related to a photo-induced plasma. Since the thermal model deals with the energy flow in homogeneous media, it explains only some of the phenomena. It cannot distinguish between different types of energy (electronic, elastic, electronic potential energy or kinetic energy, vibrational energy, or defect formation energy). The thermal picture must be regarded as very limited, even though it does give some useful semiquantitative descriptions of what is seen experimentally. For example, mixing in Si by various ions (O, Ar, Kr, Cs; energies 3–300 keV), studied by the broadening of Ge and B profiles (Cardenas *et al.* 1998), shows a mixing efficiency as a function of deposited energy per unit length which is fully consistent with a thermal model.

What are the factors that influence the surface temperatures of non-metallic solids under laser irradiation? First, band-to-band excitation can create electron–hole pairs with high concentrations. These electrons (with concentration n) and holes (with concentration p) recombine primarily by Auger recombination (Yoffa 1980). The rate of recombination for the main process is proportional to the n^2p and, since n and p are approximately equal, the rate is proportional to n^3 (Beattles & Landsberg 1959). Secondly, the free carriers can be heated by the laser beam. The cross-section is proportional to the concentrations of electrons (n) and holes (p) in the conduction band and to the cube of laser intensity (Epifanov 1974). Therefore this process is more important for shorter laser pulses. Thirdly, defect creation will lead to additional optical absorption and to the generation of extra carriers. These new defects can be created both on surfaces and in the bulk. This process makes the emission yield a superlinear function of flux or laser fluence.

11.1.2 Non-uniform damage

Non-uniform damage features are often reported in observations by electron or optical microscopes of the materials subjected to high-intensity excitation. These features may be in the bulk or at the surfaces. Damage caused by energetic heavy ions is local in some materials, like MgO, and uniform in others, like α-SiO_2. Such a difference depends partly on whether the mechanism of energy deposition is localised or delocalised. For instance, the conversion of electronic kinetic energy to thermal energy is not local, whereas the relaxation of the lattice from an electronically excited state is local for systems in which excitons are self-trapped or where defects are created. The degree of localisation in energy absorption must be considered, particularly for laser pulses with subbandgap energies, where excitation occurs either through defect excitation or through resonant multiphoton excitation.

The nucleation of defect clusters and precipitates can provide other routes to energy localisation. In semiconductors and in oxides like MgO, laser ablation appears to involve vacancy multiplication: vacancies already on a surface act as nuclei for further vacancy formation (Okano *et al.* 1993; Dickinson *et al.* 1993). Vacancies can form at the edge of a cluster of vacancies in the outer surface layer, or on the second layer which has been exposed. The relative probabilities of these two processes, together with the rates of subsequent surface migration processes, determine the way in which damage evolves. Metal colloids can show interesting phenomena. For instance, femtosecond laser pulses with energies near the interfacial plasmon resonance can change the colour of a glass containing Ag nanoparticles (Kaempfe *et al.* 1999), apparently through ultrafast particle deformation, leading to non-sperical Ag paricles of uniform orientation.

11.1.3 The early stages of energy deposition

High-intensity excitation causes dramatic changes, perhaps only transiently, from the initial ground-state system. One way to think of the highly-excited region is as a *hot zone*, in which elastic encounters have deposited a large amount of energy as nuclear kinetic energy. This view corresponds to a common description for collision cascades. The atoms have very large amplitudes of motion, since the energy deposited can be as much as 10 eV per atom. As we discussed in Chapter 2, it is not enough to assume the nuclei simply move on the electronic ground-state adiabatic energy surface. As Flynn & Averback (1988) noted, some of the energy goes to electronic excitation, even in metals. In non-metals, the hot zones will certainly involve electronic excitation as well as large nuclear motions. The situation is very similar to the catastrophic lattice disorder induced along an energetic heavy ion track in one of the materials in which excitons are self-trapped. Along these tracks, the energy density is sufficient to excite every one of the molecular units. In materials in which excitons are not self-trapped, analogous excitation effects can occur involving defect sites.

The situation is somewhat different under laser irradiation above the bandgap energy. Even for a 10 mJ laser pulse of 100 fs duration, there are about 10^{16} photons incident in the first 10 fs. The penetration depth is about 1000 layers, so about 10^{13} photons are absorbed within each layer in 10 fs, about 1/100 of the number of atoms. The density of excitation is much less than for heavy-ion tracks, at least if the absorption is delocalised. The incoming photons also excite the free electrons which are generated, and this energy turns to heat within 100 fs. So the density of excitation under laser irradiation is not as high as under ion irradiation, but the energy tends to be converted to heat rapidly. However, very high local excitation densities can occur by the localisation of electronic excitation energy at a defect or new phase. This is partly because energy can be absorbed even by excited states of defects, and partly because the defect may trap electron–hole pairs during the lifetime of the excited state. In such cases, laser irradiation will create non-uniform, localised high-density excitation.

Photons with below bandgap energies are absorbed only by defects or by multiphoton excitation, which has a very small absorption coefficient. The absorption coefficient should increase by a variety of mechanisms during laser irradiation, both within a pulse and after a sequence of pulses. Resonant multiphoton absorption through defect levels increases the number of electron–hole pairs (Matthias & Green 1990; Ang *et al.* 1999). The trapping of carriers by surface defects and the direct excitation of surface defects cause defect clusters on surfaces to evolve, giving further sources of optical absorption. Both processes increase the number of free electrons, and hence free-electron heating. The excitation of defects on surfaces leads to localisation of the electronic excitation energy. Thus the properties of surface layers should alter dramatically at the beginning of a laser pulse on the absorption of energy sufficient to cause ablation.

When the incident light excites free carriers, rather than valence electrons, the energy is diverted to other channels. If the diversion is from a bond-breaking process to the heating of a region containing many atoms, the extent of materials modification will be reduced. Thus, diamond can be ablated by two-photon absorption of 248 nm photons with 500 fs pulses at above 0.6 J/cm^2 (peak intensity 1.1 TW/cm^2), but the ablation rate is low because of the free-carrier absorption (Preuss & Stuke 1995).

11.1.4 Processing by high-intensity excitation

A typical excitation density in heavy-ion tracks is 10 keV/nm. If distributed over a radial distance of 10 nm, this figure is equivalent to about 10 eV per atom. Similarly, a 100 mJ laser pulse can deposit an energy of order 10 eV per atom. Such an energy will be larger than the vaporisation energy of the material. Material can be removed either directly, as with laser irradiation, or after chemical etching, as for heavy-ion irradiation. Thin film synthesis by laser ablation, and the drilling of holes have become common techniques, even though the mechanisms are not fully understood.

We emphasise that electronic excitation has an important role in the high-intensity excitation of all non-metallic solids. There is usually energy localisation by self-trapping or associated with defects. Furthermore, there are ways by which high-intensity excitation can cause lattice modification, even if the bandgap energy is too small to cause the change directly. Which processes are important will depend on the nature of the excited states and on the branching between the various channels, including self-trapping and defect production, which determine the final state. High-intensity excitation is potentially extremely important for materials modification, but its potential can only be realised by understanding more fully the excited-state processes and their control.

11.1.5 Sputtering by ion beams

In standard sputtering, a particle beam removes target atoms. The process is controlled by the choice of the energy, angle of incidence, and species of the incident particles. The incident particles give energy to target atoms through collisions, and subsequent collisions redistribute energy, cause mixing, and may lead to charge transfer. Sputtering lies outside the field of this book, but its parallels with laser annealing and laser ablation are helpful. Much of sputtering can be understood on the assumptions that classical collisions occur between ions, and that there is an energy barrier U to the emission from the surface of particles. There is a similar barrier for electron loss. The barrier for atom or ion emission appears to be related to the target sublimation energy, subject to obvious questions about whether one should use an average value, or whether explicit attention to steps or edges on the surface is needed. The idea that there is an energy barrier breaks down in two ways. First, there can be preferential sputter from compounds. Secondly, the model offers only a limited description of behaviour when the energy available is not much larger than that needed to remove atoms.

Ion irradiation of materials leads primarily to neutral emissions. The neutral yield from a monatomic solid is given by the following universal equation (Yamamura & Bohdanski 1985):

$$Y = 0.042(\alpha S(E)/U)[1-(E_{th}/E)^{1/2}],$$

where E is the ion energy, $S(E)$ is the elastic stopping power, U is the sublimation energy, E_{th} is the threshold energy for sputtering, and α is a parameter which depends on the ion and material combination. The equation is modified from that of Sigmund (1969) through a best fit to experimental results for monatomic solids by Matsunami *et al.* (1984). Apart from α and the factor in the square bracket, Y is in accord with Sigmund theory. The threshold energy for sputtering is significantly smaller than that for knock-ons in the bulk.

There are empirical rules describing the charge states of ions sputtered from elemental targets. For those bombarding species which readily form cations, such as Cs, one might expect charge transfer to lead to a high yield of negative ions. No systematic trend with target electron affinities is seen. For O, a bombarding species which readily forms anions, a high yield of positive ions is expected. There is a trend showing the secondary ion yield falls with target ionisation potential I (Wittmack 1980):

$$\ln(\text{secondary ion yield}) = A - I/B,$$

with B about 0.8 eV. There is a strong dependence on the column of the periodic table from which the target element comes. Most metals give a high positive ion yield (and/or a low negative ion yield), the exceptions being Os, Ir, Pt, Au, which are relatively low-yield cases. There is no strong dependence on mass for metals (Stuart & Wertheim 1962) but there is a correlation with sublimation energy; Pulker (1984) noted the threshold energy is about 4 × sublimation energy. Most anionic species (including C, Si, H) give high negative ion yields. There seem to be exceptions rather than rules for models involving local electron density and surface polarisation.

For oxides, the raw data from several references show enhancement of metal over oxide at the surface to varying extents (Table 11.1). In no case is oxygen enriched at the surface. Large oxygen losses are favoured by four circumstances: (i) the metal atom has a high mass. This, which may follow from classical collision theory, suggested to Kelly (1980) a relationship with evaporation, and hence to the vapour pressure for the oxide at 3000 K; (ii) there is a high oxidation rate. In this case, the Knotek–Feibelman process is usually possible; (iii) the system can go readily to a lower oxidation state under irradiation, so the cation can adopt several charge states and starts in a high state; and (iv) there is a high nominal charge of the cation.

Other oxides maintain their O/M ratio without significant change. The guidelines appear to be these: (i) the system is a stoichiometric, closed-shell system, having only one oxidation state. Systems in their lowest oxidation state are usually unaffected. However, we remark that even alumina and magnesia can lose oxygen under electron or laser irradiation; (ii) systems which amorphise under irradiation tend to be stable in composition; and (iii) if there is a

Table 11.1. *Sputtering behaviour of oxides. Data are from Kelly & Lam (1973), Kelly (1980), Malherbe et al. (1986). The different sources indicate slight but not significant differences for MnO and TiO_2 (marked #). Columns are in order of increasing O/M atomic ratio. We note (marked ##) that ion-beam milling of NiO preferentially removes oxygen (Ostyn & Carter 1982), resulting in topotactically-aligned Ni.*

Enhancement	MO	M_3O_4	M_2O_3	MO_2	M_2O_5	MO_3
Large metal enhancement (factor >2)				HfO_2 ZrO_2	Nb_2O_5 Ta_2O_5	
Modest metal enhancement (1.25–1.75)	MnO#, NiO##, CdO, CuO	Co_3O_4		TiO_2# UO_2 WO_2 MoO_2	V_2O_5	WO_3 MoO_3
Negligible change (0.95–1.25)	BeO, ZnO, CoO, FeO, MgO, VO, NbO	Fe_3O_4	V_2O_3 Cr_2O_3 Ti_2O_3 Al_2O_3 Fe_2O_3	SiO_2 SnO_2		

metallic oxide on the surface, again the composition tends to be stable, suggesting that electric fields play a role in oxygen loss.

11.2 Laser annealing

Laser annealing was conceived as a means to anneal damage in implanted semiconductors in a way which did not harm heat-sensitive regions or exceed thermal budgets. As a technique, it had limited success, but it can be regarded as a precursor of other rapid thermal treatments and as a stimulus for techniques like laser ablation.

It is helpful to compare three methods. Our examples will be for silicon. Laser annealing concerns materials with low bandgaps, since wide gap materials allow many alternative processes. *Pulsed laser anneal* usually exploits a ruby laser (1.785 eV) or a Nd:YAG laser (1.169 or 1.497 eV), with pulses of a few tens of nanoseconds; 1.5–2 J/cm^2 is a typical energy per pulse. Above a threshold of about 1 J/cm^2 the surface appears to melt and then resolidify. Dopants may be redistributed in these events. Above a threshold of about 7 J/cm^2, catastrophic surface damage occurs. In *continuous laser anneal,* a scanned argon laser (2.41 or 2.54 eV) is usually used, with a power of 260–270 kW/cm^2. The energy is sufficient to raise the surface to the melting temperature, but insufficient to supply the latent heat of melting. *Electron beam heating* can easily supply an energy density comparable with that for pulsed laser anneal, e.g. with a current density of 1000 A/cm^2 of 10 keV electrons. Penetration is much deeper unless a low angle of incidence is adopted; there are also problems of pulse duration and control of power density.

We shall use simple models to describe the physics of the most important processes (Poate & Mayer 1982; we shall also follow the unpublished analysis of Stoneham 1978), since we have to describe a highly inhomogeneous, time-dependent, and non-equilibrium system. If 1.5 J/cm^2 is put into the first 10 μm of Si, the energy input is about 5 eV per Si atom. The latent heat of melting is about 0.5 eV/atom. There is energy sufficient to create a very high density of electron–hole pairs, probably best called an electron–hole plasma. We shall continue to use the word 'melt', since many of the properties are easily described in this way, but emphasise that the electron and ion temperatures may be neither well defined nor equal. What is clear is that there is a transient state with different properties, and certainly a different density. The electron–hole plasma will have several consequences. It will cause the energy to be deposited close to the surface, perhaps within 50 nm. It will affect the thermal conductivity. Liquid silicon, conventionally prepared, is a metallic liquid, denser than solid Si, and with a thermal diffusivity at the melting point of about 1 cm^2/s.

Views of the transient state tend to follow one of two extreme models. In the 'plasma annealing' picture (van Vechten *et al.* 1979; van Vechten 1980), disorder develops before electronic excitation has been transformed into vibrational excitation. In the 'thermal melting' picture (e.g. R F Wood *et al.* 1984), thermalisation between electrons and ion motion is largely complete at an early stage. It is clear from our earlier chapters that the duration of the excitation pulse is critical. A very short pulse which excites electrons will certainly change the forces and force constants on the individual ions. It is not surprising that observations on Si and some III–Vs within, say, one lattice period following a short pulses of, say, 100 fs (similar to a lattice vibrational period) do show signs of initial breakdown of cubic symmetry (Shank *et al.* 1983; Tom *et al.* 1988; Sokolowski-Tinten *et al.* 1995; Shumay & Höfer 1996). This initial behaviour is likely to precede full thermalisation, as can be seen very easily by looking at the dynamics of a single oscillator under a suddenly-applied force. However, longer time observations, such as those following nanosecond pulses, should show reasonably complete thermalisation. These qualitative pictures are supported by self-consistent molecular dynamics studies of Silvestrelli *et al.* (1997).

The pulse in many laser annealing studies lasts for perhaps 50 ns, and the total time for which the Si remains at a temperature of the order of the melting temperature is less than 1 μs. However, the temperature gradients (10^6–10^7 K/cm) and cooling rates (10^9–10^{10} K/s) are very high. The melt front penetrates at about 1000 cm/s (slow compared with the velocity of sound) to a depth of perhaps 1μ. Since liquid Si is more dense that c-Si, the outer surface also recedes by about 100 nm. Recovery by solidification follows at about 100 cm/s. Resolidification may be limited by the same sort of activated atomic processes which limit a-Si to c-Si recrystallisation. In laser annealing, it is more likely that heat transfer (removal of latent heat) is significant.

Given these typical times (τ, say, which we take as 100 ns), we can estimate characteristic diffusion distances $(D\tau)^{1/2}$ for a diffusion constant D. For diffusion in the liquid phase, distances are of order 100 nm; in the solid phase, there is negligible diffusion (a distance of 0.1 nm for $D = 10^{-10}$ cm^2/s). Even in the liquid, dopant concentrations will not fully even out during the melt period. Moreover, if we use the effective cooling time (Chapter 2) to assess the nucleation of clusters, we find that precipitate nucleation and growth is unlikely to be important. However, temperatures may become relatively uniform: with a thermal diffusivity of 1 cm^2/s, the characteristic length is of order 3 μm. The electron and ion temperatures may remain different, of course.

The relative concentrations of dopants in the solid and liquid will be affected by the distribution (segregation) coefficient k (the ratio at the interface of the concentration in the solid to that in the liquid phase). For Si in equilibrium processes, larger k values (0.1–1, indicating high dopant levels in the solid) are found for B, O, Ge, P, As, and C. Smaller k values (less than 1/1000, so that most dopant remains in the liquid) are found for In, Tl, Bi, Zn, S, Cu, Ag, and Au. There is evidence that the trends in distribution coefficient k are the same in non-equilibrium laser annealing processes and conventional equilibrium experiments. The concentration profiles to be expected in simple models include a characterisitic distance $d = D/v$, with D the diffusion constant in the liquid and v the solidification velocity. For laser anneal, d is about 10 nm. The distribution coefficient is not the only factor determining the profile, of course, since there can be critical solubility limits. Further, if a high dopant concentration builds up ahead of the solidification front, constitutional supercooling may set in, leading to an instability and to cellular growth.

Constitutional supercooling is one of a number of mechanisms which can lead to non-uniformities at the surface after the anneal. Other mechanisms involve laser beam inhomogeneity, lateral redistribution of dopants, convection, surface-tension gradients (the Marangoni effect), and ultrasound produced by beats between laser modes.

The disadvantages of the original concept include the lack of spatial resolution, the non-uniformity of surface after treatment, and problems associated with the relatively rigid oxide layers. These problems have tended to outweigh the gains. One can treat devices unstable at high temperatures, like Li-drifted Si or Ge, since only a small region is heated. One can maintain an external atmosphere to avoid loss of volatile constituents, like As in GaAs or Hg in Cd/HgTe. It is possible to generate very high dopings, with substitutional dopants at well above their normal solubilities; this is of value for some classes of solar cells, ultraviolet-sensitive photodiodes, particle detectors, and junction field-effect transistors (JFETs). Ohmic contacts can be created. But, overall, the method has proved too crude a tool to be taken up widely. Perhaps this is because the electronic excitation is used largely as a heat reservoir and a means to transport energy, with some initial rapid effects in reducing elastic constants as excitation develops. The atomic transport which anneals involves liquid-phase diffusion, whereas later rapid thermal anneals appear to involve recombination-enhanced diffusion in the solid phase. Laser annealing uses energy indiscriminately: it does not select specific bonds, nor even (via electron–hole pairs) specific defect species.

11.3 Laser damage

Intense laser beams of subbandgap photon energies damage materials. This is important, and limits the uses of laser materials under strong laser beams. This has led to much work elucidating mechanisms and searching for routes to the improvement of the damage threshold (see the reviews of R M Wood (1986) and S C Jones *et al.* (1989)). Now that so-called intrinsic damage thresholds are available, it is possible to compare these values with theory. Table 11.2 gives values for damage thresholds in a number of materials. For diamond-turned MgF_2, the threshold is lower near defect sites. These damage thresholds can be compared with the ablation threshold discussed in the next section.

There are two basic theories of laser damage in defect-free solids. One is *avalanche breakdown* (Sparks *et al.* 1981); the other is *electron heating* (Schmid *et al.* 1977). The avalanche theory assumes the presence of seed electrons, their acceleration under the electric field of the laser beam, and multiplication of their number by impact ionisation. The electron heating theory assumes that the initial electrons are supplied by multiphoton ionisation. Schmid *et al.* (1977) suggested that these electrons are heated by a polaron-heating mechanism, whereas Epifanov (1974, 1976) has developed a theory of free-electron heating. Many workers rightly remark on the importance of surface preparation. For single pulses of 248 nm light (5 eV) incident on CaF_2 (bandgap 12

Table 11.2. *Damage thresholds (in J/cm²) for fluorides with differing surface treatments. These results are all for 14 ns pulses of 248 nm (M Reichling 1999, private communication).*

Surface treatment	LiF	CaF_2	SrF_2	BaF_2	MgF_2
Cleaved, on terraces		>60	>60	>60	
Cleaved, on step edges		20	20	20	
Standard polished		20	15	15	
Diamond turned	40	30		15	>60
Chemo-mechanically polished		40			

eV), for example, fluences of 60 J/cm² can be withstood without damage on extended terraces free from surface defects. However, on cleaved surfaces, or on polished surfaces, which will normally have steps and damage such as dislocations and dislocation debris, damage occurs at 20 J/cm², and is localised at features such as steps (Stenzel *et al.* 1997; Gogoll *et al.* 1996; Johansen *et al.* 1995). The laser damage threshold of SiO_2 at 308 nm can be lowered from 3 J/cm² to 1 J/cm² by Xe/O implantation, and raised to 8 J/cm² by Ar implantation (Alvisi *et al.* 1999). There can also be mechanically-loosened surface fragments, which can be identified by special methods (Johansen *et al.* 1996). Cleavage-induced defects in MgO are especially important for 5 eV optical excitation in the 2–10 J/cm² fluence range (Dickinson *et al.* 1993).

X A Shen *et al.* (1987) tested the concept of free-electron heating for SiO_2 and NaCl by using two laser beams: an ultraviolet laser beam to cause band-to-band excitation by multiphoton ionisation, and an infrared laser beam to heat the electrons in the conduction band. They found that, given the free-electron concentration constant, the energy deposited in the crystal measured by the photoaccoustic method agreed very well with the equation for free-electron heating obtained by Epifanov (1974) (see Section 1.2.1). Shen *et al.* showed also that the specimens irradiated with laser pulses at the damage threshold were heated to close to their melting temperatures, and argued that the cause of laser damage was the melting due to free-electron heating. Because of these studies, the generally accepted picture of laser-induced bulk damage in defect-free wide-gap materials involves the heating the conduction electrons generated by multiphoton ionisation, and the transfer of the energy to the lattice. The damage processes may be of several types, possibly including the generation of dislocations, fracture, and defect formation. Defects can develop during the pulse before the temperature approaches the melting point. Even if heating is a major cause of damage by melting, defect formation by excitation may also play a role.

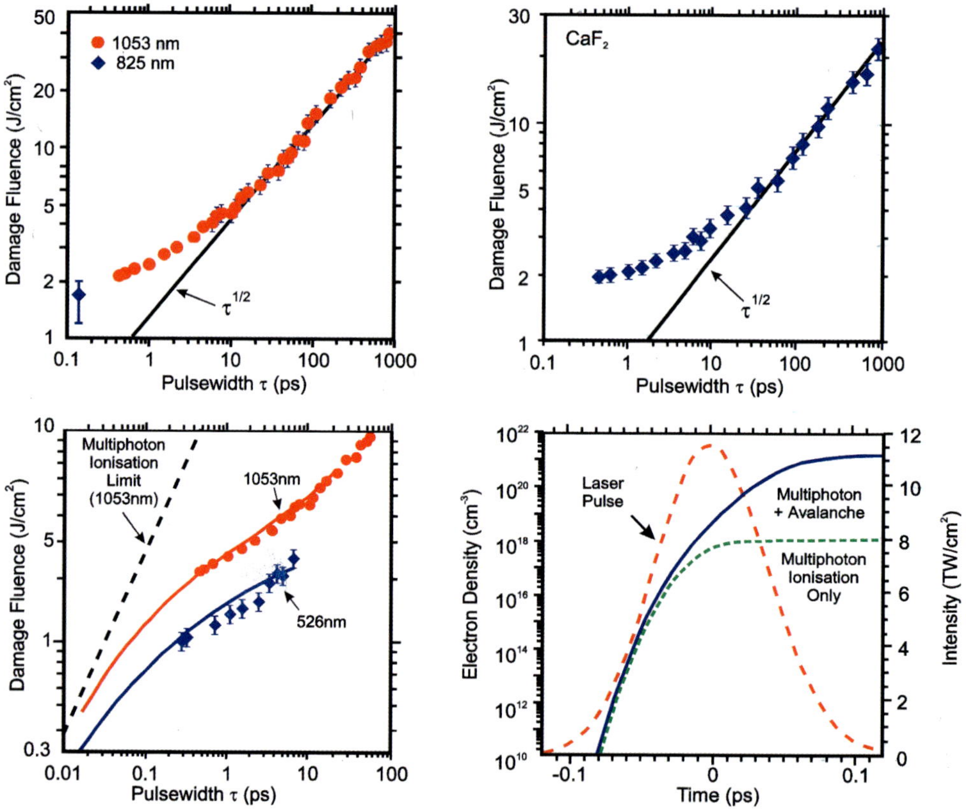

Fig. 11.1. The dependence of the laser damage threshold of SiO_2 and CaF_2 on the pulse-width.

One experimental test of the theory is the dependence on pulse-width t. If damage is caused by free-electron heating, the damage threshold will be proportional to $t^{1/2}$, since the energy remaining in the irradiated region is diffusion-limited (Bliss 1971). B C Stuart *et al.* (1996) showed that the laser-damage threshold of SiO_2 and CaF_2 at room temperature is proportional to $t^{1/2}$ for $t > 1$ ps, and to $t^{3/4}$ otherwise, for two wavelength laser pulses of 825 and 1053 nm (Fig. 11.1). They suggested a theory combining multiphoton ionisation and avalanche breakdown to explain the $t^{3/4}$ dependence. Perry *et al.* (1999) used deviations from the $t^{1/2}$ dependence as part of their evidence for the involvement of plasmon excitations.

Whichever the mechanism, results for very short (25 fs) pulses show that there are fast effects of the electron–electron interaction (M Li *et al.* 1999). There is an ultrafast decay of the plasma energy, typically 100 fs, and this decay must be recognised in analysing the breakdown process.

The dependence on pulse length applies both to single pulses and to the individual pulses in a sequence of many pulses. In particular, there are differences between sequences of pulses each longer than 1 ps and sequences each shorter than 1 ps. This implies there are processes in the solid with a timescale of the order of 1 ps, which is characteristic of some of the lattice relaxation and cooling processes. The precise timescale will vary from system to system, of course; 1 ps is representative.

Unambiguous measurements of relaxation times have been made by Chase (1994a). He used two successive laser pulses separated by time t. When t is larger than the relaxation time, the effect of the second pulse is the same as that of the first pulse. Otherwise, the cumulative effects of two pulses is more than doubled. Chase found that the relaxation time was 2 ns for laser-induced atomic emission from ZnS, as was that from borosilicate glass. These relaxation times are probably related to the lifetimes of excited states. The same approach may be used to determine relaxation times of non-linear laser-induced processes.

The damage of surfaces by laser pulses longer than 1 ps occurs over the entire area irradiated. The remaining surface is smooth, and looks as if a thin layer has been evaporated. On the other hand, the damage by shorter laser pulses is confined to a small region at the peak of Gaussian irradiance distribution. The damage looks like a shallow fracture and a pitted crater characteristic of a thin layer removed by ablation. The main difference must be associated with the lack of time for full equilibration during the short pulse, which is only a few lattice vibrational periods long. For longer pulses, the irradiated regions can establish thermal equilibrium more completely, and there is time to create states which absorb energy more readily, such as electrons and holes, transient defects. The damaged part will be heated to a high temperature in the later parts of the pulse, and can evaporate to leave a smooth surface. Ihlemann *et al.* (1992) compared the surfaces of SiO_2 ablated by 20 ns and 500 fs excimer laser pulses, and found that the surfaces ablated by 20 ns laser pulses were smooth, although those ablated by the 500 fs pulse showed clear indications of melting.

For irradiation of various silicas with the 7.9 eV (157 nm) F_2 excimer laser, Herman *et al.* (1997, 1999) found that ablation begins at 0.4 J/cm^2 for Ge-doped fused silica and at 1.0 J/cm^2 for Corning fused silica; at 193 nm, ablation only starts after several dozen incubation pulses and at around 4 J/cm^2. These data show the role of Ge (possibly for charge localisation) and also give evidence for one-photon defect generation. There is also a change in refractive index which scales as $(NF)^{0.57}$, with N the number of pulses and F the single pulse fluence.

For other oxide ceramics, the laser-induced modifications can take several forms which are not strictly damage. A nanosecond pulsed laser induces Te segregation in TeO, for example, leading to optical changes (W Y Lee *et al.*

1986). Sapphire can be amorphised (Pedreza *et al.* 1995) or partly transformed to another phase, γ-alumina (Cao *et al.* 1994). Excimer laser irradiation of $SrTiO_3$ (Yoon & Miyayama 1999) leads to O loss and results in n-type surface conductivity.

11.4 Laser ablation

Intense laser beams induce *laser ablation*, in which a large amount of material can be removed from the surface. The commonest use of laser ablation is as a source of material for deposition and film growth on a substrate. Material is ablated from a target, by some powerful excitation, and the ablated material is deposited on a substrate, where it can develop into a well-structured film. The approach has been particularly effective for non-stoichiometric oxides, such as oxide superconductors. Laser ablation can be used as a means of controlled machining as well, and sometimes holes are drilled by sharp laser beams. Electron microscopy often shows signs of melting on the surfaces from which material has been ablated. Laser ablation, like laser annealing, is indiscriminate. Laser annealing produces something akin to a melt; laser ablation causes something akin to evaporation. If laser annealing is ineffective, why is laser ablation so much more successful? The answer is that laser annealing attempts to improve an implanted surface, whereas laser ablation simply removes material from a target surface (which can be extensively damaged) to put down on a separate substrate. The laser-irradiated material is merely a source, not the product.

The effects of laser irradiation of materials depend strongly on the laser fluence. At low fluence, emissions of neutral atoms or molecules dominate for most materials. As the intensity increases, emissions of ions start to be observed, as the number of neutrals emitted from the surface increases. Electron emission is accompanied by ion emission. The emitted electrons ionise the neutrals accumulated near the surface, producing a plasma-like plume. Ions emitted from the plasma back to the surface produce further emissions of neutrals, ions, and electrons. The temperature of the surface is raised as a result, so that melting and vaporisation of the surface can occur. All of these processes interact and have cumulative effects, leading to catastrophic surface behaviour and the removal of a massive number of atoms from the surface. Electronic processes can include collective effects, as shown in the evidence (Perry *et al.* 1999) of behaviour dominated by solid-state plasma formation for pulses shorter than about 10 ps. The problem remains as to whether the initial step of laser ablation involves an electronic process, or whether the laser ablation is simply a consequence of heating. In this section we shall deal with the role of electronic processes which might occur early in laser ablation.

11.4.1 Energy absorption and transfer

Laser ablation occurs in all materials, irrespective of the presence of self-trapping. It occurs even when the photon energy is below the bandgap energy. Laser ablation is seen under irradiation by either single or multiple laser pulses. In either case, a surface undamaged before laser irradiation is changed catastrophically after laser ablation. The mechanisms of energy absorption and transfer vary, depending on several factors. These include the presence and absence of self-trapping, the magnitude of the photon energy relative to the bandgap energy, and way that the surface morphology varies during a laser pulse.

When the photon energy is less than the bandgap, energy absorption initially is mainly from multiphoton excitation or from excitations at defect sites. This initial optical absorption coefficient is extremely small. On the other hand, when the photon energy exceeds the bandgap, generation of electron–hole pairs dominates. The optical absorption coefficient is high, and almost all photons are absorbed within 1000 layers. In both cases, other sources of energy absorption build up as irradiation proceeds during a single laser pulse. These other mechanisms include free–electron absorption, which raises the electron temperature, and absorption by defects generated in the initial stage of irradiation. The concentration of electron–hole pairs is governed by Auger recombination (Yoffa 1980). The rate of recombination for the main process is proportional to n^2p, where n and p are the concentrations of conduction band electrons and valence band holes, and, since n and p are approximately equal, the rate is proportional to n^3 (Beattles & Landsberg 1959). Free carriers generated will be heated by the laser beam. The cross-section is proportional to n and p and to the cube of laser intensity (Epifanov 1974).

The energy imparted to the electronic system is gradually transferred to the lattice through electron–lattice interaction within a picosecond and by the relaxation of excited states, which results in generation of transient defects and occurs in a picosecond time range. The energy imparted to the electronic system is transferred into the bulk by the diffusion of electrons and holes. The surface lattice temperature increases gradually with a time delay of 100 ps (von Allmen 1987).

Although the surface temperature may be above the melting point at the end of a laser pulse, all the electronic processes mentioned contribute to heating and laser ablation. For instance, the rate of temperature rise depends strongly on the number of electrons in the conduction band, and so on the number of surface defects generated. Similarly, defect generation by electronic excitation usually occurs on one of the sublattices, and therefore involves lattice decomposition. Decomposition can lower the temperature at which degradation by heating will occur. Since ablation involves so many processes, it is natural that its yield is a superlinear function of laser fluence. When the yield is expressed as a power of a fluence, the power does not necessarily indicate the multiplicity of multiphoton excitation, but is simply a combination of various processes involved in laser ablation.

We know from our earlier discussions (Chapter 9) that excitation of halides and some oxides can lead to the production of point defects and the emission of individual surface atoms. Similarly, individual atom emissions from semiconductor surfaces induce defect-related multiple excitations. As the incident fluence increases, the pattern of degradation changes. Surface damage will evolve during the laser pulse, since pre-existing and laser-induced defects will affect the absorption of photons. Both electronic and thermal processes contribute to the accumulation of surface defects. Electrons, ionised from defects or impurities, can gain further energy from the light pulse, and will be emitted from the free surface. This leads to what can be a substantial electric field near the surface. The target surface changes topography. This is not a major issue, unless there is a systematic stoichiometry problem, or unless large particles (like grains) are flung off and are able to land on the target.

The consequences of excitation can depend in subtle ways on the means of excitation. This is clear from the effects of laser and electron beams on $NaNO_3$ (Webb *et al.* 1993; Shin *et al.* 1995). Laser excitation with 5 eV light has effects which show a threshold intensity. Thus, for $NaNO_3$ irradiated with fluences below 80 J/cm^2, the nitrate ion seems to dissociate, perhaps via electron attachment. NO is the sole neutral product, and its production requires an activation energy of 0.12 eV. The laser damages the surface, and there is some diffusion-limited process (of charge or energy) which does not seem to depend on substrate temperature. Above 100 J/cm^2, there is a fluorescent plume. Neutral products are NO, O_2, Na, and NO_2. The behaviour of the oxygen species is complicated, apparently because of reactions with the electrons. For $NaNO_3$ irradiated with 1–3 keV electrons, there is thermally-assisted emission, mainly of NO (presumably from nitrate dissociation) and O_2. Again, there are complex processes involving oxygen. A substantial amount of H is released. There are emission peaks near the core excitons of O (in OH) and Na. Water release seems purely thermal.

Obviously, there are two interactive factors that influence the laser ablation process: electronic excitations at the surface, and temperature rise. Electronic excitation leads directly to atomic emission, and is particularly important as a source of the temperature rise when the photon energy is higher than the bandgap energy. If the photon energy exceeds the bandgap, then the spatial distribution of the energy absorbed depends on the size of the laser beam (and so on the intensity) and on the absorption coefficient. Normally, there will be absorption which is confined to a shallow surface layer, yet reasonably homogeneous across the illuminated surface. The contribution that electronic excitations make in this case is controversial.

11.4.2 Gaps less than the laser photon energy

Most laser ablation experiments for semiconductors have used photon energies larger than the bandgap. Ichige *et al.* (1988) obtained a simple relation between the ablation threshold and the bonding strength of the material. The simplicity suggests that the phenomenon of ablation is strongly linked to the nature of cohesion of the material. There are several ways to rationalise this link. One view is a phase change when there is a dense electron–hole plasma which weakens covalent bonds and so enhances the emission (van Vechten 1982; Combescot & Bok 1982). According to this view, however, the ablation needs a long time delay. It seems unlikely that such a phase change occurs. The most common interpretion of laser ablation is that it is induced by heating of the surface layers. It is often suggested that the melt threshold coincides with the ablation threshold (Stritzker *et al.* 1981; Dreyfus *et al.* 1986). We shall show experimental observations that suggest the involvement of electronic processes in the early stage of laser ablation.

When atomic emissions are measured with submonolayer sensitivity, all semiconductors show an increase in yield when the fluence rises above a certain threshold. Above this threshold fluence, continued irradiation at the same fluence induces an increase in the yield, suggesting the generation of sites effective for atomic emission as irradiation proceeds. The idea has been put forward that emission of atoms from sites neighbouring a vacancy creates effective sites for emission. In this model, a vacancy cluster on the surface evolves from a vacancy as its neighbouring atoms are ejected; this leads to still more effective sites for atomic emission. This vacancy model of laser ablation (N Itoh *et al.* 1993) suggests that surface vacancies enhance the emission rate and make a substantial contribution to laser ablation. The ablation rates of a number of semiconductors, determined with submonolayer sensitivity, are within 20–30% of those determined macroscopically (N Itoh *et al.* 1995).

Clearly, the laser pulse that causes laser ablation can raise the surface temperature to above the melting temperature (Vivet *et al.* 1995). It is not clear whether any electronic process, including the vacancy mechanism, plays a role in this ablation. STM experiments of Xu *et al.* (1996) indicated that the emission surfaces depends strongly on morphology. If the vacancy mechanism plays a role in laser ablation, it should be possible to create surfaces with extremely small numbers of surface vacancies, and this should enhance the laser ablation threshold. At present, it is clear only that the localisation of electronic excitation energy under intense laser irradiation causes atomic emissions.

11.4.3 Gaps in excess of the laser photon energy: Effects of pre-existing defects

The irradiation of a wide-gap-material's surface with photons of energy less than the bandgap will remove material from the surface if the fluence exceeds some threshold. Ablation in this fluence range is measured in terms of the thickness of the ablated layer. The input energy for creation of a surface pit is typically 10 eV per atom removed. Smooth surfaces are seen in electron microscopic observations of surfaces ablated by nanosecond laser pulses, whereas those following picosecond laser pulses often show debris on pit surfaces (Ashkenasi *et al.* 1997, 1998), suggesting that the temperature rise was not sufficient to evaporate the surface layers. The ablation threshold is smaller for picosecond pulses than for nanosecond pulses. It is often true that the ablation rate is enhanced on repeated irradiation with laser pulses of the same fluence. A delay of a few picoseconds in the onset of atomic emissions due to 120 fs laser pulses has been measured by Rosenfeld *et al.* (1998). The delay is smaller for SiO_2 than for MgO.

Although the full mechanism of laser ablation in wide-gap materials has still to be established, we can identify the key processes for understanding ablation when the gap exceeds the photon energy. Such materials have very small optical absorption coefficients at the beginning of laser pulse. A substantial amount of laser energy has to be absorbed for laser ablation to take place. In fact, the fraction of the incident laser beam transmitted falls as the fluence increases and approaches the ablation threshold (Preuss & Stuke 1995). Obviously, the sources of optical absorption build up during irradiation, so that the optical absorption coefficient changes during a laser pulse. The sources of additional optical absorption include defects generated by band-to-band multiphoton excitations or by excitation of existing surface defects, and absorption by free electrons generated by multiphoton excitations or by defect excitations. Excitation of free electrons raises the lattice temperature through the electron–lattice interaction. As discussed in Chapter 6, defect generation by electronic excitation occurs in one of the sublattices, which implies decomposition of the material. The temperature rise due to free-electron heating occurs within a picosecond; defect generation, acompanied by lattice relaxation, requires more than one picosecond. It is likely that ablation by nanosecond laser pulses is influenced more by defect generation than is ablation by picosecond laser pulses. The formation of sharp edges with a small amount of debris may be ascribed to evaporation of decomposition products. The nature of the defects involved in laser ablation has been studied extensively in MgO and is discussed in next section.

Laser ablation can be achieved using photons with energies less than the bandgap. In such cases, there is no one-photon absorption by the perfect solid: absorption involves multi-photon processes or defects. 'Defects' in this context includes free surfaces, grain boundaries, and dislocations, as well as point defects and features (like steps and edges) on surfaces. Dislocations can have several roles: they can be localised absorbers of subbandgap light; they can be routes for electronic conduction; they may be routes for rapid ionic transport. For some purposes (see Section 11.4.4) it is sufficient to know that energy absorption occurs preferentially at surfaces and grain boundaries. But there are clear parallels between point defect processes and line defect processes, and these point to new features, such as the role of cleavage defects.

Comparisons (R M Williams *et al.* 1999) of the effects of 3 ns pulses (266 nm photons; 1–15 mJ/cm^2) and 500 fs pulses (265 nm photons; 0.3–1.1 mJ/cm^2) show interesting similarities and differences. The initial detection thresholds for desorption were about the same, despite the big differences in peak intensities. It appears that the number of photons is more important, and implies defect absorption is dominant. However, there were significant differences in species removed. For the nanosecond pulses, only Mg^+ and MgO^+ were appreciable. For the femtosecond pulses, there were also Mg^{2+} and Mg_2O^+ ions, and H^+; presumably, the hydrogen came from surface or grain boundary hydroxyl.

The irradiation of MgO by a laser beam with a fluence in the range 2–10 J/cm^2 causes the emission of many atoms from the surface, drilling a hole. In this fluence range, the surface becomes Mg-rich early in the laser pulse, so that photons incident in the later stage of the laser pulse can be absorbed strongly, causing heating. Heating can also occur by excitation of free electrons produced by ionising defects and metal colloids on the surface. The depth of a hole created by a 3.7 J/cm^2 laser pulse is about 1000 Å. If all photons were absorbed by a layer of 1000 Å, the absorbed energy per MgO molecule would be about 200 eV. Since vaporisation requires 0.5 eV per molecule, there is additional energy available to increase the number of photon absorbers. The laser ablation threshold of MgO, as detected by etch depth measurements, is about 3 J/cm^2 (Dirnberger *et al.* 1993). The formation of colour centres and of microcracks on the surface is observed even for fluences between 1–3 J/cm^2. There is an incubation period for fluences below twice the ablation threshold: no visible indications of ablation are seen for the first five laser pulses. This incubation period can be ascribed to the buildup of defects which can act subsequently as the photon absorbers (Campbell *et al.* 1999).

The study of Dickinson *et al.* (1993) concentrated on laser pulses of 248 nm (5 eV) light incident on cleaved single-crystal MgO (bandgap 7.9 eV). There was clear evidence for cleavage-induced defects and for the production of defects during laser irradiation at fluences of 3–10 J/cm^2. Cleavage appears to induce some of the same defects as does polishing or abrasion. The amount of cleavage damage can be changed by various treatments. Annealing prior to cleaving enhances cleavage damage (only 1.2 J/cm^2 for plume formation, rather than 2 J/cm^2), presumably because it dissolves the defect aggregates which impede dislocation motion during cleavage. Chemical polishing after cleavage strongly reduces damage, apparently because it removes the more highly-damaged, near-surface material; 3.4 J/cm^2 is needed for plume formation, irrespective of whether there has been an anneal.

Cleavage-induced defects are relatively sparse, so there are wide variations in response from site to site. Extensive melting is seen in regions where plastic deformation has occurred, for example along the lines where material has been undercut from the main body of the sample. Some pulses lead to a fluorescent ablation plume; others do not. Electron microscopy shows conical holes at the sites which provided a plume. These holes are very similar to the pits which are produced by etchants intended to show dislocations, although the number of laser-induced holes is much less than the number of dislocations shown by standard etch-pit methods. There are also ‘proto-holes’ which show signs of local melting. These ‘proto-holes’ are populated by small spheroidal particles enriched in Mg atoms. These particles are typically 100 nm in diameter, and there is evidence of scavenging, so that large droplets are observed associated with fewer small droplets. Presumably the local excess of Mg (reduction of MgO) results from loss of oxygen and some subsequent motion of simple vacancies, or of F^0 or F^+ centres.

Analysis of the temperature distributions expected for line defects permits an interpretation of the distribution of the droplets. The observed distribution is consistent with heating due to absorption at a line defect at an angle of about 45° to the (001) surface. The natural candidate is a screw dislocation. However, the energy balance suggests that the features need far more energy than is absorbed directly by the dislocation. At least at lower fluences (1–2 J/cm^2), the energy input to the dislocations is primarily from the non-radiative recombination of electrons and holes generated away from to dislocation core, perhaps at F^0 or F^+ centres. At higher fluences, free-carrier heating may play a role.

Dickinson *et al.* discussed two deformation mechanisms in more detail. One was cleavage step formation. Adjacent parallel but non-coplanar cracks overlap. The shear stresses in the intervening regions cause dislocation motion and substantial deformation. When the intervening material fails, what remains is an undercut edge with a high dislocation density. Point defects produced during the deformation lead to strong photoluminescence in this undercut region. In MgO, for which interstitials have a very high formation energy, one expects there to be vacancy-related defects. The 410 nm blue luminescence, characteristic of mechanical damage (Y Chen *et al.* 1975) may well be an oxygen vacancy complex. A second process, fracture-related dislocation motion, leads to rows of laser-generated pits along $\{100\}$ shallow steps. Dislocation loops are nucleated ahead of the advancing crack tip. The intense stress field drives them ahead of the crack tip, leading to long dislocation segments close to the fracture surface. The long segments, hindered by lattice defects, break into shorter loops along the original segment. These give rise to the parallel, shallow, surface steps which can be seen either as laser-induced pits or by traditional etch methods. The significant feature is that the ablation features give information about dislocation processes and the cleavage phenomenon which is complementary to that obtained by traditional techniques.

Under electron irradiation, the damage of MgO shows pronounced effects associated with OH^- (Kinoshita *et al.* 1998). The radiation tolerance appears to be related to the hydroxyls next to cation vacancies. The main resulting damage is interstitial loops, with few defect clusters or bubbles. Structural vacancies may suppress the nucleation or growth of loops, and so improve irradiation resistance.

11.4.4 Mesoscopic modelling of laser ablation

Another implication of the role of dislocations is that excitation processes cannot be modelled fully at either the macroscopic scale (integrating heat transport equations) or an atomistic level, such as molecular dynamics. It is crucial to include simulations at some intermediate scale. This has been recognised in the work of Ribeiro *et al.* (1997a,b, 1998a,b), who make mesoscopic calculations by what is, in effect, a finite difference calculation. Their model for MgO excited by subbandgap light recognised a number of important ideas. First, optical absorption of subbandgap light will occur primarily in special regions: at the surface itself, at grain boundaries, and at defects. Secondly, the carriers which are excited can move, be excited further by the laser light, recombine at the defect sites, or be emitted from the surface. Thirdly, there will be carrier motion, either diffusive or drift under the electric fields which evolve. Excitation from one site and recombination at another is a means of energy transport. Fourthly, there will be evaporation of electrons and of ions, and this

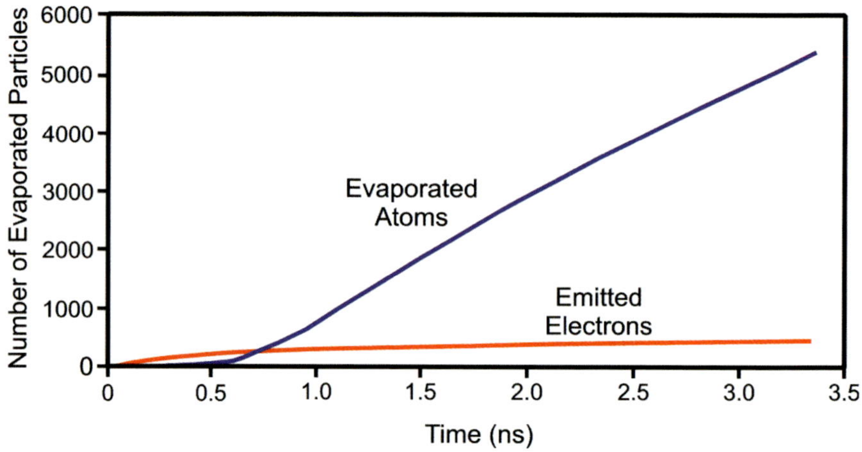

Fig. 11.2. Time dependence of electron and atom emission predicted for MgO (Ribeiro *et al.* 1998b).

will depend on the local temperatures and binding energies. Reasonable estimates of most of the parameters are possible, and can be improved by linking the parameters to fuller atomistic models.

What emerges from this approach (Figs. 11.2–11.5) is that, in the first few picoseconds, electrons are generated in special regions, such as at grain boundaries, in sufficient numbers to start additional strong absorption. Irrespective of the precise numbers, within reasonable limits, electron evaporation occurs rapidly, long before atom evaporation is significant. This is partly because, for electrons with high work functions, the electrons build up and absorb more of the light, and so heat rapidly so as to overcome the higher work function. The loss of electrons leads to electric fields, which are a major factor in carrier motion. The fields may well be large enough to cause dislocation motion. After a few tens or hundreds of nanoseconds, the numbers of atoms emitted exceed the numbers of exoelectrons, which are limited by the electric field. The predicted number of evaporated atoms has a very similar dependence on laser fluence as that observed (Ishibashi *et al.* 1994). A further conclusion from the mesoscopic modelling is that congruent removal of species is expected, since there is little dependence of emission probability for binding energies above a relatively modest value.

As evaporation proceeds, there is preferential loss from the grain boundaries. This leads to changes in surface roughness, and ultimately to the buildup of the characteristic cone structures (Ribeiro *et al.* 1997a). The extent of roughening depends on the assumed thickness of the grain boundary. ‘Thickness’ means, of course, the width of the zone in which energy absorption occurs, and will

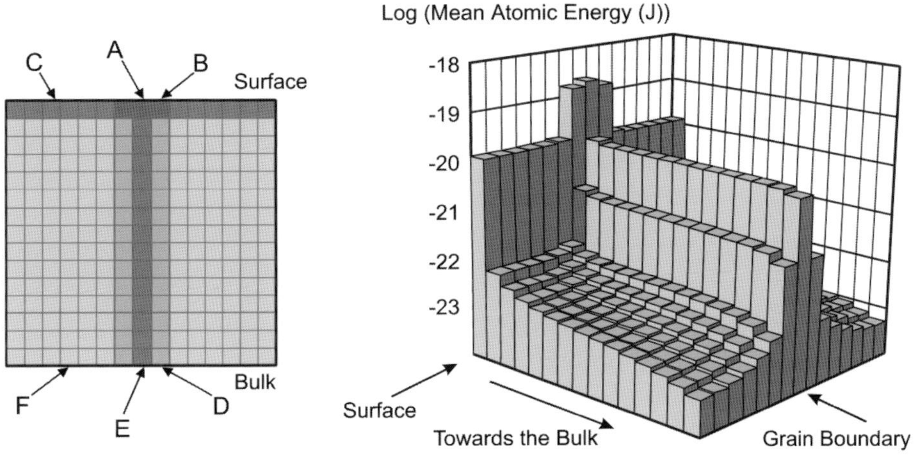

Fig. 11.3. Distribution of the mean atomic energy near a grain boundary in MgO after 3 ns (Ribeiro *et al.* 1998b).

include any defect atmosphere surrounding the boundary. The effective width will include contributions from energy absorbed by ionised centres or by free electrons formed as a result of initial excitation. Completely different behaviour is seen when laser irradiation is replaced by that of an electron beam. The energy from the electron beam is not concentrated at defects, and the rastered beam smooths the MgO surface (Bullough *et al.* 1990).

The formation of films by laser ablation involves three phases: the initial excitation, just described; processes involving the motion of ions and their interactions as they travel from the target to the substrate; and processes on the substrate. We shall not discuss these later stages, except to remark that it is possible that some of the particulates which spoil otherwise good laser ablation films are caused by grains (or large parts of grains) being forced off by strong light absorption in grain boundaries at the backs of grains, on the side away from the surface.

11.5 Lithography and nanolithography

11.5.1 Lithography by electron beams

Structures can be created on oxide surfaces using intense electron beams. Typical energies are perhaps 100 keV, so band-to-band transitions are easily excited, but there is insufficient energy to create displacement damage (460 keV would be needed for MgO). The main features are these. First, the energy is absorbed mainly in band-to-band transitions, generating electron–hole pairs,

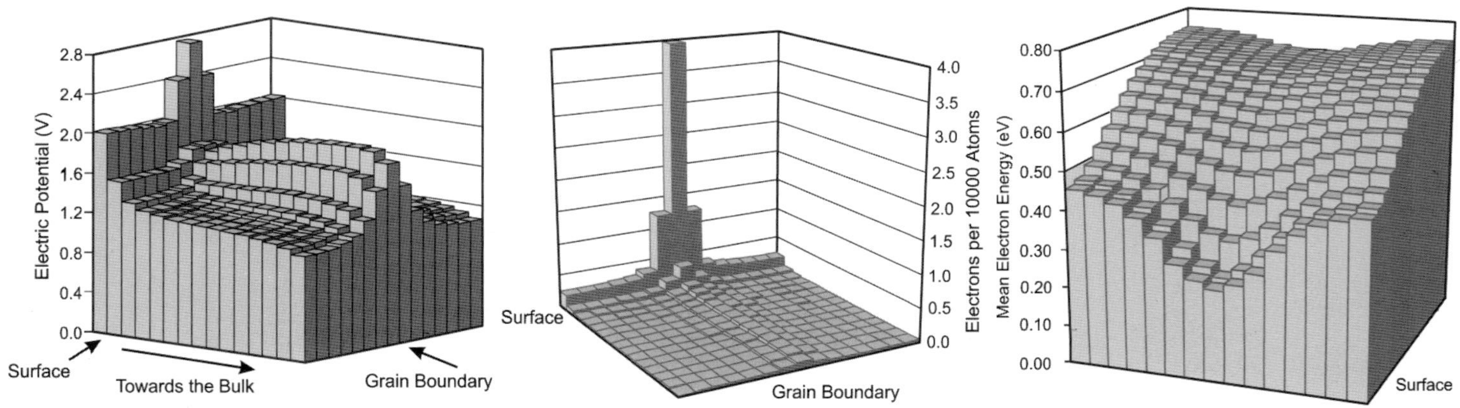

Fig. 11.4. Electrical potential, electron density, and mean electron energy near to agrain boundary in MgO (Ribeiro *et al.* 1998b).

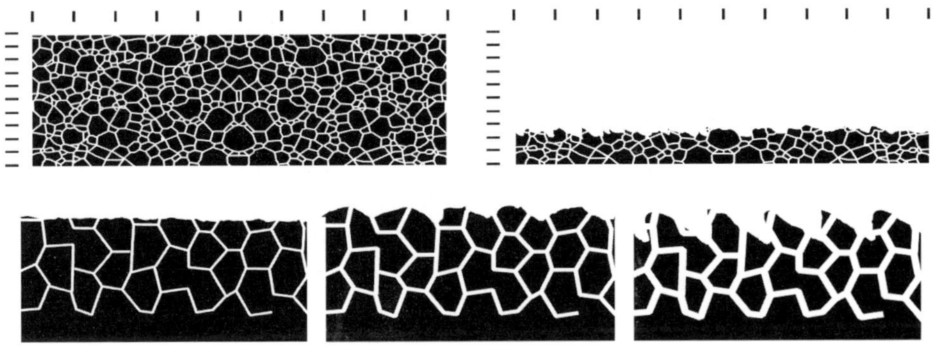

Fig. 11.5. Change in surface morphology for MgO with grain structure ((Ribeiro *et al.* 1998b).

or in plasmon excitation (Chapter 1). There is a relatively low probability of core excitation. Unlike optical excitation, there is no special need for the presence of defects. Secondly, elastic sputtering of surface atoms by electrons occurs at a threshold energy much less than that for defect production in the bulk (Cherns 1979). Transmission sputtering is more significant at the exit surface than the front surface, because of the preferred forward ejection of recoils by high-velocity electrons. Thirdly, energy is accumulated more gradually than under pulsed laser irradiation. There will be more time for diffusion processes to lead to the nucleation and growth of pits or holes under electron excitation.

11.5.2 Nanolithography of MgO and other oxides

Electron irradiation of MgO creates holes on the exit surface more efficiently than the entrance surface, for both 100 keV (Turner *et al.* 1990; Humphreys *et al.* 1990) and 200 keV electrons (Kizuka & Tanaka 1995, 1997). The removal rate from a vacuum-deposited film of MgO by the 200 keV electrons is 0.06 nm/s at 3 A/cm^2; the removal rate from MgO microcrystals by 100 keV electrons is 0.5 nm/s at 8×10^3 A/cm^2, which means one MgO unit is removed for every few thousand electrons. This difference between films and microcrystals probably stems partly from defects and partly from the difference in electron energy. Both groups show that the removal rate is proportional to dose, and that the stoichiometry does not change as atoms are removed. Turner *et al.* were able to rule out a Knotek–Feibelman mechanism. Using high-resolution transmission electron microscopy, Kizuka & Tanaka (1995) revealed preferential ejection of surface atoms at steps, and suggested that the emission was the sputtering of atoms at special sites on the surface for which the threshold energy was lower than at perfect surface sites. Preferential sputtering of such atoms with low threshold energies is highly likely, since a substantial part of electron-induced sputtering arises from the collision cascades which are generated within the oxide when the energy of the primary recoils exceeds a threshold energy.

We believe that vacancy sites or special sites on the initial surface act as nuclei for material removal, as noted above. In addition to the actual sputtering process, there will be other processes, such as surface restucturing and diffusion; the electric fields generated will also influence the way the holes develop. There are interesting similarities and differences between nanolithography and material removal by STM excitation (e.g. Heinzelmann *et al.* 1988). In both cases an electron beam is involved, and there is an electric field which affects what happens, but the roles and relative importances are different.

Nanolithography of silica

Like MgO, SiO_2 can be removed from the exit surface (Bullough *et al.* 1990; see Y Chen *et al.* 1994). The removal rate for O is faster than that for Si. In this case, removal is ascribed to the Knotek–Feibelman mechanism, but the possibility of elastic sputtering remains. It is likely that the creation of excitons and self-trapping play a role. Y Chen *et al.* (1994) showed by measuring the electron energy loss spectrum due to the Si K edge that SiO_2 is reduced to Si by irradiation with intense electron beams. The efficiency is not very high: 0.022 nm/s for 4.4×10^2 A/cm^2. The reduction to Si appears to be a photolytic reaction, perhaps due to the bond scission by the formation of self-trapped excitons or of neighbouring excitons. Neighbouring excitations are possible under high current densities or because of the low collision mean free paths of low energy electrons. G S Shen & Humphreys (1999) concluded that, for electron probe irradiation of a-SiO_2 (100 keV electrons; $10^5 - 10^9$ A/cm^2) both a surface and a volume mechanism operate; possibly initial O loss, perhaps O diffusion in the irradiated volume, then Si loss.

Nanolithography of alumina

Broers (1988) observed what he termed the direct sublimation of alumina when seeking resolution limits for electron beam lithography. He noted the emission of secondary electrons with energies up to 20 eV. Humphreys *et al.* (1990) observed for α-Al_2O_3 that an oxygen bubble forms from inside, and then bursts out. They suggest a Knotek–Feibelman mechanism. The Al L edge disappears, whereas the O K edge remains and the molecular O_2 edge emerges. Al capping seems to occur.

In the more complex oxide, β-Al_2O_3 (with prototypical form $Na_2O.11Al_2O_3$)

electron irradiation causes front-surface indentations grow at rates up to 25 nm/s. Unlike α-Al_2O_3, β-Al_2O_3 does not show the Al capping. Mochel *et al.* (1983) showed that it was possible to draw 2 nm holes and lines.

11.5.3 Nanolithography and analogous phenomena in other inorganic materials

Nanofabrication of Si by electron beams

Nanoholes are created at the exit surface for Si irradiated by intense beams of 140–200 keV electrons (Takeda *et al.* 1997a,b). The irradiated area contains several holes of 2–3 nm diameter and 5–10 nm deep with a separation of approximately 6 nm. The multiple holes are unlike other experiments, where a single hole is generated. The removal rate of material from the hole sites is approximately 0.02 nm/s for a beam current of 200 A/cm^2; this is similar to the removal rate for MgO. Hole formation does not occur above 400 °C and is prevented by oxidation of the surface. The holes appear to nucleate at the exit surface, where the loss of Si is ascribed to the elastic encounters (electron-induced sputtering). Nucleation probably starts at vacancies: the atoms around vacancies are emitted with higher probabilities. Surface diffusion may play a role as well, possibly enhanced by excitation. Takeda *et al.* (1997a,b) suggest that processes similar to spinoidal decomposition determine some of the features, such as the way that the holes are distributed.

Nanolithography of halides

Deep nanoholes can be created using convergent nanosized electron beams for many halides (Broers 1988), including NaCl (Broers *et al.* 1978; Isaacson & Murray 1981), LiF, MgF_2 (Broers *et al.* 1981) and AlF_3 (Isaacson & Murray 1981; Kratschner & Isaacson 1986). Broers (1988) suggested the mechanism is direct sublimation. For NaCl and AlF_3, holes of 0.5 nm or less were possible. Patterned AlF_3 films were demonstrated as masks for writing on silicon nitride (Kratschner & Isaacson 1986), and MgF_2 was tested as a mask for the lithography of Au. However, the doses were large (0.1 C/cm^2) for MgF_2, and the films did not stand up to etch conditions well. G S Shen & Humphreys (1999) concluded that there is a volume mechanism operating for a-AlF_3 under electron probe irradiation (100 keV electrons; 10^5-10^9 A/cm^2). Possibly a fluorine bubble develops below an Al cap, which then fails.

11.6 Irradiation with heavy ions

The high electronic charges of heavy ions mean that they have large cross-sections for elastic and electronic encounters. This means that heavy ions deposit energy along their paths with high energy density. The maximum energy deposition through elastic encounters takes place at energies of 1–10 keV/nucleon, whereas the maximum deposition by electronic encounters occurs at energies of order 1000 keV/ nucleon. At 1–10 keV/nucleon, the mean free path between knock-ons can be as short as the interatomic spacing: essentially all atoms along their paths are displaced, producing a so-called *displacement spike*. Ions in this energy range produce mainly surface effects because of their short range. When the heavy ion has an energy 100–1000 times larger, traces of heavy damage are seen in the bulk of the material, at the end of the ion paths. Fission fragments generated in nuclear fuels produce these so-called *'fission tracks'*.

The biggest cross-sections for electronic encounters are realised using ion accelerators of gigaelectron volt energy range. We now discuss the materials modifications induced by gigaelectron volt heavy ions. We will note that dense electronic excitation in a material with strong electron–lattice coupling can amorphise the material. We have already discussed amorphisation, crystallisation, and interface mixing under dense elastic encounters. Here, we shall discuss the similarities and differences when there is dense energy deposition by elastic and by electronic encounters.

11.6.1 Processes taking place near GeV heavy-ion paths in the first few femtoseconds

The primary interaction of heavy ions with non-metals creates election–hole pairs along their paths, by direct collisions with incident ions through the energetic electrons (delta rays) generated by incident ions. The kinetic energy given to the electrons is such as to distribute the electrons further from the ion path than the holes (Fig. 11.6). The number of electron–hole pairs is reduced by Auger recombination, and the energy released generates energetic electrons. These energetic free electrons lose their energy by electron–lattice interaction within 1 ps. In some materials, as discussed in Chapter 4, electron–lattice coupling can lead to lattice relaxation in the electronic excited state or to self-trapping of excitons. This will happen over 1–10 ps, and will convert the potential energy of the electron–hole pairs to lattice energy. Thus, after a few picoseconds, in materials for which excitons are not self-trapped, there will be electron–hole pairs and enhanced lattice vibrations along the ion paths. In the materials in which excitons are self-trapped, there will be a distorted lattice and enhanced lattice vibrations. The average distance between electron–hole pairs can be smaller than the exciton Bohr radius. Unlike laser beams, where the

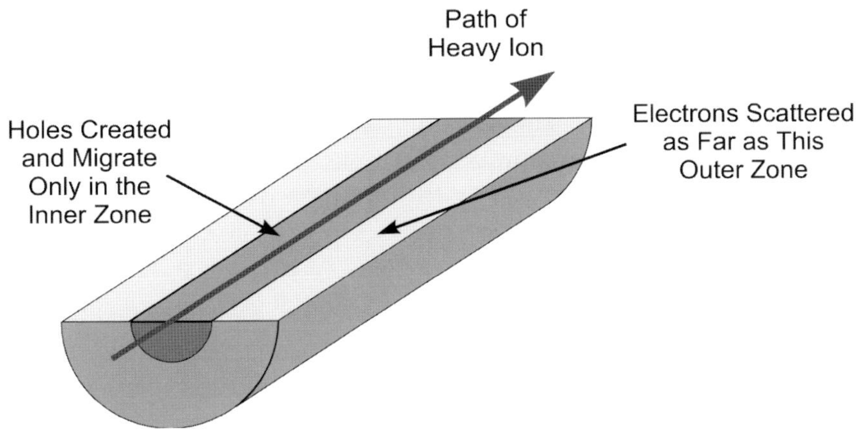

Fig. 11.6. Processes near gigaelectronvolt heavy-ion paths: the first few femtoseconds.

density of electron–hole pairs increases only slowly, heavy ions with high stopping power generate dense electron–hole pairs within a few femtoseconds.

The energies possessed by these electron–hole pairs and the enhanced lattice vibrations are taken away from the ion path by ambipolar electron–hole diffusion and by thermal diffusion, respectively. Any defects produced along ion paths should be determined by processes which take place within a few picoseconds. As we discuss later, three main driving forces have been suggested: the electric field due to the inequivalent electron and hole distributions, heat, and the self-trapping of excitons. In materials in which excitons are not self-trapped, the localisation of energy at defect sites plays a role similar to self-trapping.

The region of a material excited by electronic encounters, with the associated lattice relaxation in its excited state, is similar in certain respects to the material excited to high energy density by elastic encounters. The dense elastic encounters displace atoms from their sites to various extents, with an average atomic vibrational energy of over 1 eV. The situation is rather similar to one in which self-trapped excitons are generated at most sites along the path of a heavy ion: in both cases, most atoms are displaced and there is enhanced vibration. There are significant differences, however: in collision cascades, all atomic species are displaced, whereas it is processes on the anion sublattice which dominate in most processes involving self-trapping.

We shall concentrate on inorganic systems. In organics, such as polymers and diamond-like carbons, one major consequencc at high doses is the loss of H and the creation of relatively-conducting, possibly graphitic, tracks. For 2

MeV Ar^+ ions, low doses of $10^{10}/cm^2$ show initial beam-induced structural effects. At doses of $10^{11}/cm^2$, scission or cross-linking become evident; at $10^{12}/cm^2$, the role of oxygen becomes important; at doses above about $10^{12}/cm^2$, carbonisation occurs. This carbonisation is associated with an insulator–metal transition as the random carbonaceous tracks overlap to form a connected network (Wasserman 1986).

11.6.2 Registration of heavy-ion tracks

Fleisher *et al.* (1965) realised that heavy ions with an energy above 1 MeV per nucleon leave tracks in several materials. This phenomena is known as *track registration*. It has been shown that ions with more than a critical electronic stopping power can register tracks, which shows that electronic encounters cause the registration process. Since the stopping power is proportional to the square of the atomic number of the incident ion, the density of excitation along paths of heavy ions can be so high that all molecules within a certain region along the tracks are excited. At these excitation levels, one expects defect phenomena which are completely different from those which are associated with isolated excitons.

The critical stopping power for track registration is clearly an important parameter. The values obtained experimentally are shown in Table 11.3. For some materials shown in the table, continuous tracks are not observed. Instead, only fragmented (intermittent) tracks are seen, or only isolated defect creation in some cases. The lowest critical stopping power is found for materials like SiO_2, mica, and the high-T_c oxide superconductors. In these materials, the tracks observed are usually continuous. X-ray diffraction and Raman studies show that the continuous tracks in SiO_2 are amorphous, and are accompanied by a large volume expansion (Duraud *et al.* 1988 and references therein). The expansion is anisotropic (Benyagoub *et al.* 1992): quartz expands to the direction normal to the ion beam and shrinks in the direction parallel to the ion beam. Tracks in quartz are very different from those in LiF, again an ionic crystal, although excitons are self-trapped in both materials. In LiF, the tracks contain lithium colloids (Schwartz *et al.* 1998). The differences between LiF and quartz stem at least partly from the fact that quartz is more easily amorphised.

Oxides like MgO and Al_2O_3 show much larger critical stopping powers, and their tracks tend to be fragmented. No permanent effects of the irradiation of energetic heavy atomic ions have been observed for Si and Ge. Dunlop *et al.* (1998) showed that continuous tracks are generated by irradiation with 30 MeV fullerene ions, which give an extremely high excitation density (>10 eV/Å^3).

Table 11.3. *Threshold stopping power for track registration.*

System	Threshold stopping power (keV/nm)	Self-trapped exciton?	Reference (stopping power)
Continuous tracks	1–10	Yes	
SiO_2	2	Yes	Meftah *et al.* (1993)
$Y_3Fe_5O_{12}$	4	Probably yes	Toulemonde *et al.* (1994)
Mica	5	Possibly yes	Toulemonde *et al.* (1994)
$LiNbO_3$	7	Probably yes	Meftah *et al.* (1993)
LiF	<10	Yes	Schwartz *et al.* (1998)
Fragmentary tracks	>20		
MgO	20		Canut *et al.* (1995)
Al_2O_3	21		Canut *et al.* (1995)
Bi	31		Wang *et al.* (1996)
Si	>28		Dunlop *et al.* (1998)

This may mean that there is an even higher critical stopping power, or perhaps that recombination-enhanced diffusion eliminates the damage effectively. For metals, only defect formation is observed, and the critical electronic stopping power is very high, e.g., above 40 keV/nm for Fe (Quéré 1992; Dunlop & Lesueur 1991). In metals and semiconductors, defects are annealed under energetic heavy-ion irradiation. The annealing of defects in metals takes place at a stopping power similar to that for defect generation, whereas that in semiconductors takes place at a much lower stopping power (Huber *et al.* 1998; Iwase *et al.* 1987, see also Dunlop *et al.* 1994). It is probable that, in semiconductors, diffusion enhanced by electronic excitation is important in defect annealling.

11.6.3 Effects of heavy-ion irradiation on surfaces and interfaces

The irradiation of insulators with energetic heavy ions causes sputtering which is purely electronic in origin. Early studies measured ion emission, which is only one part of the emitted particles (Qiu *et al.* 1983). Later measurements of all the emitted particles from yttrium iron garnet show that sputtering is non-stoichiometric and has a threshold fluence, which is much larger than that for track registration (Meftah *et al.* 1996). Until measurements are made for other materials, one should be cautious in making comparisons between threshold fluences for sputtering and track registration. The non-stoichiometry of the emitted particles suggests that the sputtering is related to lattice relaxation under dense electronic excitations. Beranger *et al.* (1996) showed that the irradiation with heavy ions of MgO which contained Na precipitates reduced the size of the precipitates, implying that Na reacts with oxygen released by irradiation. It has been shown that the irradiation of SiO with heavy ions produces Si precipitates. This suggests that SiO is decomposed into Si and SiO_2 (Rodichev *et al.* 1996).

The effect of heavy-ion irradiation seems to be more dramatic for smaller grains. Hemon *et al.* (1998) found that yttrium oxides changes from cubic to a monoclinic phase on irradiation with lead ions of stopping power 38 keV/nm, and that the change is more efficient for smaller grains. Similarly Berthelot *et al.* (1998) have shown that the shapes of the relatively large nanometric grains of tin oxides are changed by irradiation with lead ions, whereas the smaller ones disappear. These results have been interpreted in terms of a thermal spike model. Multiple ionisation may play a role as well. One of the interesting applications of such phenomena is the alteration of the structures of nanoparticles embedded in materials with higher threshold stopping power.

11.6.4 The Coulomb explosion model

The earlier models of track registration did not take account of the formation of excitons, but emphasised space charge buildup. Fleischer *et al.* (1965; see also Fleischer *et al.* 1975) suggested that electrons are ejected from the ion paths, forming a *positive* space charge along the path and a *negative* space charge around it. Then he suggested a Coulomb explosion occurred, creating defects along the track.

Evidence in support of the model comes from work on the dichalcogenides MX_2 (M = Ti, Mo, W, Nb, Ta, X = S, Se, Te; Morgan & van Vliet 1970). Tracks are formed when the conductivity is less than $10^4\ \Omega^{-1}\ cm^{-1}$, and are not formed for conductivities above $10^5\ \Omega^{-1}\ cm^{-1}$; the position is less clear at intermediate conductivities. Further, changing the conductivity of WTe_2 by doping with Nb converts track-forming to non-track-forming as the conductivity rises above about $10^5\ \Omega^{-1}\ cm^{-1}$. Yet there are some difficulties in this Coulomb explosion model. The positive charge left along the path remains as positive holes, and the secondary electrons ejected cause further ionisation and produce electron–hole pairs. All these processes happen within 0.1 ps, and are followed by thermalisation of the conduction electrons within 1 ps. The passage of a heavy ion leaves non-uniformly distributed electrons in the conduction band and holes in the valence band. Very crude estimates of timescales can be made using equilibrium information about mobilities, and since we expect the energetic electrons to have mean free paths which are large compared with the excited region. But, given the relatively high electron mobilities in oxides under normal conditions, the non-uniformity should be recovered in a very short time, probably faster than a picosecond, since the electron and hole are separated by only 10 nm in a dielectric medium in which the electron has a mobility of 20 $cm^2/(V\ s)$. The question remains that of whether non-uniform charge distribution remains long enough for Coulomb explosion to take place. This means one should examine other possible roles of electronic carriers in suppressing track formation.

11.6.5 Thermal spike model

We have noted that a purely thermal (melting) model explains the appearance of new phases following displacement cascades in silicates (Meldrum *et al.* 1998). Can a purely thermal process explain track registration? The large energy deposition will lead to a temperature rise along the ion path, and this might explain track formation. The thermal model can indeed explain the relation between track radius and stopping power for many materials (Toulemonde *et al.* 1992). However, simulations suggest that the maximum temperature is attained 10 ps after the energy is converted to lattice phonons, and falls afterwards. There is a question as to whether melting can occur in this time range. The critical stopping power, explained in terms of the melting threshold, cannot explain the large difference between MgO and SiO_2, for example. Nor is it easy to understand the situation in Si, where the melt would be metallic.

11.6.6 Excitonic model

One crucial question is whether the formation of dense electron–hole pairs or excitons is important in track registration. Electron–hole pairs are created along the path of a heavy ion. Many will be converted to excitons in a short time. Some of the excitons (or indeed the electrons and holes) may diffuse away from the path; others will be trapped. It is the energy deposited locally to the lattice on electron–hole (or exciton) recombination which causes modification. The greatest impact of the energy deposition will be in those cases where electron–lattice coupling is strong, e.g. in those materials for which excitons are self-trapped. Self-trapping itself means efficient energy localisation. This suggests a reason for the low critical stopping powers in SiO_2 and mica. Experimentally, the average density of excitation within the radius of the track for SiO_2 is as large as one electron–hole pair per SiO_2 unit (N Itoh 1996; N Itoh & Stoneham 1998). We may compare these systems with ones like MgO and Al_2O_3, for which excitons are either not self-trapped or self-trapped only shallowly. In these oxides, the critical stopping powers are ten times larger than those for materials in which excitons are self-trapped. Since the excitons are not self-trapped, any excitons are localised only at defect sites near the ion path, and the others will simply diffuse away. The defect-initiated photolytic processes due to dense electron–hole pairs are now spread over a wider range. This is a possible explanation of the fragmental tracks. In this model, the trends with conductivity for the dichalcogenides noted above presumably reflect other electronic processes involving free carriers.

None of these models is well established. The processes probably comprise a temperature rise, the effects of a local electric field, and electronic changes such as bond weakening because of the electronic excitation energy. It is a characteristic highly non- equilibrium situation. Two points deserve comment. First, the formation of a highly-excited region along an ion path and the role of the excitonic mechanism are analogous to the hot zone generated in a collision cascade. The energy input along the ion path can be understood within the excitonic model. This is consistent with the thermal spike model, which explains the stopping power dependence of track radius quantitatively (Toulemonde *et al.* 1992). We can therefore bring together a microscopic (exciton) picture with an appropriate macroscopic model (thermal spike) for track registration in those materials for which excitons are self-trapped. The same argument does not hold for MgO or Al_2O_3. Secondly, the critical stopping power for defect generation in metals is even higher than in MgO or Al_2O_3. The thermal spike model fits the experimental data well, but must be related to a different microscopic picture. The essence of the concept of the excitonic model is a change in the energy surface which might be described as weakening bonds. This concept is harder to apply to metals, although it is closely linked to the ideas of Flynn & Averbach (1988) for the collision cascade. Their point is that the electronic system in a metal can act as an energy sink or as a means for redistributing energy.

We also remark that an energetic heavy ion passing through a solid is accompanied by a so-called wake potential associated with plasmon excitation, and this might be correlated with defect creation. The same concept has been suggested by Hoheisel *et al.* (1988) as the cause of laser ablation of metals. A further comment concerns the low critical stopping power of high-T_c oxides. This is not in accordance with the excitonic model, since the bandgap is small. Instead, we must recognise another factor. In oxide superconductors, very little energy is needed to move the oxygens around (in the same way, one must recognise that in silica, which has the lowest critical stopping power, very little energy is needed to generate another 'phase'). When a system needs very little energy to change structure, there may be routes to excitation-induced modifications that need not involve the excitonic mechanism in its standard form.

Excitonic ideas also point to an explanation of the luminescence associated with track formation by heavy energetic ions in quartz (Constantini *et al.* 1997). The luminescence which results (at 1.65 eV, 2.5 eV, and 2.8 eV) does not depend on either the stopping power or the fluence ϕt. However, the intensity of the luminescence decreases exponentially with fluence, varying as $\exp(-A\phi t)$. The decay cross-sectional area A increases as the stopping power is raised above the 2.5 keV/nm threshold for amorphous track formation. The value of A rises to be about an order of magnitude larger than the track core cross-section. What appears to be happening is that ion-induced defects interact with self-trapped excitons, enhancing non-radiative recombination channels.

Summary

At high densities, individual exciton events cannot be identified unambiguously. The electronic excitation at these high densities can be equivalent to the excitation of every atom within some region. Rebonding and transformation to some other structure or configuration should be relatively easy. Depending on the excitation mechanism, the energy can be given primarily as ionic kinetic energy (of which a part will become electronic excitation energy), or primarily as electronic excitation energy, which partly becomes ionic kinetic energy. Electronic excitation will give transient charge redistribution. The resulting electric fields can influence subsequent ionic motions, especially in an inhomogeneous system (fission track; end of collision cascade; ablation of a nanocrystal).

High-intensity excitation is possible using lasers, electrons, or fast ions. For lasers, there is a critical distinction between energies less than the bandgap and energies greater than the bandgap. For energies above the bandgap (as in laser annealing) the energy is absorbed close to the surface, and does not depend strongly on defects or impurities. For energies below the bandgap, energy absorption is dominated by defects, including interfaces.

Chapter Twelve

Applications of materials modification by excitation

12.1 Aims of materials modification

Most of our book aims at a coherent and (where possible) quantitative understanding of materials modification following electronic excitation. Yet this understanding is only one part of the picture. Another major component is practical: how can the consequences of this controllable excitation be exploited? We may categorise most areas of major practical importance under a small number of headings:

Removal of surface material, especially in a controlled way;
Modification of the surface layers;
Modification of the bulk in a selected way;
Altering the absolute rate of a bulk or surface process;
Switching from one process to another: altering branching ratios. In this category, we might also include the *manipulation of excitation through materials design,* even though the material itself is not modified by this excitation. For example, the xerographic process relies on the manipulation of electron and hole excitations in Se. The photochemical splitting of water using colloidal particles (typically TiO_2/RuO_2 with Pt) (e.g. Grätzel 1981) relies on the appropriate use of the energy stored following optical excitation, which ideally leaves the particle itself unaltered.

Creation and control of damage

Damage may be intentional or unintentional. Damage avoidance is achieved by the minimisation of those processes by which excitation causes harmful modifications, or simply by arranging for the damage to occur only in unimportant regions.

There is also a further classification which is helpful. Most of the applications can be grouped according to the way in which excitation causes change. In all of the methods, electrons and holes produced may recombine radiatively, giving luminescence, a phenomenon exploited in display technologies. There are four main categories:

Class I Excitation into selected electronic states

These selected states have very specific properties, and so exploit very special properties of the material. The key processes are the ones which we have discussed in Chapters 4–10.

In sophisticated applications, the photon phase may prove critical (Judson & Rabitz 1992; Roberts 2000; Assion *et al.* 1998). This is a standard idea in nuclear magnetic resonance, or in laser methods which exploit coherence. Since such applications have yet to be fully exploited in materials modification, we mention some which show promise. First, there are approaches which exploit quantum interference between two continuous wave excitations (Brumer & Shapiro 1986) to alter the balance between two different sets of products on excitation. Related to this idea are those processes in which a laser driving one transition can 'turn off' absorption associated with another transition (see, e.g., Moseley & Dunn (1995)). Secondly, there are approaches which exploit two suitably-phased ultrafast laser pulses to control the decay channel in an excited state (Tanner & Rice 1988; C J Bardeen *et al.* 1997). In this case, the first pulse raises the system to energy surface A, and the system cools through a crossing of two energy surfaces, A and B. The second pulse (which can be chirped from the first pulse) transfers population from the unwanted energy surface (A, say) to the desired surface (B, say). The idea is to reinforce the fraction in a chosen state after cooling through the intersection. The actual pulse sequences can be complex, making it appropriate to use genetic algorithm methods (Judson & Rabitz 1992; Assion *et al.* 1999). A third possible class of process exploits the coherent phonons produced by light pulses (see, e.g., Garrett *et al.* (1996)). One can envisage cases in which phonon coherence might be exploited to control the flow of vibrational energy usefully. There is every reason to consider mixed cases, in which both photons and phonons have roles. Finally, one should be aware of some of the systems being considered for quantum computing, since they may have applications in modification. Such systems include bistable systems (Chapter 7), for which one control parameter (essentially the barrier height) modifies dramatically the ease with which a move can be made from one minimum to another (Landauer 1999).

Potentially, there are many approaches which take advantage of phase selectivity as well as spectroscopic selectivity. Many of the methods used to manipulate spin states in nuclear magnetic resonance have analogues for electronic excitation. Such methods can achieve coherent population transfer among quantum states (see, e.g., Bergmann *et al.* (1998)).

Class II Energy input into a small but mesoscopic zone

In special cases, the photon phase may matter, as when an interference pattern in written in an optical fibre. The energy generates an excited state, often with free carriers, electrons, and holes. These may recombine to produce local heating (perhaps surface heating, or heating near grain boundaries, or in damaged regions). This approach helps to keep the process within a thermal budget. Such budgets are especially important as microelectronic device dimensions fall, yet the consequences of defects become larger. The situation is especially demanding when defect production is a natural result of processing, as in ion implantation. The approximate validity of the '+1' model helps the thermal budget, in that most Si vacancies appear to recombine with interstitials, leaving substitutional dopants and Si interstitials. One consequence is that the thermal budget is reduced even more. A major role of Rapid Thermal Processing (RTP, using optical methods instead of a thermal furnace) is thermal budget reduction.

Class II processes can involve the generation of significant local electric fields. Charge redistribution is especially important when X-rays or gammas are involved, or when an electron beam is used.

The continuing drive to ultimate miniaturisation means that issues of electronic excitation make contact with nanotechnology, i.e. the excitation of features consisting of very small numbers of atoms. We know of very little experimental work on the effects of excitation in this area. Clearly (as in the photographic process) there will be special behaviour of defects, defect stability, and defect processes in systems which involve only a few hundred atoms (as in quantum dots which show optical shifts from confinement) or a few tens of thousands of atoms (which show the Coulomb blockade). A single defect process can give rise to major changes. For instance, electric fields may be especially important, since the removal of one charged ion creates a field which is large enough to affect electron–hole recombination dramatically.

Another development driving materials modification of chosen small regions is photonics, the optical methods which require refractive index changes on length scales of optical wavelengths. One example relates to microfabrication for optical data storage, where two-photon polymerisation initiators have been proposed for three-dimensional data storage (Cumpston *et al.* 1999). Another example might be the creation of local refractive index changes so as to choose a strength of coupling to an optical field. This approach might be used to ensure emission of a particular polarisation, or even to make local laser action possible (Faist *et al.* 1997).

Class III Energy input to selected species from an ion beam

The selectivity is controlled either by the energy transfer (depending mainly on the target atom mass) or by a nuclear reaction (as in transmutation doping). We shall not discuss straightforward versions of methods such as ion implantation, since the important features are not directly associated with electronic excitation.

Class IV Excitation of a gas phase species, with substrate heating
The heating, in particular, can be confined to chosen regions. Normally, we regard gas phase excitation as outside our coverage, but the juxtaposition of excitation and substrate heating should be mentioned.

In the present chapter, we shall give examples of applications of materials modification by electronic excitation. Our coverage is not complete. Nor shall we suggest that these applications are inevitably successful. As in all areas of innovation, great ambitions can end in failure or partial failure. Moreover, the main achievement of the exploitation of technically-advanced approaches is sometimes improvements in the competing, lower-technology, methods. As in the case of, say, smart materials, there is a need to find the right situation to be smart about. Nonetheless, we have no doubt that certain major trends are creating an environment in which processes based on electronic excitation will grow in importance. These trends include the moves to ultimate miniaturisation and ultimate speed in microelectronics, and the emergence of photonics (the use of laser, optical fibre, and other specialised optical approaches) as a major industrial technology.

The underlying solid-state processes of materials modification are typically energy localisation, charge localisation, energy transfer, energy storage, and charge transfer. Alkali halides exhibit a wealth of processes and provide an important check of the basic ideas. The halides are especially important because of the range of closely-related materials, the great control of defects and impurities, and the many critical and accurate experiments. The complexity of excited state phenomena in alkali halides is, of course, sometimes greater than is needed in applications; other, applications involve processes which are not seen in the halides.

In halides and oxides, there are a few basic processes which are of wide interest. First, energy localisation is often associated with self-trapping (e.g. the exciton in quartz). Likewise, self-trapping can be a crucial component of charge localisation and transport. Secondly, we need to recognise the different consequences of core excitation and valence excitation (e.g. the processes leading to ion emission from the KCl surface, Chapter 9). Core excitation also allows one to examine the dynamics in the first picoseconds following excitation (cf. the work on diamond dynamics, Chapter 3). Thirdly, we must recognise that energy transfer can be from one group of lattice vibrations to another, rather than an exchange between electrons and vibrational motion (Chapter 3). The cooling transitions are the means by which equilibrium is recovered from excess local vibrations. Fourthly, we will have to understand the selectivity of the processes, and how branching ratios are determined.

In the examples which we shall mention, electronic excitation is exploited. In some cases (for example, certain forms of annealing) thermal excitation could be an alternative. In such cases, electronic excitation is essential only if it brings advantages such as control in space, or a reduced thermal budget. As an example, laser-based prototyping is expanding extremely rapidly. This process allows manufacturers to bypass conventional prototype stages. Instead, they can use computer-aided design and photonic methods. One important approach is selective laser sintering, in which a CO_2 laser is used to make a three-dimensional model from a mix of materials (nylon, polycarbonates, foundry sands, and metals (Kaplan 1998)).

In other examples, even for systems for which a large amount of information is available, we may doubt that we have the understanding to model the processes fully; photography is the classic example. To the technologist, understanding provides the basis for optimisation and improvement, and even partial understanding can be of value. To the scientist, technology poses questions which may need more interesting science than we have identified from ideal systems. In the area of defect processes, there are still many gaps in our knowledge.

The fundamental question for a technology in a business activity is 'Will it help financially?' With a new class of processes, such as materials modification by excitation, the opportunities may be taken up only slowly. There may be quite a few cases which fail to be successful, except for niches, like laser annealing. There may be others which fail on cost, simply because a photon from a laser is more expensive than hoped. But costs and needs are changing. In microelectronics, the needs of ultimate miniaturisation, ultimate speed, and low-power operation are all pushing towards new materials and new processes with lower thermal budgets. The exploitation of photonics in communications, or the ideas of quantum computing, if ever realised, will be science-based technologies which will require further processes and materials. Business opportunities may well follow technical opportunities, just as the compact disc required the semiconductor laser. The cost of lasers continues to fall, and their energy range continues to rise. New light sources are on the horizon. Given this mix of growing demand and improving technical capability, understanding the physical phenomena is crucial. Only then will one be informed so as to grasp the opportunities; only then can one avoid problems.

12.2 Modifying surface and near-surface regions

We shall give examples which illustrate the range of opportunities, rather than catalogue applications. In doing so, we shall comment on the mode of energy or charge localisation, which is often the central issue. As mentioned, we shall not discuss ion implantation in any detail, since excitation is usually a secondary factor. Implantation is usually done to dope a material, to influence another surface process, like oxidation, or to improve mechanical properties. The underlying processes are mainly alloying in ways which may not be possible by equilibrium routes, and stress and dislocation control. Avoiding interfaces is yet another option, and optical materials with graded refractive indices are under investigation. It has still to be established whether or not there are novel responses of these inhomogeneous systems under electronic excitation.

Sputtering for SIMS

In surface analysis by secondary ion mass spectroscopy (SIMS), the question of preferential sputter is critical, for the material removed must be representive of the material of the sample. Energy localisation is determined by collision sequences and by selection of the incident ion and its energy. Charge localisation is indirect: from the choice of ion (such as Cs or O), from its specific charge state, or from the extent of electron loss from the surface. The charge of the atoms or ions removed can be altered by optical ionisation. Selectivity of sputtering can be influenced by the incident particle mass and charge. Some such selectivity is observed, and there are empirical databases, but the mechanisms are not well understood.

Sputtering for film deposition and growth on a substrate

The aim is slightly different, in that one usually wants to achieve a specific stoichiometry on the target. This would be straightforward if the ablated material simply matched the stoichiometry of the target material. However, it is almost as acceptable to have consistent and reproducible ablation products, since adjustment of the composition can be done separately. Adjustments are possible by using several source materials, or by exploiting further reactions in the gas phase.

Energy localisation will match energy absorption. For optical excitation above the bandgap, energy will be absorbed near the surface. This absorption may itself cause changes of state which affect subsequent absorption. Thus, a melt will often absorb differently from a solid; a plasma generated above the surface may reflect light. If the excitation is above the bandgap, energy will be absorbed primarily at defects or grain boundaries and the like. Charge localisation may occur following electron loss from the surface.

Photochemical methods are commonly used to deposit reaction products in chosen places on a surface. Laser-assisted chemical vapour deposition (CVD) is an example. The principal reactions occur in the gas phase, so these methods, important though they are, lie outside the scope of this book. Nevertheless, these methods should draw our attention to the fact that, in laser ablation, chemical reactions can occur in the gas phase.

Smoothing and cleaning of surfaces

The gentler laser and electron methods can prove effective at smoothing surfaces. The main reason seems to be that atom removal occurs preferentially near vacancies or surface steps (Chapter 9), so that incomplete layers of atoms are removed before new vacancies form in deeper layers. Energy localisation is needed at special surface sites. The process is not so dramatic as the (less-effective) quasimelting in laser annealing. Laser excitation can also adjust surface stresses, primarily through rendering defects or dislocations more mobile.

The cleaning of surfaces by excitation can be achieved in several ways. One way is to remove certain unintended easily-vaporised materials, such as organics. The removal of dust particles exploits another feature, especially if there is some liquid (capillarity) contribution to sticking. An excimer laser can both volatilise the liquid, which provides a pressure impulse, and also remove electrons, so as to give a Coulomb contribution to removal. Successful removal of oxide prior to soldering has been demonstrated (Yonezawa *et al.* 1998) using a pulsed ArF excimer laser. Use of too intense radiation leads to altered surface morphology and, if carried out in air, to further oxidation.

One should be aware that excimer lasers do cause erosion in parts of equipment which are not expected to be affected, and that this erosion generates dust. In some cases, the dust migrates to the laser windows, and special cleaning measures are needed. Analogous problems occur in the plasmas which are used in 35–40% of semiconductor manufacturing processes (Selwyn *et al.* 1989). Plasmas are used especially for fine-line pattern transfers which need the deposition or etching of thin films. An example might be the use of plasmas to deposit <0.5 μm dielectric between metal lines for charge storage or for isolation. Similar processes have been found in processing flat-panel displays and magnetic storage discs. The dust is formed primarily in the vacuum conditions of plasma processing. The dust particles both grow and accrete. They can carry substantial charges (negative charges of 10^4 electrons are reported), and these charges determine both where the particles reside in the plasma and the way they grow.

Sputtering or ablation for control of shape or profile

Laser control of myopia by photorefractive keratectomy uses an excimer laser operating at 193 nm (6.4 eV), in excess of bond energies (Gartry 1997). It yields smooth surfaces with a clear edge: the profile is good; thermal damage is limited to 100–300 nm outside the chosen region, and the cut can be controlled in depth very precisely. There appears to be chemical decomposition as well as heating. The debris cloud has velocities as high as 2000 m/s.

Reactive ion etching and removal of oxide from silicon can use either Ar or SF_6 plasmas. Again, there appear to be two underlying processes. One is *physical sputtering* which needs a bias above 25 eV for Ar or Cl_2, or above 80 eV for He plasma (the energy relates to the bulk or surface threshold displacement energy). *Chemical sputtering* is found for SF_6 and F-containing plasmas. There can be effects on the underlying Si (see, e.g., Y H Lee (1989, 1992)). Chemical etching is used to generate ultrasharp tips for optical fibres to be used in near-field microscopy, which enables resolution beyond the traditional diffraction limits. Tips have already been etched to 70 nm; this allows powers in the micro-watt range, and hence increased scan rates.

There are two common classes of process for the direct writing of features by exploiting both laser excitation and chemistry. In *photolytic deposition* a chemical process in the gas phase is stimulated, and the reactive species modify the cold substrate. Thus, in processing III–Vs, metalorganics like $(CH_3)_3Ga$ (trimethyl gallium) are photolysed. In *pyrolytic deposition,* heating of the substrate provokes thermally-driven reactions with an ambient gas. Often, processes combine both ideas. In these cases, patterns can be written readily, often without using such high temperatures as in alternative methods.

Laser marking of Si is one case for which localised melting is effective, assisted by an electronic component (Scaroni & McKee 1997). It is possible to generate a series of dots which make a symbol using a YAG laser at 1.064 μm. The photon energy is close to the bandgap, and so absorption is temperature-sensitive, and also sensitive to local heating. This allows control of the melting process. The depth heated is about 2 μm, rather than the 250 μm expected for the usual absorption depth at that wavelength.

Ablation and surface marking

Lasers are used to remove polymeric photoresists after reactive ion etching or high-dose implants (see, e.g., Livshits *et al.* (1997)). The method combines laser ablation with plasmaless reactive chemistry. Separate excitation is used for the gas and for the surface. The gas is typically oxygen and nitrous oxide, perhaps with some halogen; a key ingredient is atomic oxygen to aid normal ablation. This route can have advantages over dry or wet stripping. Dry stripping includes plasma methods, or various ozone-based methods; plasma ashing often leaves residues. Wet stripping involves the use of Piranha (sulphuric acid plus peroxide) or organic solvents, but it is generally avoided whenever practical.

Another example of surface marking, although with less obvious application, is the creation of a nanohole array at the exit surface of a thin film of Si, following uniform electron-beam irradiation at 140–200 keV (Takeda *et al.* 1997b). The holes are deeper than 10 nm and 2–3 nm across. The process is not Frenkel defect production. Since it proved sensitive to surface treatment, it may involve mobile surface vacancies. Surface marking could also include localised bleaching of colour, an exploitation of the bleaching processes observed in sunlight. HeCd lasers (325 nm) are reported to be effective.

Laser cleaning and etching approaches use the vacuum ultraviolet, usually from discharge in hydrogen or in high-pressure rare gases, where the emission is essentially continuous from transient molecules. The photons cause C–C bond breaking and oxygen molecule dissociation (Allen *et al.* 1995). Metal contaminants also seem to be removed.

Lithography

The general principles of lithography involve the modification of a barrier coating such that parts can be removed selectively. The usual organic materials used can be cross-linked (rendered less soluble) or undergo scission (rendered more soluble) where they are illuminated or excited by electrons. These alternatives give rise to positive and negative lithographies. The resolution depends both on the molecules used and on the optical wavelength.

Pressures towards miniaturisation are moving lithography to shorter and shorter wavelengths: 248 nm is appropriate for 0.25 μm feature size and perhaps for 0.18 μm; 193 nm is needed for 0.15 μm features and the 1-Gbit DRAM chip. Plans are to move from 193 nm (ArF; for 0.13 μm features, perhaps even 0.1 μm) perhaps through 157 nm (F_2; possibly satisfactory for 0.08 μm features), 134 nm (ArK; for 0.07 μm features) to the extreme ultraviolet for 13 nm features. There are, of course, many issues, such as the changes with time of the refractive index of lithography optical materials (see Hawryluk *et al.* (1997); Pellegrini (1996); B Martin (1997)). At shorter wavelengths, the large energy input per photon will lead to major challenges of stability for lithography equipment (see also Section 12.4).

We remark that diamond-like carbons can be formed from hydrocarbons by a combination of cross-linking and selective removal of hydrogens, such as by collisions with low-energy nitrogen ions. The processes are also related to radiation- and photo-induced polymerisation. The cross-linking can also affect adhesion, partly because reactive radicals may be produced, and partly because much of the energy used to remove polymers from a metal surface may be that associated with breaking weak intermolecular bonds so as to pull a polymer strand through the network.

The effects of cross-linking on adhesion are clearly distinct from the electrostatics-based mechanisms of Chapter 10, where excitation can be used to manipulate charge states. Such electrostatic methods are used, often without a basic understanding of the mechanisms, for non-misting windows or non-misting mirrors for dental use. The same basic ideas encompass anodic bonding and, apparently, improving adhesion of tyres on ice. Indeed, essentially the same process underlies the optical poling methods noted in Chapter 8.

Excitation to produce localised luminescence

In plasma displays, only a monochrome display is possible when the light comes from the gas phase. Polychrome displays are possible if one uses phosphors excited by the plasma (Deschamps & Doyeaux 1997). Secondary electron emission is an important factor, and favours the use of MgO over alumina or silica.

12.3 Modifying the bulk solid

Once again, we shall only note the nuclear methods in passing, such as the transmutation doping of Si, since the excitation is of a type outside our main theme. Nor shall we do more than mention the subsurface implantation of oxygen into silicon to create an oxide layer, the so-called SIMOX process, since excitation is secondary, and classical dynamics is a major part of the explanation of what is seen. Some of the most significant applications are ones which we have already discussed in earlier chapters. Solar energy collectors are another type of system for which colloids are generated. Clearly, one way to achieve this is an analogue of the photographic process. In practice, ion implantation is a more straightforward route. The conventional photographic process requires a surface or interface (Chapter 5), but it and its analogues are most naturally treated as bulk processes.

Photographic processes

Photography extends well beyond the applications of the domestic camera. It includes the cosmic ray emulsions, which were used to identify elementary particles and, arguably, could be extended to include the writing of features like gratings in optical fibres and optical storage media. Other applications include the writing of graphic images directly onto polymers by using colourants which selectively change colour when exposed to laser light (*Photonics Spectra*, Aug 1997). In traditional silver halide photography, and indeed in several of the newer image storage devices for use with X-ray systems, there is a well-defined energy localisation step. Charge localisation also has a role in traditional photography (see Chapter 5).

In optical storage devices, there are several issues, such as: contrast; information density; read rate; write rate; need for refreshing the data; fatigue and other limits on the number of read–write cycles. There are questions of how best to organise the stored image, and whether or not to use a hologram. We shall not discuss these devices in detail (see, e.g., Sibley & Pooley (1974)), but remark that those based on a defect creation mechanism seem to suffer from fatigue, and so would be appropriate only for situations which are either read-only, or which do not need frequent write operations.

Photochromics

Sunglasses are the best-known photochromics. The darkening process in strong light is achieved by creating an oxide glass containing a suspension of silver-halide-enriched particles containing traces of the cuprous ion, Cu^+. The small particles, typically 10 nm across, do not scatter light significantly. Exposure to ultraviolet light creates Cu^{2+} and Ag atoms, which aggregate to form Ag colloids. Both the cupric ion and the colloids absorb in the visible. This non-equilibrium state persists whilst the light is strong, and the equilibrium state is recovered when the light is removed.

Sunglasses illustrate several distinct points. The first is that they are a system which responds to an external stimulus in a desirable way. The glass is a *'smart'* material. There are many, often exaggerated, claims for smart materials. Only in a handful of cases (examples being sunglasses and temperature compensators for pendulum clocks) are such claims demonstrable. There are two problems. The first is that most materials respond to an external stimulus in a rather weak way; the second is finding something important to be smart about. In fact, most so-called smart materials are merely *'alert'*, in that they provide a signal which allows some powered device to achieve the response. For example, there is talk of the 'controlled home', with windows that respond to house temperature (a recent approach is that of Bechinger *et al.* (1996)). A smart materials (hardware-based) solution is possible, but one has far greater flexibility if a solution based more on sensor materials and software is used.

A second point is that the changes take the system from what is an equilibrium state, or close to it, to a distinctly non-equilibrium state. It is an important feature of insulators that they can exist for extended periods with charges out of equilibrium. There is often no defined Fermi energy, since the populations have not achieved a Fermi distribution. The non-equilibrium state can last for archaeological periods in the absence of light; indeed, this is the basis of some of the most important thermoluminescence methods of dating archaeological materials and authenticating art ceramics (McKeever 1985; Zimmerman 1971).

When there is a non-equilibrium system, the issue of the kinetics of recovery can arise. In some cases, the observed kinetics can carry important information, as in the recovery of donor–acceptor pairs in semiconductors (see Chapter 2 or, e.g., Stoneham (1975) Chapter 25). The information relates to the dependence of the individual transition rates on spacings in the randomly-doped system.

A third point is the role of charge transfer. Optically-induced reactions of the form $A^0 + B^0 \rightarrow A^+ + B^-$ are common. Excitation redistributes charge. The redistribution can be over space, over species, or in more complicated forms. Some consequences are desirable; others will be undesirable. In most cases, there will be a change of colour, giving *photochromism*. Many of the classes of system showing photochromic behaviour were surveyed by Amodei (1971), who listed performance criteria and key property values. The inorganic systems are mainly based on colour centres or on transition metals. The organic systems are more varied. The main classes are: *cis–trans isomerization*, a characteristic molecular restructuring; *heterolytic cleavage*, in which a covalent bond is broken to produce a pair of ions, one positively charged, the other negatively charged; *homolytic cleavage*, to produce a pair of radicals; *tautomeric processes*, in which an atom (often hydrogen) is transferred; and *triplet-state formation*.

Since the colour change of photochromism can be reversed optically, there are possibilities for information storage. Information storage systems have been proposed in two distinct variants. One is to use a change in colour or dichroism of a region associated with a charge transfer. The region will be at least similar in size to the optical wavelength. An aim for recordable memories is to have recording, reading, and erasing operations operating at different wavelengths. Buse *et al.* (1998) have proposed a non-volatile holographic memory in $LiNbO_3$ based on $Mn^{(2+/3+)}$ and $Fe^{(2+/3+)}$. An alternative is to exploit *hole-burning*, in which essentially individual ions might be accessed; this exploits the inhomogeneous broadening of lines which have a narrow homogeneous width (Castro *et al.* 1978; Moerner *et al.* 1988; Holliday & Wild 1993; see also Stoneham 1969).

Optical memory technologies are extremely varied, and are adapted to similarly varied needs. As examples from an enormous literature, cation-deficient spinel ferrites have been suggested for high-density, write-once, read-many-times systems (Bouet *et al.* 1999); in these systems, blue light induces a phase transition to the corundum phase. Stoichiometric $LiTaO_3$ has been proposed for near-ultraviolet dynamic holography (Furukawa *et al.* 1999), where electron transfer processes are involved.

Display devices

Some of the early displays were *dark-trace* displays, writing a dark line on a bright background. Such dark-line displays have advantages; for example, because of better readability in brightly-lit conditions. In essence, these displays replaced the phosphor of a cathode ray tube by a material (KCl, or a bromide sodalite, $Na_6Al_6Si_6O_{24}.2NaBr$) which formed F centres under the electron beam. The F centres bleach in the ambient viewing light on a timescale which suits displays which are not changed too often. Selective erase facilities are possible. The colouring processes are presumably similar to those of Chapter 5. These displays have not become standard; for a review, see Sibley & Pooley (1974).

Charge transfer

We have remarked that charge transfer can redistribute charge in two main ways. *Redistribution over space* normally requires an anisotropic stimulus. In the windows of caves for the examination of spent nuclear fuels, there is a 'Compton Wind', in which electrons are moved by Compton scatter down a γ flux. In the photoelectric instabilities in laser-excited sapphire (Chapter 11; Basun *et al.* 1984a,b,c) the charge sinks, such as grain boundaries, must play a role. In radiation-enhanced electrical breakdown, it is the applied field which provides the anisotropy. In most cases, redistribution involves only a carrier moving; however, as discussed in Chapter 7, there are cases where a defect can be moved by some recombination-enhanced process. There are also cases where traps immobilise electronic carriers to such an extent that ionic motion is the main contribution to eliminating fields. This is so for Ce-doped glasses, for instance, where alkali ions or protons are the most mobile charged species.

In all cases, there is *redistribution over species or sites*. In the glasses which act as windows for spent nuclear fuel examination, there is redistribution over Ce in the glasses: sites which are most stable for 3+ ions lose an electron and sites most stable as 4+ gain an electron. Ce is special, in that neither the 3+ nor the 4+ state absorbs in the visible, so the presence of Ce combats darkening. For the same reason, Ce-based glasses are used to protect optical devices for use in space. The redistribution of charge is the basis of the colour of many gemstones (Nassau 1983). Likewise, the colouring of glasses often exploits transfers of electrons between transition metals which are stable in more than one charge state against carrier loss.

We shall not discuss conventional phosphors here (see Blasse & Grabmeier (1994); Struck & Fonger (1991)), except to remark that a number of systems exploit charge transfer. In X-ray phosphors, the charge-state changes can take several forms. The aim is to create changes using X-rays, and to read out the image by either heating or optically-scanning the material subsequently. CsI, for instance, is potentially a good X-ray material, since the high atomic number favours absorption (a more spectacular, if exotic, application is the use of a single CsI crystal the size of a household brick as part of an electromagnetic calorimeter in experiments to test the most fundamental symmetries, including *CP* violation and particle–antiparticle physics). Na-doped CsI is used as a phosphor. The Na normally present in the neutral state (Na^+ replacing Cs^+) can capture either an electron (largely through its electronegativity) or a hole (largely through elastic interactions (Ong *et al.* (1979)). X-ray excitation will generate electrons and holes, which are trapped at the amphoteric Na, and later released; recombination gives a characteristic blue luminescence.

Using change to measure excitation: detectors

Detectors are of many types, whether they are photodetectors or radiation detectors. Each type raises a specific question for its successful operation. We note four groups of detectors.

For Type 1 detectors, the question is 'How can one ensure charge created is collected?' Electron–hole pairs are created; they are then separated in an electric field, and a current measured. It is in situations like this that special materials, like diamond, can be excellent. Not only does diamond resist damage, but it is not sensitive to normal daylight, and it has an exceptionally high carrier mobility, and hence fast response. There can be subtleties in the operation of this type of detector if not all the energy generates electron–hole pairs, and this can have consequences if special effects are sought, like evidence for neutrinos with mass. As an example, for Si, there is a possible role of plasmon decays which do not lead to electron–hole pairs or charge separation (Choy & Stoneham 1993).

For Type 2 detectors, the question is 'How can one make sure light is emitted?' The typical system is a scintillator, such as CsI:Na, discussed above. For Type 3 detectors, the issue is 'How can one make sure that electrons are emitted?' This class includes photodetectors. As we have remarked in Chapter 9, there are many possible complications, and these are not all well understood. In Type 4 detectors, one asks 'How can one make sure damage is created or charge transferred so that it can be interrogated later?' X-ray fluorescence is one case where this approach might be invoked. Thermoluminescent dosimeters are very widely used, although it is uncertain as to which processes occur in some of the effective ones, like LiF:Mg,Ti (see, e.g., Sibley & Pooley (1974)). Another example of a Type 4 detector uses the predose effect in quartz or feldspars for archaeological and accident dosimetry (Zimmerman (1971); see also McKeever (1985); Halperin (1990); Martini *et al.* (1986)). In quartz, this involves a series of processes. Some are historical processes, due to natural radioactivity (the archaeological case) or irradiation due to some accident involving radioactivity. The other processes are carried out in the laboratory, using established series of irradiations and temperature cycles. In these sequences, there is charge exchange involving moving alkali and electrons around between non-bridging oxygens and other traps, such as Ge, in a systematic way (D Stoneham, N Itoh & A M Stoneham, 1996, unpublished).

12.4 Damage and deterioration

Electronic excitation can be damaging as well as beneficial. This is a major reason for emphasising understanding the underlying processes, since understanding gives the opportunity to design so as to minimise degradation, and thus to exploit the benefits. We shall mention a range of examples to illustrate the types of damage processes. Some common features emerge. First, we note that accelerated ageing tests will mislead if the degradation process is different in operation under accelerated conditions. This is the most important reason for understanding metastable defects or those defects which can emerge by evolution (like thermal donors) during processing. Secondly, we shall see that high electric fields are characteristic of a number of these cases. Thirdly, we note that interfaces are a rich source of problems, from their ability to act as defect or impurity sinks. As an example, in rapid thermal annealing, there is evidence that P_b centres are exposed at the Si/SiO_2 interface, as passivating hydrogens are removed (Hurley *et al.* 1998). Fourthly, the damage may be in a part of the system other than the active device. Thus, the early degradation of blue GaN lasers was caused in part by the ultraviolet light causing degradation of the plastic packaging (Mills 1998). Finally, in several cases, we shall see that the same phenomena can be used to advantage.

Defects in lithography optics

Vitreous silica is a standard material for lithographic optics. Its wide bandgap, and its ease of manufacture and the experience of many years give it major advantages. The quality of the silica is important, possibly because of the effects of Al and H. Defect production can be useful in optical fibres, since it allows gratings to be written; in lithography optics, the effects are wholly undesirable. The effect which is driving moves to crystalline materials (CaF_2, for instance) is compaction. There is a slow and slight change in dimensions as the optical system is used over periods of months or years (see Chapter 8). The effect may be associated with point defects or impurities; it may be associated with altered bonding patterns and changes in network topology. The drive to miniaturisation in microelectronics is forcing moves to shorter wavelength photons (in 1998 248 nm lithography was imminent; 193 nm lithography for 130 nm technology is scheduled for 2003), and these are more able to cause deterioration. Ideally, the coatings for projection optics should last 10 years without degradation. This corresponds to 10^{11} pulses, each of 1 mJ/cm^2. Trials in 1998 included reported successful coatings exposed to 10^9 pulses of 193 nm light at 15 mJ/cm^2.

Nanosecond to femtosecond laser-induced breakdown in dielectrics

The nature of optical damage depends on the total energy input and on the timescale. Intensities for very short pulses can be extremely high. As an example, B C Stuart *et al.* (1996) examined the effects of pulses from 140 fs to 1 ns at wavelengths of 1053 nm and 526 nm. If the behaviour were diffusion-dominated, there should be a characteristic square root of time in the scaling with fluence. The damage morphology observed departs from this expectation, but can be modelled successfully. It appears that ablation occurs for times <10 ps, and conventional melting, boiling, and fracture follow at times >50 ps. The roles of multiphoton ionisation, collisional (avalanche) ionisation, and Joule heating can be identified.

Electrical degradation in oxide layers

Perhaps the main source of $1/f$ noise (Kirton & Uren 1989) from oxide on silicon comes from border states. The timescales span milliseconds to hours. Charge transfer is involved, especially between Si and oxide traps. The traps can cause threshold shifts (e.g. Tanner 1996). Excitation processes leading to the release of H from these traps are among the precursors of electrical breakdown.

Catastrophic mirror blowout

Light impinging on a III–V mirror surface can cause photostimulated desorption of the Group V element. The mechanism is uncertain, and can be supresssed by an antireflection coating. The process starts at microscopic heterogeneities and results in the formation of a region rich in the Group III element. This metal-rich layer absorbs incident light more effectively, and forms an opaque region which melts locally (Ga has a very low melting point). Such melting can be observed at inclusions and precipitates. The temperature gradient between the absorbing (and so hot) front of the drop and the colder rear side means that one has melting at the front of the drop and solidification at the back. The Group V droplet diffuses down the temperature gradient in the liquid. This is known as PIZM, photo-induced zone migration. The metallic droplet is a 'dead' region, and it leaves behind it a trail of material containing defects.

We remark first that other systems, including halides, lose the anion selectively under excitation, especially under an electron beam. Secondly, there is an important role for small inhomogeneities, as in optical damage of lasers and many other cases. Thirdly, defects are formed where regrowth occurs. This is fairly general, observed in laser annealing, for example, and in other cases where nucleation is involved. Etchback and regrowth are standard tools, and are successful only when there is good control and when care is taken.

Recombination-enhanced processes

Processes enhanced by recombination were discussed in Chapter 7. In semiconductor lasers, there are high concentrations of minority carriers ($>10^{18}/\mathrm{cm}^3$) and high recombination rates ($>10^{26}/(\mathrm{cm}^3/\mathrm{s})$). In these conditions, recombination-enhanced processes can flourish. Not surprisingly, longer-wavelength lasers degrade less.

One form of degradation consists of *dark-line defects* (Kimerling 1990, p. 336), regions of reduced gain or of non-radiative recombination which can be imaged as dark regions by photoluminescence or electroluminesce. The defect comprises a network of dislocations. These appear to grow from a single dislocation which threads from the substrate, through the active region of the laser, during the crystal grwoth phase of device fabrication. The dislocations have $\langle 100 \rangle$ dipoles and Burgers' vector $(a/2)\langle 110 \rangle$; they show interstitial-type contrast in electron microscopy. There are prismatic loops behind the dipole front, with vacancy-type contrast. The growth of the dark-line defects is a non-conservative climb process in which atoms are added. The Ga interstitials, trapped at heterostructure interfaces (AlGaAs/GaAs) during crystal growth, become mobile during operation through recombination-enhanced diffusion. The precipitation of these interstitials is one component of dislocation growth. However, climb needs both extra Ga and extra As. The extra As appear to leave As vacancies, which themselves condense to form the vacancy loops. The overall reaction can be described as $\mathrm{Ga_i} \rightarrow \mathrm{V_{As}}$. Just how this occurs is unclear. The original suggestion (Petroff & Kimerling 1976) is complicated. We favour another view, namely that antisite defects are involved (Stoneham 1984, unpublished). If so, $\mathrm{Ga_i}$ would generate $\mathrm{As_i}$ and $\mathrm{Ga_{As}}$. The climb is then easy to understand. The harder part is how the $\mathrm{Ga_{As}}$ generate the vacancy loops.

Gradual degradation

Slow damage processes occur at the heterostructure interfaces between active layers and cladding layers (Kimerling 1990, p. 339). Small strained regions or growth imperfections lead to the accumulation of defects (gettering). In operation, recombination causes these centres to become mobile, and they generate dislocation loops in the active region. New defect formation does not seem important. This process is slow, since there is a homogeneous nucleation barrier to loop formation, so there are analogies with photochemical processes in alkali halides (Chapter 5). The loops cause non-radiative recombination, degrading device gain. The process is less important for high-quality interfaces and low bandgaps.

Another example in which excitation can affect contamination and gettering concerns Fe impurity. The key reaction is probably $[Fe_iB_s] \rightarrow Fe_i^+ + B_s^-$ and its reverse. Electron capture can be regarded as having two effects: it cancels the electrostatic interaction between the Fe cation and the B anion, and it deposits some vibrational energy. These two effects together stimulate breakup of the pair, a critical step.

Degradation of avalanche devices

Carrier multiplication is a feature of devices such as avalanche photodiodes (APD) and devices based on impact ionisation (IMPATT). A region of high electric field accelerates carriers to above-bandgap energies, leading to impact ionisation. The large concentrations of minority carriers ensure that recombination rates are high, with the potential for degradation. Degradation is especially a problem at heterointerfaces. In APD, the high concentration of minority carriers can be isolated from the heterointerface by suitable design. In IMPATT, the region of high field is often adjacent to the Schottky metal contact; defects associated with the metal–semiconductor interface may diffuse into the depleted region and affect device operation. For example, accelerated tests on a Pt/GaAs IMPATT indicated a long mean device lifetime, more than 10^6 hours at the designed operating temperature. The ageing (actually due to an interface reaction between Pt and GaAs) suggested a 1.6 eV activation energy. In operation, devices were short-lived, and the effective activation energy was only 0.3 eV; the process involved recombination-enhanced defect motion.

Enhanced migration of interstitials

Most interstitials in semiconductors show enhanced diffusion. Thus, in Si, apart from the likely case of self-interstitial motion, Kimerling *et al.* (1989) list evidence for enhanced diffusion of neutral interstitial B and O and of positively-charged C, Li, Fe, and Al (Al is doubly-charged). Chapter 7 discusses other situations and open questions; for muons, the evidence is especially clear. C is a special problem because hydrocarbon contamination is so common. A reactive-ion etch (RIE) plasma, or rapid thermal processing (RTP) lamps, or a reflection high-energy electron diffraction (RHEED) gun during molecular beam epitaxy (MBE) can all lead to in-diffusion of interstitial C. The device may then show instabilities when the motion of this C is stimulated in operation. There can be long-range motion of C during hole injection into n-Si.

Radiation-induced processes

Even though damage by particle beams is mainly outside our remit, it is important to mention some of the parallels. For example, electronic excitation occurs in collision cascades, even in metals. This affects nuclear plant lifetimes: behaviour on a femtosecond timescale (collision cascades) can affect performance on the decades (of years) timescale (Stoneham *et al.* 1998a). One component is *energy transfer from nuclei to electrons.* As regards localisation, the collision cascade sets one of the scale lengths; the electron mean free path is another length scale. The behaviour is highly non-equilibrium.

We have already mentioned mechanisms which may involve more than one type of radiation. We emphasise that, when several radiation sources are present, their effects need not be additive. This is especially true for mixed γ and n^0 damage in optical fibres. It will certainly be true if optical excitation accompanies sputtering.

When radiation is causing intentional damage, recombination-enhanced processes can have unexpected effects. As we noted in Chapter 7, implantation of GaAs with H, D, or T to give electrical isolation (Steeples *et al.* 1980) shows non-monotonic behaviour with isotope. This is associated with competition between recombination at the hydrogenic impurity and recombination-enhanced annealing of radiation damage. In plasma processing, hydrogen can also affect dislocation motion in Si. The effect is a reduction in activation energy from 2.3 eV to 1.2 eV, and appears to be due to H, not to the plasma providing excitation (Y Yamashita *et al.* 1997).

12.5 Changing rates and altering processes

It is in this final section that we feel that the potential for materials modification by excitation is barely appreciated. The examples we can give are few in number, and often exploratory. For instance, it is clear that excitation can speed the oxidation of silicon by transforming the oxygen molecule into something more effective. Excitation may transform the first few monolayers of oxide, and offer control of one of the most important systems as it faces exceptionally demanding technical challenges in the next few years. Yet, since photochemistry has many well known examples of chemical reaction manipulation, and since photosynthesis is a reminder of what is possible, we believe that far more impressive examples than those we mention here will emerge before long. For example, it may prove possible to exploit coherent control of processes (see Castella & Zimmermann (1999)). Or one might be able to exploit the many bistabilities of excited states in non-metals, although maintaining coherence among states strongly-coupled to the lattice and with level-crossings is challenging.

As a first example, we cite laser microchemical technology for the real-time editing of first-run silicon devices, that is to say prototype integrated circuits. The process aids the development of new integrated circuits. The first examples of some initial design may have hardware failures or vulnerable regions which need editing and optimisation before the design is finalised. One way is to use *focused ion beams,* but this can be time consuming and tricky. A *laser microchemical* tool can do the job better. Now one can use an ambient gas and locally heat the circuit component so that either etching or deposition occurs (so it is not the gas which is excited). The laser can scan through a window, heating localised regions by a few hundred degrees so that a surface reaction occurs (not ablation) to deposit or remove material. The advantages of lasers over ion beams are: first speed, secondly the purity of the product material, thirdly that the system can be external to the process chamber (whereas an ion beam requires a vacuum), fourthly a lower energy input and no unwanted implanted species, and fifthly flexibility, e.g. to produce long interconnects (up to a few mm), or to remove large areas of passivating oxide, or to adjust the device while it is actually operating. This range of benefits is typical of the promise of materials modification by excitation.

As a second example, we note the enhanced growth of diamond films (Mistry & Roy 1996). The approach uses pulsed lasers of several wavelengths to heat a small spot on a surface, whilst CO_2 and nitrogen shielding gas pass over it. The lasers are standard machining lasers (usually chosen from KrF excimer, XeCl excimer, Nd:YAG, or CO_2 lasers). The growth rate can be as high as 0.36 cm/h, and there appears to be some control over crystal size and orientation. The advantage lies less in the rate, good though that is, than in the ability to put diamond down where you want it. This process is a natural generalisation of laser-assisted CVD. What it shows is that there are ways to take state-of-the-art methods to new systems. There are continuing opportunities for novel excitation-based methods.

A third example concerns chemically-amplified resists in lithography. We have mentioned the standard resist ideas, in which the important chemistry is driven directly by the photons absorbed. Yet it is a new technique which is used for all positive and negative resists for 248 nm and 193 nm applications

(D Seeger 1997; Ito & Wilson 1984). 'Chemically-amplified' means that the photon stimulates a photoacid generator to form an acid which catalyses changes in the resist material. The catalyst molecule can result in many reactions per photon. Reductions in critical dimensions lead to some significant technical challenges.

12.6 Closing comments

The number of processes which involve materials modification by excitation is substantial. Many of the processes involve excitation as only a minor part of a complex series of steps. Sometimes, the excitation process and its consequences are not understood. If we look at the major examples, ordered by class of material, we see that they are the following.

Halides

Photography is the major example. Other imaging systems, whether for medical applications, or particle detection, have extended the role of halide-based image formation. Halide-based photography itself is being challenged by electronic approaches, although not for all applications. The other key examples for halides concern *lack* of materials modification for light below their wide bandgaps. The uses of CaF_2 for lithography optics, of MgF_2 for ultraviolet windows, or of fluoride glasses for optical fibres all rely on a sufficient understanding of damage processes to avoid problems.

Oxides

The commonest applications for oxides appear to involve charge transfer. In many cases, this is to achieve a change in colour, or to achieve a lack of colour. It is the amorphous oxides (including glasses) which have most evident uses, especially silicon dioxide. Both the thin-film form (oxide on silicon) and the fibre form have crucial technological roles. For the thin films, being able to minimise changes under electrical stress is important. In the fibre, controlled changes to create gratings and the like are very important. The resistance of oxides to modification is an asset in many applications, such as coatings, or fibres used to transmit laser power.

Semiconductors

The applications for semiconductors concern processing: enhancement, acceleration, or adding extra control. The enhancements range from smoothing surfaces to improvements in lithographies. The acceleration aspect is inevitably in competition with heat, and the move to smaller features is demanding lower thermal budgets. Rapid thermal processing using, for example, tungsten halogen lamps, is used for oxidation, for silicide formation, for dopant activation after implants, and for a range of other steps. In rapid thermal processing, it is highly likely that there is an excitation component, in addition to simple heat supply. The nature of this enhancement is often obscure, but it is likely that enhanced diffusion is important. The nature of electronic excitation, whether by photons or electrons, is such that there can be some local control: local enhancement of reactions or processes is possible. It is this control which should provide opportunities which are not yet obvious.

Summary

We have defined four broad classes of materials modification by electronic excitation. Most emphasis in practice has been on Classes II (spatially-localised energy input) and IV (excitation from the gas phase). There has been significant work on Class III (selective energy transfer from an ion beam), but relatively little work on Class I (selected excited states). Class I offers opportunities which will become increasingly important as control on a fine scale becomes critical.

Halide crystals with closed-shell cations (NaCl, CaF_2) Lithography; X-ray detectors; dosimetry; exploit resistance to excitation.

Amorphous halides and halide glasses Resistance to change under excitation in fibres.

Halide crystals in which the cation can change charge state (AgCl, $PbCl_2$) Photography.

Oxide crystals with closed shell cations (MgO, ZnO, Al_2O_3, usually SiO_2) Ablation; secondary electron emission; dosimetry.

Oxide glasses with closed-shell cations, like borosilicate or aluminosilicate glasses Lithography; Bragg gratings; exploit resistance to excitation and to breakdown.

Oxide crystals with cations which can have several charge states ($Fe_{1-x}O$ and other transition metal oxides) Colour changes; display devices; optical memories.

Oxide glasses which include transition metals Colour changes; optical memories.

Semiconductor crystals (III–V, Group IV) with relatively wide bands and small gaps Chemical sputtering; diamond-like carbon (tribological coatings); process enhancement possible, but more often some degradation results (which can be useful in electrical isolation). Surface texture control. Adjustment of oxidation.

Amorphous semiconductors Examples include the Xerox process.

Other systems We are well aware that we have concentrated on inorganic systems in our book. Organic systems and mixed organic–inorganic systems provide a wealth of further opportunities for the exploitation of electronic excitation.

References

Abakumov, V N & I N Yassievich 1976 *Sov Phys JETP* **44** 345
Abriou, D, F Creuzet & J Jupille 1996 *Surf Sci* **352** 499
Abromov, A 1979 *Sov Phys Solid State* **21** 47
Adair, M, C H Leung & K S Song 1985 *J Phys* **C18** L909
Affatigato, M, K Tang, R F Haglund & C H Chen 1994 *Appl Phys Lett* **65** 1751
Agranovich V V & V M Kirsanov 1971 *Sov Phys Solid State* **12** 2147
Agullo-Lopez, F & P D Townsend 1980 *Phys Status Solidi B* **97** 575
Ahrenkiel, R K & R S van Heyningen 1966 *Phys Rev* **B144** 576
Akazawa, H, J Takahashi, Y Utsumi, I Kawashima & T Uris 1992a *Appl Phys Lett* **60** 974
Akazawa, H, Y Utsumi, J Takahashi & T Uris 1992b *IEICE Trans Electron* **E75-C** 781
Akazawa, H, M Nagase & Y Utsumi 1994 *Nucl Instrum Methods* **B91** 644
Al Jammal, Y & P D Townsend 1973 *J Phys* **C6** 955
Al Jammal, Y, D Pooley & P D Townsend 1973 *J Phys* **C6** 247
Albaugh, K B 1986 *Materials Letters* **4** 465
Albert, J 1998 *MRS Bulletin* **23(10)** 36
Albert, J, C Jouanin & C Gout 1977 *Phys Rev* **B 16** 925 4619
Albert, J, B Malo, K O Hill, F Bilodeau, D C Johnson & S Thériault 1995 *Appl Phys Lett* **67** 3529
Alcala, R, N Koumvakalis & W A Sibley 1975 *Phys Status Solidi A* **30** 449
Alig, R C & S Bloom 1975 *Phys Rev Lett* **35** 1522
Allen, A M T & J D Comins 1992 *J Phys Condens Matter* **4** 2701
Allen, A M T, J D Comins & P J Ford 1985 *J Phys* **C18** 5783
Allen, L R, J M Grant, T Nguyen, K Valiev, L Velikov & B Meshman 1995 *Sol State Technol* (May 1995) 77
Allmen, M von 1987 *Laser-Beam Interactions with Materials*; Springer-Verlag, Berlin
Allnatt, A & A B Lidiard 1993 *Atomic Transport in Solids;* Cambridge University Press, Cambridge
Alonso P J , L E Halliburton, E E Khonke & R B Bossoli 1983 *J Appl Phys* **54** 5369
Althaus, H L, G Weiser & S Nagal 1978 *Phys Status Solidi B* **87** 117
Aluker, E, V V Gavrilov & S A Chernov 1992 *Phys Status Solidi B* **171** 253
Alvarez Rivas, J L 1980 *J Phys (France)* **41** C6-353
Alvisi, M, G de Nunzio & M R Perrone 1999 *Thin Solid Films* **B38** 269
Amer, N M, A Skumanich & W B Jackson 1983 *Physica* **B117–118** 897

Amodei, J J 1971 p 533 *et seq* of *Handbook of Lasers with Selected Data on Optical Technology* (edited R J Pressley); CRC Press, Cleveland
Anderson, P W 1958 *Phys Rev* **109** 1492
Anderson, P W 1975 *Phys Rev Lett* **34** 953
Ang, L K, Y Y La & R M Gilgenbach 1999 *Appl Phys Lett* **74** 2912
Angus, J C 1986 *Thin Solid Films* **142** 145
Angus, J C & F Janssen 1988 *J Vac Sci Technol* **A6** 1778
Anthony, T R 1983 *J Appl Phys* **54** 2419
Antonini, A, A Manara & P Lensi 1978 p 316 of *The Physics of SiO_2 and Its Interfaces* (edited by S T Pantelides); Pergammon, Oxford
Arimoto, O, K Kan'no, K Nakamura & Y Nakai 1984 *J Phys Soc Jpn* **53** 70
Arutyunyan, V V, A K Babayan & V A Gevorkyan 1995 *Sov Phys Solid State* **37** 443
Asami, K, T Naka & M Ishiguro 1986 *Phys Rev* **B34** 5658
Ashby, M F 1972 *Acta Metall* **20** 887
Ashkenasi, D, A Rosenfeld, H Varel, M Wahmer & E E B Campbell 1997 *Appl Surf Sci* **120** 65
Ashkenasi, D, A Rosenfeld, H Varel, M Wahmer & E E B Campbell 1998 *Appl Surf Sci* **129** 128
Ashley, J C 1991 *J Appl Phys* **69** 674
Assion, A, T Baumert, M Bergt, T Brixner, B Keifer, V Seyfried, M Strehle & G Gerber 1998 *Science* **282** 919
Assmann, S, S Schweitzer & J-M Spaeth 1999 *Phys Status Solidi* **212** 325
Atkinson, A 1985 *Rev Mod Phys* **57** 437
Auciello, O & R Kelly 1984 *Ion Bombardment Modification of Surfaces*; Elsevier, Amsterdam
Audebert P, Ph Daguzan, A Dos Santos, J C Gauthier, S Guizard, G Hamoniaux, K Krastev, P Martin, G Petite & A Antonnetti 1994 *Phys Rev Lett* **73** 1990
Ausnees, A, M Elango, A Kikas, E Nommiste, J N Anderson, R Nyholm & I Martinson 1990 *Solid State Commun* **76** 1383
Averyanov, V L, A V Kolobov, B T Kolomiets & V M Lyubin 1980 *Phys Status Solidi A* **57** 81
Avouris, Ph 1996a *Chem Phys Lett* **257** 148
Avouris, Ph 1996b *Surf Sci* **363** 368
Avouris, Ph, R E Walkup, A R Rossi, H C Akpati, P Nordlander, T-C Shen, G C Abeln & J W Lyding 1996 *Surf Sci* **268** 1590
Awano, T, T Namba, M Ikezawa T Matsuyama & H Yamoaka 1989 *J Phys Soc Jpn* **58** 2570
Awazu, K & K Watanabe 1994 *Nucl Instrum Methods* **B91** 422
Awazu, K, H Onuki, S Iijima & K Watanabe 1994 *J Non-Cryst Solids* **179** 276
Azuma N, T Miyazaki, K Fueki, I Sakaguchi & H Hirano 1986 *J Amer Ceram Soc* **1** 19
Bachheimer, J P 1986 *J Phys* **C19** 5509
Baetzold, R C 1973 *Photo Sci Eng* **17** 78
Baetzold, R C 1975 *Photo Sci Eng* **19** 11
Baetzold, R C 1987 *Phys Rev* **B36** 9182
Baetzold, R C 1997 *J Phys Chem* **101** 8180
Baetzold, R C & K S Song 1991 *J Phys Condens Matter* **3** 2499
Baetzold, R C & K S Song 1993 *Phys Rev* **B48** 14907
Baglin, J E E, G J Clark & J Böttiger 1984 *Mat Res Soc Symp Proc* **25** 179
Bagnell, D M, Y F Chen, Z Zhu, T Yao, Y Shen & T Goto 1998 *Appl Phys Lett* **73** 1038
Bagus P S & Schaefer H F III 1972 *J Chem Phys* **56** 224

Baird, R J & R J Haeberle 1986 *J Vac Sci Technol* **4** 532

Bajic, S & R V Latham 1988 *J Phys* **D21** 200

Balanzat, E, S Bouffard, A Bourell, E Dooryhee, S Hemon & J M Ramilon 1996 *Nucl Instrum Methods B* **116** 159

Ballamy, W C & L C Kimerling 1978 *IEEE Trans Electron Devices* **ED-25** 746

Balzar, D, H Ledbetter, P W Stephens, E T Park & J L Routbort 1999 *Phys Rev* **B59** 3414

Balzer, R, H Peisl & W Weidelich 1968 *Phys Lett A* **27** 31

Balzer, R, H Peisl & W Weidelich 1969 *Phys Status Solidi* **31** K29

Baraff, G A & M Schluter 1978 *Phys Rev* **41** 892

Baraff, G A, E O Kane & M Schlüter 1979 *Phys Rev Lett* **37** 1504

Baraff, G A, M Schlüter & G Allan 1983 *Phys Rev Lett* **50** 739

Baranova, G K, A V Gorbunov & Yu I Koval' 1987 *Sov J Opt Technol* **55** 115

Bardeen, C J, J Che, K R Wilson, V V Yakovlev, P Cong, B Kohler, J L Krause & M Messina 1997 *J Phys Chem* **101** 3815

Bardeen, J 1947 *Phys Rev* **71** 717

Barerra, R G & C B Duke 1976 *Phys Rev* **B13** 4477

Barnett, R N, U Landman & C L Cleveland 1988a *J Chem Phys* **88** 4421

Barnett, R N, U Landman, C L Cleveland & J Jortner 1988b *J Chem Phys* **88** 4429

Barnett, R N, U Landman & A Nitzan 1989 *J Chem Phys* **91** 5567

Barniol, N, F Perez-Murano & X Aymerich 1992 *Appl Phys Lett* **61** 462

Barriac, C, P Pinard & F Davoine 1966 *C R Acad Sci Paris* **266** 423

Bartholomé, S A, A Dubus, J Devooght, J-P Ganachaud & C Attard 1997 *Nucl Instrum Methods, Phys Res* **B125** 13

Bartholomé, S A, A Dubus, R Renoud, C Attard & J-P Ganachaud 1998 *Supplement à la Revue 'Le Vide: science, technique et applications'* **287** 519

Bartram, R H & A M Stoneham 1975 *Solid State Commun* **17** 1593

Bartram, R H & A M Stoneham 1983 *Semicond & Insulators* **5** 297

Bartram, R H & A M Stoneham 1985 *J Phys* **C18** L549

Bastiani, S, A Rousse, J P Geindre, P Audebert, C Quoix, G Harmoniaux, A Antonetti & J C Gautier 1997 *Phys Rev* **E56** 7179

Basun, S A & A A Kaplyanskii, 1989 *Cryst Latt Def & Amorph Mat* **18** 137

Basun, S A, A A Kaplyanskii & S P Feofilov 1984a *Sov Phys JETP* **60** 1182

Basun, S A, A A Kaplyanskii & S P Feofilov 1984b *JETP Lett* **37** 586

Basun, S A, A A Kaplyanskii, S P Feofilov & A S Furman 1984c *JETP Lett* **39** 189

Basun, S A, A A Kaplyanskii & S P Feofilov 1985 *J de Physique Colloque* **C7** 46 267

Basun, S A, A A Kaplyanskii & S P Feofilov 1987a *Fiz Tverd Tela* **29** 1284

Basun, S A, A A Kaplyanskii & S P Feofilov 1987b *Izv Akad Nauk Sci USSR (ser fiz)* **51** 2172

Basun, S A, A A Kaplyanskii & S P Feofilov 1987c *J Lumin* **38** 120

Basun, S A, A A Kaplyanskii, S P Feofilov & A S Furman 1986 *Sov Phys Solid State* **28** 521

Bauer, W & A Sosin 1964 *Phys Rev* **135** A521

Bauer, Ph, D Dufor, C Jaouen, G Marchal, J Pacaud, J Grilh & J C Jousset 1997a *Phys Rev* **B81** 116

Bauer, P, C Dufour & C Jaouen 1997b *J Appl Phys* **81** 116

Beales, K J, J E Midwinter, G R Newns & C R Day 1974 *Post Office Elec Eng J* **67** 80

Beattles, A B & P T Landsberg 1959 *Proc Royal Soc A* **249** 16

Beaumont, J H, J V Gee & W Hayes 1970 *J Phys* **C5** 1475

Bechinger, C, S Ferrer, A Zaban, J Sprague & B A Gregg 1996 *Nature* **383** 608

Becker, R S, J A Golovchenko & B S Swartzentruber 1987 *Nature* **325** 419

Behr, A, H Peisl & W Waidelich 1967a *Phys Lett* **24A**, 379
Behr, A, H Peisl & W Waidelich 1967b *Phys Status Solidi* **21** K9
Behrisch, R 1983 *Sputtering by Particle Bombardment* Vol 2; Springer; Berlin; see especially G Betz & G K Wehner pps 11 *et seq*, on Sputtering of multicomponent materials
Bel'kov, V V, J Hirschinger, V Novák, F-J Niedernostheide, S D Ganichev & W Prettl 1999 *Nature* **397** 398
Bellamy, W C & L C Kimerling 1978 *IEEE Trans Electron Devices ED-25*
Bennebroek, M T, A Arnold, O G Poluektov, P G Baranov & J Schmidt 1996 *Phys Rev* **B53** 15607
Bennewitz, R, M Reichling & E Matthias 1997 *Surf Sci* **387** 69
Bennewitz, R, D Smith & M Reichling 1999 *Phys Rev* **B59** 8237
Benninghoven, A, F G Rödenauer & H W Werner 1987 *Secondary Ion Mass Spectrometry: Basic Concepts, Instrumental Aspects, Applications and Trends*; John Wiley, New York
Benyagoub, A, S Loffler, M Rammensee, S Klaumunzer & G Saemann-Ischenko 1992 *Nucl Instrum Methods* **B65** 228
Beranger, M, P Thevenard, R Brenier, B Canut, S M M Ramos, A Brunelle, S Della Negra, Y Le Beyec, E Balanzat & T Tombrello 1996 *Phys Rev* **B53** 14773
Bergh, A A 1971 *IEEE Trans Electron Devices* **18** 166
Bergmann, K, H Theuer & B W Shore 1998 *Rev Mod Phys* **70** (3)
Berthelot, A, S Hemon, F Gourbilleau, C Dufour, E Dooryhee & E Paumier 1998 *Nucl Instrum Methods* **146** 437
Betzig, E & R J Chichester 1993 *Science* **262** 1422
Beuneu, F & P Vajda 1996 *Phys Rev Lett* **76** 4544
Beuneu, F, P Vajda & G Jaskierowicz 1997 *Phys Rev* **B55** 11263
Bezina, B 1992 *Nucl Instrum Methods B* **72** 202
Bezina, B 1995 *J Mol Structure* **349** 269
Biegelsen, D K & M Statzmann 1986 *Phys Rev* **B33** 3006
Biegelsen, D K & R A Street 1980 *Phys Rev Lett* **44** 803
Bier, M & R D Astumian 1993 *Phys Rev Lett* **71** 1649
Biersack, J P & L G Haggmark 1980 *Nucl Instrum Methods* **124** 257
Bilodeau, T G, K J Ewing, G M Nau & I D Aggarwal 1995 *J Appl Phys* **77** 5524
Birkhoff, R D 1958 *Encyclopedia of Physics*, Vol 34; Springer, Berlin
Bishop, S G, U Storm & P C Taylor 1977 *Phys Rev* **B15** 2278
Blasse, G 1981 *Comments Inorg Chem* **1** 245
Blasse, G 1994 *Chem Mater* **6** 1465
Blasse, G & B C Grabmeier 1994 *Luminescent Materials*; Springer, Berlin
Bliss, E S 1971 *Opto-electronics* **3** 99
Block, D & A Wasiela 1979 *Solid State Commun* **28** 455
Block, D, A Wasiela & Y Merle d'Aubigne 1978 *J Phys* **C11** 4201
Blumen, A & J Klafter 1983 *Phil Mag* **B47** L5
Bochove, E J & J F Walkup 1990 *Am J Phys* **58** 131
Bonch-Bruevich, V L & E G Landsberg 1968 *Phys Status Solidi* **29** 9
Bonn, M, M J P Brugmans, A W Kleyn, R A van Santen & H J Bakker 1996 *Phys Rev Lett* **76** 2440
Boroin, M P 1973 *J Amer Ceram Soc* **56** 254
Boscaino, R, M Cannas, F M Gelardi & M Leone 1999 *J Phys Condens Matter* **11** 721
Bosio, C & W Czaja 1993 *Europhys Lett* **24** 197

Böttiger, J, J E E Baglin, V Brusic, G J Clark & D Anfiteatro 1984 *Mat Res Soc Symp Proc* **25** 203

Bouet, L, P Teilhades, I Pasquet, C Bonningue, S LeBrun & A Rousset 1999 *Jap J Appl Phys* (Part 1) **38** 1826

Bourgoin, J & J W Corbett 1978 *Radiat Effects* **36** 157

Bouzidi, S, T Angot, V Langais, J-M Debever, R Sporken, J L Longueville & P A Thiry 1994 *Surf Sci* **307–309** 1038

Boyd, I W & R B Jackman 1992 *Photochemical Processing of Electronic Materials*; Academic Press, London

Bozack, M J, W J Choyke, L Muelhoff & J T Yates 1986 *J Appl Phys* **60** 3750

Brailsford, A D & R Bullough 1972 *J Nucl Mat* **44** 121

Brailsford, A D & R Bullough 1973 p 267 of *Irradiation Embrittlement & Creep in Fuel Cladding and Core Components*; British Nuclear Engineering Society, London

Brethean, T, J Castaing, T Rabier & P Vessengière 1979 *Adv Phys* **28** 135

Brewer, P D, J J Zinck & G L Olson 1990 *Appl Phys Lett* **57** 2526

Bridges, F, G Davies, J Robertson & A M Stoneham 1990 *J Phys Condens Matter* **2** 2875

Brinciotti, A, M Piacentini & N Zema 1994 *Radiat Eff Defects Solids* **128** 89

Broer R & Nieuwpoort W C 1981 *Chem Phys* **54** 291

Broer R & Nieuwpoort W C 1988 *Theor Chim Acta* **73** 405

Broer R, Van Oosten A B & Nieuwpoort W C 1991 *Rev Solid State Sci* **5** 79

Broers, A N 1988 *IBM J Res Dev* **32** 502

Broers, A N, J J Cuomo, J Harper, W Molzen, R B Laibowitz & M Pomeranz 1978 p 343 of *Electron Microscopy 1978* Vol III (edited by J M Sturgess); Microsc Soc of Canada

Broers, A N, J J Cuomo & W Krakow 1981 *IBM Tech Disclosure Bull* **24** 1534

Brower, K L 1988 *Phys Rev* **B38** 9657

Brower, K L & S M Meyers 1990 *Appl Phys Lett* **57** 162

Brown, F C, B R Sever, N Kristianpolar & J-P Stott 1987 *Physica Scripta* **35** 582

Brown R, B Lounis & M Orrit 1997 *Europhysics News* Sept/Oct 174

Brückner, R 1986 p 109 of *Current Topics on Non-Crystalline Solids* (edited by M D Baró & N Clavaguera); World Scientific, Singapore

Brumer, P & M Shapiro 1986 *Farad Soc Disc* **82** 177

Brune, H, J Wintterlin, J Trost & G Ertl 1993 *J Chem Phys* **99** 2128

Bucksbaum, P H 1998 *Nature* **396** 217

Budhani, R C, S Prakash, H J Doerr & R F Bunshah 1986 *J Vac Sci Technol* **A4** 3023

Bullough, T J, C J Humphreys & R W Devenish 1990 *MRS Proc* **157** 323

Bulsara, A R & L Gammaitoni 1996 *Physics Today* (March) 39

Bunton, P H, R F Hauglund, Jr, D Liu & N H Tolk 1991 *Surf Sci* **243** 227

Bunton, P H, R F Hauglund, Jr, D Liu & N H Tolk 1992 *Phys Rev* **B45** 4566

Burns, R G 1981 *Ann Rev Earth Planet Sci* **9** 345

Burt, M G 1981 *J Phys* **C14** 3269

Burton, B 1977 *Diffusional Creep of Polycrystalline Materials*; Trans Tech, Zürich

Buse, K, A Adibi & D Psaltis 1998 *Nature* **393** 665

Buttman, M, J Nakamura & H Kawano 1994 *Appl Surf Sci* **78** 421

Cabrera, N & N F Mott 1948 *Rep Prog Phys* **12** 163

Cade, P E, A M Stoneham & P W Tasker 1984 *Phys Rev* **B30** 4621

Cade, P E, A M Stoneham & P W Tasker 1986 *Phys Rev* **B33** 4166

Cahn, R W & E Lifshin 1993 eds. *Concise Encyclopaedia of Materials Characterisation*; Pergamon, Oxford
Caldas, M J, R J Baierle, E Molinari & S Ossicini 1997 *Mater Sci Forum* **11** 258
Call, P J, W Hayes & M N Kabler 1975 *J Phys* **C8** L60
Campbell, E E B, D Ashkenasi & A Rosenfeld 1999 *Mater Sci Forum* **301** 123
Canut, B, A Benyagoub, G Marest, A Meftah, N Moncoffre, S M M Ramos, F Studer, P Stuvenard & M Toulemonde 1995 *Phys Rev* **B51** 12194
Cao, S, A J Pedraza, D H Lowndes & L F Allard 1994 *Appl Phys Lett* **65** 2940
Carbucicchio, M, A Valenti, G Battaglin, P Mazzoldi & R Dal Maschio 1985 *Radiat Eff Insulators* **3** 187
Cardenas, J, B G Svensson & M Petravic 1998 *J Appl Phys* **84** 4809
Cartier, E & F R McFeely 1991 *Phys Rev* **B44**, 10689
Castella, H & R Zimmermann 1999 *Phys Rev* **B59** R7801
Castro, G, D Haarer, R M Macfarlane & H P Trommsdorf 1978 US Patent No 4,101,976
Catlow, C R A & A M Stoneham (eds.) 1989 *J Chem Soc Faraday Trans* **11** 85
Catlow, C R A, K M Diller & L W Hobbs 1980 *Phil Mag* **42** 123
Catlow, C R A, J Corish, P W M Jacobs & A B Lidiard 1981 *J Phys* **C14** L121
Caturla, M J, T Diaz de la Rubia & G H Gilmer 1995 *J Appl Phys* **77** 3121
Cavenett, B C 1981 *Adv Phys* **30** 475
Cazaux, J 1995 *Microsc Microanal Microstruct* **6** 345
Cazaux, J 1999 *J Appl Phys* **85** 1137
Chadi, D J & K J Chang 1988 *Phys Rev Lett* **61** 873
Chakarov, D & B Kasemo 1998 *Phys Rev Lett* **81** 5181
Champion, J V & S J Dodd 1995 *J Phys* **D28** 398
Champion, J V & S J Dodd 1996 *J Phys* **D29** 862
Champion, J V, S J Dodd & J M Alison 1996 *J Phys* **D29** 2689
Chandler, D 1984 *J Phys Chem* **88** 3400
Chandler, D & P Woylnes 1981 *J Chem Phys* **74** 4078
Chase, L L 1994a *Nucl Instrum Methods* **B91** 597
Chase, L L 1994b p 53 *et seq* of *Laser Ablation: Principles and Applications* edited by J C Miller; Springer, Berlin
Che, M & A J Tench 1983 *Adv Catal* **32** 1
Chelikowsky, J R & M Schlüter 1977 *Phys Rev* **B15** 4020
Chen, C Y & M A Kastner 1986 *Phys Rev* **B33** 1073
Chen, G S 1999 *J Vac Sci Technol* **A17** 403
Chen, L F & K S Song 1990 *J Phys CondensMatter* **2** 3507
Chen, L F, Y Cai & K S Song 1994 *Nucl Instrum Methods* **B91** 614
Chen, X H, J C Polanyi & D Rogers 1997 *Surf Sci* **376** 77
Chen, Y, M M Abraham, T J Turner & C M Nelson 1975 *Phil Mag* **32** 99
Chen, Y, F W Clinard, B D Evans, E H Farnum, R H French, R Gonzalez, J J O'Dwyer, F W Wiffen & X F Zong 1994 *J Nucl Mat* **217** 32
Cheng, C C *et al.* 1994 *J Vac Sci Technol* **A12** 2630
Cheng, S, H G Berry, R W Dunford, H Esbensen, D S Gemmel, E P Kanter & T Lebrun 1996 *Phys Rev* **A54** 3182
Cheng, Z Y & C H Ling 1998 *J Appl Phys* **83** 5289
Cherns, D 1979 *Surf Sci* **90** 336
Cherynskov, S V 1991 *J Non-Cryst Solids* **134** 218
Chou, N, A D Marwick, R D Goldblatt, Leping Li, G Coleman, J E Heidenrich & J R Paraszczak 1992 *J Vac Sci Technol* **A10** 248
Choy, T C & A M Stoneham 1993 *Europhys Lett* **23** 99

Chu, S N G, S Nakahara, M E Twigg, L A Koszi, E J Flynn, A K Chin, B P Segner & W D Johnston 1988 *J Appl Phys* **63** 611
Cichy, H & E Fromm 1991 *Thin Solid Films* **195** 147
Cielo, P 1984 *J Appl Phys* **56** 230
Cini, M 1978 *Phys Rev* **B17** 2788
Cini, M 1985 *Phys Rev* **B32** 1951
Ciraci, S & I P Batra 1983 *Phys Rev* **B28** 982
Clark, G J, J E E Baglin, F M d'Heurle, C W White, G Farlow & J Narayan, 1984 *Mat Res Soc Symp Proc* **27** 55
Clifford, F K & D T Tuma 1982 *Sensors and Actuators* **3** 238
Clinard, F W & L W Hobbs 1986 p 388 in *Physics of Radiation Effects in Crystals* (edited by R A Johnson & A N Olrov); Elsevier Science, Amsterdam
Cobden, D H & B A Muzykantskii 1995 *Phys Rev Lett* **75** 4274
Cobden, D H, M J Uren & M J Kirton 1990 *Appl Phys Lett* **56** 1245
Collot, P, G Gautherin, B Agius, S Rigo & F Rochet 1985 *Phil Mag* **B52** 1057
Combescot, M & J Bok 1982 *Phys Rev Lett* **48** 413
Compton, W D 1957 *Phys Rev* **107** 1271
Constantini, J M, F Brisard, G Biotteau & L Protin 1997 *J Non-Cryst Solids* **216** 36
Cooke, D W, B L Bennett, E H Farnum, W L Hults, K E Sickafus, J F Smith, T Smith, N Taylor & P Tiwan 1996 *Appl Phys Lett* **68** 1663
Cope, R G & A W Penn 1968 *Brit J Appl Phys* **1** 161
Cordischi, D, D Indovina & M Occhiuzi 1978 *J Chem Soc Faraday Trans* (I) **74** 456
Coulson, C A 1956 *Electricity*; Oliver and Boyd, London
Cowern, N E B, E J H Collart, J Politiek, P H L Bancken, J G M van Berkum, K Kyllesbech Larsen, P A Stolk, H G A Huizing, P Pilcher, A Burenkov & D J Gravesteijn 1997 *Mat Res Soc Symp Proc* **469** 265
Cox, P A & V Henrich 1994 *The Surface Science of Metal Oxides*; Cambridge University Press, Cambridge
Cox, P A & A A Williams 1986 *Surf Sci* **175** L782
Cox, S F J & R L Lichti 1997 *J Alloys Compounds* 253–254 414
Craciun, V & I W Boyd 1993 pps 363 *et seq. Proceedings of the 1st International Rapid Thermal Processing Conference* (edited by R B Fair & B Lojek)
Cumpston, B *et al.* 1999 *Nature* **398** 51
Dagata, J A, J Schneir, H H Harary, C J Evans, M T Postek & J Bennet 1990 *Appl Phys Lett* **56** 2001
Daguzan, Ph, S Guizard, K Krastev, P Martin & G Petite 1994 *Phys Rev Lett* **73** 2352
Daguzan, Ph, P Martin, S Guizard & G Petite 1995 *Phys Rev* **B52** 17099
Danilov, S V, N G Kotelevskaya, V I Skorospelova, S A Stepanov, N A Khalyavka & Y G Yagmurov 1990 *Sov J Opt Technol* **1991** 58 633
Dapkus, P D & C H Henry 1976 *J Appl Phys* **47** 4061
Davidson, E R & W T Borden 1972 *J Phys Chem* **87** 4783
Davis, E A & S F J Cox (eds.) 1996 *Protons and Muons in Materials Science*; Taylor & Francis, London
Davis, E A & N F Mott 1970 *Phil Mag* **22** 903
Day, H C & D R Allee 1993 *Appl Phys Lett* **62** 2691
Day, H C & D R Allee 1996 *Nanotechnol* 7 106
De Raedt, B, H Sprik & H Klein 1984 *J Chem Phys* **80** 5719
Dean, P J & W J Choyke 1977 *Adv Phys* **26** 1
Dearnaley, G, A M Stoneham & D V Morgan 1970 *Rep Prog Phys* **33** 1129
Dearsch, H, J Stuke & J Beichler 1980 *Appl Phys Lett* **38** 456

Debeau, J, T Pochet, L M Hamet, B Equer & A Karar 1991 *Nucl Instrum Methods* **B54** 458
Debeau, J, L A Hamet & T Pochet 1996 *Phys Rev* **B53**, 10740
Degraeve, R, P H Roussel, G Groeseneken & H E Maes 1996 *Microelectronics & Reliability* **36** 1639
Del Buono, G S, P J Rossky & T H Murphrey 1992 *J Phys Chem* **96** 7761
Delbecq, C J, W Hayes & P H Yuster 1961 *Phys Rev* **121** 1043
Delbecq, C J, E Hutchinson & D Schoemaker 1969 *Phys Rev* **187** 1103
Delbecq, C J, D Schoemaker & P H Yuster 1973a *Phys Rev* **B3** 473
Delbecq, C J, D Schoemaker & P H Yuster 1973b *Phys Rev* **B9** 1913
Delbecq, C J, D Schoemaker & P H Yuster 1974 *Phys Rev* **B7** 3933
Delbecq, C J, E Hutchinson, P H Yuster & S A Marshall 1984 *J Chem Phys* **79** 1408
DelFatti, N, P Langot, R Tommasi & F Vallee 1999 *Phys Rev* **B59** 4576
Dellin, T, D A Tichenov & E H Barssis 1977 *J Appl Phys* **48** 1131
Denee, P B 1969 *J Appl Phys* **40** 5396
den Hartog, H W, J C Groote, J R W Weerkamp, J Seinen & H Datema 1993 p 410 in *Defects in Insulating Materials* (edited by O Kanert & J-M Spaeth); World Scientific, Singapore
Depas, M, T Nigam & M M Heyns 1996 *IEEE Trans Electron Devices* **43** 1499)
Depinna, S P & B C Cavenett 1982 *Phil Mag* **B46** 71
Deschamps, J & H Doyeaux 1997 *Physics World* (June) 39
Deshmukh, S, L M Hitchcock, E W Roth & G P Reck 1994 *Diamond & Related Mat* **3** 195
Devine, R A B 1987 *Phys Rev* **B35** 9783
Devine, R A B 1989 *Phys Rev* **B39** 5132
Devine, R A B 1990 *Nucl Instrum Methods* **46** 244
Devine, R A B 1994 *Nucl Instrum Methods* **91** 387
Devochkin, O V & L A Ivanov 1987 *Fiz Khim Obrab Mater* **6** 93
Dexter, D L 1956 *Phys Rev* **101** 48
Dexter, D L, C C Klick & G A Russell 1956 *Phys Rev* **100** 603
Dickinson, J T, L C Jensen, R L Webb, M L Dawes & S C Langford 1993 *J Appl Phys* **74** 3758
Dickinson, J T, S C Langford J J Shin & D L Dpering 1994 *Phys Rev Lett* **73** 2630
Dickinson, J T, J-J Shin & S C Landford 1996 *Appl Surf Sci* **96–98** 316
Dickmann, K 1997, in Cleo '97, quoted *Photonics Spectra* May 1997 p 133
Diener, J, M Ben-Choring, D I Kovalev, S D Ganichev & F Koch 1995 *Phys Rev* **B52** R8617
DiMaria, D J 1991, p 65 of *Insulating Films on Semiconductors* (edited by W Eccleston & M Uren); Adam Hilger (now Inst of Physics Publishing), Bristol
DiMaria, D J 1996 *Appl Phys Lett* **68** 3004
DiMaria, D J & J W Stasiak 1989 *J Appl Phys* **65** 2342
Dionne, G F 1973 *J Appl Phys* **44** 5361
Dirnberger, L *et al.* 1993 *Appl Surf Sci* **69** 216
Ditchfield, R & E G Seebauer 1999 *Phys Rev Lett* **82** 1185
Ditmire, T, J W G Tisch, E Springate, M B Mason, N Hay, R A Smith, J Marangos & M H R Hutchinson 1997 *Nature* **386** 54
Donckers, M C J M, O G Poluektov, J Schmidt & P G Baranov 1992 *Phys Rev* **B45** 13061
Donovan, E P, F Spaepen, J M Poate & D C Jacobson 1989 *Appl Phys Lett* **55** 1516
Dorssen, G E van, H D Roper H A Padmore A D Smith & G N Greaves 1995 *Rev Sci Instr* **66** 1480

Dow, J D & D Redfield 1972 *Phys Rev* **B5** 594
Dragnea, B & B Bourguinon 1999 *Phys Rev Lett* **82** 3085
Dresendorfer, P V 1989 *Nucl Instrum Methods* **B40** 1291
Dreyfus, R W 1993 *Surf Sci* **283** 177
Dreyfus, R W, R Kelly & R E Walkup 1986 *Appl Phys Lett* **49** 1478
Dreyfus, R W, F A McDonald & R J von Gutfeld 1987a *Appl Phys Lett* **50** 1491
Dreyfus, R W, F A McDonald & R J von Gutfeld 1987b *Appl Phys Lett* **50** 1059
Dubeau, J, L A Hamel & T Pachet 1966 *Phys Rev* **B53** 10740
Duckett, S W 1968 *Phys Rev* **166** 302
Duffy, D M & A M Stoneham 1983 *J Phys* **C16** 4087
Duffy, D M & P W Tasker 1986 *Phil Mag* **A53** 113
Duffy, D M, J H Harding & A M Stoneham 1995a *Acta Met Mater* **43** 1559
Duffy, D M, J H Harding & A M Stoneham 1995b *J Appl Phys* **76** 2791
Duley, W W 1984 *Phil Mag* **B49** 159; see also *J Chem Soc Farad* I **80** 1173
Dunlop, A & D Lesueur 1991 *Proc Int Conf Irrad Metals, Siofok*
Dunlop, A, D Lesueur, P Legrand, H Dammak & J Dural 1994 *Nucl Instrum Methods B* **90** 330
Dunlop, A, G Jaskierowicz, & S Della-Negra 1998 *Nucl Instrum Methods B* **146** 302
Dunstan, D J 1984 *Phil Mag* **B49** 191
Duraud, J P, F Jollet, Y Langevin & E Dooryhee 1988 *Nucl Instrum Methods B* **32** 248
Dusastre, V & D E Williams 1998 *J Phys Chem* **B102** 6732
Dyke, J P van, & G A Samara 1975 *Phys Rev* **B11** 4935
Dyson, F J 1955 *Phys Rev* **98** 349
Ebinger, H D & J T Yates 1998 *Phys Rev* **B57** 1976
Edamatsu, K, M Ikezawa, S Sato, T Kono & T Sagawa 1983 *J Phys Soc Jpn* **52** 1521
Edamatsu, K, M Ikezawa, H Tokailin, T Takahashi & T Sagawa 1986 *J Phys Soc Jpn* **55** 314
Edamatsu, K, M Sumita, S Hirota & M Hirai 1993 *Phys Rev* **B47** 6747
Edwards, A H 1993 *Phys Rev Lett* **71** 3190
Eernisse, E P & C B Norris 1974 *J Appl Phys* **45** 5196
Eigler, D M & E K Schweizer 1990 *Nature* **344** 524
Ejiri, A, K Nakagawa & A Hatano 1990 *Physica Scripta* **41** 95
Elango, M, J Pruulmann & A P Zhurakovskii 1983 *Phys Status Solidi B* **115** 399
Eley, D D & M A Zammitt 1971 *J Catal* 21 377
Elliott, D J & P D Townsend 1971 *Phil Mag* **23** 249
Elliott, S R 1978 *Phil Mag* **38** 325
Elliott, S R 1979 *Phil Mag* **B39** 349
Elliott, S R 1986 *J Non-Cryst Solids* **81** 71
Elliott, S R & K Shimakawa 1990 *Phys Rev* **B42** 9766
Elliott, S R, T Rayment & S Cummings 1982 *J Phys (Paris)* **43** C9–35
Elman, B, E S Koteles, P Melman & C A Armiento 1989 *J Appl Phys* **66** 2104
Emin, D 1977 in *Amorphous and Liquid Semiconductors*; University of Edinburgh Press, Edinburgh
Emin, D 1980 *J Non-Cryst Solids* **35–36** 969
Emin, D & M-N Bussac 1994 *Phys Rev* **B49** 14290
Emin, D & T Holstein 1976 *Phys Rev Lett* **36** 323
Emtsev, V V, U Dedek, P Erhart, P D Kervalishvili, M A Margaryan, D S Poloskin & H Zillgen 1997 *Mater Sci Forum* **258–263** 575
Engemann, D & Fisher 1977 *Phys Status Solidi B* **79** 195
Englman, R 1972 *The Jahn–Teller Effect in Molecules and Crystals*; Wiley, New York

Englman, R 1979 *Non-Radiative Decay of Ions and Molecules in Solids*; North Holland, Amsterdam

Epifanov, A S 1974 *Sov Phys JETP* **40** 897

Epifanov, A S, A A Manenkov & A M Prokorov 1976 *Sov Phys JETP* **43** 377

Esaid, M, J L Brebner, J Albert & K Awazu 1998 *J Appl Phys* **84** 4193

Eshita, T K Tanimura & N Itoh 1984 *Phys Status Solidi B* **122** 489

Estle, T L, K H Chow, S F J Cox, E A Davis, B Hitti, R F Kiefl, R L Lichti & C Schwab 1997 *Mater Sci Forum* **258–263** 849

Faist, J, F Capasso, C Sirtori, D L Sivco, A L Hutchinson & A Y Choy 1997 *Phys Rev Lett* **79** 2994

Faist, J *et al.* 1999 *Nature* **387** 777

Falco, J, J P von der Weid, M A Aegerter, T Iida & T Nagapka 1980 *J Phys* **C13** 993

Fallgren, H F, H B Neilsen & E Matthias 1986 *Appl Phys Lett* **49** 770, 930

Faraday, B J & W D Compton 1965 *Phys Rev* **A138** 893

Farlow, G, A Blose, J Sister, B Feldott, B Sounsberry & L Slifkin 1983 *Radiat Effects* **75** 1

Farlow, G C, B R Appleton, L A Boatner & C J McHargue 1985 *Mat Res Soc Symp Proc* **45** 137

Fatemi, F K , D J Fatemi & L A Bloomfield 1995 *Phys Rev Lett* **77** 4895

Feenstra, R M & T C McGill 1982 *Phys Rev* **B25** 6329

Feher, G & F Kip 1955 *Phys Rev* **98** 337

Feibelman, P J & M L Knotek 1978 *Phys Rev* **B18** 6531

Feit, M 1971 *Phys Rev* **B3** 1223

Feynman, R P 1955 *Phys Rev* **97** 660

Feynman R P 1963 *Lectures on Physics*; Addison Wesley, New York

Feynman, R P & A R Hibbs 1965 *Quantum Mechanics and Path Integrals*; Mcgraw Hill, New York

Fiebig, M, K Miyano, Y Tomioka & Y Tokura 1998 *Science* **280** 1925

Finnis, M W, P Agnew & A J E Foreman 1990 *Phys Rev* **B44** 967

Fiori, C & R A B Devine 1984 *Phys Rev Lett* **52** 2081

Fischetti, N V & D J DiMaria 1988 p 375 of *The Physics and Technology of Amorphous* SiO_2 (edited by R A B Devine); Plenum, New York

Fisher, A 1985 p 159 of *Amorphous Semiconductors* (edited by M H Brodsky); Springer, Berlin

Fisher, A J, W Hayes & D L Wallace 1989 *J Phys Condens Matter* **1**, 5567

Fisher, A J, W Hayes & A M Stoneham 1990a *Phys Rev Lett* **64** 2667

Fisher, A J, W Hayes & A M Stoneham 1990b *J Phys Condens Matter* **2** 6707

Fleischer, R L, 1993, p 468 *et seq* of *Concise Encyclopaedia of Materials Characterisation* (edited by R W Cahn & E Lifshin); Pergamon, Oxford

Fleisher, R L, P B Price & R M Walker 1965 *J Appl Phys* **36** 3465

Fleisher, R L, P B Price & R M Walker 1975 *Nuclear Tracks in Solids*; University of California Press, Berkeley

Fleury, L, A Gruber, A Drabenstedt, J Wrachtrup & C von Borczynskoski 1997 *J Phys Chem* **B101** 7933

Flower, H M 1993 p 134 of *Concise Encyclopaedia of Materials Characterisation* (edited by R W Cahn & E Lifshin); Pergamon, Oxford

Flynn, C P 1972 *Defects and Diffusion in Solids*; Oxford University Press, Oxford

Flynn, C P & R S Averback 1988 *Phys Rev* **B38** 7118

Flynn, C P & G Jaccucci 1982 *Phys Rev* **B25** 6225

Flynn, C P & A M Stoneham 1970 *Phys Rev* **B1** 3966

Fokker, P A, W D Koster, J I Dijkhuis, H W de Wijn L Lu, R S Meltzer & W M Yen 1997 *Phys Rev* **B56** 2306
Foot, J D, E A Colbourn & C R A Catlow 1988 *J Phys Chem Solids* **49** 1225
Fowler, W B 1966 *Phys Rev* **151** 657
Fowler, W B (ed.) 1968 *Physics of Color Centers*; Academic Press, New York
Franchy, R 1998 *Rep Prog Phys* **61** 691
Frederickson, A R & A S Karakashian 1995 *J Appl Phys* **77** 1627
Freeman, G R (ed.) 1987 *Kinetics of Inhomogeneous Processes*; J Wiley, New York
Frisch, H L & J R Patel 1967 *Phys Rev Lett* **18** 784
Frumar, M 1997 *J Non-Cryst Solids* **213–214** 215
Frumar, M, A P Firth & A W Owen 1984 *Phil Mag* **B50** 463
Fu, C-R & K S Song 1999 *Phys Rev* **B59** 2529
Fu, C-R, L F Chen & K S Song 1999 *J Phys Condens Matter* **11** 5517
Fujimaki, M, T Watanabe, T Katoh, T Kasahara, N Miyazaki & Y Ohki 1998 *Phys Rev* **B57** 3920
Fujimaki, M, T Katoh, T Kasahara, N Miyazaki & Y Ohki 1999 *J Phys Condens Matter* **11** 2589
Fujita, M, H Nakagawa, K Fukui, H Matsumoto, T Miyanaga & M Watanabe 1991 *J Phys Soc Jpn* **60** 4393
Fujiwara, H, T Suzuki & K Tanimura 1997 *J Phys Condens Matter* **9** 923
Fukutome, H 1983 *Prog Theor Phys* **47** 1156; **49** 22; **50** 1433
Furukawa, Y, K Kitamura, K Niwa, H Hatano, P Bernasconi, G Montemezzani & P Gunter 1999 *Jap J Appl Phys (Part 1)* **38** 1816
Gadzuk, J W 1991 *Phys Rev* **B44** 13466
Gale, G M, G Gallot, F Hache, N Lascoux, S Bratos & J C Leicknam 1999 *Phys Rev Lett* **82** 1068
Galii, P V & N A Tsai 1985 *Sov Phys Solid State* **27** 1922
Gammaitoni, L, P Hänngi, P Jung & F Marchesoni 1998 *Rev Mod Phys* **70** 223
Ganjoo, A, Y Ikeda & K Shimakawa 1999 *Appl Phys Lett* **74** 2119
Garcia, R, M Calleja & F Perez-Murano 1998 *Appl Phys Lett* **72** 2295
Garrett, G A, T F Albrecht, J F Whittaker & R Merlin 1996 *Phys Rev Lett* **77** 3661
Gartry, D S 1997 *Proceedings of the Royal Institution* **68** 233 Oxford University Press, Oxford
Gayen, S K, V P Tomaselli & R R Alfano 1995 *Photonics Spectra* (May) 144
Gazecki, J, G A Sai-Halasz, R G Elliman & A Kellock 1985 *Appl Surf Sci* **22–23**, 1034
Geguzin, Y E & M Krivoglaz 1973 *Migration of Macroscopic Inclusions in Solids*; Consultants Bureau, New York
Gehan, R, A J Amaratunga & J Robertson 1996 *Phys Rev* **B53** 16306
Georgiev, M & N Itoh 1990 *J Phys Condens Matter* **2** 10021
Gerlach, B & H Lowen 1991 *Rev Mod Phys* **63** 63
Gilbert, T L 1966 NATO Summer School (Ghent; unpublished; discussed in Fowler (1968))
Gillen, K T & R L Clough 1992 *Polymer* **33** A 358
Glachant, A, P Soukiassian, S T Kim, S Kapoor, A Papageorgopoulis & Y Baros 1991 *J Appl Phys* **70** 2387
Gladden, L F, S R Elliott & G N Greaves 1988 *J Non-Cryst Solids* **106** 189
Glasstone, S, K J Laidler & H Eyring 1941 *Theory of Rate Processes*; McGraw Hill, New York
Godwin, P D, A P Horsfield, A M Stoneham, S J Bull, I J Ford, A H Harker, D G Pettifor & A P Sutton 1996 *Phys Rev* **B54** 15785

Gogoll, S, E Stenzel, H Johansen, M Reichling & E Matthias 1996a *Nucl Instrum Methods* **B116** 279
Gogoll, S, E Stenzel, M Reichling, H Johansen & E Matthias 1996b *Appl Surf Sci* **96–98** 332
Goldberg, L S & H Meistrich 1968 *Phys Rev* **172** 877
Goldberg, R D, J S Williams & R G Elliman 1999 *Phys Rev Lett* **82** 771
Goldman, A 1977 *Phys Status Solidi B* **81** 9
Goldstein, H 1959 *Fundamental Aspects of Radiation Shielding*; Addison-Wesley, New York
Gong, W L, L M Wang & R C Ewing 1996 *Phys Rev* **B54** 3800
Gonzalez, R, M A Monge, J E Santiuste Munoz, R Paneja, Y Chen, E Kotomin, M M Kukla & A I Popov 1999 *Phys Rev* **B59** 4786
Goodwin, I 1997 *Physics Today* (Oct) 85
Gosar, P 1977 *Phys Status Solidi B* **79** 175
Gosnell, T R, A J Sievers & R H Silsbee 1984 *Phys Rev Lett* **52** 303
Gotoh, T, S Takagi & G Tominaga 1989 p 327 in *Desorption Induced by Electronic Transitions DIET IV* (edited by G Betz & P Varga) Springer, Berlin
Govolov, V V, Yu A Matveets, A M Sanov & V S Letokhov 1992 *JETP Lett* **55** 440
Grais, K I & A M Bastawros 1982 *J Appl Phys* **53** 5239
Grätzel, M 1981 *Acc Chem Res* **14** 376
Grazhulis, V A, V V Kveder & Yu A Ossipyan 1981 *Sov Phys JETP* **53** 618
Green, T A, M E Riley & M A Coltrin 1989a *Phys Rev* **B39** 5397
Green, T A, M E Riley, P M Richards, G I Loubriel, D R Jennison & R T Williams 1989b *Phys Rev* **B39** 5407
Green, T A, G M Loubriel, P M Richards, L T Hudson, P M Savundararaj, R G Albridge, A V Barnes & N H Tolk 1989c p 201 of *Desorption Induced by Electronic Transitions DIET IV* (edited by G Betz & P Varga) Springer, Berlin
Grein, C H & S John 1990 *Phys Rev* **B41** 7641
Griffin, P B, M Cao, P Van der Voorde, Y-L Chang & W M Greene 1998 *Appl Phys Lett* **73** 2986
Griffith, J E, Y Qiu & T A Tombrello 1982 *Nucl Instrum Methods* **198**, 607
Grigorjeva, L G, D K Millers, E A Kotomin, V G Artyushenko & A V Nomoev 1989 *Opt Spectr USSR*, **67** 608
Grigorjeva, L G, E A Kotomin, D K Millers & R I Eglitis 1995 *J Phys Condens Matter* **7** 1483
Grimes, R W, C R A Catlow & A M Stoneham 1989 *J Amer Ceram Soc* **72** 1856
Grimes, R W, C R A Catlow & A L Shluger (eds) 1992 *Quantum Mechanical Cluster Calculation in Solid State Studies*; World Scientific, Singapore
Griscom, D L 1975 p 209 of *Radiation Damage Processes in Materials* (edited by C H D Dupuy); Noordhoff, Groningen
Griscom, D L 1979 p 98 of *Proceedings of the Third International Frequency Control Symposium*; Electronic Industries Association, Washington, DC
Griscom, D L 1989 *Phys Rev* **B40**, 4224
Griscom, D L 1991 *J Ceram Soc Jpn* **99**, 923
Griscom, D L 1992a *J Non-Cryst Solids* **149** 137
Griscom, D L 1992b *J Electron Mater* **21** 763
Griscom, D L 1993 *J Non-Cryst Solids* **161** 45
Griscom, D L 1995a *J Appl Phys* **77** 5008
Griscom, D L 1995b *J Appl Phys* **78** 6996
Griscom, D L & E J Friebele 1991 *Phys Rev* **B43** 7427

Grobunov, S V, V Yu Vaklovlev V Yu Ivanov & A V Kruzhalov 1990 *Sov Phys Solid State* **32** 2942

Gruber, A, A Drabenstedt, C Tietz, L Fleury, J Wachtrup & C von Borczyskowski 1997 *Science* **276** 2012

Grunthaner, P J, M H Hecht, F J Grunthaner & N M Johnson 1986 *Mater Sci Report* **1** 65

Guillot, G, A Nouailhat & P Pinard 1975 *J Phys Soc Jpn* **39** 398

Guizard, S, P Martin, Ph Daguzan, G Petite, P Audebert, J P Geindre, A Dos Santos & A Antonnetti 1995 *Europhys Lett* **29** 401

Guizard, S, P Martin, G Petite, P D'Olivera & P Meynardier 1996 *J Phys Condens Matter* **8** 1281

Gusev, V E & O B Wright 1998 *Phys Rev* **B57** 2878

Gutfield, R J von, F A McDonald & R W Dreyfus 1986 *Appl Phys Lett* **49** 1478

Guzzi, M, M Martin, M Mattaini, F Pio & G Spinolo 1987 *Phys Rev* **B35** 9407

Haart, L G J de, *et al.* 1985 *J Solid State Chem* **59** 291

Haassen, P 1975 *Phys Status Solidi A* **28** 145

Haglund, R F & N Itoh 1994 p 11 *et seq* of *Laser Ablation: Principles and Applications* (edited by J C Miller); Springer, Berlin

Hagon, J P, M Jaros & A M Stoneham 1985 *J Phys* **C18** 4957

Haken, H 1976 *Quantum Field Theory of Solids*; North Holland, Amsterdam

Halas, N J & J Bokor 1989 *Phys Rev Lett* **62** 1679

Hall, T P P, A Leggeat & J W Twidell 1969 *J Phys* **C2** 1590

Hall, T P P, A E Hughes & D Pooley 1976 *J Phys* **C9** 439

Hallen, A, N Keskitalo, L Josyula & B G Svensson 1999 *J Appl Phys* **86** 214

Halliburton, L E, N Koumvakalis, M E Markes & J J Martin 1981 *J Appl Phys* **52** 3565

Halperin, A 1990 *J Phys Chem Solids* **51** 303; see also A Halperin & E W Scucov 1991 *J Phys Chem Solids* **52** 1039; A Halperin & E W Scucov 1993 *J Phys Chem Solids* **54** 43

Halperin, A & S Katz 1988 *J Phys Chem Solids* **49** 577

Halperin, A, M G Jani & L E Halliburton 1986 *Phys Rev* **B34** 5702

Hamanaka, H, K Tanaka & S Iizima 1977 *Solid State Commun* **23** 63

Hamilton, J 1988 *Adv Phys* **37** 359

Hamou, A, G Fleury & C Viger 1985 *Thin Solid Films* **123** 87

Han, B Y, K Nakayama & J H Weaver 1999 *Phys Rev* **B60** 13846

Hand, D P & P St J Russell 1990 *Opt Lett* **15** 102

Hanson, R C 1962 *J Phys Chem* **66** 2376

Happek, U, C E Mungen, W von der Osten & A J Sievers 1994 *Phys Rev Lett* **72** 3903

Harding, J H 1985 *Phys Rev* **B32** 6861

Harding, J H 1990 *Rep Prog Phys* **53** 1403

Harding, J H & A M Stoneham 1982 *J Phys* **C15** 4649

Harding, J H, M J L Sangster & A M Stoneham 1987 *J Phys* **C20** 5281

Harper, P, J Hodby & R A Stradling 1973 *Rep Prog Phys* **36** 1

Harrison, W A 1989 *Electronic Structure and Properties of Solids: The Physics of the Chemical Bond*; Dover, New York

Harrison, W A 1998 *Phys Rev* **B57** 9727; see also *Phys Rev* **B58** 15914

Hart, E J, & M Anbar 1970 *The Hydrated Electron*; John Wiley, New York

Härtel, H & F Lüty 1963 *Z Phys* **177** 369

Hattori, K, A Okano, Y Nakai & N Itoh 1992 *Phys Rev* **B45** 8424

Hautala, J, W D Ohlsen & P C Taylor 1988 *Phys Rev* **B38** 11048
Hawryluk, A M, N M Ceglio & D A Markle 1997 *Solid State Technol* (July) 151
Hayashi, I 1981 *Proc 15th Int Conf on the Physics of Semiconductor*
Hayashi, T, T Yanase, M Matsumoto, K Kan'no, K Toyoda & Y Nakai 1992 *J Phys Soc Jpn* **61** 1098
Hayes, W & T J L Jenkin 1985 *J Phys* **C18** L849
Hayes, W & T J L Jenkin 1986 *J Phys* **C19** 6211
Hayes, W & T J L Jenkin 1988 *J Phys* **C21** 2391
Hayes, W & A M Stoneham 1974 p 185 of *Crystals with the Fluorite Structure* (edited by W Hayes); Oxford University Press, Oxford
Hayes, W, & A M Stoneham 1985 *Defects and Defect Processes in Nonmetallic Solids*; John Wiley, New York
Hayes, W, R F Lambourn, G Rangarajan & I M Ritchie 1973 *J Phys* **C6** 27
Hayes, W, R F Lambourn & J P Stott 1974 *J Phys* **C7** 2429
Hayes, W, I B Owen & G I Pilopenk 1975 J *Phys* **C8** L407
Hayes, W, I B Owen & P J Walker 1977 *J Phys* **C10** 1751
Hayes, W, M J Kane, O Salminen & A I Kuznetsov 1984a *J Phys* **C17** L383
Hayes, W, M J Kane, O Salminen, R L Wood & S P Doherry 1984b *J Phys* **C17** 2943
Hebard, A F, A T Fiory, S Nakahara & R H Eick 1986 *Appl Phys Lett* **48** 520
Heggie, M & R Jones 1983 *Inst Phys Conf Ser* **67** 45
Heifets, E N & A L Shluger 1992 *J Phys Condens Matter* **4** 8311
Heinrich, M, C Domke, Ph Ebert & K Urban 1996 *Phys Rev* **B53** 10894
Heinzelmann, H, D Anselmetti, R Wiesendanger, H-J Guntherodt, E Kaldis & A Wisard 1988 *Appl Phys Lett* **53** 2447
Heitz, R, M Veit, N N Ledentsov, A Hoffmann, D Bimberg, V M Ustinov & P S Kop'ev 1997 *Phys Rev* **B56** 10435
Heller, E J 1975 *J Chem Phys* **62** 1544
Heller, E J 1981 *J Chem Phys* **75** 2923
Heller, E J 1991 *J Chem Phys* **94** 2723
Hemley, J, H K Mao, P M Bell & B O Mysen 1986 *Phys Rev Lett* **57** 747
Hemon, S, Ch Dufour, F Gourbilleau, E Paumier, E Dooryhee & S Begin-Colin 1998 *Nucl Instrum Methods B* **146** 443
Henisch, H K 1949 *Metal Rectifiers*; Oxford University Press, Oxford
Henke, B L, J Liesegang & S D Smith 1979 *Phys Rev* **B19** 3004
Henry, C H 1980a p 19 of *Relaxation of Elementary Excitations* (edited by R Kubo & E Hanimera); Springer, Berlin
Henry, C H 1980b *J Appl Phys* **51** 3051
Henry, C H & P D Dapkus 1976 *J Appl Phys* **47** 4067
Henry, C H & D V Lang 1977 *Phys Rev* **B15** 989
Henry, C H & R A Logan 1977 *J Appl Phys* **48** 3902
Henry, C H, P M Petroff, R A Logan & F R Merritt 1979 *J Appl Phys* **50** 3721
Herman, P R, K Beckley, B Jackson, K Kurosawa, D Moore, T Yamanashi & J Yang 1997 *SPIE* **2992** 86
Herman, P R, G Goodno, X Gu, B Kalbfleish, J Long, M Lukacs, R Majoribanks, R J Dwayne Miller, M Nantel, S Ness & A Oertl 1999 *SPIE* **3618**
Herring, C 1950 *J Appl Phys* **21** 437
Herschel, J F W 1839 *Proc Roy Soc* **4** 131
Hickmott, T W 1964 *J Appl Phys* **35** 2118, 2679
Higashi, G S & M Kastner 1981 *Phys Rev* **B24** 2295
Hill, K O, B Malo, F Bilodeau & D C Johnson 1993 *Ann Rev Mat Sci* **23** 125

Hill, P C, G R Atkins, J Canning, G C Cox & M G Sceats 1993 *Appl Opt* **34** 7689
Himpsel, F-J, & W Steinman 1975 *Phys Rev Lett* **35** 1025
Hirabayashi, I, K Morigaki & S Nitta 1980 *Jap J Appl Phys* **19** L357
Hirabayashi, I, K Morigaki & S Nitta 1981 *J Phys Soc Jpn* **50** 2961
Hirai, M 1990 *J Phys Chem Solids* **51** 737
Hirai, M & S Wakita 1983 *Semicond & Insulators* **5** 231
Hirai, M, Y Suzuki, H Hattori, T Ebara & E Kitamura 1987 *J Phys Soc Jpn* **56** 2948
Hirota, S, K Edamatsu, Y Kondo & T Itoh 1995 *Phys Rev* **52** 7779
Hirsch, P B 1980 *J de Physique* **42** C3–149
Hirsch, P B 1981, p 257 of *Defects in Semiconductors* (edited by J Narayan & T Y Tan); North Holland, Amsterdam
Hirsch, P B 1983 *Inst Phys Conf Ser* **67** 1
Hirsch, P B 1985 p 333 of *Dislocations and Properties of Real Materials*, Institute of Materials, London
Hirschenfelder, J 1983 *Ann Rev Phys Chem* 34 1
Hirth, J P & J Lothe 1982 *Theory of Dislocations*, 2nd edition; McGraw Hill, New York
Hizhnyakov, V V, V G Plekhanov, V V Shepelev & G S Zavt 1981 *Phys Status Solidi B* **108** 531
Ho, W 1994 *Surf Sci* **299–300** 996
Hobbs, L 1994 *Nucl Instrum Methods Phys Res* **B91** 30
Hobbs, L W, A E Hughes & D Pooley 1973 *Proc Roy Soc A* **332** 167
Hodby, J W, J A Borders & F C Brown 1967 *Phys Rev Lett* **19** 952
Hodby, J W, J G Crowder & C C Bradley 1974 *J Phys* **C7** 3033
Hoffman, A, M Petravic, G Comtet, A Heurtel, L Hellner & G Dujardin 1999 *Phys Rev* **B59** 3203
Hodgson, E R 1991 *J Nucl Mater* **179–181** 383
Hodgson, E R 1992 *J Nucl Mater* **191–194** 552
Hodgson, E R 1994 *J Nucl Mater* **212–215** 1123
Hodgson, R T, R Tsu, J A van Vechten & E J Yoffa 1979 *Bull Amer Phys Soc* **24** 315
Hoheisel, W, K Jungmann, M Vollmer, R Weidenauer & F Trager 1988 *Phys Rev Lett* **60** 1649
Hohne, M & M Stasiw 1968 *Phys Status Solidi B* **28** 247
Holliday, K & U P Wild 1993 p 149 of *Molecular Luminescence Spectroscopy,* Part 3 (edited by S G Schulman); Wiley, New York
Holstein, T 1959 *Ann Phys* **8** 343
Hong Wang, G K Ng & A A Hopgood 1998 *J Phys* **D31** 3168
Hopman, H J, J Verhoeven, J J Scholtz & R Fastenau 1997 *Appl Surf Sci* **111** 270
Hori, T 1997 *Gate Dielectrics and MOS ULSIs: Principles, Technologies and Applications*; Springer, Berlin
Horsfield, A P, P D Godwin, D G Pettifor & A P Sutton 1996 *Phys Rev* **B54** 1573
Hoshi, J, M Saidoh & N Itoh 1975 *Cryst Latt Defects* **6** 15
Hosono, H, K-I Kawamura, H Kawazoe & J Nishi 1996 *J Appl Phys* **80** 3115
Hosono, H, H Kawazoe & N Matsunami 1998 *Phys Rev Lett* **80** 317
Hosono, H , M Mizuguchi, H Kawazoe, T Ichimura, Y Watanabe, Y Shinkuma & T Ogawa 1999a *J Appl Phys* **85** 3038
Hosono, H, M Mizuguchi, H Kawazoe & T Ogawa 1999b *Appl Phys Lett* **74** 2755
Howard, R E & R Smolouchowski 1959 *Phys Rev* **116** 314
Howe, A T 1999 *Photonics Spectra* (Jan) 126
Huang, K & A Rhys 1950 *Proc Roy Soc A* **204** 406

Huber, H, W Assman, S A Karamian, H D Mieskes, H Nolte, E Gazis, M Kokkoris, S Kossionides, R Vlastou, R Groumltzschel, A Muumcklich & W Prusseit 1998 *Nucl Instrum Methods* **B146** 309
Hughes, A E 1966 *Proc Phys Soc* **88** 449
Hughes, A E 1983 *Radiat Effects* **74** 57
Hughes, A E & S C Jain 1979 *Adv Phys* **28** 717
Hughes, A E & A B Lidiard 1989 UKAEA Report AERE R-13319 *The Growth of Colloids in Irradiated NaCl: A Comparison of Theory and Experiment*
Hughes, A E, D Pooley, H U Rahman & W A Runciman 1967 UKAEA Report AERE-R 5604
Hughes, R C 1973 *Phys Rev Lett* **30** 1333
Hughes, R C 1977 *Phys Rev* **B15** 2012
Hughes, R C 1978 *Solid State Electron* **21** 251
Huisinga, M, M Reichling & E Matthias 1997 *Phys Rev* **B55** 7600
Huisinga, M, V E Puchin & M Reichling 1998 *Nucl Instrum Methods B* **141** 528
Humphreys, C J, T J Bullough, R W Devenish, D M Maher & P S Turner 1990 *Scanning Microsc Suppl* **4** 185
Hunsche, S, T Starcczewski, A l'Huillier, A Persson, C-G Wahlstrom, B van Linden van den Heuvell & S Svanberg 1996 *Phys Rev Lett* **77** 1966
Hurley, P, C Leveugle, A Mathewson, D Doyle, S Whiston & J Prendergast 1998 MRS Spring Meeting, to appear in *MRS Symp Proc* **510** (edited by S Ashok *et al.*) *Defect and Impurity Engineered Semiconductors and Devices II*
Husinski, W, P Wurz, K Mader, E Wolfrum, B Strehl, G Betz, R F Huglund Jr, A V Barnes & N H Tolk 1988 *Nucl Instrum Methods B* **33** 824
Hutchinson, M H R, T Ditmire, E Springate, J W G Tisch, Y L Shao, M B Mason, N Hay & J P Marangos 1998 *Phil Trans Roy Soc* **356** 297
Hutchinson, P W & P S Dobson 1975 *Phil Mag* **32** 745
Ichige, M, Y Matsumoto & A Namiki 1988 *Nucl Instrum Methods* **B33** 820
ICRU 1979 Report 31 *Average Energy Required to Produce an Ion Pair*; International Commission on Radiation Units and Measurements, Washington DC
Imai, H & H Hirashima 1994 *J Non-Cryst Solids* **179** 202
Imai, H, K Arai, H Imagawa, H Hosono & Y Abe 1988 *Phys Rev* **B38** 12772
Ioselevich, A S & E I Rashba 1985 *Sov Phys JETP* **88** 1873
Irene, E A & R Ghez 1987 *Appl Surf Sci* **30** 1
Irene, E A & E A Lewis 1987 *Appl Phys Lett* **51** 767
Irsigler, P, D Wagner & D J Dunstan 1983 *J Phys* **C16** 6605
Isaacson, M & A Murray 1981 *J Vac Sci Technol* **19** 1117
Isenor, N R & M C Richardson 1971 *Appl Phys Lett* **18** 224
Ishibashi, H, S Arisaka, K Inoshita & T Kobayashi 1994 *Jap J Appl Phys* **33** 4971
Ishikawa, K, J Kanasaki, Y Nakai & N Itoh 1996 *Surf Sci* **349** L153
Ismailov, K M & V Murk 1992 *Sov Phys Solid State* **34** 165
Isoya, J, J A Weil & R F C Claridge 1978 *J Chem Phys* **69** 4876
Ito, H & C G Wilson 1984 p 11 of *Polymers in Electronics*, American Chemical Society Symposium Series 242 (edited by T Davidson); American Chemical Society
Itoh, C, K Tanimura & N Itoh 1988 *J Phys* **C21** 4693
Itoh, C, K Tanimura & N Itoh 1989 *Phys Rev* **B39** 11183
Itoh, C, T Suzuki & N Itoh 1990 *Phys Rev* **B41** 3479
Itoh, M 1988 *J Phys Soc Jpn* **57** 372
Itoh, M 1995 *SPIE* **2362** 202

Itoh, M, N Ohno & S Hashimoto 1992 *Phys Rev Lett* **69** 113
Itoh, M, N Ohno & Y Uzawa 1994 *J Phys Soc Jpn* **63** 825
Itoh, N 1972 *Cryst Latt Defects* **3** 115
Itoh, N 1976 *Nucl Instrum Methods* **132** 211
Itoh, N 1982 *Adv Phys* **31** 491
Itoh, N 1987 *Nucl Instrum Methods B* **27** 155
Itoh, N 1996 *Nucl Instrum Methods B* **116** 33
Itoh, N 1997 *Nucl Instrum Methods B* **122** 405
Itoh, N 1997 *Mater Sci Forum* **239–241** 509
Itoh, N & T Nakayama 1982 *Phys Lett (a)* **92** 471
Itoh, N & M Saidoh 1969 *Phys Status Solidi* **33** 649
Itoh, N & M Saidoh 1973 *J Phys (France)* **34** C-9 101
Itoh, N & A M Stoneham 1998 *Nucl Instrum Methods B* **146** 362
Itoh, N & K Tanimura 1986 *Radiat Effects* **98** 269
Itoh, N & K Tanimura 1988 *Radiat Effects* **98** 435
Itoh, N & K Tanimura 1990 *J Phys Chem Solids* **51** 715
Itoh, N, B S H Royce & R Smoluchowski 1965 *Phys Rev* **A138** 1766
Itoh, N, A M Stoneham & A H Harker 1977 *J Phys* **C10** 4197
Itoh, N, A M Stoneham & A H Harker 1979b *J Phys Soc Jpn* **49** 1364
Itoh, N, T Nakayama & T Tombrello 1985 *Phys Lett (a)* **108** 480
Itoh, N, T Eshita & R T Williams 1986 *Phys Rev* **B34** 4230
Itoh, N, K Tanimura & C Itoh 1988 p 135 of *The Physics and Technology of Amorphous* SiO_2 (edited by R A B Devine); Plenum, New York
Itoh, N, K Tanimura, A M Stoneham & A H Harker 1989a *J Phys Cond Mat* **1** 3911
Itoh, N, A M Stoneham & A H Harker 1989c *Surf Sci* **217** 573
Itoh, N, T Tanaka & F W Clinard 1991 *J Nucl Mat* **183** 25
Itoh, N, A Okano, K Hattori, J Kanasaki & Y Nakai 1993 *Nucl Instrum Methods* **B82** 310
Itoh, N, J Kanasaki, A Okano & Y Nakai 1995 *Ann Rev Mat Sci* **25** 97
Itoh, N, A M Stoneham & K Tanimura 1998 Transient Defects and Electronic Excitation, in R Devine (ed.) *Structure and Imperfections in Amorphous and Crystalline Silicas*; John Wiley, New York
Itoh, N, K Kanasaki & J Xu 1999 *Prog Surf Sci* **61** 1
Ivanov, V Yu, E K Fel'dback, S V Gorbunov, A V Kruzhalov & I L Kuusman 1988 *Sov Phys Solid State* **30** 1572
Ivanov, V Yu, V A Pustovarov, S V Gorbunov & A V Kruzhalov 1994 *Sov Phys Solid State*
Iwai, S, T Tokizaki, A Nakamura, T Shibata, K Tanimura, A L Shluger & N Itoh 1994 *J Lumin* **60–61** 720
Iwai, S, T Tokizaki, A Nakamura, K Tanimura, N Itoh & A Shluger 1996 *Phys Rev Lett* **76** 1691
Iwase, A, S Sasaki, T Iwata & T Nihira 1987 *Phys Rev Lett* **58** 2450
Jackson, K A 1988 *J Mat Res* **3** 1218
Jacobs, P W M & E A Kotomin 1992 *Phys Rev Lett* **69** 1411
Jacobs, P W M, E A Kotomin, A Stashans & E V Stefanovich 1992 *J Phys. Condens Matter* **4** 7531
Jacobs, S D, A L Hrycin & K A Cerqua 1986 *Thin Solid Films*, **144**, 69
Jacobson, S & B Jänsson 1983 *Thin Solid Films* **107** 89
Jaenicke, W 1977 *Adv Electrochem and Electrochem Eng* **10** 91
Jain, U & A B Lidiard 1977 *Phil Mag* **35** 245
Jakas, M M & D E Harrison 1984 *Phys Rev* **B30** 3577

Jakas, M M & D E Harrison 1985 *Phys Rev* **B32** 2752
James, T H 1987, p 57 of *Pioneers of Photography: Their Achievements in Science and Technology* (edited by E Ostroff); The Society for Imaging Science and Technology, Springfield, VA
Janssen, G J M & W C Nieuwpoort 1988 *Int J Quant Chem Symp* **22** 679
Jeffery, S, C J Sofield & J B Pethica 1998 *Appl Phys Lett* **73** 172
Jette, A N, T L Gilbert & Das T P 1969 *Phys Rev* **184** 884
Jiang, Z & X C Zhang 1998 *Opt Lett* **23** 1114
Joannopoulos, J D, M Schlüter & M L Cohen 1975 *Phys Rev* **B11** 2186
Johansen, H, S Gogoll, E Stenzel, M Reichling & E Matthias 1995 *Rad Eff Defects Solids* **136** 151
Johansen, H, S Gogoll, E Stenzel, M Reichling & E Matthias 1996 *J Appl Phys* **80** 4982
John, S, C Soukoulis, M H Cohen & E N Economou 1986 *Phys Rev Lett* **57** 1777
John, S, M Y Chou, M H Cohen & C M Soukoulis 1988 *Phys Rev* **B37** 6963
Johnson, W L, Y-T Cheng, M van Rossum & M-A Nicolet 1985 *Nucl Instrum Methods* **B7–8** 657
Jones, R 1980 *Phil Mag* **B42** 213
Jones, R & G M S Lister 1990 *Phil Mag* **B61** 881
Jones, S C, A H Fischer, P Braunlich & P Kelly 1988 *Phys Rev* **B37** 755
Jones, S C, P Braunlich, R T Casper, X-A Shen & P Kelly 1989 *Opt Eng* **28** 1039
Jonsson, J, B Ranby, D Mary, C Laurent & C Mayoux 1995 *IEEE Trans Dielectrics & Electrical Insulation* **2** 107
Jonsson, J, B Ranby, C Laurent & C Mayoux 1996a *IEEE Trans Dielectrics & Electrical Insulation* **3** 148
Jonsson, J, B Ranby, F Massines, D Mary & C Laurent 1996b *IEEE Trans Dielectrics & Electrical Insulation* **3** 859
Joosen, W, S Guizard, P Martin, G Petite, P Agostini, A Dos Santos, G Grillon, D Hulin & A Antonetti 1992 *Appl Phys Lett* **61** 2260
Jortner, J 1997 *J Chem Soc Faraday Trans* **108** 1
Jortner, J, N R Kestner, S A Rice & M H Cohen 1965 *J Chem Phys* **43** 2614
Judson, R S & H Rabitz 1992 *Phys Rev Lett* **68** 1500
Kabler, M N 1964 *Phys Rev* **136** A1296
Kabler, M N 1972 p 291 of *Point Defects in Solids* (edited by J H Crawford & L M Slifkin); Plenum Press, New York
Kabler, M N 1975 p 171 of *Proceedings for NATO Advanced Study Insitute on Radiation Damage* (edited by C H S Dupuy); Noordhoff International, Leyden
Kabler, M N & D A Patterson 1967 *Phys Rev Lett* **19** 652
Kac, M 1943 *Am Math J* **65** 609
Kaempfe, M, T Rainer, K J Berg, G Seifert & H Graener 1999 *Appl Phys Lett* **74** 1200
Kamei, T, P Stradins & A Matsuda 1999 *Appl Phys Lett* **74** 1707
Kanamori, J & A Kotani (eds.) 1988 *Core-Level Spectroscopy in Condensed Systems*; Springer, Berlin
Kane, E O 1966 *Phys Rev* **147** 335
Kanasaki, J, T Ishida, K Ishikawa & K Tanimura 1998 *Phys Rev Lett* **80** 4080
Kanasaki, J, K Iwata & K Tanimura 1999 *Phys Rev Lett* **82** 644
Kannan, S, J Li, R L Lehman & G H Sigel 1993a *Appl Optics* **32** 6684
Kannan, S, M E Fineman, J Li & G H Sigel 1993b *Appl Phys Lett* **63** 3440
Kanney-Wallace, G A 1981 *Adv Chem Phys* **47** 535
Kan'no, K, K Tanaka & T Hayashi 1990 *Rev Solid State Sci* **4** 383
Kan'no, K, M Shirai, T Matsumoto & I Akimoto 1997 *Mater Sci Forum* **239–241** 569

Kantorovich, L & M J Gillan 1997 *Surf Sci* **374** 373
Kantorovich, L N, E A Kotomin, V N Kuzovkov, I Tale & Yu R Zakis 1991 *Models of Processes in Wide-Gap Solids with Defects*; Zinatne, Riga
Kantorovich, L N, E N Heifets, A Livshicz, M Kluja & P Zapol 1993 *Phys Rev* **B47** 875
Kanzaki, H 1984 *Phys Rev* **B29** 3573
Kanzaki, H & T Mori 1984 *Phys Rev* **B29** 3573
Kanzaki, H & Y Tadakuma 1994 *J Phys Chem Solids* **55** 631
Kanzaki, H & Y Tadakuma 1997 *J Phys Chem Solids* **58** 221
Kanzaki, H, S Sakugari & K Suemoto 1971 *Solid State Commun* **9** 999
Kanzig, W 1955 *Phys Rev* **99** 1890
Kanzig, W & T O Woodruff 1958 *J Phys Chem Solids* **9** 70
Kao, C-T, L G Rowan & L Slifkin 1992 *Phys Rev* **B46** 2809
Kaplan, H 1998 *Photonics Spectra* (Oct) 60
Kaplyanskii, A A 1991 *J Lum* **48–49** 1
Karasawa T & M Hirai 1975 *J Phys Soc Jpn* **39** 999
Karlsson, U O, F J Himpsel, J F Morar, F R McFeely, D R Reiger & J A Yarmoff 1986 *Phys Rev Lett* **57** 1247
Kasabov, J, E Atanassova, J Vassileva & D Dimirov 1992 *Solid State Commun* **35** 1621
Kassim, H A, J A D Matthew & B Green 1978 *Surf Sci* **74** 109
Kastner, M, D Adler & H Fritsche 1976 *Phys Rev Lett* **37** 1504
Kawai, T, K Kobayashi, M Kurita & Y Makita 1971 *J Phys Soc Jpn* **30** 1101
Kazanskii, S A 1985 *JETP Lett* **41** 224
Kazansky, P G & V Pruneri 1997 *Phys Rev Lett* **78** 2956
Keimer, B 1998 *Science* **280** 1904
Keller, F J & R B Murray 1966 *Phys Rev* **150** 670
Keller, F J & F W Patten 1969 *Solid State Commun* **7** 1603
Kelly, R 1980 *Surf Sci* **100** 85; see also R Kelly 1977 *Radiat Effects* **32** 91
Kelly, R & R W Dreyfus 1988 *Nucl Instrum Methods B* **32** 341
Kelly, R & N Q Lam 1971 *Radiat Effects* **10** 247
Kelly, R & N Q Lam 1973 *Radiat Effects* **19** 39
Kerssen, J, W Degruijter & J Volger 1973 *Physica* **70** 375
Kesternich, W 1999 *J Appl Phys* **85** 748
Kesternich, W, F Scheuermann & S J Zinkle 1993 *J Nucl Mater* **206** 68
Kesternich, W, F Scheuermann & S J Zinkle 1995 *J Nucl Mater* **219** 190
Khoo, G S, C K Ong & N Itoh 1993 *J Phys Condens Matter* **5** 1187
Kikas, A & M Elango 1985 *Phys Status Solidi B* **130** 211
Kim, K S & D Turnbull 1973 *J Appl Phys* **44** 5237
Kimerling, L C 1978 *Solid State Electr* **21** 1391
Kimerling, L C 1990 *Rev Solid State Sci* **4** 335
Kimerling, L C, H J Leamy & K A Jackson 1980 p 242 of *Laser and Electron Beam Processing of Electronic Materials* (edited by C L Anderson, G K Celler & G A Rozgonyi) Electrochemical Society
Kimerling, L C, M T Asom, J L Benton, P J Drevinsky & C E Caefer 1989 *Matter Sci Forum* **38** 141
Kimmel, G A & T M Orlando 1994 *J Chem Phys* **101** 3282
Kimura, K, C Andou & K Nakajima 1998 *Phys Rev Lett* **81** 5438
Kinchin, G H & R S Pease 1955 *Rep Prog Phys* **18** 1
King, S, M R Wilby & I W Boyd 1996 *Mater Sci Eng* **B37** 162
Kingery, W D 1983 p 3 of *Ceramic Powders* (edited by P Vincenzini); Elsevier, Amsterdam

Kingery, W D, H K Bowen & D R Uhlmann 1976 p 677 *et seq* of *Introduction to Ceramics* John Wiley, New York

Kink, R A, G G Liidja Ch B Lushchik & T A Soovik 1967 *Iz SSSR, ser Fiz* **31** 1982

Kink, R, T Avarmaa, V Kisand, A Lõhmus, I Kink & I Martinson 1998 *J Phys Condens Matter* **10** 693

Kinoshita, C, T Sonoda & A Manabe 1998 *Phil Mag* **78** 657

Kirin, M, A Lushchik, Ch Lushchik, I Martinson, V Nagirnyi & E Vasil'chenko 1998 *J Phys Condens Matter* **10** 3509

Kirm, M, G Zimmerer, E Feldbach, A Lushchik, C Lushchik & F Savikhin 1999 *Phys Rev* **B60** 502

Kirton, M J & M J Uren 1989 *Adv Phys* **38** 367

Kiryukin, V, Y J Yang, F C Chou, M A Kastner & R J Birgeneau 1999 *Phys Rev* **B59** R6581

Kiss, Z 1970 *Physics Today* (Jan) 42

Kislov, A N, A N Varaksin & V G Mazurenko 1991 *Fizika Tverdogo Tela* **33** 2932

Kitaura, M & H Nakagawa 1996 *J Electron Spectrosc Related Phen* **79** 171

Kittel, C 1964 *Quantum Theory of Solids*; J Wiley, New York

Kittel, C 1996 p 327 of *Introduction to Solid State Physics*, 7th edition; J Wiley, New York

Kizuka, T 1999 *Phil Mag Lett* **79** 417

Kizuka, T & T Tanaka 1995 *Phil Mag* **A71** 631

Kizuka, T & T Tanaka 1997 *Phil Mag* **A76** 289

Kizuka, T & T Tanaka 1999 *Jap J Appl Phys* (1) **38** 1595

Kizuka, T, M Iijima & N Tanaka 1998 *Phil Mag* **A77** 413

Klein, C A 1968 *J Appl Phys* **39** 2029

Klick, C C 1960 *Phys Rev* **120** 760

Knights, J C & D K Biegelsen 1977 *Solid State Commun* **22** 133

Knotek, M L 1980 *Surf Sci* **91** L17

Knox, R S 1963 *Theory of Excitons*, Supplement of Solid State Physics

Koch, E, V Petrova-Koch & T Muschik 1993 *J Lumin* **57** 271

Kolobov, A V & G J Adriaenssens 1994 *Phil Mag* **B69** 21

Kolobov, A V & G E Bedel'baeva 1991 *Phil Mag* **B64** 21

Kolobov, A V & S R Elliott 1995 *J Non-Cryst Solids* **189** 297

Kolobov, A V, H Oyanagi, Ka Tanaka & Ke Tanaka 1997 *Phys Rev* **B55** 726

Kolodziej, J, P Czuba, P Piatkowski, A Poradzisz, Z Postawa & M Szymonski 1992a *Nucl Instrum Methods* **B65** 507

Kolodziej, J, P Czuba, V Kempter, P Piatkowski, A Poradzisz, Z Postawa & M Szymonski 1992b *Surf Sci* **260** 295

Kolodziej, J, P Czuba, P Piatkowski, Z Postawa & M Szymonski 1996 *Surf Sci* **363** 229

Kometer, R & N Schwentner 1996 *J Chem Phys* **104** 6967

Kometer, R & N Schwentner 1997 *J Chem Phys* **106** 51

Komiyama, S, T Masumi & T Kajita 1979 *Phys Rev* **B20** 5192

Kondo, M & K Morigaki 1993 *J Non-Cryst Solids* **164–166** 227

Kondo, Y, M Hirai & M Ueta 1972 *J Phys Soc Jpn* **33** 151

Kondo, Y, S Hoshina, S Hirota, I Goto & Y Kon'no 1993 *Phys Rev Lett* **70** 810

Kondo, Y, I Goto & N Sakaida 1997 *Phys Rev* **B55** 9534

Kosloff, R 1988 *J Phys Chem* **92** 2087

Kotani, A & Ogasawara H 1997 *J Electron Spectrosc Relat Phenom* **60** 257

Kotomin, E A & A V Doktorov 1982 *Phys Status Solidi B* **114** 287

Kotomin, E A & V V Kuzovkov 1992 *Rep Prog Phys* **55** 2079

Kotomin, E A & V V Kuzovkov 1996 *Modern Aspects of Diffusion-Controlled Reactions*; Elsevier, Amsterdam

Kotomin, E A & A I Popov 1998 *Nucl Instrum Methods* **B141** 1

Kotomin, E A, V Tale, P Tale, P Butlers & P Kulis 1989 *J Phys Condens Matter* **1** 6777

Koumvakalis, N & W A Sibley 1976 *Phys Rev* **B13** 4509

Kratschner, E & M Isaacson 1986 *J Vac Sci Technol* **B4** 361

Kremerman, V A, M Lax, S G Demos, D M Calistru & R R Alfano 1997 *Phys Rev Lett* **78** 374; *Phys Rev* **B56** 14391

Krivolgaz, M A 1969a *Theory of X-ray and Thermal Neutron Scattering by Real Crystals*; Plenum, New York

Krivolgaz, M A 1996b *Diffuse Scattering of X-rays and Neutrons by Fluctuations;* Springer, Berlin

Krylova, I V & A G Petrukhin 1996 *(Sov Phys) Semiconductors* **30** 231

Kubo, T, A Okano, J Kanasaki, K Ishikawa, Y Nakai & N Itoh 1993 *Phys Rev* **B49** 4931

Kubota, S, H Suzuki, J Ruan, F Shiraishi & Y Takami 1986 *Nucl Instrum Methods* **A242** 291

Kuiper, P, G Kruizinger, J Ghijsen, G A Sawatzky & H Verwiej 1989 *Phys Rev Lett* **62** 221

Kulis, P A, Z A Rachko, M J Springs, I A Tale & J L Jansons 1986 *Electronic Processes and Defects in Ionic Crystals* (in Russian); Latvian University Press, Riga

Kumazaki, Y, Y Nakai & N Itoh 1987 *Phys Rev Lett* **59** 2883

Kunz, A B 1982a *Phys Rev* **B26** 2056

Kunz, A B 1982b *Phys Rev* **B26** 2070

Kunz, A B & M P Guse 1977 *Chem Phys Lett* **45** 18

Kurtz, R L, B Stockbauer, R Nyholm, S A Flodstrom & F Senf 1987 *Phys Rev* **B35** 7749

Kurtz, R L, R Stockbauer & T E Madey 1988 p 109 in *Desorption Induced by Electronic Transitions DIET III* (edited by R H Stulen & M L Knotek); Springer Berlin

Kuusmann, I L, G G Liidja & Ch B Lushchik 1976 *Trudi Ist Phys Estonian Acad Sci* **46** 5

Kuvalovskii, V D, V G Lysenko & V B Timofeev 1986 *Sov Phys Uspekhi* **28** 735

Kuznetsov, A I, V N Abramov, N S Rooze & T I Savikhina 1978 *Sov Phys JETP Lett* **28** 602

Kuznetsov, A I, T E Jaessar, B R Namozov & T V Uibo 1982 *Opt Spectr* **52** 438

Kuznetsov, A I, B R Namozov & V V Murk 1985 *Sov Phys Solid State* **27** 1819

Kuznetsov, A I, V N Abramov, V V Murk & B R Namozov 1991 *Sov Phys Solid State* **33** 1126

Kuzovkov, V V & E A Kotomin 1988 *Rep Prog Phys* **51** 1479

Kuzukawa, Y, A Ganjoo, K Shimakawa & Y Ikeda 1999 *Phil Mag* **B79** 249

Kyrukhin, V *et al.* 1997 *Nature* **386** 813

Lamatsch, H *et al.* 1971 *Phys Status Solidi* **48** 311

Landau, L D 1933 *Phys Z Sowjetunion* **3** 664

Landau, L D 1936 *Phys Z Sowjetunion* **10** 154; also L D Landau & E M Lifshitz 1972 *Course of Theoretical Physics: Physical Kinetics*; Pergamon, Oxford

Landau, L D & E M Lifshitz 1959 p 323 of *Course in Theoretical Physics: Statistical Physics*; Pergamon, Oxford

Landau, L D & E M Lifshitz 1984 Chapter XIV of *Course of Theoretical Physics: Electrodynamics*; Pergamon, Oxford
Landauer, R 1999 p 77 of *Feynman and Computation* (edited A G Hey); Perseus Books, Reading, Mass
Landis, G A & P R Younger 1979 *Trans IEEE Components, Hybrids and Manufacturing Technology* **2** 350, see also Younger, P R *et al.* 1980 *Proc 14th IEEE Photovoltaic Specialists Conference* and Younger, P R 1980 *J Non-Cryst Solids* **38** 39
Landman, U, D Scharf & J Jortner 1985 *Phys Rev Lett* **54** 1860
Landry, R J, E Snitzer & R H Bartram 1971 *J Appl Phys* **42** 3827
Lang, D V 1982 *Ann Rev Mat Sci* **12** 377
Lang, I G 1977 *Sov Phys JETP* **72** 2152
Lang, N D 1992 *Phys Rev* **B45** 13599
Langer, J M 1990 *Rev Solid State Sci* **4** 167
Langreth, D C 1971 *Phys Rev* **B3** 3120
Lannoo, M & A M Stoneham 1968 *J Phys Chem Sol* **29** 1987
Laredo, E, W Paul, S E Wand, L G Rowman & L Slifkin 1983 *Phys Rev* **B27** 2470
Lazneva, E F 1991 *Rad Eff Defects Solids* **115** 257
Lee, W Y, H Coufal, C R Davis, V Jipson, G Lim, W Parrish, F Sequeda & R E Davis 1986 *J Vac Sci Technol* **4** 2988
Lee, Y H 1989 *J Vac Sci Technol* **A7** 903
Lee, Y H 1992 *J Vac Sci Technol* **A10** 1318
Leforestier, C *et al.* 1991 *J Comput Phys* **94** 59
Lenahan, P M & P V Dressendorfer 1984 *J Appl Phys* **55** 3495
Lengel, G, J Harper & M Weimer 1996 *Phys Rev Lett* **76** 4725
Lester, S D, T S Kim & B G Streetman 1987 *J Appl Phys* **62** 2950
Letant, S, & G Fishman 1998 *Phys Rev* **B58** 15344
Leung, C H, G Brunet & K S Song 1985 *J Phys* **C18** 4459
Leung, C H, C G Zhang & K S Song 1992 *J Phys Condens Matter* **4** 1489
Lewis, J T, J L Kolopus, E Sonder & M M Abraham 1973 *Phys Rev* **B7** 810
Li, M, S Menon, J P Niberger & G N Gibson 1999 *Phys Rev Lett* **82** 2394
Li, N, T Yoshinobu & H Iwasaki 1998 *Jap J Appl Phys* **37** L995
Li, X, R D Beck & R L Whetten 1992 *Phys Rev Lett* **68** 3420
Liang, W Y & A D Yoffe 1968 *Phys Rev Lett* **20** 59
Lidiard, A B 1998 *Z Phys* **206** 219
Liebrant, G W R, G Hoogers & F H P M Habraken 1992 *Phys Rev Lett* **68** 1947
Liedtke, A *et al.* 1989 *J Non-Cryst Solids* **114** 522
Liijda, G & V L Plekhanov 1973 *J Lumin* **6** 71
Lindhard, J 1954 *K Dansk Vidensk Selsk Mat-Fys Medd* **28** (8)
Lindhard, J, M Scharf & H Schiott 1963a *Mat Phys Med Dan Vid Selsk* **33(14)** 1
Lindhard, J, V Neilsen, M Scharff & P V Thomsen 1963b *Mat Phys Med Dan Vid Selsk* **33(10)**
Lindholm, E & S Lundqvist 1985 *Physica Scripta* **32** 220
Lindner, R, M Reichling, E Matthias & H Johansen 1999 *Appl Phys B* **68** 233
Ling, C H 1994 *J Appl Phys* **76** 581
Liu, F-X, J Y Qian, X L Wang, L Liu & H Ming 1997 *Phys Rev* **B56** 3066
Liu, L, V E Henrich, W P Ellis & I Shindo 1996 *Phys Rev* **B54** 2236
Livshits, B, O Teher-Zahav, E Iskevitch & M Genut 1997 *Solid State Technology* (July) 197
Lockwood, D J, Z H Lu & J-M Baribeau 1996 *Phys Rev Lett* **76** 539
Long, J P, S S Goldberg & M N Kabler 1992 *Phys Rev Lett* **68** 1014

Long, X C & S R J Brueck 1999 *Appl Phys Lett* **74** 2110
Lorenzi, G de, & G Jacucci 1986 *Phys Rev* **B33** 1993
Lundt, H & G Weiser 1983 *Solid State Commun* **48** 827
Lushchik, A Ch & Ch B Lushchik 1986 *Sov Phys Solid States* **28** 765
Lushchik, A, E Feldbach, A Frorip, K Ibragimov, I Kuusmann & Ch Lushchik 1994 *J Phys Condens Matter* **6** 2357
Lushchik, Ch B 1982, p 505 *et seq* of *Excitons* (edited by E I Rashba & M D Sturge); North Holland, Amsterdam
Lushchik, Ch B & A Ch Lushchik 1989 *Decay of Electronic Excitations with Defect Formation in Solids*; Nauka, Moscow
Lushchik, Ch, I Vitol & M Elango 1969 *Sov Phys Solid States* **10** 2166
Lushchik, Ch, A Lushchik & E Vasil'chenko 1981 p 323 in *Defects in Insulating Crystals* (edited by V M Tuchkevic & K K Shvarts); Zinatne, Riga
Lushchik, Ch B, J Kolk, A Ch Lushchik, N E Lushchik, M M Taiirov & E A Vasilchenko 1982 *Phys Status Solidi* **114** 103
Lüty, F 1968 in *Physics of Color Centers* (edited by W B Fowler); Academic Press, New York
Lüty, F 1971, unpublished Abstract H158 of the Reading UK Colour Centre meeting
Lüty, F 1983 *Semicond Insulators* **5** 245
Lyding, J W, K Hess & I C Kizilyally 1996a *Appl Phys Lett* **68** 2526
Lyding, J W, T-C Chen, G C Abein, C Wang & J R Tucker 1996b *Nanotechnol* **7** 128
Lyo, I-W & P Avouris 1991 *Science* **253** 171
Lyubin, V M & V K Tikhomirov 1989 *J Non-Cryst Solids* **114** 133
Macaulay-Newcombe, R G & D A Thompson 1984 *Nucl Instrum Methods B* **1** 176
MacLean, S G & W W Duley 1984 *J Phys Chem Soc* **45** 223
Maeda, K & S Takeuchi 1983 *J de Physique 44 Colloque Suppl* **9** C4–375
Mainwood, A M & A M Stoneham 1994 *J Phys Condens Matter* **6** 4917
Mainwood, A M & A M Stoneham 1997 *J Phys Condens Matter* **9** 2453
Maki, A G & R Forneris 1966 *Spectrochem Acta* **A23**, 867
Makimura, M, K Tanimura, N Itoh, T Tokizaki & A Nakamura 1994 *J Phys Condens Matter* **6** 4581
Malherbe, J B, S Hofmann & J M Sanz 1986 *Appl Surf Sci* **27** 355
Malshe, A P, A M Ozkan, W D Brown & P A Mollian 1998, *MRS Proceedings* Vol 526 (edited by R K Singh, D H Lowndes, D P Chrisey, E Fogasrassy & J Narayan); Materials Research Society, Warrendale, PA
Marchetti, A P & D S Tinti 1981 *Phys Rev* **B24** 7361
Margulis, W, F C Garcia, E N Hering, L C Guedes Valente, B Lesche, F Laurell & I C S Carvalho 1998 *MRS Bulletin* **23(10)** 31
Markvardt, T 1980 *Solar Cells* **2**
Markvardt, T & P T Landsberg 1981 *Tblisi Conference on the Physics of Semiconductors*
Marquardt, C L, R T Williams & M N Kabler 1971 *Solid State Commun* **9** 2285
Marrone, M J, F W Patten & M N Kabler 1973 *Phys Rev Lett* **31** 467
Martin, B 1997 *GEC J Technol* **14** 2
Martin, M, W Mader & E Fromm 1994 *Thin Solid Films* **250** 61
Martin, P, S Guizard, Ph Daguzan, G Petite, P D'Oliveira, P Meynendiev & M Perdrix 1997 *Phys Rev* **B55** 5799
Martini, M, G Spinolo & A Vedda 1986 *J Appl Phys* **60** 1705
Martini, M, G Spinolo & A Vedda 1987 *J Appl Phys* **61** 2486
Martini, M, A Paleari, G Spinolo & A Vedda 1995 *Phys Rev* **B52** 138

Mary, D, C Laurent, C Mayoux, J Berdala & R Clavreul 1996 *Rev de l'Electricité et de l'Electronique (Cables d'Énergie et Science des Isolants)* **1** 102
Mashkov, V A, W R Austin, L Zhang & R A Leisure 1996 *Phys Rev Lett* **76** 2926
Masri, P M & A M Stoneham 1984 *J Electron Mater* **14** 205
Masri, P M, A M Stoneham & A H Harker 1983 *J Phys* **C16** L613
Matsui, Y, R Nagoyoshi, M Nakamura, M Okayama & Y Hamakawa 1992 *Jap J Appl Phys* **31** 1972
Matsumoto, T, T Kawata, A Miyamoto & K Kan'no 1992 *J Phys Soc Jpn* **61** 4229
Matsumoto, T, K Ichinose & K Kan'no 1993 *J Phys Soc Jpn* **62** 1860
Matsumoto, T, M Shirai & K Kan'no 1995a *J Phys Soc Jpn* **64** 291
Matsumoto, T, M Shirai & K Kan'no 1995b *J Phys Soc Jpn* **64** 987
Matsunami, N, Y Yamamura, Y Itikawa, N Itoh, Y Kazumata, S Miyagawa, K Morita, R Shimizu & H Tawara 1984 *Atomic Data Nucl Data Tables* **31** 1
Mattheis, L F 1972 *Phys Rev* **B5** 290
Matthias, E & T A Green 1990 p 112 in *Desorption Induced by Electronic Transition* (edited by G Betz & P Varga); Springer, Berlin
Matthias, E, H B Nielsen, J Reif, A Rosen & E Westin 1987 *J Vac Sci Technol* **B5** 1415
Mattson, S M & G R Rossman 1987 *Phys Chem Min* **14** 94
McCune, R C & P Wynblatt 1983 *J Amer Ceram Soc* **66** 111
McHague, C, D L Joslin & C M White 1994 *Nucl Instrum Methods Phys Res* **B91** 549
McKeever, S W S 1985 *Thermoluminescence in Solids*; Cambridge University Press, Cambridge
Meftah, A, F Brisard, J M Constantini, E Dooryhee M Hage-Ali, M Hervieu, J P Stoquert, F Studer & M Toulemonde 1993 *Phys Rev* **B48** 920
Meftah, A, M Djebara, J P Stoquert, F Studer & M Toulemonde 1996 *Nucl Instrum Methods B* **107** 242
Meldrum, A, S J Zinkle & L A Boatner 1998 *Nature* **395** 56
Menzel, D & R Gomer 1964 *J Chem Phys* **41** 3311
Mesarwi, A, A Ignatiev & J S Liu 1988 *Solid State Commun* **65** 319
Meshulach, D & Y Silberberg 1998 *Nature* **396** 239
Midwinter, J E 1979 *Optical Fibers for Transmission*; J Wiley, New York
Mieghem, P van 1992 *Rev Mod Phys* **64** 755
Migus, A, Y Gauduel, J L Martin & A Antonetti 1987 *Phys Rev Lett* **58** 1559
Mihaychuk, J G, N Shamir & H M van Driel 1999 *Phys Rev* **B59** 2164
Mihokova, E, V Nagirnyi, M Nikl, A Stolovich, G P Pazzi, S Zazubovich & V Zepelin 1996 *J Phys Condens Matter* **8** 4301
Milinchuk, V K & V I Tupikov (eds.) 1989 *Organic Radiation Chemistry Handbook* Ellis Horwood, Chichester
Miller, A J, R G Leisure, V A Mashkov & F L Galeener 1996 *Phys Rev* **B53** R8818
Miller, H C 1981 *Physica* **104(C)** 188
Miller, R J D 1991 *Ann Rev Phys Chem* **42** 581
Mills, A 1998 *III-Vs Review* **11(1)** 46, citing D Steigerwald & M Osinski
Misewich, J A, T F Heinz & D M Newns 1992 *Phys Rev Lett* **68** 3737
Mistry, N B & J Roy 1996 *Photonics Spectra* (Jun) 34
Mitchell, I V, J S Williams, D K Sood, K T Short, S Johnson & R G Elliman 1984 *Matt Res Symp* **25** 189
Mitchell, J W 1957 *Rep Prog Phys* **20** 433
Miura, K 1995 *Solid State Commun* **94** 243
Miura, K, K Sigiura, R Souda, T Aizawa, C Oshima & Y Ishizawa 1991 *Jap J Appl Phys* **30** 809

Miyagawa, S 1973 *J Appl Phys* **44** 5617
Miyake, T, E S Gillman, I Oodake & H Petek 1996 *Surf Sci* **357–358** 155
Miyano, K, T Tanaka, Y Tomioka & Y Tokura 1997 *Phys Rev Lett* **78** 4257
Miyazaki, T, N Azuma, S Yoshida & K Fueki 1988 *Radiat Phys Chem* **32** 695
Miyoshi, T, K Nitta, H Ohkumi, K Shiraishi & N Matsuo 1998 *Jap J Appl Phys* **37** 4336
Mizuno, B, K Tanimura & N Itoh 1986 *J Phys Soc Jpn* **55** 3528
Mochel, M E, C J Humphreys, J M Mochel & J A Eades 1983 *41st Annual Meeting of the Electron Microscopy Society of America* p 100
Moerner, W E, W Lenth & G C Bjorklund 1988 *Topics in Current Physics*, vol 44 (edited by W E Moerner); Springer, Berlin
Moisson, J M & M Bensussan 1983 *Surf Sci* **126** 294
Momose, Y, Y Tamura, M Ogino, S Okazaki & M Hirayama 1992 *J Vac Sci Technol* **A10** 229
Monnier, R, K S Song & A M Stoneham 1977 *J Phys* **C10** 4441
Montant, S, A LeCalvez, E Freysz, A Ducasse, V Nazabel, E Fargin & G LeFlem 1999 *Appl Phys Lett* **74** 2623
Moreth, B & G Weiser 1979 *J Lumin* **18–19** 297
Morgan, D V & D van Vliet 1970 *Contemp Phys* **11** 173
Mori, T 1998 *Phys Rev* **B58** 12543
Morigaki, K 1988 *Jap J Appl Phys* **27** 163
Morigaki, K 1999 *Physics of Amorphous Semiconductors*: World Scientific, Singapore
Morigaki, K, I Hirabayashi, M Nitta & K Shimakawa 1980 *Solid State Commun* 33 851
Morigaki, K, M Yamaguchi & I Hirabayashi 1993 *J Non-Cryst Solids* **164–166** 571
Morimoto, Y, S Nozawa & H Hosano 1999a *Phys Rev* **B59** 4066
Morimoto, Y, Y Yasuda & S Nasu 1999b *J Appl Phys* **85** 7385
Morita, S Y Fukano T Uchihashi Y Sugawara Y Yamanishi & T Oasa 1994 *Appl Surf Sci* **75** 151
Moseley, P T, A M Stoneham & D E Williams 1991 pps 46–60 of *Techniques and Mechanisms in Gas Sensing* (edited by P T Moseley, J O W Norris & D E Williams), Adam Hilger (now Institute of Physics Publishing), Bristol
Moseley, R & M Dunn 1995 *Physics World* (Jan) 30
Moser, F, R S van Heyningen & S Lyu 1969 *Solid State Commun* **7** 1609
Moses, W W, S E Derenzo, M J Weber, F Cerina & A Ray-Chaudhuri 1994 *J Lumin* **59** 89
Motakabbir, K A & P J Rossky 1989 *Chem Phys* **129** 253
Mott, N F 1969 *Contemp Phys* **10** 125
Mott, N F & E A Davis 1979 p 244 of *Electronic Processes in Non-Crystalline Materials*, Oxford University Press, Oxford
Mott, N F & R W Gurney 1948 *Electronic Processes in Ionic Crystals*; Oxford University Press, Oxford
Mott, N F & M J Littleton 1938 *Trans Faraday Soc* **34** 485
Mott, N F, E A Davis & R A Street 1975 *Phil Mag* **32** 961
Mott, N F, S Rigo, F Rochet & A M Stoneham 1989 *Phil Mag* **B60** 189
Muller, P, S Spenke & J Teltow 1970 *Phys Status Solidi* **41** 81
Murayama, K 1983 *J Non-Cryst Solids* **59–60** 983
Murayama, K & M A Bosch 1982 *Phys Rev* **B25** 6542
Murdin, B N, C J G M Langerak, S-C Lee, I Gailbraith, G Strasser, E Gornick, W Heiss, M Helm & C R Pidgeon 1997 *Phys Rev* **B55** 5171
Murk, V V 1997 *Mater Sci Forum* **239–241** 537

Murk, V V, A Kuznetsov, B Namonov & K Ismailov 1994 *Nucl Instrum Methods* **B91** 327
Murk, V V, B Namonov & N Yaroshevich 1995 *Radiat Measurements* **24** 371
Murray, R B & P G Bethers 1969 *Phys Rev* **177** 1269
Nagahara, L A, T Thundat & S M Lindsay 1990 *Appl Phys Lett* **57** 270
Nagesha, K & L Sanche 1998 *Phys Rev Lett* **81** 5892
Naguib, H M & R Kelly 1975 *Radiat Effects* **25** 1
Nakayama, K & J H Weaver 1999 *Phys Rev Lett* **82** 980
Nakayama, M, A Soumura, K Hamasaki, H Takeuchi & H Nishimura 1997 *Phys Rev* **B55** 10099
Narayan, J, R A Weeks & E Sonder 1978 *J Appl Phys* **49** 5977
Nassau, K 1983 *The Physics and Chemistry of Color*; J Wiley, New York
Navinsek, O B 1965 *J Appl Phys* **36** 1678
Navrotsky, A 1987 *Diffusion & Defect Data* **53–54** 61
Neidhart, T, M Sporn, M Schmid & P Varga 1995 *Nucl Instrum Methods* **B101** 127
Nelson, R L 1966 *Disc Farad Soc* **41** 322
Nelson, R L & A J Tench 1964 *J Chem Phys* **40** 2736
Nelson, R L & A J Tench 1967 *Trans Farad Soc* **63** 3039
Nelson, R L, A J Tench & B J Harmsworth 1967 *Trans Farad Soc* **63** 1427
Nelson, R L, J W Hale, B J Harmsworth & A J Tench 1968 *Trans Farad Soc* **64** 2525
Neria, E & A Nitzan 1993 *J Chem Phys* **99** 1109
Nicklaus, E 1979 *Phys Status Solidi A* **53** 217
Nie, S, D T Chiu & R N Zare 1994 *Science* **266** 1018
Nikitenko, V A & A I Tereshchenko 1979 *Opt Spectrosc* **47** 671
Ning, Yu, P C McIntyre, M Nastasi & K E Skafus 1995 *Phys Rev* **B52** 17518
Nishii, J, H Yamanosa, H Hosono & H Kawazoe 1995 *Opt Lett* **20** 1184
Nishimura, H 1983 *J Phys Soc Jpn* **52** 3233
Nishimura, H 1989 p 56 in *Defect Processes Induced by Electronic Excitation in Insulators* (edited by N Itoh); World Scientific, Singapore
Nishimura, H, M Sakata, T Trujimoto & M Nakayama 1995 *Phys Rev* **B51** 2167
Nisoli, M, S de Silvestra & O Svelto 1996 *Phys Rev Lett* **77** 3463
Nistor, S V, E Goovaerts & D Schoemaker 1993 *Phys Rev* **B48** 9575
Nistor, S V, E Goovaerts, M Stefan & D Schoemaker 1998 *Nucl Instrum Methods B* **141** 538
Niwano, M, Y Takakuwa, H Katakura & N Miyamoto 1991 *J Vac Sci Technol* **A9** 212
Noda, K, K Uchida, T Tanifuji & S Nasu 1981 *Phys Rev* **B24** 3376
Noda, K, Y Ishii, H Matsui & H Watanabe 1985 *Radiat Effects* **133–134** 205
Noda, K, Y Ishii, H Matsui & H Watanabe 1986 *Radiat Effects* **97** 297
Norgett, M J & A M Stoneham 1973 *J Phys* **C6** 238
Norgett, M J, A P Pathak & A M Stoneham 1977 *J Phys* **C10** 555
Nuritidinov, I, M A Musaeva & V M Reiterov 1999 *Optics & Spectroscopy* **86** 219
Oates, W A, A M Mainwood & A M Stoneham 1978 *Phil Mag* **38A** 607
Öbergs, P K Sitch, R Jones & M I Heggie 1995 *Phys Rev* **B51** 13138
O'Dwyer, J J 1964 *Theory of Electrical Breakdown*; Oxford University Press, Oxford
Oguri, K, K Fujita, M Takahashi, Y Omori, A Tonegawa, N Honda, M Ochi, K Takayama & Y Nishi 1998 *J Mat Res* **13** 3368
Ohanian, H C 1983 *Am J Phys* **51** 1020
Ohashi, H, E Ishiguro, T Sasano & K Shobatake 1996 *Appl Phys Lett* **68** 3713
Ohmukai, M, Y Takigawa & K Kurosawa 1999 *Appl Surf Sci* **137** 78

Ohnishi, A, K Kan'no, Y Iwabuchi & N Mori 1994 *Nucl Instrum Methods B* **91** 210
Ohnishi, A, K Kan'no, Y Iwabuchi & N Mori 1996 *J Electron Spectrosc Rel Phenom* **79** 159
Ohwaki, T & Y Taga 1990 *J Vac Sci Technol* **A8** 2173
Okano, A, A Y Matsuura, K Hattori, N Itoh & J Singh 1993 *J Appl Phys* **71** 3158
Olmstead, M A, N M Amer, S Kohn, D Fournier & A C Boccara 1983 *Appl Phys* **A3** 2141
Ong, C K, K S Song, R Monnier & A M Stoneham 1979 *J Phys* **C12** 4641
Ong, C K, A H Harker and A M Stoneham 1993 *Interface Science* **1** 139
Orrit, M & J Bernard 1990 *Phys Rev Lett* **65** 2716
Ossipyan, Yu A 1981 *Cryst Res Technol* **16** 239
Ossipyan, Yu A 1983a *Inst Phys Conf Ser* **67** 11
Ossipyan, Yu A 1983b *J de Phys 44 Colloq* **44** C4 103
Osten, von der, W 1984 p 293 of *Polarons and Excitons in Polar Semiconductors and Ionic Crystals*; Plenum Press, New York
Ostroff, E (ed.) 1987 *Pioneers of Photography: Their Achievements in Science and Technology*; The Society for Imaging Science and Technology, Springfield, VA
Ostyn, K M & C B Carter 1982 *Surf Sci* **121** 360
Ourmazd, A, D W Taylor, J Aschler & J Bevk 1987a *Phys Rev Lett* **59** 213
Ourmazd, A, W T Tsang, J A Rentschler & D W Taylor 1987b *Appl Phys Lett* **50** 1417
Overeijnder, H, R R Tol & A E de Vries 1979 *Surf Sci* **90** 265
Palmberg, P W & T N Rhodin 1968 *J Phys Chem Solids* **29** 1917
Palmer, R E & P J Rous 1992 *Rev Mod Phys* **64** 383
Panchratov, O & M Sheffler 1995 *Phys Rev Lett* **75** 701
Pandey, K *et al.* 1989 *Phys Rev* **B39** 12565
Pantelides, S T & M Long 1978 p 339 of *Physics of* SiO_2 *and Its Interfaces* (edited by S T Pantelides); Pergamon, New York
Pantelides, S T, D J Mackeish & A B Kunz 1974 *Phys Rev* **B10** 5203
Papaléo, R M *et al.* 1996 *Phys Rev* **B53** 2305
Parks, C C, Z Hussain, D A Shirley, M L Knotek, G Loubriel & R A Rosenberg 1983 *Phys Rev* **B28** 4793
Parrinello, M & A Rahman 1984 *J Chem Phys* **80** 860
Pascucci, M R & L Hobbs 1980 *J Phys (Paris)* **41** C6 237
Pasccuci, M R & L Hobbs 1983 *Radiat Effects* **74** 219
Pässler, R 1977 *Phys Status Solidi B* **83** K55, K111, K163
Patel, J R, L R Testardi & P E Freeland 1976 *Phys Rev* **B13** 3548
Patel, J R, L R Testardi & P E Freeland 1977 *Phys Rev* **B15** 4124
Patten, F W & M J Marrone 1966 *Phys Rev* **142** 513
Pedreza, A J, M J Desilva & R A Kumar 1995 *J Appl Phys* **77** 5176
Peierls, R E 1979 *Surprises in Theoretical Physics;* Princeton University Press, Princeton
Pellegrini, J C 1996 *Sol State Technol* (Oct) 87
Pells, G P 1991 *J Nucl Mater* **184** 177
Pells, G P & D C Phillips 1979 *J Nucl Mater* **80** 207
Pells, G P & A Y Stathopoulos 1983 *Radiat Effects* **74** 181
Penn, D R 1987 *Phys Rev* **B35** 482
Perry, M D, B C Banks, M D Feit, V Yanovsky & A M Rubenchik 1999 *J Appl Phys* **85** 6803
Petite, G, Ph Daguzan, S Guizard & P Martin 1995 p 223 of *Supplément à la Revue Le Vide: Science, Technique et Applications* No 275

Petite, G, P Martin, S Guizard & R Deblock 1998 p 133 of *Supplément à la Revue Le Vide: Science, Technique et Applications* No 287
Petrenko, V F 1998 *J Appl Phys* **84** 261
Petrenko, V F *et al.* 1977 *Sov Phys Solid State* **19** 154
Petroff, P M 1979 *J de Physique 40 Colloque* **C6-201**
Petroff, P M & R L Hartman 1974 *J Appl Phys* **45** 3899
Petroff, P M & L C Kimerling 1976 *Appl Phys Lett* **29** 461
Pfeiffer, G, M A Paesler & S C Agrawal 1991 *J Non-Cryst Solids* **130** 111
Pham, T A, A Daunois, J-C Merle, J Le Moigne & J-Y Bigot 1995 *Phys Rev Lett* **74** 904
Phillips, J C 1978, in *Thin Films* (edited by J M Poate, K N Tu & J W Mayer); Electrochemical Society, New York
Phillips, J C 1979 *Phys Rev Lett* **42** 153
Phipps, C R, T P Turner, R T Harrison, G W York, W Z Ozborne, G K Anderson, X F Coris, L C Haynes, H S Steele & K C Spicochi 1988 *J Appl Phys* **64** 1083
Pian, T R, M M Traum, J S Kraus, N H Tolk, N G Stoffel & G Margaritondo 1983 *Surf Sci* **128** 13
Pilloud, J J & C J Jaccard 1975 *Solid State Commun* **17** 907
Pilloud, J J & C J Jaccard 1979 *Phys Status Solidi B* **92** 233
Platikanova, V & J Malinowski 1978 *Phys Status Solidi A* **47** 683
Poate, J M & J W Mayer 1982 *Laser Annealing of Semiconductors*; Academic, New York
Poborchii, V V, A V Kolobov & Ka Tanaka 1999 *Appl Phys Lett* **74** 215
Poindexter, E & P Caplan 1983 *Prog Surf Sci* **14** 211
Pointer, D 1997 reply to letter in *Europhotonics (*Feb/Mar)
Polman, A, J S Custer, P M Zangwijn, A M Molenbroek, P F A Alkemade 1997 *J Appl Phys* **81** 150
Pomerantz, D I 1968 US Patent 3 397 273 (taken out by P R Mallory & Co)
Pong, W, & J Smith 1973 *Phys Rev* **B7** 5410
Pong, W, & J Smith 1974 *Phys Rev* **B9** 2674
Poole, R T, J G Jenkin, J Liesegang & R C Leckey 1975 *Phys Rev* **B11** 5179
Pooley, D 1965 *Solid State Commun* **39** 241
Pooley, D 1966 *Proc Phys Soc* **87** 245
Pooley, D & W A Runciman 1970 *J Phys* **C3** 1815
Popp, R D & R B Murray 1972 *J Phys Chem Solids* **33** 601
Portnyagin, A S, I I Mil'man & V S Kortov 1991 *Fiz Tver Tela* **33** 2258
Portnyagin, A S, I I Mil'man & V S Kortov 1992 *Fiz Tver Tela* **34** 1444
Postawa, Z and M Szymonski 1989 *Phys Rev* **B39** 12950
Postawa, Z, J Rutkowski, A Poradzisz, P Czuba & M Szymonski 1987 *Nucl Instrum Methods B* **18** 574
Postawa, Z, R Maboudian, M El-Maazawi, M H Ervin, M C Wood & N Winograd 1992 *J Chem Phys* **96** 3298
Postawa, Z, J Kolodziej, P Czuba, P Piatkowski, A Poradzisz, M Szymonski & J Fine 1993 p 299 of *Desorption Induced by Electronic Transition DIET V* (edited by A B Burns, E B Stechel & D R Jennison) Springer, Berlin
Powell, C J 1985a *Surf Interf Analysis* **7** 263
Powell, C J 1985b *J Vac Sci Technol* **A3** 1338
Powell, C J 1994 *Surf Sci* **299–300** 34
Presby, H & W L Brown 1974 *Appl Phys Let* **24** 511
Preuss, S & M Stuke 1995 *Appl Phys Lett* **67** 338

Price, D F, R M More, R S Walling, G Guethlein, R L Shepherd, R E Stewart & W E White 1995 *Phys Rev Lett* **75** 252
Primak, W 1984 *J Appl Phys* **55** 3315
Priolo, F & E Rimini 1990 *Mater Sci Rep* **5** 319
Prokes, S M & O J Glembocki 1994 *Phys Rev* **B49** 2238
Prokhorov, A M, V I Konov, I Ursu & I N Mhailescu 1990 *Laser Heating of Metals*; Institute of Physics Publishing, Bristol
Puchin, V E, A L Shluger & N Itoh 1993a *Phys Rev* **B47** 10760
Puchin, V E, A L Shluger, K Tanimura & N Itoh 1993b *Phys Rev* **B47** 6226
Puchin, V E, A L Shluger & N Itoh 1995 *J Phys Condens Matter* **7** L147
Puchin, V E, A L Shluger & N Itoh 1996 *Phys Rev* **B52** 6254
Puchina, A V, V E Puchin, M Huisinga, R Bennwitz & M Reichling 1998 *Surf Sci* **402–404** 678
Pulker, H K 1984 *Coatings on Glass*, Elsevier, Amsterdam
Purdy, A E, & R B Murray 1975 *Solid State Commun* **16** 1293
Purdy, A E, R B Murray K S Song & A M Stoneham 1977 *Phys Rev* **B15** 2170
Qin, G G & Y Q Jia 1993 *Sol State Commun* **86** 559
Qin, G G, H Z Song, B R Zheng, J Lin, J Q Duan & G Q Yao 1996 *Phys Rev* **B54** 2548
Qiu, Y, J B Griffith, W J Meng & T A Tombrello 1983 *Radiat Effects* **70** 231
Quéré, Y 1992 p 49 of *Materials Modelling: From Theory to Technology* (edited by C A English, J R Matthews, H Rauh, A M Stoneham & R Thetford); Institute of Physics Publishing, Bristol
Rabin, H & C C Klick 1960 *Phys Rev* **117** 1005
Rachko, Z A & J A Valbis 1979 *Phys Status Solidi B* **93** 161
Rakshani, C Hogarth & Abidi 1976 *J Non-Cryst Solids* **20** 25
Ralston, J, A L Moretti & R K Jain 1987 *Appl Phys Lett* **50** 1817
Ramaker, D E 1983a *J Vac Sci Technol A* **1** 1137
Ramaker, D E 1983b p 70 in *Desorption Induced by Electronic Transitions* (edited by N H Tolk, M M Traum, J C Tully & T E Madey); Springer, Berlin
Ramaker, D E, C T White & J S Murday 1982 *Phys Lett A* **89** 211
Ramos, M M D, A M Stoneham, A P Sutton & J B Pethica 1990 *J Phys Condens Matter* **2** 5913
Ramos, M M D, A M Stoneham & A P Sutton 1994 *Synthetic Metals* **67** 137
Ramsden, J J 1984 *Proc Roy Soc* **A392** 427
Rasagni, G, F Vernier, J P Palmari, N Mayani, N Rasagni & A Llebaria 1983 *Opt Comm* **46** 294
Rascon, A & J L Alvarez-Rivas 1983 *J Phys C* **16** 241
Rashba, E 1957 *Opt Spectr* **2** 75
Rashba, E 1982 pp 543–80 of *Excitons* (edited by E Rashba & M Sturge); North-Holland, Amsterdam
Ray, A R & C A Hogarth 1990 *Int J Electr* **69** 97
Redhead, P E 1964 *Can J Phys* **42** 886
Rees, G J, H G Grimmeis, E Janzen & B Skarstem 1980 *J Phys* **C13** 6157
Reichling, M 1995 *Nucl Instrum Methods B* **101** 108
Reichling, M 1998 p 532 *et seq Experimental Methods in the Surface Sciences*, Vol 30: *Laser Ablation and Deposition* (edited by J C Miller & R F Hauglund); Springer, Berlin
Reichling, M, R M Wilson, R Bennewitz, R T Williams, S Gogoll, E Stenzel & E Matthias 1996 *Surf Sci* **366** 531

Reichling, M, A Bodemann & N Kaiser 1998a *Thin Solid Films* **320** 264

Reichling, M, M Uisinga, D Ochs & V Kempter 1998b *Surf Sci* **402–404** 145

Reif, J 1989 *Opt Eng* **28** 1122

Rentzepis, P M, R P Jones & J Jortner 1973 *J Chem Phys* **59** 766

Repzka, E, G Mitroaica, S Lefrant & L Taurel 1973 *Solid State Commun* **13** 1499

Revcolevschi, A & J Livage 1981 p 94 of *Treatise on Materials Science and Technology* vol 20 (edited by H Herman); Academic, New York

Rezvani, G A & R J Friauf 1993 *Phys Rev* **B47** 9215

Rhoderick, E H 1980 *Metal Semiconductor Contacts*; Oxford University Press, Oxford

Ribeiro, R M, M M D Ramos & A M Stoneham 1997a *Appl Surf Sci* **109–110** 158

Ribeiro, R M, M M D Ramos & A M Stoneham 1997b p 489 of *Advances in Materials and Processing Technologies* (edited by M Andritschky); Institute of Materials, Universidade do Minho, Braga, Portugal

Ribeiro, R M, M M D Ramos & A M Stoneham 1998a *Comput Mat Sci* **10** 33

Ribeiro, R M, M M D Ramos & A M Stoneham 1998b *Mat Res Symp Proc* **526** 21

Rice, S A 1958 *Phys Rev* **112** 804

Ridley, B K 1982 *Quantum Processes in Semiconductors*; Oxford University Press, Oxford

Ristein, J & G Weiser 1986 *Phil Mag* **B54** 533

Ristein, J & G Weiser 1988 *Sol State Commun* **66** 361

Ristein, J, P C Taylor, W D Ohlsen & G Weiser 1990 *Phys Rev* **B42** 11845

Robbins, D J 1980 *J Electrochem Soc* **127** 2694

Roberts, F P, G H Jenks & C D Bopp 1976 *Battelle North West Labs Report* 1944

Roberts, G 2000 *Phil Trans Roy Soc* **A358** 345

Robertson, J 1983 *Adv Phys* **32** 361

Robertson, J 1996 *Phys Rev* **B53** 16302

Robertson, J & E P O'Reilly 1987 *Phys Rev* **B35** 2946

Robins, L H & M A Kastner 1984 *Phil Mag* **B50** 29

Robins, L H & M A Kastner 1987 *Phys Rev* **B35** 2867

Rochet, F, S Rigo, M Froment, C d'Anterroches, C Maillot, H Roulet & G Dufour 1986 *Adv Phys* **35** 237

Rodichev, D, Ph Lavallard, E Dooryhee, A Slaoui, J Perriere, M Gandais & Y Wang 1996 *Nucl Instrum Methods B* **107** 259

Roessler, D M & W C Walker 1967 *Phys Rev* **159** 733

Roessler, D M, W C Walker & E Loh 1969 *J Phys Chem Sol* **30** 157

Roosbroek, W van 1965 *Phys Rev* **A139** 1702

Rosenberg, R A & C-R Wen 1988 *Phys Rev* **B37** 5841

Rosenfeld, A, D Ashkenasi, H Varel, M Wahmer & E E B Campbell 1998 *Appl Surf Sci* **129** 76

Rossky, P J & J Schnittke 1988 *J Phys Chem* **92** 4277

Rothschild, M, C Arnone & D Egrlich 1986 *J Vac Sci Technol* **B4** 310

Rothwarf, A 1973 *J Appl Phys* **44** 752

Roy, A, A V Kolobov & K Tanaka 1998 *J Appl Phys* **83** 4956

Rudra, J K, W B Fowler & F J Feigel 1985 *Phys Rev Lett* **55** 2614

Rühle, M & S L Sass 1984 *Phil Mag* **A49** 759

Rundquist, A, C Durfe, Z Chang, G Taft, E Zeek, S Backus, M Murnane, H C Kapteyn, I Christov & V Stoev 1997 *Appl Phys* **B65** 161

Rundquist, A, C Durfe, Z Chang, C Herne, S Backus, M Murnane & H C Kapteyn, 1998 *Science* **280** 1412

Ryan, J F 1998 Chapter 8 of *Hot Electrons in Semiconductors* (edited by N Balkan); Oxford University Press, Oxford
Saeta, P N & B J Greene 1993 *Phys Rev Lett* **70** 3588
Sai-Halasz, G A & G Gazecki 1984 *Appl Phys Lett* **45** 1067
Saidoh, M & N Itoh 1973 *J Phys Chem Solids* **34** 1165
Saiki, K, Y Sato, K Ando & A Koma 1987 *Surf Sci* **192** 1
Sakaguchi, K & T Suita 1952 *Tech Rep Osaka Univ* **2** 177
Sakuragi, S & H Kanzaki 1977 *Phys Rev Lett* **38** 1302
Salaam, G P, M Persson & R E Palmer 1994 *Phys Rev* **B49** 10655
Salehpour, M A, S Satpathy & G P Das 1991 *Phys Rev* **44** 8880
Salminen, O, P Rihola, A Ozols & T Viitala 1996 *Phys Rev* **B53** 6129
Sanche, L 1997 *IEEE Trans Dielectrics & Electrical Insulation* **4** 507
Sangster, M J L & A M Stoneham 1980 *Phil Mag* **B43** 609
Sangster, M J L & A M Stoneham 1984 *J Phys* **C17** 6093
Sawatzky, G A 1977 *Phys Rev Lett* **39** 504
Sawatzky, G A & A Lenselink 1980 *J Chem Phys* **72** 3748
Sa-Yakanit, V & H R Glyde 1987 *Comm Cond Mat Phys* **13** 35
Scaroni, J & T McKee 1997 *Solid State Technol* (July) 245
Sceats, M G & D S B Poole 1991 *Austral Conf Opt Fibre Technol* p302
Sceats, M G, G R Atkins & S B Poole 1993 *Ann Rev Mat Sci* **23** 381
Schmid, A, P Braunlich & P K Rol 1975 *Phys Rev Lett* **35** 1382
Schmid, A, P Kelly & P Braunlich 1977 *Phys Rev* **B16** 4569
Schmidt, J A, R D Arce, R H Buitrago & R R Koropecki 1997 *Phys Rev* **B55** 9621
Schmidt, J A, R D Arce, R R Koropecki & R H Buitrago 1999 *Phys Rev* **B59** 4568
Schober, H R, C Gaukel & C Oligschleger 1997 *Prog Theor Phys Suppl* **126** 67
Schoemaker, D 1966 *Phys Rev* **149** 693
Schoemaker, D 1973 *Phys Rev* **B7** 786
Schoemaker, D & L Kolopus 1970 *Phys Rev* **B2** 1148
Schofield, S A, R G Wyatt & P G Wolynes 1996 *J Chem Phys* **105** 340
Scholz, C & P Ehrhart 1993 *Mat Res Soc Symp Proc* **279** 427
Scholz, F, H Rabus & G Um 1996 *Appl Phys Lett* **69** 2974
Schreiber, M & W Schafer 1984 *Phys Rev* **B29** 2238
Schwaller, P, S Berner, T Greber, J Osterwalder & H Berger 1999 *Appl Phys Lett* **74** 1877
Schwartz, K, G Wirth, T Steckenreiter, O Geiss & M Krämer 1998 *Phys Rev* **B58** 11232
Schwarz, B J, E R Bittner, O V Prezhdo & P J Rossky 1996 *J Chem Phys* **104** 5942
Schweitzer, S & J-M Spaeth 1999 *J Phys Condens Matter* **11** 1723
Scott, P L & C D Jeffries 1962 *Phys Rev* **127** 32
Seah, M P & W A Dench 1979 *Surf Interf Analysis* **1** 2
Seeger, D 1997 *Solid State Technol* (Jun) 115
Seeger, K 1973 *Semiconductor Physics*; Springer, Vienna
Seitz, F 1940 *Modern Theory of Solids*; McGraw Hill, New York
Seitz, F 1954 *Rev Mod Phys* **26** 7
Selloni, A, R Car & M Parrinello 1987 *J Phys Chem* **91** 4947
Selwyn, G S, J Singh & R S Bennett 1989 *J Vac Sci Technol* **A7** 2758
Sempolinski, D R & W D Kingery 1980 *J Amer Ceram Soc* **63** 11
Seth, U & R Chaney 1975 *Phys Rev* **B12** 5923
Shank, C V, R Yen & C Hirliman 1983 *Phys Rev Lett* **50** 454; **51** 900
Sharma, J & R Smoluchowski 1965 *Phys Rev* **A137** 259

Sharma, R R & A M Stoneham 1976 *J Chem Soc Farad II* **72** 913
Shen, G S & C J Humphreys 1999 *J Appl Phys* **85** 148
Shen, T-C & P Avouris 1997 *Surf Sci* **390** 35
Shen, X A, P Braunlich, S C Jones & P Kelly 1987a *Phys Rev Lett* **59** 1605
Sheng, X A, J C Jones, P Braunlich & P Kelly 1987b *Phys Rev* **B36** 2831
Shibata, T, S Iwai, T Tokizaki, K Tanimura, A Nakamura & N Itoh 1994 *Phys Rev* **B49** 13255
Shigemura, H, Y Kawamato, J Nishii & M Takahashi 1999 *J Appl Phys* **85** 3413
Shih, A, J Yatter, P Pehrsson, J Butler, C Hor & R Abrams 1997 *J Appl Phys* **82** 1860
Shikama, T & G P Pells 1983 *Phil Mag A* **47** 369
Shikama, T, M Narui, Y Endo, A Ochiai & H Kayano 1992 *J Nucl Mater* **191–194** 544
Shikama, T, S J Zinkle, K Shikama, L L Snead & E H Farnum 1998 *J Nucl Mater* **258–263** 1867
Shikama, T, K Yasuda, S Yamamoto, C Kinoshita, S J Zinkle & E R Hodgson 1999 *J Nucl Mater* **271–272** 560
Shimakawa, K & S R Elliot 1995 *Electronic, Optoelectronic and Magnetic Thin Films* (edited by J M Marshall, N Kirov & A Vavrek); Wiley, New York
Shimakawa, K, S Imnami, T Kato & S R Elliot 1992 *Phys Rev* **B46** 10062
Shimakawa, K, A Kolobov & S R Elliott 1995 *Adv Phys* **44** 475
Shimizu, T, H Kidoh, M Matsumoto, A Morimoto & M Kumeda 1989 *J Non-Cryst Solids* **114** 630
Shimizu-Iwayama, T, M Ohshima, T Niimi, S Nakao, K Saitoh, T Fujia & N Itoh 1993 *J Phys Condens Matter* **5** L375
Shin, J J, S C Langford, J T Dickinson & Y Wu 1995 *Nucl Instrum Methods* **B103** 284
Shirai, M & K Kan'no 1998 *J Phys Soc Jpn* **67** 2112
Shluger, A L 1988 *J Phys* **C21** L431
Shluger, A L & N Itoh 1990 *J Phys Condens Matter* **2** 4119
Shluger, A L & E Stefanovich 1990 *Phys Rev* **B42** 9664
Shluger, A L & A M Stoneham 1993 *J Phys Condens Matter* **5** 3049
Shluger, A, E Kotomin & L Kantorovich 1982 *Solid State Commun* **42** 749
Shluger, A L, N Itoh, V E Puchin & E Heifets 1991a *Phys Rev* **B44** 1499
Shluger, A L, R W Grimes & C R A Catlow 1991b *J Phys Condens Matter* **3** 3125
Shluger, A L, R W Grimes, C R A Catlow & N Itoh 1991c *J Phys Condens Matter* **3** 8027
Shluger, A L, A H Harker, R W Grimes & C R A Catlow 1992a *Phil Trans Roy Soc* **341** 221
Shluger, A L, L N Kantorovich, E N Heifets, E K Shidlovskaya & R W Grimes 1992b *J Phys Condens Matter* **4** 7417
Shluger, A L, E N Heifets, J D Gale & C R Catlow 1992c *J Phys Condens Matter* **4** 5711
Shluger, A L, A H Harker, V E Puchin, N Itoh & C R A Catlow 1993 *Model Sim Mater Sci Eng* **1** 673
Shluger, A L, V E Puchin, T Suzuki, K Tanimura & N Itoh 1995 *Phys Rev* **B52** 4017
Shockley, W 1961a *Czech J Phys* **11** 81
Shockley, W 1961b *Solid State Electronics* **2** 35
Sholtz, F, H Rabus & G Ulm 1996 *Appl Phys Lett* **69** 2974
Shumay, I L & U Höfer 1996 *Phys Rev* **B53** 15878
Shur, D & G Rosenman 1999 *J Phys D: Appl Phys* **32** L29

Shur, D, G Rosenman, Ya E Krasik & V D Kugel 1996 *J Appl Phys* **79** 3669
Sibley, W A & D Pooley 1974 p 46 of *Treatise on Materials Science and Technology*, vol 5; Academic Press, New York
Siebert, W *et al.* 1987 *Phys Status Solidi B* **140** 311
Siegbahn, K *et al.* 1967 *ESCA: Atomic, Molecular and Solid State Structure Studies by Means of Electronic Spectroscopy*; Almqvist and Wiksells, Stockholm
Sieger, M T, D A Luh, T Miller & T C Chiang 1996 *Phys Rev Lett* **77** 2758
Sigmund, P 1969 *Phys Rev* **184** 383
Sigmund, P 1972 *Rev Roum Phys* **17** 823, 969, 1079
Sigmund, P 1974 *Appl Phys Lett* **25** 169 (*erratum* 1975 *Appl Phys Lett* **27** 52)
Silin, A R, L N Skuja & A N Trukhin 1980 *J Non-Cryst Solids* **38–39** 195
Silva, S R P, J Robertson, G A Rusliand & J Amaratunga 1996 *Phil Mag* **B74** 369
Silvermann, Scott 1996 *Solid State Technol* (September) 113; see also p 50
Silvestrelli, P L, A Alavi, M Parrinello & D Frankel 1997 *Phys Rev* **B56** 3806
Simmons, J G & R R Verderber 1967 *Proc Roy Soc* **A301** 77
Sinke, W C, A Polman, S Roorda & P A Stolk 1989 *Appl Surf Sci* **43** 128
Sitch, P, R Jones, S Oberg & M I Heggie 1994 *Phys Rev* **50** 17717
Sjodin, T, H Petek & H L Dai 1998 *Phys Rev Lett* **81** 5664
Skuja, L 1994a *J Non-Cryst Solids* **167** 229
Skuja, L 1994b *J Non-Cryst Solids* **179** 51
Slagle, O D 1984 *J Amer Ceram Soc* **67** 169; also Hanford reports HEDL S/A 1079, 1538, 77–83 describing work presented at meetings of the American Ceramic Society
Slater, N B 1959 *Theory of Unimolecular Reactions*; Cornell University Press, Ithaca
Smirnov, V N 1986 *Sov Opt Tech* **1** 113
Smith, D R & A J Tench 1968 *Chem Comm* 1113
Smith, G & R G J Strens 1976 *The Physics & Chemistry of Minerals and Rocks* (edited by R G J Strens); Wiley, New York, p 583
Smith, R A 1978 *Semiconductors*; Cambridge University Press, Cambridge
Snipe, B C, M A Rezaei, W Ho, S Gao, M Persson & B I Lundqvist 1997 *Phys Rev Lett* **78** 4410
Snow, E S, P M Campbell & P J McMarr 1993 *Appl Phys Lett* **63** 749
Snyder, K C & W B Fowler 1993 *Phys Rev* **B48** 13238
Snyder, L C 1971 *J Chem Phys* **55** 95
Sofield, C J & A M Stoneham 1995 *Semiconcl Sci Technol* **10** 215
Sokolowski-Tinten, K, J Bialkowski & D von der Linde 1995 *Phys Rev* **B51** 14168
Sokolowski-Tinten, K, J Bialkowski, A Cavalleri, D von der Linde, J Meyer-ter-Vehn & S Anisimov 1998 *Phys Rev Lett* **81** 224
Solomon, I 1998 Royal Society Claude Bernal Lecture (unpublished)
Sonder, E 1975 *Phys Rev* **B12** 1516
Sonder, E & W A Sibley 1972 p 201 in *Point Defects in Solids* (eds J H Crawford & L M Slifkin); Plenum, New York
Sonder, E & L C Templeton 1967 *Phys Rev* **164** 1106
Sonder, E, K F Kelton, J C Pigg & R A Weeks 1978 *J Appl Phys* **49** 5971
Song, H Z & X M Bao 1997 *Phys Rev* **55** 6988
Song, K S 1967 *J Phys Chem Sol* **28** 2003
Song, K S & R C Baetzold 1992 *Phys Rev* **B46** 1960
Song, K S & L F Chen 1994 *Radiat Eff Defects Solids* **128** 35
Song, K S & C H Leung 1987 *J Phys Soc Jpn* **56** 2113
Song, K S & A M Stoneham 1976 *Solid State Commun* **18** 367

Song, K S & R T Williams 1993 *Self-Trapped Excitons*; Springer Berlin
Song, K S, A M Stoneham & A H Harker 1975 *J Phys* **C8** 1125
Song, K S, C H Leung & A M Stoneham 1981 *Solid State Commun* **37** 949
Sood, D K, W M Skinner, J Gazecki & J S Williams 1985 *Mat Res Soc Symp* **45** 219
Souda, R 1999 *Phys Rev Lett* **82** 1570
Soukoulis, C M, A D Zdetsis, E N Economou 1986 *Phys Rev* **B34** 2253
Spaeth, J M, W Meise & K S Song 1994 *J Phys Condens Matter* **6** 1801
Sparks, M, D L Mills, R Warren, Y Holstein, A A Maraduduin, L J Sham & D F King 1981 *Phys Rev* **B24** 3519
Spitzer, R C, W P Ambrose & A J Sievers 1986 *Phys Rev* **B34** 7307
Staebler, D L & C R Wronski 1977 *Appl Phys Lett* **31** 292
Starnberg, H I, P Soukassian, M H Bakshi & Z Hurych 1988 *Phys Rev* **B37** 1315
Stathis, J & E Cartier 1994 *Phys Rev Lett* **72** 2745
Stathis, J H & M A Kastner 1987 *Phys Rev* **B35** 2972
Stathis, J, S Rigo & I Trimaille 1991 *Solid State Commun* **79** 119
Stathis, J, S Rigo, I Trimaille & M S Crowder 1992 *Mat Sci Forum* **83–87** 1421
Steeples, K A, G Dearnaley & A M Stoneham 1980 *Appl Phys Lett* **36** 981
Steer, M J, D J Mowbray, W R Tribe, M S Skolnick, M D Sturge, M Hopkinson, A Cullis, C R Whitehouse 1996 *Phys Rev* **B54** 17738
Stefanov, B B & K Raghavachari 1997 *Phys Rev* **B56** 5035
Stenzel, E, S Gogoll, J Sils, M Huisinga, H Johansen, G Kästner, M Reichling & E Matthias 1997 *Appl Surf Sci* **109–110** 162
Stephens, R B 1980 *J Appl Phys* **51** 6197
Stesmans, A 1993 *Phys Rev Lett* **70** 1723; *Phys Rev* **B48** 2410
Stesmans, A & K Vanheusen 1991 *Phys Rev* **B44** 11353
Stevels, J M 1944 *Progress in the Theory of the Physical Properties of Glass*; North Holland, Amsterdam
Stirniman, M J, W Li & S J Siebener 1995 *J Chem Phys* **103** 451
Stock, D M, G H Gilmer, M Jaraiz & T Diaz de la Rubia 1995 *Nucl Instrum Methods* **B102** 207
Stockbro, K, C Thirstrup, M Sakurai, U Quaade, Ben Yu-Kuang Hu, F Perez-Murano & F Grey 1998 *Phys Rev Lett* **80** 2618
Stoffel, N G, R Riedel, E Colavita, Taglauer & N H Tolk 1985 *Phys Rev* **B32** 6805
Stokstad, R G, P M Jacobs, I Tserruya, L Sapir & G Mamane 1986 *J Mater Res* **1(2)** 231; and also Stokstad, R G, P M Jacobs, I Tserruya, L Sapir & G Mamane 1986 *Nucl Instrum Methods Phys Res* **B16**, 465
Stoneham, A M 1965 *Proc Phys Soc* **86** 1163
Stoneham, A M 1966 *Proc Phys Soc* **89** 909
Stoneham, A M 1969 *Rev Mod Phys* **41**
Stoneham, A M 1974 *J Phys* **C7** 2476
Stoneham, A M 1975 *Theory of Defects in Solids*; Oxford University Press, Oxford
Stoneham, A M 1977 *Phil Mag* **36** 983
Stoneham, A M 1978 UKAEA Report AERE M-3001 *Laser Annealling of Semiconductors*
Stoneham, A M 1979a *Adv Phys* **28** 457
Stoneham, A M 1979b *J Phys* **C12** 891
Stoneham, A M 1979c *Hyperfine Int* **6** 211
Stoneham, A M 1980 *Solar Cells* **2** 43
Stoneham, A M 1981a *J Amer Ceram Soc* **64** 54
Stoneham, A M 1981b *Rep Prog Phys* **44** 1251
Stoneham, A M 1985a AERE Report M-3523 *Electronic Processes in Ionic Semiconductors*

Stoneham, A M 1985b *Phil Mag* **B51** 161
Stoneham, A M 1987a *Cryst Latt Def Amorph Mat* **14** 173
Stoneham, A M 1987b *Introduction to Current Issues in Condensed Matter Structure* Adam Hilger (now Institute of Physics Publications), Bristol
Stoneham, A M 1989a *J Chem Soc Farad II* **85** 505
Stoneham, A M 1989b *Ionic Solids at High Temperatures*; World Scientific, Singapore
Stoneham, A M 1989c *Physica Scripta* **T25** 17
Stoneham, A M 1989d *Nucl Instrum Methods* **B48** 389
Stoneham, A M 1990 *Rev Solid State Sci* **4** 161
Stoneham, A M 1991 pps 19–32 of *Insulating Films on Semiconductors '91* (edited by W Eccleston & M J Uren) Institute of Physics Publishing, Bristol
Stoneham, A M 1994 *Nucl Instrum Methods Phys Res* **B91** 1
Stoneham, A M 1995 *Supplément 275 à la Revue La Vide: Science, Technique et Application* p 146
Stoneham, A M 1997a *IEEE Trans Dielectrics & Electrical Insulation* **4** 604
Stoneham, A M 1997b *Radiat Eff Defects Solids* **142** 191
Stoneham, A M 1998 *Proc 3rd International Conference on Electric Charge in Solid Insulators: Supplément 287 à la Revue "Le Vide: Science, Technique et Applications"*
Stoneham, A M & R H Bartram 1970 *Phys Rev* **B2** 3403
Stoneham, A M & B A McKinnon 1998 *J Phys Condens Matter* **10** 7665
Stoneham, A M & M M D Ramos 1993 *J Solid State Chem* **106** 2
Stoneham, A M & M J L Sangster 1980 *Phil Mag* **B43** 609
Stoneham, A M & M J L Sangster 1983 *Radiat Effects* **73** 267
Stoneham, A M & M J L Sangster 1985 *Phil Mag* **B52** 717
Stoneham, A M & C J Sofield 1998 pps 79–88 of *Fundamental Aspects of Ultrathin Dielectrics on Si-based Devices* (edited by E Garfunkel, E Gusev & A Vul')
Stoneham, A M & L W Smith 1990 AERE-InTec 0206 (unpublished)
Stoneham, A M & L W Smith 1991 *J Phys Condens Matter* **3** 225
Stoneham, A M & P W Tasker 1985a *J Phys* **C18** L543
Stoneham, A M & P W Tasker 1985b *Mat Res Soc Symp Proc* **40** 291
Stoneham, A M & P W Tasker 1987 *Phil Mag* **B55** 237
Stoneham, A M, M J L Sangster & P W Tasker 1981 *Phil Mag* **B44** 603
Stoneham, A M, C R M Grovenor & A Cerezo 1987 *Phil Mag* **B55** 201
Stoneham, A M, C A English & W J Phythian 1998a *Radiat Eff Defects Solids* **144** 311
Stoneham, A M, P D Godwin, A P Sutton & S J Bull 1998b *Appl Phys Lett* **72** 3124
Stoneham, A M, A H Harker, A L Shluger, M A Szymanski & C J Sofield 2000a "Modelling silicon oxidation and the oxide dielectric: Can silicon oxidation be solved? to appear in *Physics and Control of Defects in Semiconductors* (edited by H Katayama-Yoshida); Gordon & Breach
Stoneham, A M, M A Szymanski & A L Shluger 2000b *MRS Symposium on Ultrathin Dielectrics* Vol 592B (edited by A Edwards, T Hattori, H von Bardeleben & D A Buchanan); Materials Research Society, Warrendale PA
Stookey, S S, G H Beall & J E Pierson 1978 *J Appl Phys* **49** 5114
Storchak, V, J H Brewer & G D Morris 1995 *Phys Rev Lett* **75** 2384
Storchak, V, S F J Cox, S P Cottrell, J H Brewer, G D Morris, D J Arsenau & B Hitti 1997 *Phys Rev Lett* **78** 2835
Street, R A 1976 *Adv Phys* **25** 397
Street, R A 1977 *Solid State Commun* **24** 363
Street, R A 1981 *Adv Phys* **30** 593

Street, R A 1991 *Hydrogenated Silicon*; Cambridge University Press, Cambridge
Street, R A & D K Biegelsen 1980 *J Non-Cryst Solids* **35–36** 651
Street, R A & N F Mott 1975 *Phys Rev Lett* **35** 1293
Street, R A, R J Nemanich & G A N Connel 1978 *Phys Rev* **B18** 6915
Stritzker, B, A Pospieszczyk & J A Tagle 1981 *Phys Rev Lett* **47** 356
Struck, C & W M Fonger 1991 *Understanding Luminescence Spectra and Efficiencies using W_p and Related Functions*; Springer, Heidelberg
Stuart, B C, M D Feit, S Herman, A M Rubenchik, B W Shore & M D Perry 1996 *Phys Rev* **B53** 1749
Stuart, R & G K Wertheim 1962 *J Appl Phys* **33** 2345
Stuart, R, F Wooten & W E Spicer 1964 *Phys Rev* **A4** 95
Stutzmann, M, & D K Biegelsen 1989 *Phys Rev* **B40** 9843
Stutzmann, M, W B Jackson & C C Tsai 1984 *Appl Phys Lett* **45** 1075
Stutzmann, M, W B Jackson & C C Tsai 1985 *Phys Rev* **B32** 23
Sugioka, K, S Wada, H Tashiro, K Toyoda, Y Ohnuma & A Nakamura 1995 *Appl Phys Lett* **67** 2789
Sugiyama, T, H Fujiwara, T Suzuki & K Tanimura 1996 *Phys Rev* **54** 15109
Sujimoto, T, H Nishimura & M Nakayama 1996 *Phys Rev* **B54** 16579
Sumi, H 1980a *Physica* **116B&C** 297
Sumi, H 1980b *J Phys Soc Jpn* **49** 1701
Sumi, H 1982 *Physica* **117/118C** 197
Sumi, H 1984 *J Phys Soc Jpn* **53** 3498
Sumi, H 1985 *Phys Rev* **B29** 4616
Sumi, H & Y Toyozawa 1971 *J Phys Soc Jpn* **31** 342
Sutton, A P & J B Pethica 1988 *J Vac Sci Technol* **A6** 2490
Suzuki, T, K Tanimura & N Itoh 1993 *Phys Rev* **B48** 9298
Suzuki, T, K Tanimura & N Itoh 1994 *Phys Rev* **B49** 7233
Szajman, J, J Liesegang, J G Jenkins & R C G Leckey 1981 *J Electr Spectr* 23 97
Szymanski, M, A M Stoneham & A L Shluger 2000 *The roles of charged and neutral oxidising species from ab-initio and Monte-Carlo calculations*, Proc 10th Workshop on Dielectrics in Microelectronics, Barcelona 1999
Szymonski, M 1980 *Radiat Effects* **52** 9
Szymonski, M 1983 p 216 of *Desorption induced by electronic excitation I* (edited by N H Tolk, M M Traum, J C Tully & T E Madey); Springer Berlin
Szymonski, M 1990 *Nucl Instrum Methods* **B46** 427
Szymonski, M 1993 p 495 of *Fundamental Processes in Sputtering of Atoms and Molecules* (edited by P Sigmund); Munksgaard, Copenhagen
Szymonski, M & A E de Vries 1981 *Radiat Effects* **54** 135
Szymonski, M, J Ruthowski, A Poradzisz & Z Postawa 1985 p 160 of *Desorption Induced by Electronic Transitions II* (edited by W Brenig & D Menzel); Springer Berlin
Szymonski, M, J Kolodziej, P Czuba, P Piatkowski, A Poradzisz & N H Tolk 1991a *Nucl Instrum Methods B* **58** 485
Szymonski, M, J Kolodziej, P Czuba, P Piatkowski, A Poradzisz, N H Tolk & J Fine 1991b *Phys Rev Lett* **67** 1906
Tada, T & T Ninomiya 1991 *J Non-Cryst Solids* **137–138** 997
Taglauer, E, N Tolk, R Reidel, E Colvatia, G Margaritondo, N Gershenfeld, N Stoffel, J A Kelber, G Loubriel, A S Bommanavar, M Bakshi & Z Huric 1986 *Surf Sci* **169** 267
Taiirov, M M 1983 *Fiz tverd Tela* **25** 450
Takahashi, M, T Fujiwara, T Kawachi & A J Ikushima 1997 *Appl Phys Lett* **71** 993

Takeda, S & J Yamasaki 1999 *Phys Rev Lett* **83** 320

Takeda, S, K Koto, M Hirata, T Kuno, S Iijima & T Ichihashi 1997a *Mat Sci Forum* **258–263** 553

Takeda, S, K Koto, S Iijima & Ichihashi 1997b *Phys Rev Lett* **79** 2994

Takigawa, Y, K Kurosawa, W Sasaki, M Okuda, K Yshida, E Fujiwara, Y Kato & Inoue 1990 *J Non-Cryst Solids* **125** 107

Tanaka, K 1975 *Appl Phys Lett* **26** 243

Tanaka, K 1976 *Structure and Excitons of Amorphous Solids* edited by G Lucovsky & F L Galeener); American Institute of Physics, New York

Tanaka, K 1977 *Appl Phys Lett* **26** 243

Tanaka, K 1980 *J Non-Cryst Solids* **35–36** 1023

Tanaka, K 1983 *J Non-Cryst Solids* **59–60** 925

Tanaka, K, K Ishida & N Yoshida 1996 *Phys Rev* **B54** 9190

Tanaka, K, K Kan'no & Y Nakai 1990 *J Phys Soc Jpn* **59** 1474

Tanaka, T, T Eshita, K Tanimura & N Itoh 1985 *Cryst Latt Defects Amorph Mat* **11** 221

Tanimura, K 1989 pp 177 & 178 of *Defect Processes Induced by Electronic Excitation in Insulators* (edited by N Itoh); World Scientific, Singapore

Tanimura, K & N Itoh 1981 *J Phys Chem Solids* **42** 901

Tanimura, K & N Itoh 1984 *J Phys Chem Solids* **45** 323

Tanimura, K & N Itoh 1988a *Nucl Instrum Methods Phys Res* **B32** 211

Tanimura, K & N Itoh 1988b *Phys Rev Lett* **60** 2753

Tanimura, K & N Itoh 1990 *Phys Rev Lett* **64** 1429

Tanimura, K & N Itoh 1991 *J Appl Phys* **69** 7831

Tanimura, K & N Itoh 1992 *Phys Rev* **B45** 1432, 9417

Tanimura, K & T Okada 1980 *Phys Rev* **B21** 1690

Tanimura, K, M Fujiwara, T Okada & T Suita 1974 *Phys Lett* **50A** 301

Tanimura, K, T Tanaka & N Itoh 1983 *Phys Rev Lett* **51** 423

Tanimura, K, T Tanaka & N Itoh 1984 *Nucl Instrum Methods* **B1** 187

Tanimura, K, C Itoh & N Itoh 1988 *J Phys* **C21** 1869

Tanimura, K, T Katoh & N Itoh 1989 *Phys Rev* **B40** 1282

Tanimura, K, T Makimura, T Shibata, N Itoh, T Tokizaki, S Iwai & A Nakamura 1993 *Defects in Insulating Materials* (edited by O Kanert & J-M Spaeth); World Scientific, Singapore

Tanner, D J & S A Rice 1988 *Adv Chem Phys* **70** 441

Tanner, P, cited in *Solid State Technol* 1996 (Nov) 56

Tannor, D J 1994 Nature **369** 445 (comment on Metz, Pfeiffer, Thoemke, Crim (1994), Amer Chem Soc meeting; see *J Chem Phys* **92** 6333; **94** 4928; **99** 1744; *Chem Phys Lett* **221** 347)

Tasker, P W 1983 p 457 of *Mass Transport in Solids* (edited by F Beniere & C R A Catlow); Plenum Press, New York

Tasker, P W & D M Duffy 1984 *Surf Sci* **137** 91

Tasker, P W & A M Stoneham 1977 *J Phys Chem Sol* **38** 1185

Tasker, P W, E A Colbourn & W C Mackrodt 1985 *J Amer Ceram Soc* **68** 74

Taurel, L, E Rzepka & S Lefrant 1983 *Radiat Eff Defects Solids* **72** 115

Taylor, G M 1983 *Welding Inst Res Bull* **24** 55; see also *Welding Engineer* 1969 (Jun)

Temkin, H, S Mahajan & M A DiGuiseppe 1982 *Appl Phys Lett* **40** 562

Tench, A J 1971 *Surf Sci* **25** 625

Tench, A J 1972 *J Chem Soc Farad I* **68** 1181

Tench, A J & K F J Kibblewhite 1973 *J Chem Soc Comm* **1402** 955

Tench, A J & R L Nelson 1966 *J Chem Phys* **44** 1714
Tench, A J & R L Nelson 1968 *J Colloid Interface Sci* **26** 364
Thoma, E D, H M Yochum & R T Williams 1997 *Phys Rev* **B56** 8001
Thompson, M W 1969 *Defects and Radiation Damage in Metals*; Cambridge University Press, Cambridge
Thorpe, M F 1983 *J Non-Cryst Solids* **57** 355
Tikhomirov, V K, G J Adrianessens & S R Elliott 1997 *Phys Rev* **B55** 660
Tinivella, G & A M Stoneham 1980 UKAEA Harwell Report M-3142
Tjebg, L H, A R Vos & G A Sawatzky 1990 *Surf Sci* **235** 269
Tokizaki, T, T Makimura, H Akiyama, A Nakamura, K Tanimura & N Itoh 1991 *Phys Rev Lett* **67** 2701
Tolk, N H, M N Traum, J S Kraus, T R Pian & W E Collins 1982 *Phys Rev Lett* **49** 812
Tolpygo, K B 1957 *Ukr Fiz Zhur* **2** 242
Tolpygo, K B 1961 *Uspekhi Fiz Nauk* **74** 269
Tolpygo, K B & E Ya Shaterman 1987 *Sov Phys Solid State* **29** 1793
Tom, H W K, G D Aumiller & C H Brito-Cruz 1988 *Phys Rev Lett* **60** 1438
Tombrello, T A 1983a *Nucl Instrum Methods Phys Res* **218** 679
Tombrello, T A 1983b *Int J Mass Spectr Ion Phys* **53** 307
Tombrello, T A 1984 *Mat Res Soc Symp Proc* **25** 173
Tomiki, T, G Fukudome, M Kaminao, M Fujisawa, Y Tanahara & T Futemma 1988 *J Lumin* **40–41** 379
Torres, V J B, P M Masri & A M Stoneham 1987 *J Phys* **C20** L143
Torres, V J B, A M Stoneham, C J Sofield, A H Harker & C F Clement 1995 *Interface Sci* **3** 131
Tougaard, S & P Sigmund 1982 *Phys Rev* **B2** 4452
Toulemonde, M, C Dufour & E Paumier 1992 *Phys Rev* **B46** 14362
Toulemonde, M, S Bouffard & E Studer 1994 *Nucl Instrum Methods B* **91** 108
Townsend, P D 1973 *J Phys* **C6** 961
Toyozawa, Y 1954 *Prog Theor Phys* **12** 422
Toyozawa, Y 1961 *Prog Theor Phys* **26** 29
Toyozawa, Y 1974 p 317 of *Vacuum Ultraviolet Physics* (edited by E E Koch, R Haensel & C Kunz; Pergamon, Braunschweig
Toyozawa, Y 1981 *J Phys Soc Jpn* **50** 1861
Toyozawa, Y 1990 *Rev Solid State Sci* **4** 133
Toyozawa, Y & A Sumi 1974 p 179 of *Proceedings International Conferences on Physics of Semiconductors* (edited by M H Pilkuhn); Teubner, Stuttgart
Trautmann, C, J M Constantini, A Meftah, K Schwarz, J R Stoquart & M Toulemonde 1998, *MRS Research Proceedings*, Vol 526 (edited by R K Sungh, D H Lowndes, D P Chrisey, E Fogasrassy & J Narayan) Materials Research Society, Warrendale, PA
Trotter, D M 1991 *Scientific American* (April) 56
Trotter, D M & D W Smith 1984 *Appl Phys Lett* **45** 112
Troxell, J R, A P Chaterjee, G D Watkins & L C Kimerling 1979 *Phys Rev* **B19** 5336
Trukhin, A N 1978 *Phys Status Solidi* **B 86** 67
Trukhin, A N 1979 *Sov Phys Solid State* **21** 644
Trukhin, A N 1980 *Phys Status Solidi* **B 98** 541
Trukhin, A N 1991 *Sov Phys Solid State* **33** 2888
Trukhin, A N 1992 *J Non-Cryst Solids* **149** 32
Trukhin, A N 1994 p 7 of *Polish Ceramic Bulletin 11 Ceramics* 49
Trukhin, A N 1995 *J Non-Cryst Solids* **189** 1

Trukhin, A N 1997 *Mat Sci Forum* **239–241** 531
Tsai, T E & D L Griscom 1991 *Phys Rev Lett* **67** 2517
Tsang, K L, Y Chen & H T Tohver 1984 *Phys Rev* **B30** 6093
Tsujimoto, T, H Nishimura & M Nakayama 1996 *Phys Rev* **B54** 16579
Tung, C J, J C Ashley & R H Ritchie 1979 *Surf Sci* **81** 427
Turner, P S, T J Bullough, R W Devenish, D M Maher & C J Humphreys 1990 *Phil Mag Lett* **61** 181
Uberna, R *et al.* 1999 *J Chem Phys* **108** 9259
Uchida, H, D Huang, F Grey & M Aono 1993 *Phys Rev Lett* **70** 2040
Ueda, O 1993 *Mat Sci Eng* **B20** 9
Ueda, O, H Imai, T Fujiwara, S Yamakashi, T Sugawara, & T Yamaoka 1980 *J App Phys* **51** 5316
Ueda, Y 1976 *J Phys Soc Jpn* **41** 1255
Uematsu, M & K Wada 1991 *Appl Phys Lett* **58** 2015
Ueta M, H Kanzaki, K Kobayashi, Y Toyozawa & E Hanamura 1986 *Excitonic Processes in Solids*; Springer, Berlin
Underhill, P & T E Gallon 1982 *Solid State Commun* **43** 9
Uren, M J, J H Stathis & E Carter 1996 *J Appl Phys* **80** 3915
Vail, J M 1990 *J Phys Chem Solids* **51** 589
Vail, J M, R Pandey & B Kunz 1991 *Rev Solid State Sci* **5** 241
Vaisburd, D 1998 of *Supplément à la Revue 'Le Vide: Science, Technique et Applications'* No 287 (ed. G Damamme), p 185
Vajda, P & F Beuneu 1996 *Phys Rev* **B53** 5335
Valbis, Ya A & N Itoh 1991 *Radiat Eff Defects Solids* **11** 171
Valbis, Ya A, K A Kalder, I L Kuusmann, Ch B Lushchik, A A Ratas, Z A Rachko, M E Springs & V M Tiit 1975 *Sov Phys JETP Lett* **22** 36
Van Puymbroeck, W & Schoemaker D 1981 *Phys Rev* **B23** 1670
Varley, J H O 1954 *J Nucl Energy* **1** 130
Varley, J H O 1962 *J Phys Chem Solids* **23** 985
Vechten, J A van 1975 *J Electronic Mat* **4** 1159
Vechten, J A van 1980 *J de Physique Colloque* **41** C4–15
Vechten, J A van 1982 p 49 of *Laser and Electron-Beam Interaction* (edited by B A Appleton & G K Celler); North Holland, Amsterdam
Vechten, J A van 1985 *J Vac Sci Technol* **B3** 1240
Vechten, J A van, R Tsu, F W Saris & D Hoonhout 1979 *Phys Lett* **74A** 417, 422
Venables, J A, G D T Spiller & M Hanbrücken 1984 *Rep Prog Phys* **47** 399
Vertes, A, P Juhasz, M De Wolf & R Gijbels 1988 *Scanning Microsc* **2** 1853
Verwey, J F 1970 *J Phys Chem Solids* **31** 163
Verwey, J F & N G Westerlink 1969 *Physica* **42** 293
Vineyard, G H 1957 *J Phys Chem Sol* **3** 121
Vivet, L, M F Barthe, T Gibert-Legrand & B Dubreuit 1995 *J Appl Phys* **78** 3411
Vlaston, R, R Grotzschel, A Mucklich & W Prusseit 1998 *Nucl Instrum Methods B* **146** 309
Vleck, J H van 1941 *Phys Rev* **58** 724
Vos, M H, F Rappaport, J-C Lambry, J Breton & J-L Martin 1993 *Nature* **363** 320 (see also J Maddox, same issue p 297)
Vuoristo, P, T Mäntylä & P Ketturnen 1986 *J Vac Sci Technol* **A4(6)** 2932
Wade, E, A M Stoneham & J A Kilner 1979 *Mat Res Bull* **14** 661
Wagner, M & A Koengeter 1988 *J Chem Phys* **88** 7550
Walkup, R E & R L Kurtz 1988 p 160 in *Desorption Induced by Electronic Transitions DIET III* (edited by R H Stulen & M L Knotek); Springer, Berlin

Walkup, R E, Ph Avouris & A P Ghosh 1986 *Phys Rev Lett* **57** 2227

Walkup, R E, Ph Avouris & A P Ghosh 1987 *Phys Rev* **36** 4577

Walkup, R E, Ph Avouris & A P Ghosh 1988 p 248 in *Desorption Induced by Electronic Transitions DIET III* (edited by R H Stulen & M L Knotek) Springer, Berlin

Walkup, R E, D M Newns & Ph Avouris 1993 *Phys Rev* **B48** 1858

Wallace, D S, A M Stoneham W Hayes A J Fisher and A H Harker 1991 *J Phys Condens Matter* **3** 3905

Walle, C van der & W B Jackson 1996 *Appl Phys Lett* **69** 2441

Wallis, G 1970 *J Amer Ceram Soc* **53** 563

Wallis, G & D I Pomerantz 1969 *J Appl Phys* **40** 3946

Walton, S G, B L Peko & R L Champion 1998 *Phys Rev* **B58** 15430

Wang, Z G, Ch Dufour, E Paumier & M Toulemonde 1996 *Nucl Instrum Methods B* **11** 577

Wang, Z L, J Westendorp & F W Saris 1982 *Nucl Instrum Methods* **209–210** 115

Wang, Z-L, N Itoh & N Matsunami 1994 *Appl Phys Lett* **64** 1000

Wang, Z-L, N Itoh, N Matsunami & Q T Zhao 1995 *Nucl Instrum Methods B* **100** 493

Warman, J M, M P de Haas, P Pichat, T P M Koster, E A van der Zouwen-Assink, A Mackor & R Cooper 1991 *Radiat Phys Chem* **37** 433

Wasiela, A, G Ascareli & Y Merle d'Aubigne 1973 *Phys Rev Lett* **31** 993

Wasserman, B 1986 *Phys Rev* **B34** 1926

Watanabe, H, S Fujita & S Haruno 1997 *Appl Phys Lett* **71** 1038

Watkins, G D 1965 in *Radiation Damage in Semiconductors* (edited by P Baruch); Dunod, Paris

Watkins, G D 1990 *Rev Solid State Sci* **4** 279

Webb, R L, L C Jensen, S C Langford, & J T Dickinson 1993 *J Appl Phys* **74** 2323

Webb, R L, S C Langford & J T Dickinson 1995 *Nucl Instrum Methods B* **103** 297

Weber, B, D M Stock & K Gartner 1999 *Nucl Instrum Methods B* **148** 375

Weeks, J D, J C Tully & L C Kimerling 1975 *Phys Rev* **B12** 3286

Weeks, R A 1994 *J Non-Cryst Solids* **179** 1

Weidenauer, R, M Vollmer, W Hoheisel, U Schulte & F Traeger 1989 *J Vac Sci Technol* **A7**

Weinberg, Z A & G W Rubloff 1978 *Appl Phys Lett* **32** 184

Weinberg, Z A, D R Young, D J DiMaria & G W Rubloff 1979 *J Appl Phys* **50** 5757

Welland, M E & M P Murrell 1993 *Scanning* **15** 251

Werner, B T, T Vreeland, Jr, M H Mendenhall, Y Qui & T A Tombrello 1982 *Mat Res Soc Symp Proc* **18** 163

Werner, B T, T Vreeland, Jr, M H Mendenhall, Y Qui & T A Tombrello 1983 *Thin Solid Films* **104** 163

Weyl, W A 1951 *J Phys Colloid Chem* **55** 507

Whether, N R 1964 *J Appl Phys* **33** 3729

Whipple, C 1997 *Photonics Spectra* (Oct) 40 *re* work by A Fujishima (Tokyo U)

White, R C & W C Walker 1969 *Phys Rev* **188** 1380

Whitworth, R W 1974 *Adv Phys* **24** 203

Wight, D 1977 *J Phys* **D10** 431

Wilkins, J W & J R Gabriel 1963 *Phys Rev* **132** 1950

Williams, D E & K F E Pratt 1998 *J Chem Soc Farad Trans* **94** 3943

Williams, D L, B J Ainslie J R Armitage & R Kashyap 1992 *European Conference on Optical Communications (EFOC'92)*

Williams, J S 1986 *Rep Prog Phys* **49** 491
Williams, R M, K M Beck, A G Joly, J T Dickinson & W P Hess 1999 *Proc SPIE*, Vol 3618 22–44 Laser Applications in Microelectronic and Optoelectronic Manufacturing (edited by J Dubrowski, H Helvajian, E W Kreutz & K Sugioka)
Williams, R T 1989 *Radiat Eff Defects Solids* **109** 175
Williams, R T & M N Kabler 1972 *Solid State Commun* **10** 49
Williams, R T & K S Song 1990 *J Phys Chem Sol* **51** 679
Williams, R T, M N Kabler, W Hayes & J P Stott 1976 *Phys Rev* **B14** 725
Williams, R T, C L Marquardt, J W Williams & M N Kabler 1977 *Phys Rev* **B15** 5003
Williams, R T, J N Bradford & W L Faust 1978 *Phys Rev* **B18** 7038
Williams, R T, K S Song, W L Faust & C H Leung 1986 *Phys Rev* **B33** 7232
Williams, R T, H Liu, G P Williams & K J Platt 1991 *Phys Rev Lett* **66** 2140
Williams, T M, A M Stoneham & D Harries 1976 *Met Sci* **10** 14
Wilson, B A & T P Kerwin 1982 *Phys Rev* **B25** 5276
Wilson, R A, P Hu, J P Harbison & T M Jedju 1983 *Phys Rev Lett* **50** 1490
Winterbon, K B, P Sigmund & J B Sanders 1970 *Mat Phys Med Dan Vidensk Selsk* **37(14)**
Wittmaack, K 1980 *Nucl Instrum Methods* **168** 343
Woggon, U, H Giessen, F Gindele, O Wind, B Fluegel & N Peyghambarian 1996 *Phys Rev* **B54** 17681
Wolters, D R 1981 p 180 of *Insulating Films on Semiconductors* (edited by M Schulz & G Pensl); Springer, Berlin. See also D R Wolters 1980, in *The Physics of MOS Insulators*; *Proceedings of a topical conference held in Raleigh, North Carolina*)
Wood, R F 1966 *Phys Rev* **151** 629
Wood, R F, C W White & R T Young (eds.) 1984 *Pulsed Laser Processing of Semiconductors*; Academic Press, Orlando
Wood, R M 1986 *Laser Damage in Optical Materials*; Adam Hilger, Bristol
Wooten, F 1972 *Optical Properties of Solids*; Academic Press, New York
Wurz, P & C Becker 1989 *Surf Sci* **241** 224
Wurz, P, J Sarnthein, W Husinski & G Betz 1990 p 289 of *Desorption Induced by Electronic Transitions* (edited by G Betz & P Varga); Springer, Berlin
Wurz, P, J Sarnthein, W Husinski, G Betz, P Nordlander & Y Wang 1991 *Phys Rev* **B43** 6729
Xu, J, S H Overburg & J F Wendelkin 1995 *Phys Rev* **B53** R4245
Xu, J, S H Overburg & J F Wendelkin 1996 *J Vac Sci Technol B* **14** 918
Yamaga, M, Y Hayashi & H Yoshioka 1978 *J Phys Soc Jpn* **44** 154
Yamamura, Y & J Bohdanski 1985 *Vacuum* **35** 561
Yamasaki, S, H Okushi, A Matsuda & K Tanaka 1990 *Phys Rev Lett* **65** 756
Yamasaki, S, J Isoya & K Tanaka 1991 *Phil Mag* **B63** 163
Yamashita, J & T Kurosawa 1958 *J Phys Chem Sol* **5** 34
Yamashita, Y, F Jyobe, Y Kamiura & K Maeda 1997 *Mat Sci Forum* **258–263** 313
Yan, Q, A V Barnes, N Seifert, R Albridge & N Tolk 1995 *Nucl Instrum Methods* **B100** 224
Yanagisawa, Y 1986 *J Phys Soc Jpn* **55** 3479
Yanagisawa, Y & H Huzimura 1981 *J Phys Soc Jpn* **50** 209
Yanagisawa, Y & R Huzimura 1984 *J Phys Soc Jpn* **53** 66
Yang, C Y, M A Paesler & D E Sayers 1987 *Phys Rev* **B36** 9160
Yang, K H & S W S McKeever 1990 *J Phys* **D23** 237
Yang, M H & C P Flynn 1994 *Phys Rev Lett* **73** 1809

Yang, M H & C P Flynn 1996 *J Phys Condens Matter* **8** L279
Yassievich, I N 1994 *Semic Sci Tech* **9** 1433
Yoffa, E J 1980 *Phys Rev* **B21** 2415
Yokohama, T & M Kobayashi 1985 *J Phys Soc Jpn* **54** 2329
Yonezawa, Y, T Minamikawa, A Moromito & T Shimizu 1998 *Jap J Appl Phys* **37** 4
Yoon, J W & M Miyayama 1999 *Jap J Appl Phys (1)* **38** 894
Yoshida, M & K Morigaki 1989 *J Phys Soc Jpn* **58** 3371
Yoshinari, T, T Matsuyama, H Yamoaka & K Aoyagi 1985 *Jap J Appl Phys* **24** L720
Yoshioka, H, N Sugimoto & M Yamaga 1985 *J Phys Soc Jpn* **5** 3990
Young, E M & W A Tiller 1987 *Appl Phys Lett* **50** 46
Yuste, M, L Taurel & M Rahmani 1975 *Solid State Commun* **17** 1435
Zecchina, A, M G Lofthouse & F S Stone 1975 *J Chem Soc Farad Trans (I)* **81** 2835
Zema, N, M Piacenti, P Czuba, J Kolodziej, P Piatkowski, Z Postawa & M Szymonski 1997 *Phys Rev* **B55** 5448
Zewail, A H 1996 *J Phys Chem* **100** 12701
Zhang, B L & K Raghavachari 1995 *Phys Rev* **B51** 7946
Zhang, B L & K Raghavachari 1997 *Phys Rev* **B55** 15993
Zhang, J, K Sugioka, S Wada, H Tashiro & K Toyoda 1997 *Appl Phys* **A64** 477
Zhang, L, V A Mashkov & R G Leisure 1995 *Phys Rev Lett* **74** 1605
Zhang, X, C K Ong & A M Stoneham 1994 *J Phys Condens Matter* **6** 5647
Zhdanov, V G, B T Kolomiets, V M Lyubin & K Malinovskii 1979 *Phys Status Solidi A* **52** 621
Zhong, D & A H Zewail 1997 *J Amer Chem Soc* **119**, 2305, 5978
Zhong, D & A H Zewail 1998 *J Phys Chem* **A102** 4031
Zhu, H L & R S Averback 1995 *Phil Mag* **A71** 735
Ziegler, J F & J M Manoyan 1988 *Nucl Instrum Methods Phys Res* **B35** 215
Ziegler, J F, J P Biersack & U Littmark 1985 *Stopping Power and Ranges of Ions in Solids*; Pergamon, New York
Ziman, J M 1979 *Models of Disorder*; Cambridge University Press, Cambridge
Zimmerman, J 1971 *J Phys* **C4** 3265
Zong, X-F, C-F Shen, S Liu, Z-C Wu, Yi Chen, Y Chen, J G Zhu, B D Evans & R Gonzalez 1994 *Phys Rev* **B49** 15514
Zong, X-F, C-F Shen, S Liu, Z-C Wu, Yi Chen, R Zhang, Y Chen, J G Zhu, B D Evans, R Gonzalez & C H Sellers 1995 *J Nucl Mat* **219** 176
Zong, X-F, C-F Shen, S Liu, Yi Chen, R Zhang, Y Chen, J G Zhu, B D Evans & R Gonzalez 1996 *Phys Rev* **B54** 139
Zweiful, H 1998 *Stabilization of Polymeric Materials*; Springer, Berlin

Index